Principles and Applications of Structural Biology

Principles and Applications of Structural Biology

Editor: Zandra Edmonds

R CALLISTO REFERENCE

www.callistoreference.com

Callisto Reference,
118-35 Queens Blvd., Suite 400,
Forest Hills, NY 11375, USA

Visit us on the World Wide Web at:
www.callistoreference.com

ISBN: 978-1-64116-077-3 (Hardback)

Cataloging-in-Publication Data

Principles and applications of structural biology / edited by Zandra Edmonds.
    p. cm.
Includes bibliographical references and index.
ISBN 978-1-64116-077-3
1. Molecular biology. 2. Biomolecules--Structure. 3. Molecular structure.
I. Edmonds, Zandra.
QH506 .P75 2019
572.8--dc23

Table of Contents

Preface

Structural biology is the scientific study of the molecular structure of important macromolecules like amino acids, proteins and nucleic acids. It comes under the domain of molecular biology, but builds on the technological tools of biophysics like spectrometry, crystallography, proteolysis, scattering, etc. It also uses the principles of biochemistry to study, design, alteration and evolution of macromolecules. This book discusses the modern approaches of structural biology. It also elucidates the principles and applications of structural biology in a multidisciplinary manner. Experts and students in biotechnology, biophysics, biochemistry, structural biology and bioinformatics will be assisted by this book.

All of the data presented henceforth, was collaborated in the wake of recent advancements in the field. The aim of this book is to present the diversified developments from across the globe in a comprehensible manner. The opinions expressed in each chapter belong solely to the contributing authors. Their interpretations of the topics are the integral part of this book, which I have carefully compiled for a better understanding of the readers.

At the end, I would like to thank all those who dedicated their time and efforts for the successful completion of this book. I also wish to convey my gratitude towards my friends and family who supported me at every step.

Editor

Structural importance of the C-terminal region in pig aldo-keto reductase family 1 member C1 and their effects on enzymatic activity

Minky Son[1†], Chanin Park[1†], Seul Gi Kwon[2†], Woo Young Bang[3], Sam Woong Kim[2], Chul Wook Kim[2*] and Keun Woo Lee[1*]

Abstract

Background: Pig aldo-keto reductase family 1 member C1 (AKR1C1) belongs to AKR superfamily which catalyzes the NAD(P)H-dependent reduction of various substrates including steroid hormones. Previously we have reported two paralogous pig AKR1C1s, wild-type AKR1C1 (C-type) and C-terminal-truncated AKR1C1 (T-type). Also, the C-terminal region significantly contributes to the NADPH-dependent reductase activity for 5α-DHT reduction. Molecular modeling studies combined with kinetic experiments were performed to investigate structural and enzymatic differences between wild-type AKR1C1 C-type and T-type.

Results: The results of the enzyme kinetics revealed that V_{max} and k_{cat} values of the T-type were 2.9 and 1.6 folds higher than those of the C-type. Moreover, catalytic efficiency was also 1.9 fold higher in T-type compared to C-type. Since x-ray crystal structures of pig AKR1C1 were not available, three dimensional structures of the both types of the protein were predicted using homology modeling methodology and they were used for molecular dynamics simulations. The structural comparisons between C-type and T-type showed that 5α-DHT formed strong hydrogen bonds with catalytic residues such as Tyr55 and His117 in T-type. In particular, C3 ketone group of the substrate was close to Tyr55 and NADPH in T-type.

Conclusions: Our results showed that 5α-DHT binding in T-type was more favorable for catalytic reaction to facilitate hydride transfer from the cofactor, and were consistent with experimental results. We believe that our study provides valuable information to understand important role of C-terminal region that affects enzymatic properties for 5α-DHT, and further molecular mechanism for the enzyme kinetics of AKR1C1 proteins.

Keywords: Aldo-keto reductase, Homology modeling, Molecular dynamic simulation, NADPH-dependent reduction, Steroid hormone

Background

The aldo-keto reductase (AKR) superfamily is mostly comprised of monomeric oxidoreductases that catalyze NAD(P)H-dependent reductions of a wide range of aldehydes and ketones including steroids, carbohydrates, bile

* Correspondence: cwkim@gntech.ac.kr; kwlee@gnu.ac.kr
†Equal contributors
²Swine Science and Technology Center, Gyeongnam National University of Science & Technology, Jinju 660-758, Korea
¹Division of Applied Life Science (BK21 Plus), Systems and Synthetic Agrobiotech Center (SSAC), Plant Molecular Biology and Biotechnology Research Center (PMBBRC), Research Institute of Natural Science (RINS), Gyeongsang National University (GNU), 501 Jinju-daero, Jinju 660-701, Republic of Korea
Full list of author information is available at the end of the article

acids, and prostaglandins [1,2]. The AKRs have been classified into 14 families (AKR1 to AKR14) and AKR1 family have been further divided into 6 subfamilies (AKR1A to AKR1G) [3,4]. Among the subfamilies, AKR1C enzymes are known as hydroxysteroid dehydrogenases (HSDs) which play a pivotal role in metabolism and regulation of steroid hormones such as progesterone, 5α-dihydrotestosterone (DHT), and testosterone. Pig aldo-keto reductase family 1 member C1 (AKR1C1) shows both 3α- and 20α-HSD activities and also plays a crucial role in progesterone metabolism, maintenance of pregnancy, and hormone regulation during the estrous cycle [5]. It officially named as AKRlCL1 (aldo-keto reductase family 1, member C-like 1), consists of 14 amino acid residues longer than that of general AKR1C1 [3]. The longer amino acid residues have been reported to alter enzymatic activities of several steroid hormones [3]. The structures of AKRs have the $(\alpha/\beta)_8$-barrel or TIM-barrel motif and three conserved loop regions, loop A, B, and C, which are related with steroid hormone specificity [4]. The enzymes catalyze an ordered bisequential kenetic process in which binding of cofactor is obligatory for the reaction [6,7]. The nicotinamide group of NADPH cofactor lies in *anti*-conformation with respect to the ribose group, so that 4-pro-R-hydride is transferred from the cofactor to the 3-ketosteroid substrate [2,6,8]. The hydride transfer is mediated by a highly conserved catalytic tetrad consisting of Asp50, Tyr55, Lys84, and His117, where Tyr 55 acts as the general acid/base [9-11]. Recently we have identified two paralogous pig AKR1C1s with or without C-terminal region (R320 to L337) which was truncated by a non-synonymous variation [3]. Also, the C-terminal region significantly affects the NADPH-dependent reductase activity for 5α-DHT reduction [3].

In this study, we performed molecular modeling studies combined with kinetic experiments to examine structural difference between wild-type AKR1C1 (C-type) and C-terminal-truncated AKR1C1 (T-type) for 5α-DHT. Since there was no available experimental structure of pig AKR1C1, we have carried out homology modeling to build 3D structure models of the both types, which were used for molecular dynamics (MD) simulation study. Our findings provide structural insights into important role of C-terminal region of the enzyme. It can be helpful for understanding different enzymatic properties for 5α-DHT between C-type and T-type.

Methods
Materials
The following chemicals were used in the experiments; 5α-dihydrotestosterone (5α-DHT), methylglyoxal, 9,10-phenanthrenequinone and hydrindantin were purchased from Sigma (St. Louis, MO), and Ni-NTA chelating agarose CL-6B was purchased from Peptron company (Promega corporaton, USA). Bio-Rad Bradford Protein assay kit was purchased from (Bio-Rad Laboratories, Inc (South Korea). The others, including Na2HPO4, NaH2PO4, NaCl, bovine serum albumin (BSA), and imidazole, were purchased from Sigma (St. Louis, MO).

Recombinant protein purification
In previous study [3], two types of pPROEX HTb-AKR1CL1 clones were constructed for the production of his-tagged fusion proteins for C-type and T-type. They were used for the IPTG-induced expression of each of the clones in *E. coli* BL21. In this study, the IPTG-induced proteins were subjected to the affinity chromatography

using Ni-NTA agarose, according to manufacturer's manual (Peptron, Daejon, Korea). Briefly, basal buffer for protein purification was prepared by 50 mM sodium phosphate buffer, pH8.0, and 500 mM NaCl. Imidazole (Sigma, USA) was added to required concentration according to purification manual with NTA Chelating Agarose CL-6B (Promega corporation, USA). The overexpressed cells were precipitated by centrifugation, and suspended by binding buffer including 5 mM imidazole. The collected cells were lyzed by SONICS Vibracell VCX750 Ultrasonic Cell Disruptor, which was done twice by conditions as following; 5 min by 2 sec interval of on/off and 35% amplitude during ice cooling. The supernatant to obtain water-soluble protein was collected from the cells treated by centrifugation for 30 min at 10,000 rpm, 4°C. Purification of the protein was done by NTA Chelating Agarose CL-6B (Promega corporation, USA) according to manufacturer's directions. The purified recombinant proteins were concentrated by Ultrafree-0.5 Centrifugal Filter Device (Millipore Corporation, Germany). The concentrated proteins were quantified by Bio-Rad Bradford Protein assay kit (Bio-Rad Laboratories, Inc., Korea) by OD595 nm in wavelength. The purified proteins were added with 50% glycerol and 50 mM Sodium Phosphate Buffer (pH 6.4) for long-term storage at −20°C.

Measurement of NADPH-dependent carbonyl reductase activity
The reductase activity was measured under conditions described previously [12]. Reaction mixtures included 60 mM sodium phosphate (pH 6.5), purified recombinant proteins such as C-type and T-type, 0.1 mM NADPH and 0.1 mM substrates (the reproductive steroid hormones indicated above) and were incubated in a total volume of 0.5 ml at 37°C. The assay of reductase activity was spectrophotometrically carried out by monitoring the decrease in absorbance at 340 nm with time.

Statistical analysis
To determine kinetic parameters with a Michaelis–Menten plot, the data were analyzed by nonlinear regression using GraphPad prism 6 software (GraphPad Software Inc., San Diego, CA). Enzyme concentrations of 39 and 38.9 nmol/mg were used for the calculation of turnover rates (k_{cat}) for C-type and T-type, respectively. The significant differences were analyzed by Student's t-test ($p < 0.01$ or $p < 0.05$) using the above software. The results are expressed as means ± standard errors (S.E.) of at least 3 independent experiments.

Homology modeling
The sequence of pig AKR1C1 consisting of 337 amino acids, was obtained from UniProtKB (http://www.

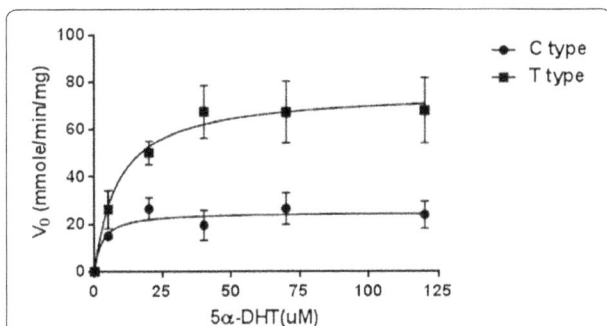

Figure 1 A Michaelis-Menten plot from measurement of the NADPH-dependent reduction of 5α-DHT by recombinant AKR1C1 (WT) or AKR1C1 (ΔC term). The data, obtained from measurement of the NADPH-dependent reduction of 5α-DHT by recombinant AKR1C1 (WT) and AKR1C1 (ΔC term), were analyzed by nonlinear regression using GraphPad prism 6 software (GraphPad Software Inc., San Diego, CA). Each spot represents the mean ± S.E. (n=3).

uniprot.org/) (accession no. Q1KLB4). In order to build a structure model of pig AKR1C1 homology modeling was conducted using Phyre2 server (Protein Homology/analogY Recognition Engine V 2.0) with intensive modeling mode [13], which the server utilizes multiple templates and *ab initio* techniques to predict 3D structure model. The generated homology model was subjected to energy minimization to refine the model as well as to reduce steric clashes. The minimization with the steepest descent algorithm for 10,000 steps was carried out by GROMACS 4.5.3 package [14,15] with CHARMM27 force field. The stereochemical quality of the model was assessed by PROCHECK [16], ProSA [17,18], and ERRAT [19]. All other analyses including multiple sequence alignment were done by Discovery Studio v3.1 (DS).

Molecular docking calculation

AKR1C1 C-type and T-type in complex with NADPH were subjected to molecular docking calculation. The structure of T-type was prepared by deleting the C-terminal region (R320 to L337) from C-type. The coordinates of the cofactor were taken from the structure of human AKR1C3 (PDB: 1S1P). The substrate, 5α-DHT, was downloaded from PubChem Compound Database (CID: 10635) [20]. Then 5α-DHT was subjected to energy minimization with CHARMm force field and implicit solvent model using DS. The binding pose of 5α-DHT

was predicted using GOLD v 5.0.1 (Genetic Optimization for Ligand Docking) [21,22] which uses genetic algorithm (GA) for docking flexible ligands in the binding site of the protein. The binding site was assigned through *Define and Edit Binding Site* tool in DS. All residues within the radius of 5 Å of the center of binding sphere were included in the calculation and the number of GA runs was set to 100. All other parameters were used as their default values. The docking poses were ranked based on GOLD fitness score and top solution was selected as initial conformation for MD simulation.

Molecular dynamics simulation

MD simulations for C-type and T-type in complex with NADPH and 5α-DHT were performed using GROMACS 4.5.3 with CHARMM27 force field. Topology files for the ligands were obtained from SwissParam server [23]. At the beginning, protonation states of the ionizable residues were set at pH7. A water box with the size of 1.5 nm from the protein surface was created to make an aqueous environment, and immersed using explicit TIP3P water model [24]. The size of the system was 6.05 × 5.78 × 5.90 nm for C-type and 6.03 × 5.77 × 5.03 nm for T-type, respectively. Several water molecules were replaced with sodium ions to neutralize the system. Energy minimization for 10,000 steps was executed using steepest descent algorithm until the maximum force lower than 1000 kJ/mol. After minimization, the systems were subjected to 100 ps NVT equilibration at 300 K and then 100 ps NPT equilibration at 300 K and 1 bar of pressure. The equilibrated systems were used in 20 ns production runs under NPT ensemble. A constant temperature and pressure were kept using V-rescale thermostat [25] and Parrinello-Rahman barostat [26,27]. During the simulation, LINCS [28,29] and SETTLE [30] algorithms were used to constrain all bond lengths and the geometry of water molecules, respectively. Short-range interactions were treated with the cut-off value of 1.2 nm and long-range electrostatic interactions were calculated by applying particle mesh Ewald (PME) method [31,32]. The periodic boundary conditions were adopted to avoid edge effects. A grid spacing of 0.12 nm was applied for fast Fourier transform calculations. We repeated the simulations two times under the same conditions except that the simulation time was 10 ns. All

Table 1 Kinetic parameters for 5α-DHT reduction measured by spectrophotometer

Enzyme	V_{max} (nmol/min per mg)	K_m (μM)	K_{cat} (s^{-1})	K_{cat}/K_m (s^{-1} M^{-1})
AKR1C1 C-type	25.74 ± 2.338	4.978 ± 3.091	0.011 ± 0.000999	2,210 ± 323.196
AKR1C1 T-type	75.5 ± 5.2	7.736 ± 2.816	0.0340 ± 0.00234	4,340 ± 830.966

Each value indicates mean ± SEM (n = 3).
Kinetic parameters were determined using data in Figure 1 through GraphPad prism 6 software (GraphPad Software Inc., San Diego, CA).

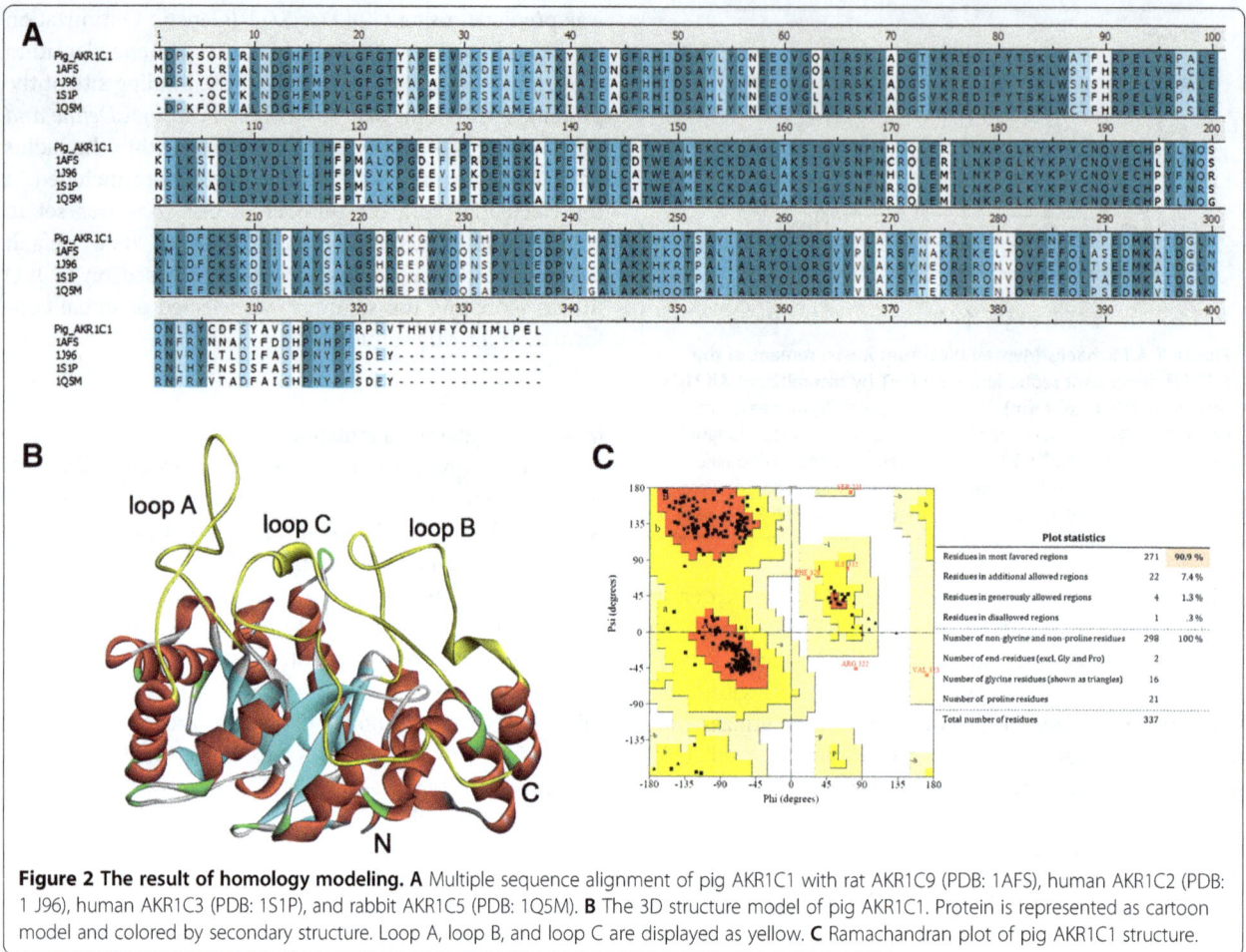

Figure 2 The result of homology modeling. A Multiple sequence alignment of pig AKR1C1 with rat AKR1C9 (PDB: 1AFS), human AKR1C2 (PDB: 1 J96), human AKR1C3 (PDB: 1S1P), and rabbit AKR1C5 (PDB: 1Q5M). **B** The 3D structure model of pig AKR1C1. Protein is represented as cartoon model and colored by secondary structure. Loop A, loop B, and loop C are displayed as yellow. **C** Ramachandran plot of pig AKR1C1 structure.

simulations were performed with the time step of 2 fs and the coordinates were saved every 1 ps for analyses.

Results and discussion

The C-terminal region in AKR1C1 alters significantly the enzymatic properties to 5α-DHT

AKR1C1 exhibits broadly enzymatic activities to various steroid hormones [33]. Among steroid hormones,

AKR1C1 originated from human, previously named as 20α-hydroxysteroid dehydrogenase, detects specifically to progesterone with high activity [33]. A variant truncated at the C-terminus of pig AKR1C1 was employed in this study. In our previous study, we found a new novel single nucleotide variant (SNV) truncated in C-terminus, where the SNV is a nonsense mutant lacking 18 amino acid residues (R320 to L337) in C-terminus [3]. During the evaluation of enzymatic activities with

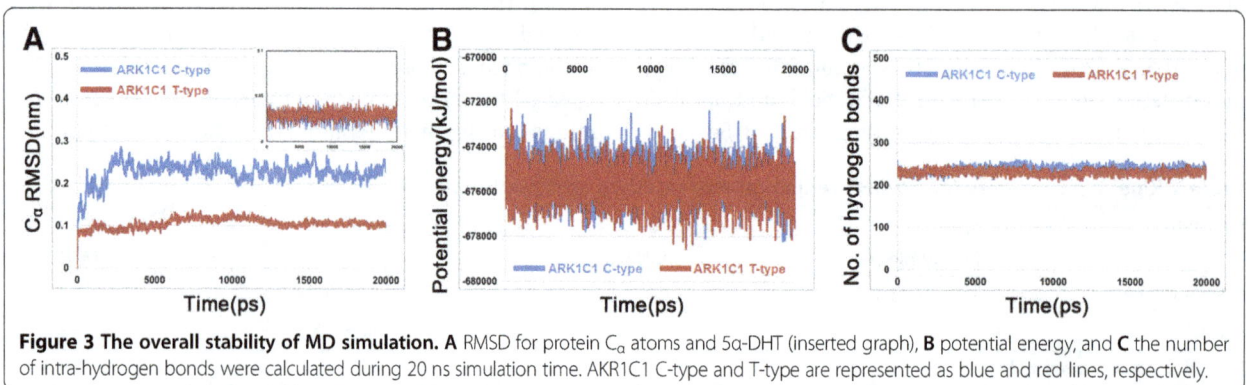

Figure 3 The overall stability of MD simulation. A RMSD for protein C$_\alpha$ atoms and 5α-DHT (inserted graph), **B** potential energy, and **C** the number of intra-hydrogen bonds were calculated during 20 ns simulation time. AKR1C1 C-type and T-type are represented as blue and red lines, respectively.

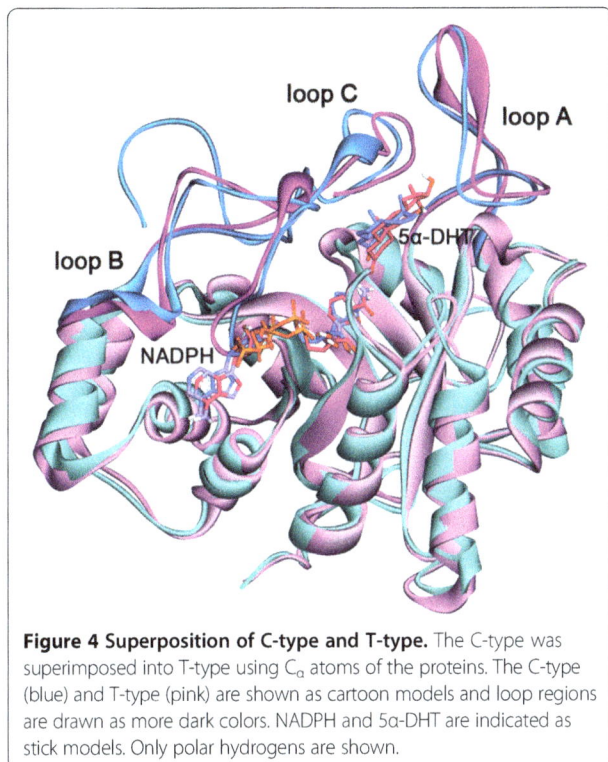

Figure 4 Superposition of C-type and T-type. The C-type was superimposed into T-type using C$_\alpha$ atoms of the proteins. The C-type (blue) and T-type (pink) are shown as cartoon models and loop regions are drawn as more dark colors. NADPH and 5α-DHT are indicated as stick models. Only polar hydrogens are shown.

with substrate 5α-DHT and cofactor NADPH. The V_{max} value of AKR1C1 T-type was 2.9 fold higher than that of C-type, but K_m value was lower 1.6 fold (Figure 1 and Table 1). Furthermore, the values of k_{cat} and catalytic efficiency of the T-type were 2.9 and 1.9 folds higher than those of C-type. These results suggest that C-terminal truncated AKR1C1 improves the values of V_{max}, k_{cat} and catalytic efficiency.

The structure prediction of pig AKR1C1 using homology modeling

Since crystal structure of pig AKR1C1 has not been determined yet, we have constructed the 3D structure model using four structures as templates; rat AKR1C9 (PDB: 1AFS), human AKR1C2 (PDB: 1 J96), human AKR1C3 (PDB: 1S1P), and rabbit AKR1C5 (PDB: 1Q5M). The multiple sequence alignment with the four templates revealed that catalytic tetrad of Asp50, Tyr55, Lys84, and His117 were conserved and they have high sequence identity and similarity with each template; 71.5% and 85.6% between Pig AKR1C1 and rat AKR1C9, 75.5% and 88.2% between Pig AKR1C1 and human AKR1C2, 75.9% and 88.8% between Pig AKR1C1 and human AKR1C3, 76.2% and 88.5% between Pig AKR1C1 and rabbit AKR1C5, respectively (Figure 2A). Since there was no proper structural information for 14 residues at the end of the C-terminal region of AKR1C1, the region was modeled by *ab initio* method. The homology model for pig AKR1C1 was refined through the energy minimization and it showed conserved loop regions which are structural features of AKR superfamily (Figure 2B). The stereochemical quality of the generated model was evaluated using three programs. Ramachandran plot obtained from PROCHECK showed that 90.9% of residues were in most favored regions and only one residue was in disallowed region (Figure 2C).

different steroid hormones, differential activities between AKR1C1 C- and T-types were shown, 5α-DHT being anyway the preferred substrate for both of them [3]. Therefore, 5α-DHT was employed for enzymatic kinetics in this study.

In order to analyze enzymatic activity of AKR1C1s, the enzymes were cloned into overexpression vector and then purified to homogeneity by affinity chromatography. The purified AKR1C1s were applied for enzymatic kinetics

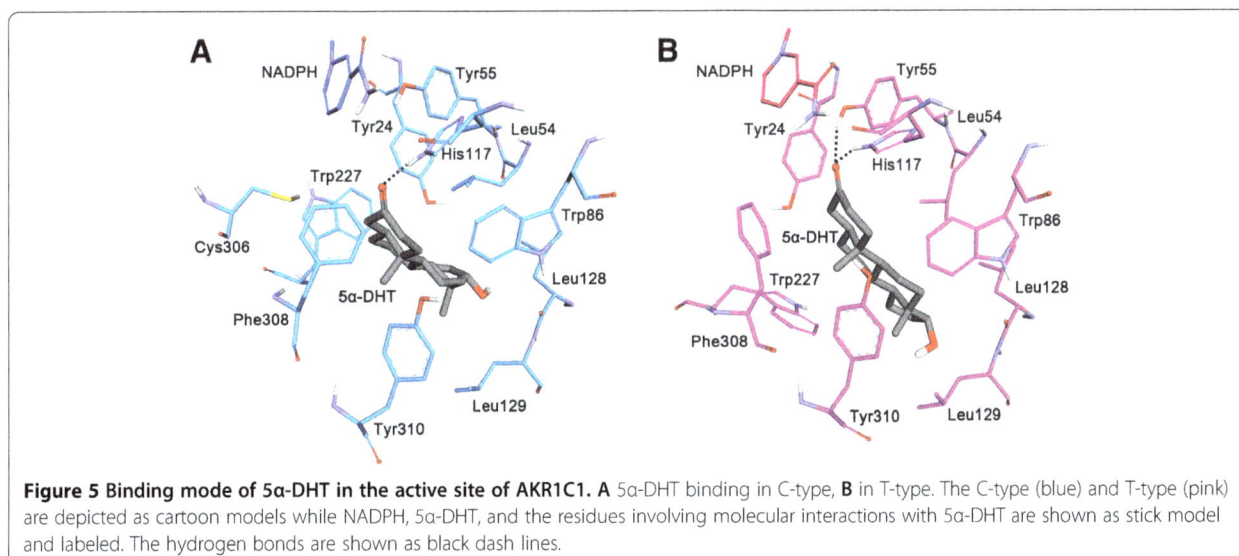

Figure 5 Binding mode of 5α-DHT in the active site of AKR1C1. A 5α-DHT binding in C-type, **B** in T-type. The C-type (blue) and T-type (pink) are depicted as cartoon models while NADPH, 5α-DHT, and the residues involving molecular interactions with 5α-DHT are shown as stick model and labeled. The hydrogen bonds are shown as black dash lines.

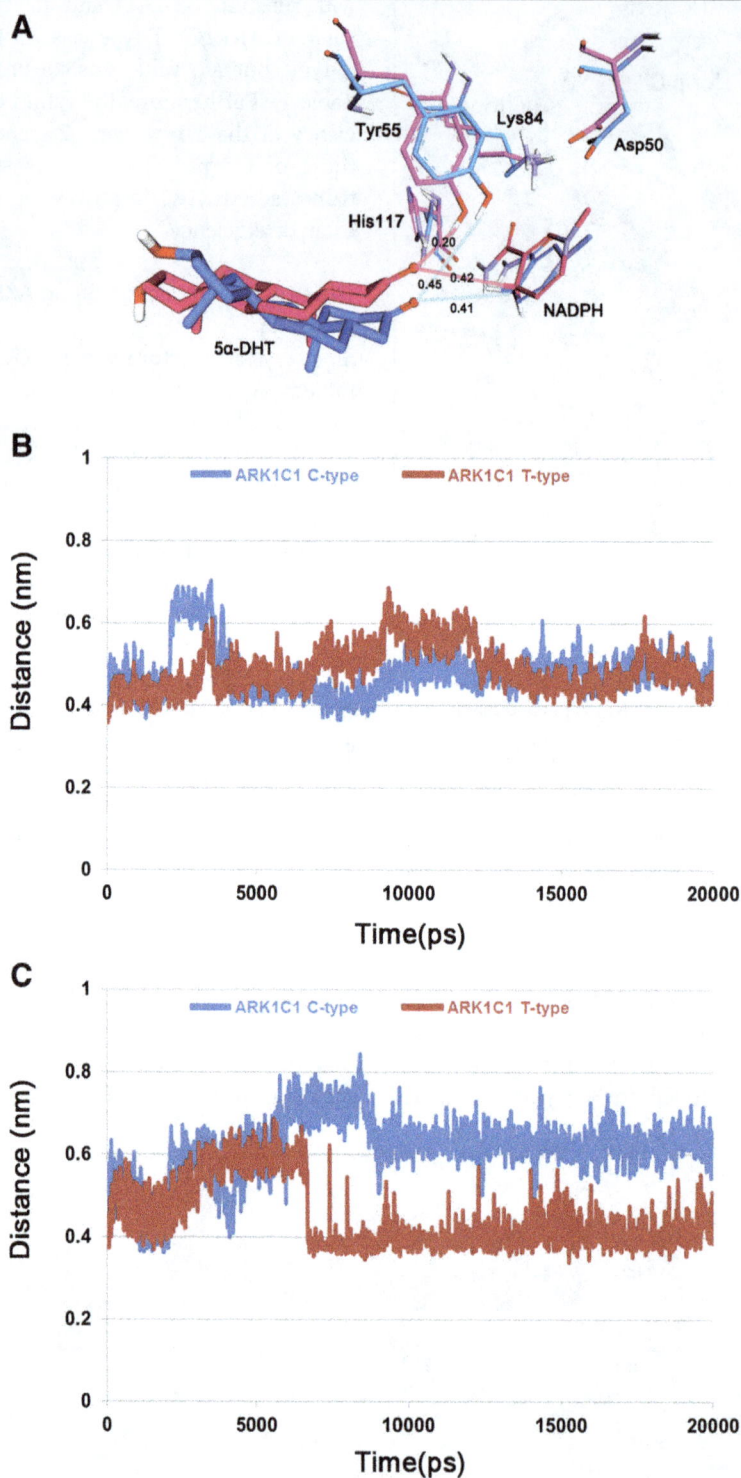

Figure 6 The measurement of key distance for reduction reaction. A The crucial distance to initiate catalytic reaction and relative position of the catalytic tetrad and cofactor in the active site of C-type and T-type. The catalytic tetrad and 5α-DHT in C-type (blue) and T-type (pink) are displayed as stick models. The key distances are given in nm. **B** The distance between C3 of 5α-DHT and C4N of NADPH. **C** The distance between C3 of 5α-DHT and OH of Tyr55. The both distances in C- and T-types were measured during 20 ns simulation time. Blue and red lines indicate C-type and T-type, respectively.

Overall quality factor scores calculated from ERRAT and ProSA were 90.49 and –11.04, respectively.

Binding mode of 5α-DHT in the active site of AKR1C1 C-type and T-type

A molecular docking study was performed to discover proper binding conformations for 5α-DHT at the active site of the C-type and T-type. The docking conformations were clustered and ranked according to their GOLD fitness scores. A conformation having high fitness score in the most populated cluster was selected as putative binding pose of each system. The docking results revealed that 5α-DHT bound to the both types of AKR1C1 in a similar manner forming hydrophobic interactions with Tyr24, Leu54, Trp86, Phe118, Leu128, Trp227, Phe308, and Tyr310 which have been reported as key residues for steroid binding [34,35]. We found that 5α-DHT formed two hydrogen bonds with Tyr55 and His177 in the C-type, whereas, in the T-type, there was additional hydrogen bond with Leu129 as well as two hydrogen bonds. The final docking poses in C-type and T-type were used as initial structures in MD simulation study to understand the effect of C-terminal region on the enzymatic activity in atomic level. We evaluated the overall stability of MD simulations by calculating C_α root-mean-square deviation (RMSD), potential energy, and the number of intra-hydrogen bonds (Figure 3). The RMSD values for each system were converged to around 0.25 nm in C-type and 0.1 nm in T-type (Figure 3A). During the whole simulation time, the RMSD value of C-type was relatively higher than that of T-type with the average value of 0.23 nm and 0.11 nm, respectively. The RMSD plot for only 5α-DHT also revealed that the substrates in both C- and T-types achieved stabilization and their average values were 0.03 nm (inserted in Figure 3A). Moreover, potential energy and the number of intra-hydrogen bonds for the systems remained constant for the simulation time (Figure 3B and C). These results indicate that the MD simulations for both systems were successfully completed and there were no abnormal behaviors in the structures throughout the simulation time. A structural comparison between the C-type and T-type was performed using their representative structures which were the closest snapshot to the average of all snapshots obtained from the last 5 ns. Although there were no significant conformational changes in both systems, they showed a difference in 5α-DHT binding in terms of hydrogen bond interactions (Figure 4). The C-type showed only one hydrogen bond interaction between oxygen atom of 5α-DHT and hydrogen atom of His117 with the distance of 0.21 nm (Figure 5A). On the other hand, oxygen atoms of 5α-DHT formed hydrogen bonds with hydrogen atoms of Tyr55 and His177 in T-type and the distances of bonds were within 0.21 nm

(Figure 5B). The residues Tyr24, Leu54, Trp86, Leu128, Leu129, Trp227, Phe308, and Tyr310 in both structures were participated in hydrophobic interactions which were similar to that observed in the initial docked structures. These further stabilized 5α-DHT binding in both active sites.

Difference in 5α-DHT binding between C- type and T-type

A major difference in 5α-DHT binding for C-type and T-type of AKR1C1 was the relative distance from Tyr55 which is important to initiate the catalytic reaction of the enzyme. The C3 ketone of 5α-DHT in the T-type was positioned much closer to the catalytic tetrad and the 4-pro-R hydride of the NADPH than in the C-type (Figure 6A). The distance between C3 position of 5α-DHT and C4 position in nicotinamide ring of NADPH were 0.41 nm in C-type and 0.42 nm in T-type. In contrast, the distance between the C3 of the 5α-DHT and hydroxyl group of Tyr55 in T-type was 0.20 nm which is much shorter than the value of 0.45 nm in C-type. The monitoring these distances during 20 ns simulation time revealed that the both distances were relatively short in T-type compared to C-type (Figure 6B and C). Superimposition of the two structures showed that 5α-DHT in T-type was sandwiched between Leu54 and Trp227 and its β-face was oriented toward Trp227, whereas in the case of C-type, the flipping of the side chain of Trp227 hindered the interaction with β-face of 5α-DHT (Figure 7). The side chain of Tyr24 also showed different conformation in the both types and that was

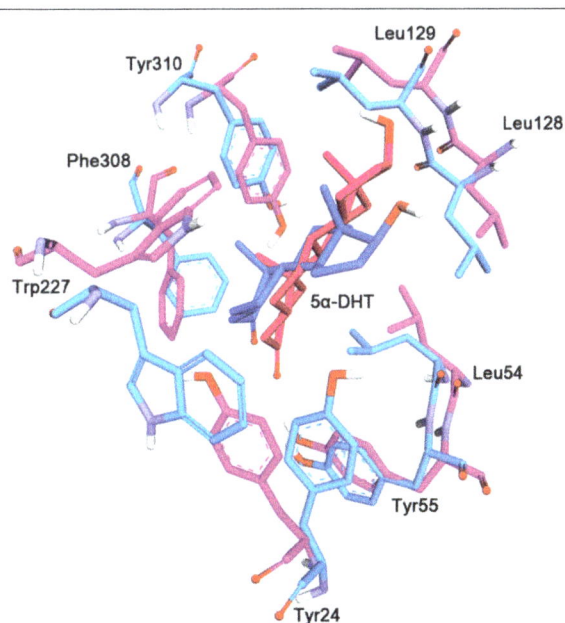

Figure 7 Comparison of the 5α-DHT binding in the active site of C-type and T-type. Interacting residues and 5α-DHT in C-type (blue) and T-type (pink) are drawn as stick models.

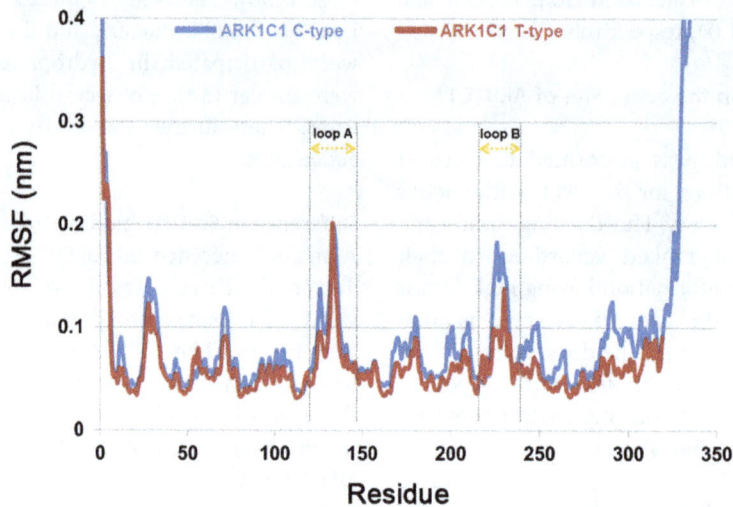

Figure 8 RMSF plot showing the atomic fluctuations by residues of C-type and T-type. RMSF values for C_α atoms of the proteins are drawn as blue and red lines, respectively.

probably due to displacement of Trp227. From the structural comparison, it appears that binding conformation of 5α-DHT in T-type was more favorable for catalytic reaction than that of C-type. In root mean square fluctuation (RMSF) plot, it was observed that the residues 226–229 in T-type exhibited higher flexibility than in C-type, while flexibilities of other residues were quite similar in the both structures except for highly flexible regions such as N- or C-terminal part of the protein (Figure 8). These differences might be explained by flipping of Trp227 in C-type. From RMSD plot calculated using all atoms of Trp227, the RMSD value in C-type showed relatively high with the average of 0.12 nm and it started to increase from 5 ns (Figure 9). In the simulation for C-type, the flipping of

Trp227 side chain was observed and 5α-DHT was gradually alienated from Tyr55 during that time. Compared to C-type, RMSD value of Trp227 in T-type was very stable, less than 0.05 nm, throughout 20 ns simulation time and the average value was 0.03 nm. This might be related to the observation that flipping of Trp227 hardly ever happened in T-type. These analyses demonstrated that the instability of Trp227 caused by flipping of the side chain might be correlated with the distance from 5α-DHT to the hydroxyl group of Tyr55. Additionally, the interaction energy between 5α-DHT and the protein was −36.42 kcal/mol in C-type and −44.98 kcal/mol in T-type. It also indicated that 5α-DHT in T-type had energetically favorable conformation.

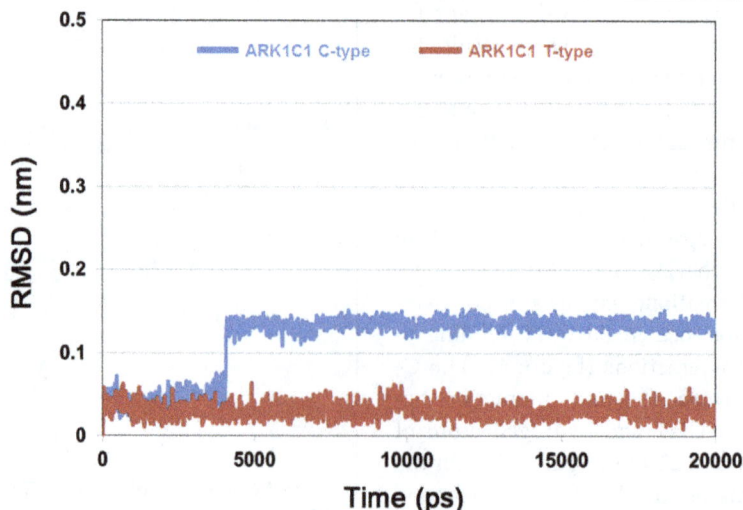

Figure 9 RMSD plot for Trp227 of C-type and T-type. RMSD values for all atoms of Trp227 are displayed as blue and red lines, respectively.

Conclusions

The study of enzyme kinetics revealed that the C-terminal region in AKR1C1 contributed significantly the enzymatic properties for 5α-DHT reduction. To gain structural insights into the difference between C-type and T-type of AKR1C1 for 5α-DHT reduction, MD simulations for both structures were carried out. Prior to the simulation, we generated homology model structure for AKR1C1 due to lack of experimentally determined structures. Then C-type and T-type in complex with 5α-DHT obtained from molecular docking study were used as initial conformations for MD simulation. Although there were no significant conformational changes in both systems during 20 ns simulation time, binding conformations of 5α-DHT were different in the active site of C-type and T-type. The structural comparisons showed that T-type formed strong hydrogen bonds with Tyr55 and His117, while only His117 was found in C-type. To initiate catalytic reaction, the C3 ketone group of 5α-DHT should be close to Tyr55 and the nicotinamide ring of NADPH which are involved in hydride transfer. The distances between these groups were monitored during 20 ns simulation time. As a result, 5α-DHT was close to the cofactor in the both structures, whereas the distance between 5α-DHT and Tyr55 in T-type was relatively much shorter than C-type. On the contrary, the flipping of the side chain of Trp227 in C-type might disrupt the interaction with β-face of 5α-DHT. The interaction energies between 5α-DHT and the proteins also indicated that T-type was energetically stable compared to C-type. Taken together, our simulation results demonstrated that binding conformation of 5α-DHT in T-type was more favorable for catalytic reaction than that of C-type. These structural explanations were also in agreement with kinetic experimental results. Our findings will be useful to understand molecular mechanism for the enzyme kinetics of AKR1C1 protein.

Competing interests
The authors declare that they have no competing interests.

Authors' contributions
MS and CP performed structural modeling and analyzed data. SGK, WYB, and SWK designed the study and carried out the experiments and statistical analysis. MS, CP, SGK, WYB, and SWK wrote the manuscript. CWK and KWL interpreted the data and correct the manuscript. All authors read and approved the final manuscript.

Acknowledgments
This work was supported by grants from Priority Research Centers Program (2011–0022965) and Management of Climate Change Program (2010–0029084) through the National Research Foundation of Korea (NRF) funded by the Ministry of Education of Republic of Korea, the project, "Search & Discovery of Utility Value from Biological Resources (2015)", from the National Institute of Biological Resources of Korean Government, the Export Promotion Technology Development Program (no. 313012–05) of Ministry of Food, Agriculture, Forestry and Fisheries, Republic of Korea. And this work was also supported by the Next-Generation BioGreen 21 Program (PJ009486) from Rural Development Administration (RDA) of Republic of Korea.

Author details
[1]Division of Applied Life Science (BK21 Plus), Systems and Synthetic Agrobiotech Center (SSAC), Plant Molecular Biology and Biotechnology Research Center (PMBBRC), Research Institute of Natural Science (RINS), Gyeongsang National University (GNU), 501 Jinju-daero, Jinju 660-701, Republic of Korea. [2]Swine Science and Technology Center, Gyeongnam National University of Science & Technology, Jinju 660-758, Korea. [3]National Institute of Biological Resources, Environmental Research Complex, Incheon 404-708, Korea.

References
1. Jez J, Bennett M, Schlegel B, LEWIS M, Penning T. Comparative anatomy of the aldo–keto reductase superfamily. Biochem J. 1997;326:625–36.
2. Penning TM. Hydroxysteroid dehydrogenases and pre-receptor regulation of steroid hormone action. Hum Reprod Update. 2003;9(3):193–205.
3. Kwon S, Bang W, Jeong J, Cho H, Park DH, Hwang J, et al. Important role of the C-terminal region of pig aldo-keto reductase family 1 member C1 in the NADPH-dependent reduction of steroid hormones. Indian J Biochem Biophys. 2013;50(3):237.
4. Hyndman D, Bauman DR, Heredia VV, Penning TM. The aldo-keto reductase superfamily homepage. Chem Biol Interact. 2003;143:621–31.
5. Seo K-S, Naidansuren P, Kim S-H, Yun S-J, Park J-J, Sim B-W, et al. Expression of aldo-keto reductase family 1 member C1 (AKR1C1) gene in porcine ovary and uterine endometrium during the estrous cycle and pregnancy. Reprod Biol Endocrinol. 2011;9(1):139-139.
6. Askonas LJ, Ricigliano JW, Penning TM. The kinetic mechanism catalysed by homogeneous rat liver 3 alpha-hydroxysteroid dehydrogenase. Evidence for binary and ternary dead-end complexes containing non-steroidal anti-inflammatory drugs. Biochem J. 1991;278:835–41.
7. Grimshaw CE, Bohren KM, Lai C-J, Gabbay KH. Human aldose reductase: rate constants for a mechanism including interconversion of ternary complexes by recombinant wild-type enzyme. Biochemistry. 1995;34(44):14356–65.
8. Penning TM, Drury JE. Human aldo–keto reductases: function, gene regulation, and single nucleotide polymorphisms. Arch Biochem Biophys. 2007;464(2):241–50.
9. Schlegel BP, Jez JM, Penning TM. Mutagenesis of 3α-Hydroxysteroid Dehydrogenase Reveals a "Push–Pull" Mechanism for Proton Transfer in Aldo–Keto Reductases†. Biochemistry. 1998;37(10):3538–48.
10. Bohren KM, Grimshaw CE, Lai CJ, Harrison DH, Ringe D, Petsko GA, et al. Tyrosine-48 is the proton donor and histidine-110 directs substrate stereochemical selectivity in the reduction reaction of human aldose reductase: enzyme kinetics and crystal structure of the Y48H mutant enzyme. Biochemistry. 1994;33(8):2021–32.
11. Grimshaw CE, Bohren KM, Lai C-J, Gabbay KH. Human aldose reductase: pK of tyrosine 48 reveals the preferred ionization state for catalysis and inhibition. Biochemistry. 1995;34(44):14374–84.
12. Tanaka M, Ohno S, Adachi S, Nakajin S, Shinoda M, Nagahama Y. Pig testicular 20 beta-hydroxysteroid dehydrogenase exhibits carbonyl reductase-like structure and activity. cDNA cloning of pig testicular 20 beta-hydroxysteroid dehydrogenase. J Biol Chem. 1992;267(19):13451–5.
13. Kelley LA, Sternberg MJ. Protein structure prediction on the Web: a case study using the Phyre server. Nat Protoc. 2009;4(3):363–71.
14. Berendsen HJ, van der Spoel D, van Drunen R. GROMACS: A message-passing parallel molecular dynamics implementation. Comput Phys Commun. 1995;91(1):43–56.
15. Van Der Spoel D, Lindahl E, Hess B, Groenhof G, Mark AE, Berendsen HJC. GROMACS: fast, flexible, and free. J Comput Chem. 2005;26(16):1701–18.
16. Laskowski RA, MacArthur MW, Moss DS, Thornton JM. PROCHECK: a program to check the stereochemical quality of protein structures. J Appl Crystallogr. 1993;26(2):283–91.
17. Wiederstein M, Sippl MJ. ProSA-web: interactive web service for the recognition of errors in three-dimensional structures of proteins. Nucleic Acids Res. 2007;35 suppl 2:W407–10.
18. Sippl MJ. Recognition of errors in three-dimensional structures of proteins. Proteins: Structure, Funct Bioinformatics. 1993;17(4):355–62.
19. Colovos C, Yeates TO. Verification of protein structures: patterns of nonbonded atomic interactions. Protein Sci. 1993;2(9):1511–9.

20. Bolton EE, Wang Y, Thiessen PA, Bryant SH. PubChem: integrated platform of small molecules and biological activities. Annu Rep Comput Chem. 2008;4:217–41.

21. Jones G, Willett P, Glen RC, Leach AR, Taylor R. Development and validation of a genetic algorithm for flexible docking. J Mol Biol. 1997;267(3):727–48.

22. Verdonk ML, Cole JC, Hartshorn MJ, Murray CW, Taylor RD. Improved protein–ligand docking using GOLD. Proteins: Structure, Funct Bioinformatics. 2003;52(4):609–23.

23. Zoete V, Cuendet MA, Grosdidier A, Michielin O. SwissParam: a fast force field generation tool for small organic molecules. J Comput Chem. 2011;32(11):2359–68.

24. Jorgensen WL, Chandrasekhar J, Madura JD, Impey RW, Klein ML. Comparison of simple potential functions for simulating liquid water. J Chem Phys. 1983;79:926.

25. Bussi G, Donadio D, Parrinello M. Canonical sampling through velocity rescaling. J Chem Phys. 2007;126(1):014101.

26. Parrinello M, Rahman A. Polymorphic transitions in single crystals: A new molecular dynamics method. J Appl Phys. 1981;52(12):7182–90.

27. Nosé S, Klein M. Constant pressure molecular dynamics for molecular systems. Mol Phys. 1983;50(5):1055–76.

28. Ryckaert JP, Ciccotti G, Berendsen HJC. Numerical integration of the cartesian equations of motion of a system with constraints: molecular dynamics of n-alkanes. J Comput Phys. 1977;23(3):327–41.

29. Hess B, Bekker H, Berendsen HJC, Fraaije JGEM. LINCS: a linear constraint solver for molecular simulations. J Comput Chem. 1997;18(12):1463–72.

30. Miyamoto S, Kollman PA. SETTLE: an analytical version of the SHAKE and RATTLE algorithm for rigid water models. J Comput Chem. 1992;13(8):952–62.

31. Darden T, York D, Pedersen L. Particle mesh Ewald: An N log (N) method for Ewald sums in large systems. J Chem Phys. 1993;98:10089-10089.

32. Essmann U, Perera L, Berkowitz ML, Darden T, Lee H, Pedersen LG. A smooth particle mesh Ewald method. J Chem Phys. 1995;103(19):8577–93.

33. Zhang Y, Dufort I, Rheault P. Characterization of a human 20alpha-hydroxysteroid dehydrogenase. J Mol Endocrinol. 2000;25(2):221–8.

34. Nahoum V, Gangloff A, Legrand P, Zhu D-W, Cantin L, Zhorov BS, et al. Structure of the Human 3α-Hydroxysteroid Dehydrogenase Type 3 in Complex with Testosterone and NADP at 1.25-Å Resolution. J Biol Chem. 2001;276(45):42091–8.

35. Bennett MJ, Albert RH, Jez JM, Ma H, Penning TM, Lewis M. Steroid recognition and regulation of hormone action: crystal structure of testosterone and NADP+ bound to 3-hydroxysteroid/dihydrodiol dehydrogenase. Structure. 1997;5(6):799–812.

2

A comparative analysis of the foamy and ortho virus capsid structures reveals an ancient domain duplication

William R. Taylor[1]* ⓘ, Jonathan P. Stoye[2] and Ian A. Taylor[3]

Abstract

Background: The *Spumaretrovirinae* (foamy viruses) and the *Orthoretrovirinae* (e.g. HIV) share many similarities both in genome structure and the sequences of the core viral encoded proteins, such as the aspartyl protease and reverse transcriptase. Similarity in the *gag* region of the genome is less obvious at the sequence level but has been illuminated by the recent solution of the foamy virus capsid (CA) structure. This revealed a clear structural similarity to the orthoretrovirus capsids but with marked differences that left uncertainty in the relationship between the two domains that comprise the structure.

Methods: We have applied protein structure comparison methods in order to try and resolve this ambiguous relationship. These included both the `DALI` method and the `SAP` method, with rigorous statistical tests applied to the results of both methods. For this, we employed collections of artificial fold 'decoys' (generated from the pair of native structures being compared) to provide a customised background distribution for each comparison, thus allowing significance levels to be estimated.

Results: We have shown that the relationship of the two domains conforms to a simple linear correspondence rather than a domain transposition. These similarities suggest that the origin of both viral capsids was a common ancestor with a double domain structure. In addition, we show that there is also a significant structural similarity between the amino and carboxy domains in both the foamy and ortho viruses.

Conclusions: These results indicate that, as well as the duplication of the double domain capsid, there may have been an even more ancient gene-duplication that preceded the double domain structure. In addition, our structure comparison methodology demonstrates a general approach to problems where the components have a high intrinsic level of similarity.

Keywords: Virus capsid structure, Foamy virus evolution, Protein structure comparison

Background

Taxonomically, the *Orthoretrovirinae* (orthoretroviruses) and *Spumaretrovirinae*[1] (spumaviruses) make up the two subfamilies of *Retroviridae*. They share many similarities, including overall genome structures with gag, pol and env genes encoding proteins for replication and life cycles involving reverse transcription and integration into the chromosomes of infected cells. However, there are also a number of differences distinguishing these viral subfamilies, including finer details of genome organisation, the absence of a Gag-Pol fusion protein in spumaviruses and the timing of reverse transcription [1].

Gag is the major structural protein of both Ortho and Foamy viruses and is responsible for many of the differences and similarities between the viral subfamilies. Ortho and Foamy viral Gags are required for particle assembly, budding from the cell, reverse transcription and delivery of the viral nucleic acid into the newly infected cell. However, there are a number of striking differences including how the Gag precursor is targeted to the cell membrane, the absence of a Major Homology Region and Cys-His box

*Correspondence: william.taylor@crick.ac.uk
[1]Computational Cell and Molecular Biology Laboratory, Francis Crick Institute, Midland Road, NW1 1AT London, UK
Full list of author information is available at the end of the article

in Foamy viruses and very different patterns of processing during viral maturation [2]. In all Ortho viruses, Gag is proteolytically cleaved to form distinct, well-studied proteins, matrix (MA), capsid (CA) and nucleocapsid (NC), found in mature virions, whilst in spumaviruses Gag processing to remove a C-terminal peptide occurs only in a fraction of the Gag molecules [3].

Structural information regarding foamy virus Gag has been limited to the crystal structure of the N-terminal Env binding region of Prototypic Foamy virus (PFV) Gag (PFV-Gag-NtD) that although maintaining some of the function of orthoretrivial MA shared no structural similarity [4]. However, more recently the solution NMR structure of the PFV Gag central CA domains has shed new light on the relationship between ortho and spumaviruses. It reveals that the CA structures of both viral subfamilies share a common protein fold, implying that their Gag proteins may be evolutionarily related [5].

However, an intriguing aspect of this relationship was an ambiguity in the degree of relatedness between the CA domains of the Gag proteins, with the Spumaretroviral CA domains, NtDCEN and CtDCEN, appearing almost equally similar to either the amino- (CA-NtD) or carboxy-terminal (CA-CtD) domains of the orthoretroviruses. With small domains that share a high degree of background similarity, particularly those composed entirely of α-helices, it is very difficult to evaluate the significance of their structural relationships as chance combinations of a few helices can give rise to an apparently convincing overlaps with a low RMSD.

In this paper, we now investigate and clarify the nature of the relationship between these capsid domains and discuss its evolutionary implications. Our work provides a demonstration of a general approach to the resolution of difficult comparison problems in which the proteins share a high intrinsic level of similarity.

Results
Full-length comparison
To investigate the structural relationship between the capsid structure of the ortho viruses (HIV, MLV, etc.), and the new structure of the foamy virus capsid [5] (PDB codes: 5m1g, 5m1h), the foamy virus structure was compared to one of the few full double domain ortho virus structures, the HIV capsid with PDB code: 3nte, using the flexible superposition program SAP [6]. Even though this program has a tolerant approach to relative domain shifts, the comparison produced a high RMSD value of 14Å over the 100 best superposed positions. The amino (N) terminal domain positions roughly corresponded but shifts in the relative orientation of the carboxy (C) terminal domain resulted in large deviations between equivalent helices. The superposed structures are shown in Fig. 1a and the domain divergence can

be seen clearly as a jump in the cumulative RMSD plot (Fig. 1b).

DALI searches
Although this initial superposition (Fig. 1) did not appear encouraging, the foamy virus structure was scanned across the Protein DataBank (PDB), using the DALI program [7] to search for any similarities.

Full chain scan
A scan of the full-length foamy structure using the DALI server[2] over the 90% non-redundant protein structure databank identified a wide selection of retroviral capsid structures. In the ranked list of structure hits, capsids were identified from position 2 to position 550. The top hits are shown in Fig.2 (See Additional file 1 for a summary of the full 550 with Z-scores over 2). Many capsids are found in the top 20 hits and although the top scoring hit is not obviously a capsid protein, it is thought to have originated from the Ty3/Gypsy retrotransposon family *gag* gene [8]. However, almost all of these are partial hits, covering little more than half the query structure. The structural alignment of the top two hits is shown in Fig. 3 coloured to emphasise the matched regions.

The result of the DALI search indicated that the Foamy virus structure shares some similarity with the capsid structure of the ortho-viruses. However, the matches consist only of a small number of helices and appears barely more convincing than other matches to proteins that seem very unlikely to have any meaningful connection to a viral capsid. The preponderance of capsid matches throughout the list of hits might seem to add some support to the relationship but may simply be a reflection of the number of capsid structures in the structure databank.

Adding confusion to the ortho/foamy relationship is the additional observation that the distribution of matches to the ortho-virus structures between the amino (N) and carboxy (C) terminal domains are mixed. For example; taking the top 10 matches, the N-terminal domain of the Foamy structure aligns with 6 C-terminal domains and 4 N-terminal domains of the ortho virsuses and the best match with the corresponding Foamy C-terminal domain aligns with an ortho N-terminal domain.

Domain scans
To clarify the domain match specificity, the two domains of the Foamy virus (1–88 and 89–180, as defined automatically [9]) were scanned separately using the DALI program. The individual domains were much more specific at matching known capsid structures[3], both in the full PDB and PDB-90 collections as can be seen from the plots in Fig. 4.

The results of these scans strengthened the identification of the relationship to the ortho capsids and supported

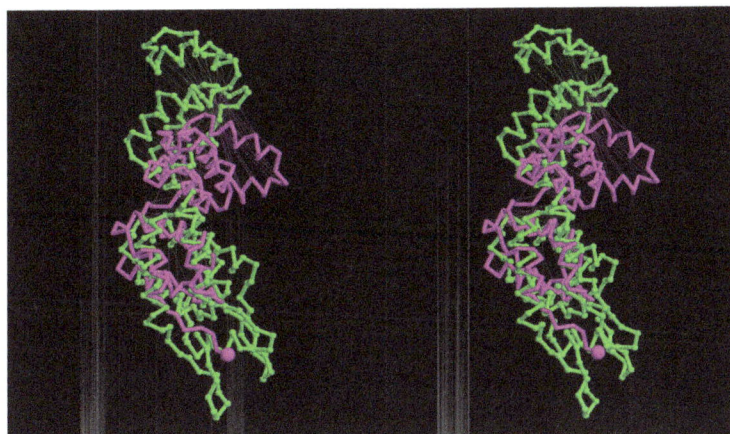

(a)

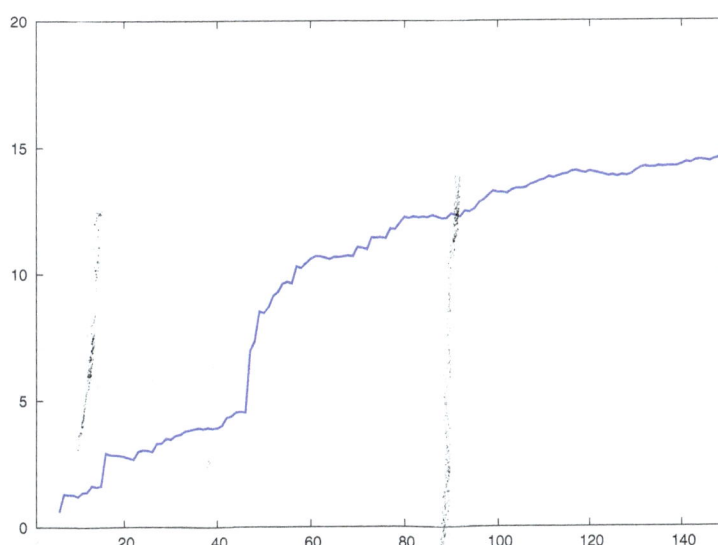

(b)

Fig. 1 Full ortho/foamy virus capsid superposition. The superposed structures are shown in part (**a**) as a stereo pair, coloured as *green* = ortho virus (HIV, PDB code: `3nte-A`) and *magenta* = foamy virus capsid. (The amino terminus is marked by a small sphere). Part (**b**) shows the cumulative RMSD plot for this superposition which plots the RMSD value (*Y*-axis) for increasingly larger sets of residues as ranked by their SAP similarity score (*X*-axis). The sharp rise in this trace marks the transition into subsets that include positions from the displaced domain

the swapped specificity for the N-terminal match of the Foamy structure with the C-terminal match of the ortho virus and *vica versa*, with all top 12 hits of each domain matching their opposed counterpart. The structure-based sequence alignments of each domain based on this equivalence are shown in Fig. 5.

Although domain transposition is not impossible in viral genomes, it is sufficiently unexpected to warrant deeper investigation, especially as it is hard to imagine how an ancestral capsid protein could tolerate such a large rearrangement and still pack to form a competent shell. We therefore undertook a more thorougher evaluation using alternative methods to assess the statistical significance of these structural similarities.

Structural alignment significance
Reversed-structure searches
For each comparison, the DALI program calculates an empirical Z-score, combining an estimation of significance with protein length normalisation. The program reports all matches over Z=2, however, when the proteins are small and especially when the structures being compared are both predominantly alpha-helical in nature, then matches over this cutoff include many functionally unrelated hits where the similarity has arisen through the fortuitous alignment of a few helices.

Therefore, to calculate a stricter cutoff on score, we created a decoy probe by reversing the alpha-carbon backbone then reconstructing the full atomic structure, using a

```
No:  Chain  Z    rmsd lali nres %id PDB  Description
 1:  4x3x-A 5.0  3.1   66   82  11 PDB  MOLECULE: ACTIVITY-REGULATED CYTOSKELETON-ASSOCIATED;
 2|  3g29-A 3.7  2.7   60   77   8 PDB  MOLECULE: GAG POLYPROTEIN;
 3|  3g0v-A 3.7  2.9   62   76   8 PDB  MOLECULE: GAG POLYPROTEIN;
 4:  2v60-D 3.6  2.2   41  998   7 PDB  MOLECULE: MULTIDRUG RESISTANCE PROTEIN MEXB;
 5:  3j38-i 3.6  2.5   40  113   3 PDB  MOLECULE: 60S RIBOSOMAL PROTEIN L10A-2;
 6|  4ph2-A 3.6  3.2   69  127   7 PDB  MOLECULE: BLV CAPSID - N-TERMINAL DOMAIN;
 7:  1iqp-E 3.6  3.8   69  326   7 PDB  MOLECULE: RFCS;
 8:  4gco-A 3.6  5.7   55  120  11 PDB  MOLECULE: PROTEIN STI-1;
 9|  3g29-B 3.6  2.8   62   77   8 PDB  MOLECULE: GAG POLYPROTEIN;
10|  3g11-B 3.6  2.9   62   75   8 PDB  MOLECULE: GAG POLYPROTEIN;
11|  3g21-A 3.6  2.8   60   77   8 PDB  MOLECULE: GAG POLYPROTEIN;
12:  2a0u-A 3.5  3.1   68  374   4 PDB  MOLECULE: INITIATION FACTOR 2B;
13:  1j7q-A 3.5  2.9   60   86   5 PDB  MOLECULE: CALCIUM VECTOR PROTEIN;
14:  2a0u-B 3.5  8.1   80  367   4 PDB  MOLECULE: INITIATION FACTOR 2B;
15:  1iqp-A 3.5  3.7   70  326   7 PDB  MOLECULE: RFCS;
16|  4ph0-C 3.5  4.6  101  199   8 PDB  MOLECULE: BLV CAPSID;
17|  4ph0-D 3.5  4.2  101  198   8 PDB  MOLECULE: BLV CAPSID;
18|  4ph2-B 3.5  3.3   69  127   7 PDB  MOLECULE: BLV CAPSID - N-TERMINAL DOMAIN;
19:  1sxj-B 3.4  3.5   65  316   3 PDB  MOLECULE: ACTIVATOR 1 95 KDA SUBUNIT;
20:  2afd-A 3.4  2.7   59   88  14 PDB  MOLECULE: PROTEIN ASL1650;
```

Fig. 2 Top structural similarities. Found by the DALI program in the 90% non-redundant PDB (PDB-90) using the full length foamy virus capsid as a query (145 residues). The columns are: the ranked number of the hit (No.), marked by a '|' for a capsid protein, otherwise ':'; the PDB entry identifier (Chain, with the chain designation after the dash); the DALI Z-score (Z) (significance estimate); the root-mean-square-deviation (rmsd) over aligned α-carbon positions; the number of aligned positions (lali); the number of residues in the matched structure (nres); the percentage sequence identity of the match (%id) followed by a description of the molecule. It can be seen from the number of matched positions (lali) that most matches are partial, covering typically less than half the query structure

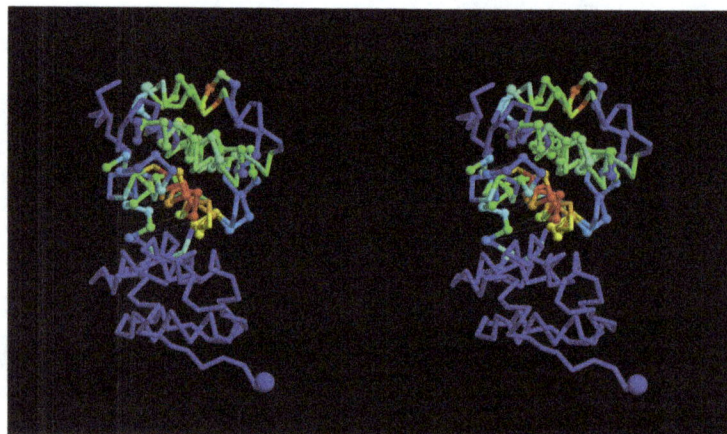

(a) 4x3x-A

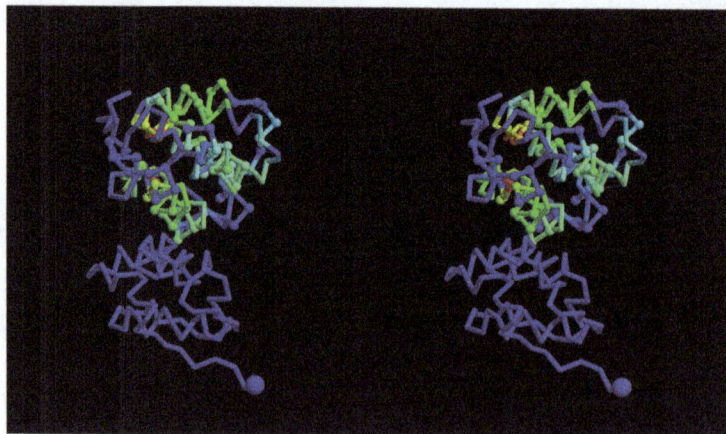

(b) 3g29-A

Fig. 3 Top hits superposed. The top two DALI hits to the full foamy virus capsid are shown as a α-carbon backbone (stereo pair) coloured using the residue similarity score calculated by SAP. (red = strong similarity, blue = none). The amino terminus of the foamy structure is marked by a large ball and the other structure is distinguished by small balls on its α-carbon atoms. **a** a cytoskeleton associated protein (fragment) of the arc/arg3.1 gene (PDB code: 4x3x-A), (which is thought to have originated from a Ty3/Gypsy retrotransposon family capsid) and (**b**) the structure of the capsid C-terminal domain of the Rous scarcoma virus (PDB code: 3g29-A)

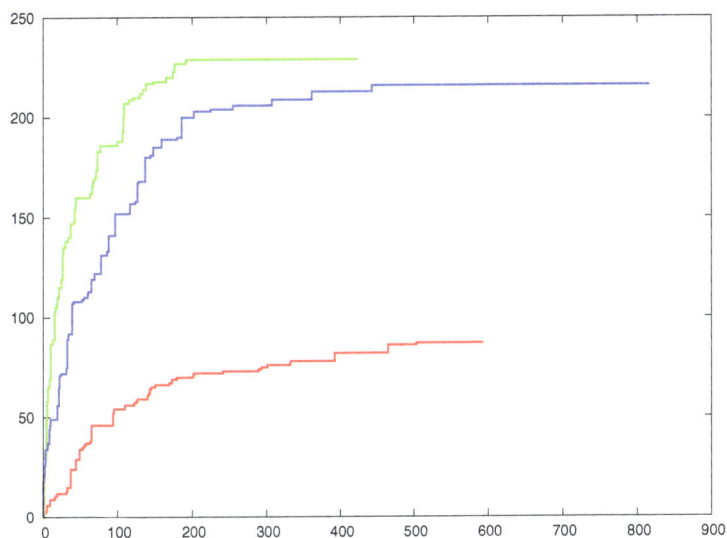

(a) Full PDB

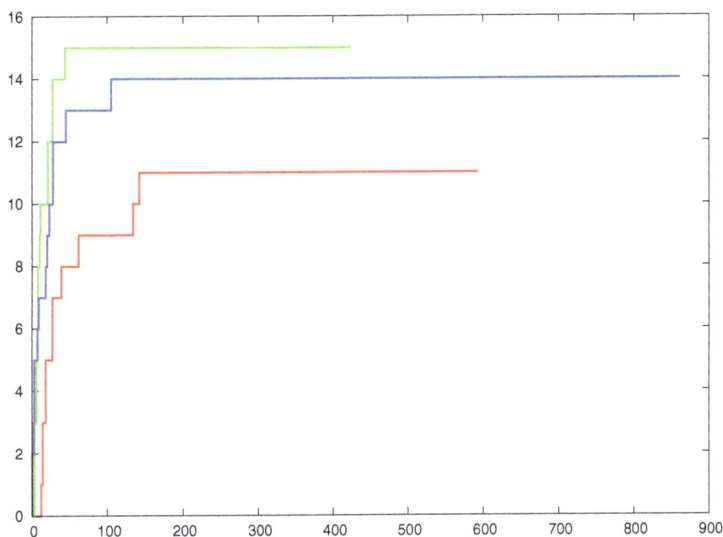

(b) PDB-90

Fig. 4 PDB capsid structure matches. The number of capsid structures identified by the DALI program in (**a**) the full PDB and (**b**) the 90% non-redundant PDB (PDB-90) is shown for queries using the full foamy capsid structure (*red*), the carboxy terminal domain (*green*) and the amino terminal domain (*blue*). The number of capsid hits (*Y*-axis) is plotted against the order of all hits ranked by Z-score down to a value of 2. A *curve* approaching the *top left corner* indicates greater specificity and the extent of a curve to the right indicates the total number of hits

simple algorithm to regenerate a full backbone[4]). Figure 6 plots the ranked DALI Z-scores for the separate (native) foamy domains. As would be expected, the larger C-terminal domain has hits with a higher significance than the smaller N-terminal domain: the former covers the range Z=2.5 to Z=5 over the true hits (magenta dots) whereas the latter tracks a similar profile running one Z-value unit lower (2–4 over true red dots). Plotting the Z-scores against the log of their rank produces almost linear traces for the hits from the PDB-90, making it easy

to compare N-domain (red/cyan dots) with C-domain (magenta/green dots) (for T/F hits) in Fig. 6.

The equivalent scans with the reversed domain structures, using both the foamy and ortho (HIV) structures (neither of which should have any particular relationship to the capsid or any other natural protein) also found hits with high Z-scores (black and blue points in Fig. 6, respectively). When compared with the native domains (Fig. 6), these decoys had a profile that tracked mostly above the N-terminal native domain but below the

```
             APSPVIPIQHIRAVTGEVPNNPRDIPMWIGRNAPAIEGVYPVTTPDLRARIINALIGGKSGIHLTAPEAVTWASAVAAIFTRTHGSFP
             GILSVLPISQIRTVIGNTPVDPKKVPLWIAKSASAIEGVMPTNTPDIRCRLVNALLPQHGGLILQPHECNSWTQIASALYTRVNGMIP
             VIGPVIPINHLRSVIGNTPPNPRDVALWLGRSTAAIEGVFPIVDQITRMRVVNALVASHPGLTLTENEAGSWNAAISALWRKAHGAAA
             AQPVVIPINVIRSVCGDTPSNPQDIPLWMGRIIPAIEGVFPIDNPNLRMRVVNALLALHPGLAITELNAQTWGQVLAVLHMRALGHTA
             QPIHHLPITHIRAVIGETPAQIRDVPLWLAQSIPALTGVVPAMDAGTLTRLVNAITARHPGLALGMNEAGSWHEAVHLIWQRTFGATA
 Nter        PIGTVIPIQHIRSVTGEPPRNPREIPIWLGRNAPAIDGVFPVTTPDLRCRIINAILGGNIGLSLTPGDCLTWDSAVATLFIRTHGTFP
             aaaaaaaaaa         aaaaaaaaaaaaaaaaa      aaaaaaaaaaaaaaaa            aaaaaaaaaaaaaaa
             :  ::  |    |::|                                              :  |   :  ::|
             aaaaaaaaaaaaaaaa      aaaaaaaaaaaaaa                         aaaaaaaaaaaaaa
 3g1gA       ---------PWAD--IMQGPS--SFVDFANRLIKAVEGSDL-ARAPVIIDCFRQKSQPQQLI--PSTL-TTPGEIIKYVLDRQK----
 3tirA       ---------PWAD--IMQGPS--SFVDFANRLIKAVEGSDL-ARAPVIIDCFRQKSQPQQLI-------TTPGEIIKYVLDRQ-----
 3g1iA       ---------PWAD--IMQGPS--SFVDFANRLIKAVEGSDL-ARAPVIIDCFRQKSQPQQLI----TLTT-PGEIIKYVLDRQ-----
 3g29A       ---------PWAD--IMQGPS--SFVDFANRLIKAVEGS---ARAPVIIDCFRQKSQPQQLI-------TTPGEIIKYVLDRQ-----
 3gOvA       ---------PWAD--IMQGPS--SFVDFANRLIKAVEGSAL-ARAPVIIDCFRQKSQPQQLI-------TTPGEIIKYVLDRQ-----
 3g29B       ---------PWAD--IMQGPS--SFVDFANRLIKAVEGSNL-ARAPVIIDCFRQKSQPQQLI-------TTPGEIIKYVLDRQ-----
 3g1iB       ---------PWAD--IMQGPS--SFVDFANRLIKAVEGSDL-ARAPVIIDCFRQKSQPQQLI-------TTPGEIIKYVLDRQ-----
 3g26A       ---------PWAD--IMQGPS--SFVDFANRLIKAVEGS---CRAPVIIDCFRQKSQPQQLI-------TTPGEIIKYVLDRQ-----
 3dtjC       ---------SILD--IRQGPK--EPFRDYVDRFYKTLR--VKNW--MTATLLVQNANPD-TILKGPGA--TLEEMMTA-CQGV-----
 3dtjB       ---------SILD--IRQGPK--EPFRDYVDRFYKTLR--VKNW--MTATLLVQNANPD-TILKGPGA--TLEEMMTA-CQGV-----
 3dtjA       ---------SILD--IRQGPK--EPFRDYVDRFYKTLR--VKNW--MTATLLVQNANPD-TILKGPGA--TLEEMMTA-CQGV-----
 3g21A       ---------PWAD--IMQGPS--SFVDFANRLIKAVEGSDL-ARAPVIIDCFRQKSQPQQLI-------TTPGEIIKYVLDRQ-----

             MHNLSAILTGIANGEGVESAYNLGMMLSNGDFNLVYGIVRGLLPGQAAVAYMQQRLDAEPSDALRAQNFIQHLHLVYEILGLNHRGQSIR
             LHALPQTLSQVTKEEGILVAYQIGMTFTGQNFPLTWGILRPLLPGQAVVAMMQGYLDQYPTDDLKAVNFASILRRVFDILGLNYMGQNIR
             QHELAGVLSDINKKEGIQTAFNLGMQFTDGNWSLVWGIIRTLLPGQALVTNAQSQFDLMGDDIQRAENFPRVINNLYTMLGLNIHGQSIR
             LHQLPALLETIVKTDGILPAYNMGMEVTQQDFSYVWGILRTLLPGQAFVLSMQNELDRLPAA QRPGMFPGLLQRTLDILGLNSRGQNIQ
             LHALSDVLKGIAQRNGVVMALEMGLMFTNDDWDLTWSVIRRCLPGQASVVTIQARLDALPNNQARIIQAGFIIREVYEVLGLDPLGRPLH
 Cter        MHQLGNVIKGIVDQEGVATAYTLGMMLSGQNYQLVSGIIRGYLPGQAVVTALQQRLDQEIDNQTRAETFIQHLNAVYEILGLNARGQSIR
             aaaaaaaaaaaaa aaaaaaaaaaaa aaaaaaaaaa      aaaaaaaaaaaaaa aaaaaa aaaaaaaaaaaaa
             |: :|::  : ::::    |     :       |::  ::::|:   :|  ||  :
             aaaaaaaaaaaaa    aaaaaaaa      aaaaaa      aaaaaaaaaaaaaa  aaaaaaaaaaaaaaaaa
 1l6nA       SPRTLNAWVKVVEEKA-IPMFSALSE--GATPQDLNTMLNTVGGHQAAMQMLKETINEEA--EIYKRWIILGLNKIVRMYS------PTS
 3j34U       SPRTLNAWVKVVEEKA-IPMFSALSE--GATPQDLNTMLNTVGGHQAAMQMLKETINEEA--EIYKRWIILGLNKIVRMY-------SPT
 4uObF       SPRTLNAWVKVVEEKA-IPMFSALSC--GATPQDLNTMLNTVGGHQAAMQMLKETINEEA--EIYKRWIILGLNKIVRMY-------SPT
 4uObG       SPRTLNAWVKVVEEKA-IPMFSALSC--GATPQDLNTMLNTVGGHQAAMQMLKETINEEA--EIYKRWIILGLNKIVRMY-------SPT
 3h4eB       SPRTLNAWVKVVEEK--IPMFSALSC--GATPQDLNTMLNTVGGHQAAMQMLKETINEEA--EIYKRWIILGLNKIVRMY-------SPT
 2jprA       SPRTLNAWVKVVEEKA-IPMFSALSE--GATPQDLNTMLNTVGGHQAAMQMLKETINEEA--EIYKRWIILGLNKIVRMY----------
 1afvB       SPRTLNAWVKVVEEKAVIPMFSALSE--GATPQDLNTMLNTVGGHQAAMQMLKETINEEA--EIYKRWIILGLNKIVRMY-------SPT
 4uObE       SPRTLNAWVKVVEEK--IPMFSALSE--GATPQDLNTMLNTV-GHQAAMQMLKETINEEA--EIYKRWIILGLNKIVRMY-------SPT
 4uObK       SPRTLNAWVKVVEEK--IPMFSALSC--GATPQDLNTMLNTVGGHQAAMQMLKETINEEA--EIYKRWIILGLNKIVRMY-------SPT
 4uObH       SPRTLNAWVKVVEEK--IPMFSALSC--GATPQDLNTMLNTVGGHQAAMQMLKETINEEA--EIYKRWIILGLNKIVRMY-------SPT
 2gonA       SPRTLNAWVKVVEEK--VIPXFSALSE--GATPQDLNTMLNTVGGHQAAXQXLKETINEEA--EIYKRWIILGLNKIVRXYS--------
 1afvA       SPRTLNAWVKVVEEKAVIPMFSALSE--GATPQDLNTMLNTVGGHQAAMQMLKETINEEA--EIYKRWIILGLNKIVRMY-------SPT
```

Fig. 5 Top domain similarity alignments. The sequence alignments are shown for the top 12 capsid domain matches found by the `DALI` program using the foamy virus capsid N and C domains separately as a query over the full PDB. The sequence of the N-terminal domain (`N-ter`) is shown at the top of the first alignment block and the sequences of the C-terminal domain (`C-ter`) at the *top* of the second block. The sequences of the ortho-virsuses aligned below these all come from the "swapped" relationship of C and N terminal domains, respectively. These alignments, which are determined by structure not sequence, exhibit no specific similarity beyond what would be expected from aligning similar secondary structures from similar sized domains. (Amino acid identities are marked by a bar and similarities by a colon). The location of alpha helices is marked by the letter 'a', taken from the PDB entries of their adjacent proteins. A selection of other foamy virus sequences are aligned above the foamy virus sequence of known structure (human) which, from the *top*, are from: simian (orangutan), squirrel-monkey, cat, simian (unspecified) and horse. (NB. no alignment is implied between the two blocks of aligned domains)

C-terminal domain. However, with the latter domain, this was only distinct in the hits to the full PDB whereas with the PDB-90, the native domain was only clearly better over the top 10 matches, half of which were to non-capsid structures.

The results with the simple reversed decoy using `DALI` suggested that the match of the foamy virus domains to the ortho virus capsid N-terminal domain may be due to chance and that the match to the C-terminal domain looks meaningful if based on the hits to the full PDB but may be only marginal based on the PDB-90 hits.

However, both the N and C terminal domains pocess a degree of internal symmetry which gives rise to a partial match with their reversed 'doppleganger' decoys. The N-terminal domain superposed on its decoy had an RMSD of 5.4/60 (Å/α-carbon s) and 5.5/24 for the C-terminal domain (Fig. 7). The higher symmetry of the smaller domain may be sufficient to explain its poor level of specificity seen in Fig. 6 and to try and resolve this ambiguity, a more diverse set of decoys was generated based on cyclic permutation and segment swapping combined with chain reversal [10].

Customised decoy comparisons

To improve the statistical analysis of the foamy/ortho capsid similarity, we employed a method based on the generation of a population of customised 'decoy' models to provide a background distribution of unrelated protein scores [10]. This method retains the advantage of the simple reversed structures where every comparison that constitutes the random pool is between two models of the same size and secondary structure composition as the pair of native structures being compared. For this study we collected 12 capsid N-terminal domains and 7 C-terminal domains, each of which were compared with the foamy N-terminal domain and the foamy C-terminal domain. (The structures are identified in Table 1 with full details in the "Methods" section).

For each domain pair to be compared, decoys were created using cyclic permutation and segment swapping with chain reversal to generate a family of customised decoys for each comparison [10]. All pairs of forward/reversed decoys were then compared, with each pair being drawn from a pool of models generated from the two native structures. This ensures that the native domains (which

(a)

(b)

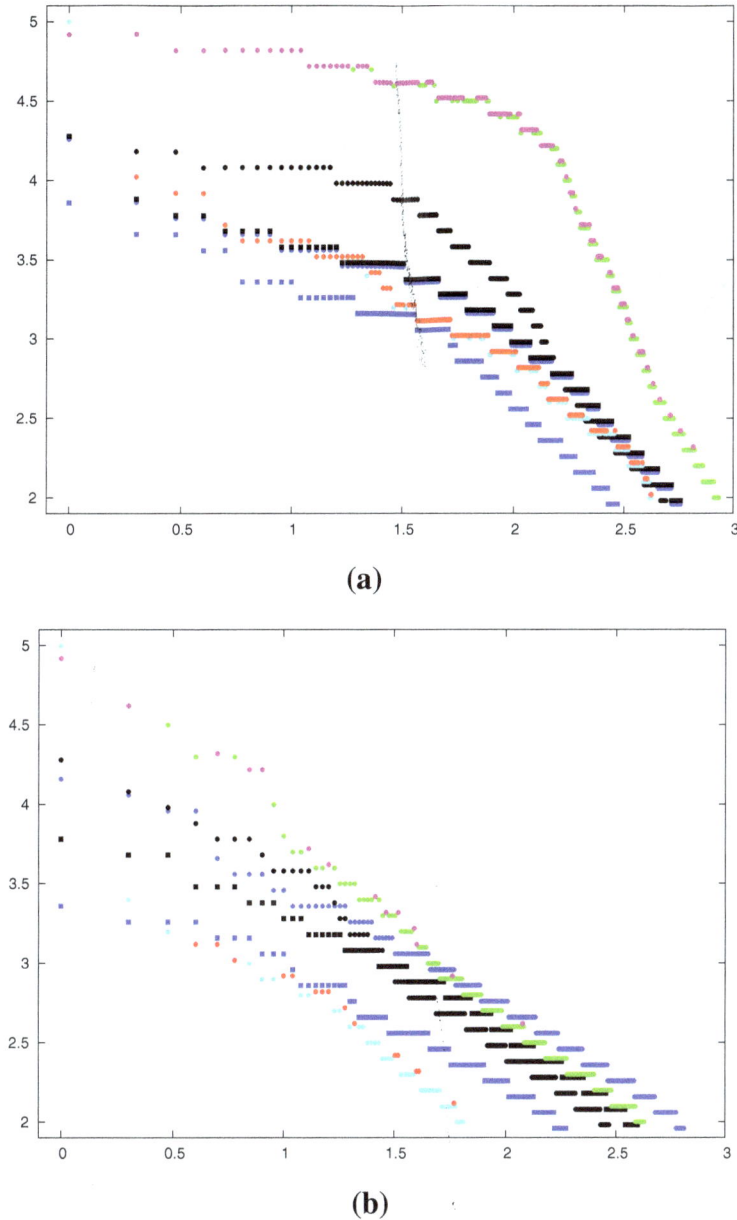

Fig. 6 DALI scores with decoys. The DALI Z-scores (Y-axis) are plotted against the $\log_{10}$ of their ranked position in the list of hits (X-axis) with the amino-terminal domain (N) as T=*red*, F=*cyan* dots and the carboxy-terminal domain (C) as T=*magenta* and F=*green* dots, where T is a true capsid hit and F is a false hit to a non-capsid protein. Four sets of decoys are compared to these, consisting of the reversed foamy capsid domains in *black* and the reversed HIV capsid domains in *dark-blue* (with a circle = N and a square = C domains in both). The DALI score for each set of hits has been slightly displaced to prevent coincident dots from being obscured. (This happens because of the integral number of residues and the DALI score being specified to only one decimal place). **a** full PBD. **b** PDB-90

may have different lengths) are always evaluated against a decoy pair with the same length combination. (See Methods section for details). All the decoy comparisons, of which there are typically 150–300 for each comparison, can then be compared to the native pair on a plot of RMSD against the number of matched residues (α-carbon atoms). An example is shown in Fig. 8(c)

for the comparison of the HIV1 structure (PDB codes: 1ak4 (N) and 1a43 (C)) domains against the foamy virus Gag domains.

Statistical analysis of the decoy comparisons

The quality of the comparisons in Fig. 8c can be quantified as a combination of their RMSD (R) and the

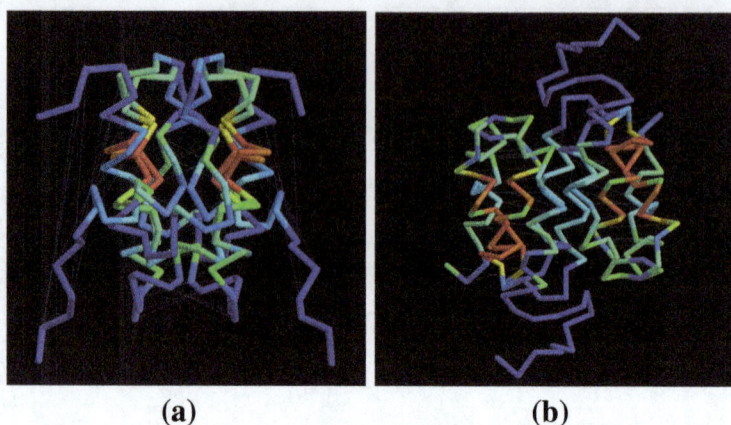

(a)   **(b)**

Fig. 7 Native/decoy similarity. When superposed using the program SAP, both N-terminal (*left*) and C-terminal (*right*) domains have some degree of similarity to their reversed decoy 'doppleganger', which is more marked for the *N* domain. The superposed structures are coloured by the SAP residue-level score as *red* = high similarity, *blue* = low. The *N* domain has roughly 60 equivalent α-carbon positions compared to only 24 in the larger C domain. **a** N. **b** C

Table 1 Ortho and foamy domain comparison Z-score statistics

a	ortho-N					
	foamy-N			foamy-C		
Virus	Pool	a-value	Z-score	Pool	a-value	Z-score
BLV6	300	0.552	**4.073**	244	0.542	3.692
BLV	251	0.550	**4.494**	184	0.400	3.669
HIV6	312	0.551	3.781	220	0.405	3.579
HIV1	312	0.573	3.703	213	0.402	3.692
HML2	264	0.777	2.166	196	0.438	**4.594**
HTLV	400	0.592	**4.030**	328	0.457	**4.013**
JSRV	225	1.063	0.896	190	0.601	3.237
MLV	326	0.751	3.044	188	0.508	3.151
MPMV	269	0.565	**3.902**	185	0.523	2.918
PSIV	285	0.621	3.731	235	0.369	**5.019**
RELIK	234	0.639	3.688	237	0.700	3.297
RSV	204	0.543	3.123	239	0.526	3.542
b	ortho-C					
BLV6	144	0.763	3.019	212	0.709	**4.046**
BLV	154	0.578	3.400	204	0.556	**4.047**
HIV1	157	0.593	3.760	174	0.705	3.362
HIV6	179	0.780	3.175	177	0.640	**4.380**
HML2	185	0.732	3.027	184	0.676	**3.900**
HTLV	156	0.685	3.847	163	0.694	2.807
RSV	155	0.448	3.754	235	0.403	**5.009**

For each amino (N) and carboxy (C) domain pair between an ortho virus structure and the foamy virus capsid structure, a **Z-score** is calculated based on the **a-value** (Equ^n. 1) derived from the comparison RMSD and length, relative to the **pool** of background decoy comparisons. The ortho **virus** identity is indicated by the code to the left, full details of which can be found in the "Methods" section. The top 12 Z-scores are high-lighted in bold, only three of which support a swapped domain match

number of matched (superposed) positions (N). However, as explained in the "Methods" section, for statistical analysis, it is easier to combine this pair of numbers as a single number, called the a-value (Equ^n. 1), which is the scaling factor that causes a theoretical curve to pass through the point (R, N).

When expressed by a single a-value all the data points in a comparison, such as Fig. 8c, can be plotted as a frequency histogram and examined to see if they approximate a Normal distribution. The distributions were found to be a good fit to unskewed Gaussians and so were treated as normal distributions (rather than extreme value distributions that have also been considered previously as a model for random structure comparison scores [10, 11]). The frequency data from the comparison of the orthoN domain from HIV1 and the foamyC domain (Fig. 8c) is shown in Fig. 9a along with a Normal distribution that has the same mean (μ) and standard deviation (σ) as the data. On this plot, the value of a (Equ^n. 1) for the comparison of the native pair of domains is also plotted (blue triangle) and from its position, a Z-score can be calculated.

In this way, the significance of all combinations of the native ortho and foamy domain superpositions were calculated, using the background distribution of 'customised' decoy comparisons based on each individual native pair. The resulting Z-scores (σ units) are collected in Table 1. The degree of similarity between the domains ranged from less than 1σ to over 5σ, with the latter (highly significant) result being obtained for both a swapped (NC) and forward (CC) combination. However, of the top 12 scores, only three now came from swapped pairings.

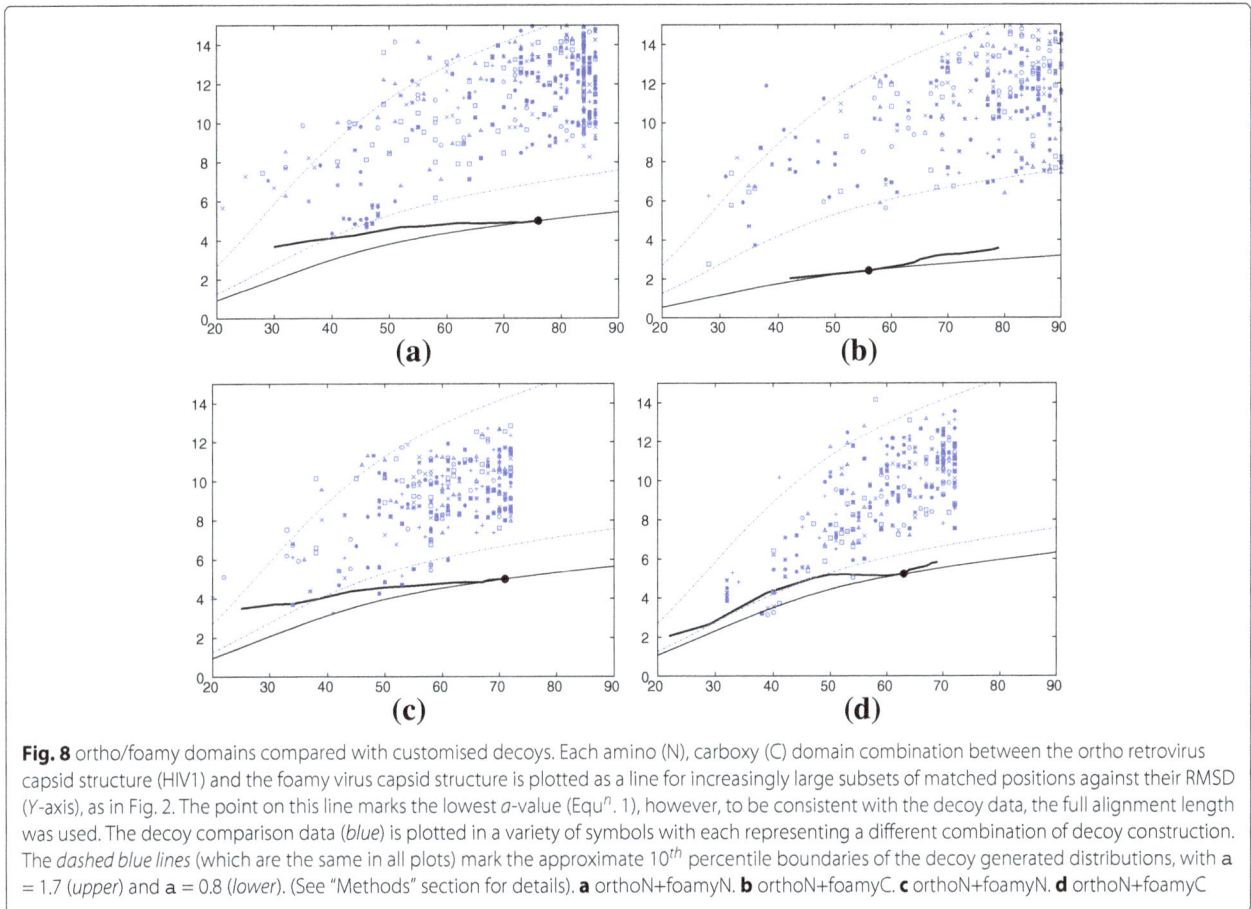

Fig. 8 ortho/foamy domains compared with customised decoys. Each amino (N), carboxy (C) domain combination between the ortho retrovirus capsid structure (HIV1) and the foamy virus capsid structure is plotted as a line for increasingly large subsets of matched positions against their RMSD (Y-axis), as in Fig. 2. The point on this line marks the lowest a-value (Equ^n. 1), however, to be consistent with the decoy data, the full alignment length was used. The decoy comparison data (*blue*) is plotted in a variety of symbols with each representing a different combination of decoy construction. The *dashed blue lines* (which are the same in all plots) mark the approximate 10th percentile boundaries of the decoy generated distributions, with a = 1.7 (*upper*) and a = 0.8 (*lower*). (See "Methods" section for details). **a** orthoN+foamyN. **b** orthoN+foamyC. **c** orthoN+foamyN. **d** orthoN+foamyC

Asymmetry statistics: To quantify the degree of bias for domains of like-type (NN, CC) to be more similar than those of mixed-type (NC, CN), the observed ranking of like and mixed pairs, based on their Z-value (Table 1), was compared to that expected by chance. The positions of all pairs in the list were shuffled a million times and the asymmetry of each arrangement was quantified as the number of like-pairs in the top half and also by their second moment: $\sqrt{\left(\left(\sum r_i^2\right)/N\right)}$, where r is the rank of the like-pair i in a list of N pairs. The chance of obtaining a distribution with more like-pairs being ranked higher can be caluclated by summing the area of the tail of each empirical distribution that lies beyond the observed value. However, these values were calculated over all pairs and neglects the principle that emphasis should be given to the more significant similarities. Rather than rely on a single significance cutoff (like 3σ) or an arbitrary cutoff (like the "3-out-of-12" mentioned above), we calculated statistics for all such cutoffs (Fig. 10a).

The majority of values in Fig. 10a lie below the 0.05 probability level for the larger sample sizes, with those for the top-half bias statistic (blue line) being more significant than the moment-based statistic (red line). While

confirming the visual trend towards a bias of higher scoring like-type domain similarities, the analysis summarised in Fig. 10a is complicated by having unequal numbers of amino and carboxy domain comparisons and also by including some closely related structures. To produce a more balanced data-set, one of each pair of the two most similar carboxy domain structures was discarded leaving five structures and for each of these, their matching amino terminal domain was also retained, leaving: BLV-1, HIV-1, HML2, HTLV-1 and RSV. Despite having a smaller set of comparisons (5N + 5C domains giving 20 rather than 38 Z-scores), the results for this reduced set indicated an equally clear bias towards towards a preferred like-domain equivalance, especially as measured by their occurrence in the upper half of the ranked list, with several having a probability below the 0.05 level and a few below the 0.005 level (Fig. 10b).

T-test statistic: An alternative to the above analysis, which still remains marginally significant, is to pool the raw comparison data for all the domain comparisons and their background distributions giving now not just a single value compared to a distribution but two distributions (Fig. 9b). For these data, a significance was calculated

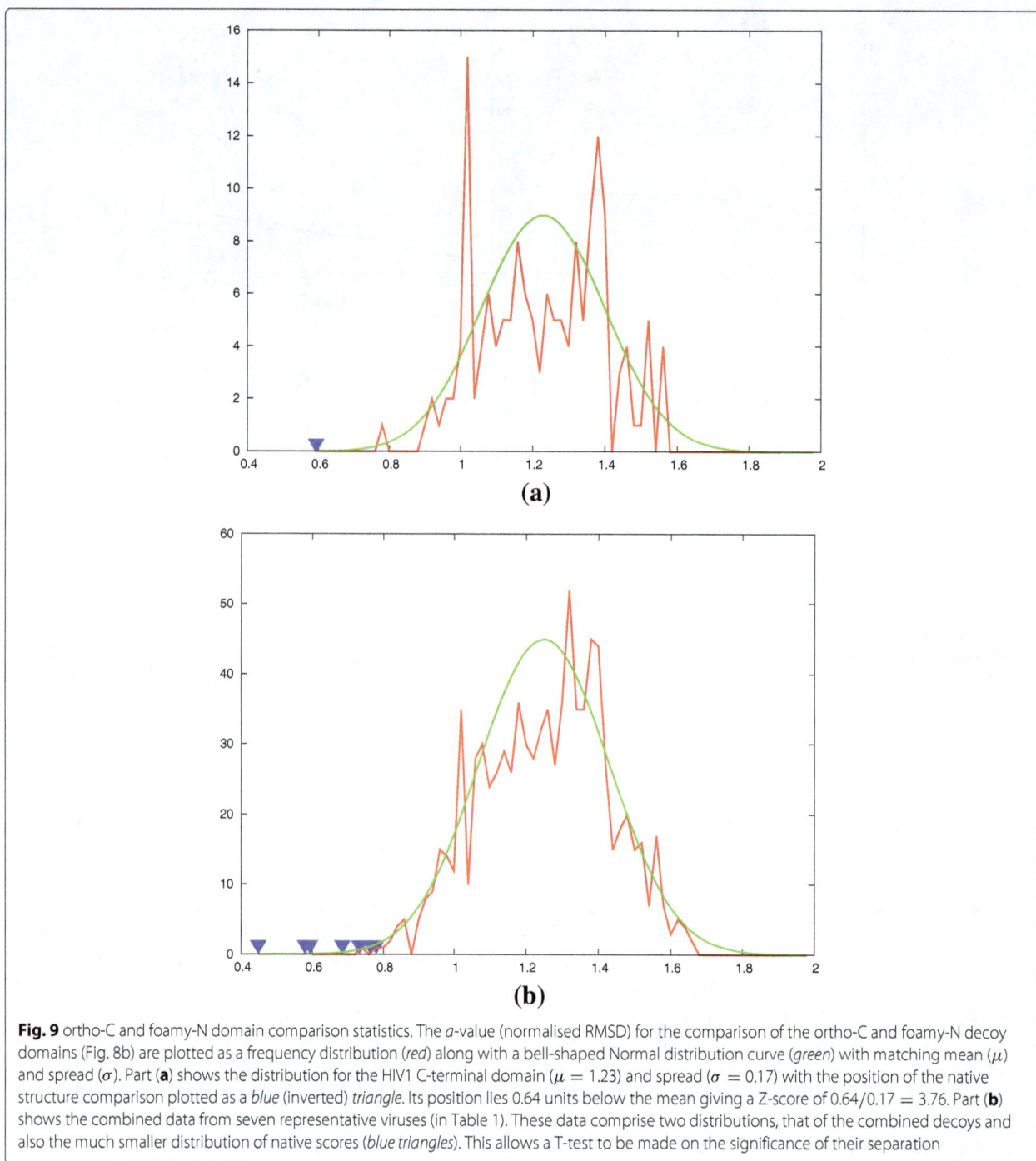

Fig. 9 ortho-C and foamy-N domain comparison statistics. The *a*-value (normalised RMSD) for the comparison of the ortho-C and foamy-N decoy domains (Fig. 8b) are plotted as a frequency distribution (*red*) along with a bell-shaped Normal distribution curve (*green*) with matching mean (μ) and spread (σ). Part (**a**) shows the distribution for the HIV1 C-terminal domain ($\mu = 1.23$) and spread ($\sigma = 0.17$) with the position of the native structure comparison plotted as a *blue* (inverted) *triangle*. Its position lies 0.64 units below the mean giving a Z-score of $0.64/0.17 = 3.76$. Part (**b**) shows the combined data from seven representative viruses (in Table 1). These data comprise two distributions, that of the combined decoys and also the much smaller distribution of native scores (*blue triangles*). This allows a T-test to be made on the significance of their separation

using Student's T-test, the values of which are given in Table 2.

From these results, it can be seen that all the four possible pairings are highly significant with probabilities ranging from 10^{-10} to over 10^{-20}. It is also clear that the two swapped pairings (NC and CN) have higher probabilities than the forward pairings (NN and CC). Combining the probabilities (*P*) as: $\Delta P = \log_{10}(P_{NN}P_{CC}) - \log_{10}(P_{NC}P_{CN})$, gives a value of

17.7 (42.7 - 25.0) which means that the swapped pairing is almost 18 orders of magnitude less likely than the forward pairing. Calculating the same statistic on the reduced 5N+5C domain data set gave a similar result but with a difference reduced 1000-fold to 15 orders of magnitude.

The unexpected swapped pairing, which was indicated originally by the DALI results, now seems less likely. The preferred, and biologically more reasonable, result is that

(a) Full dataset

(b) 5N+5C dataset

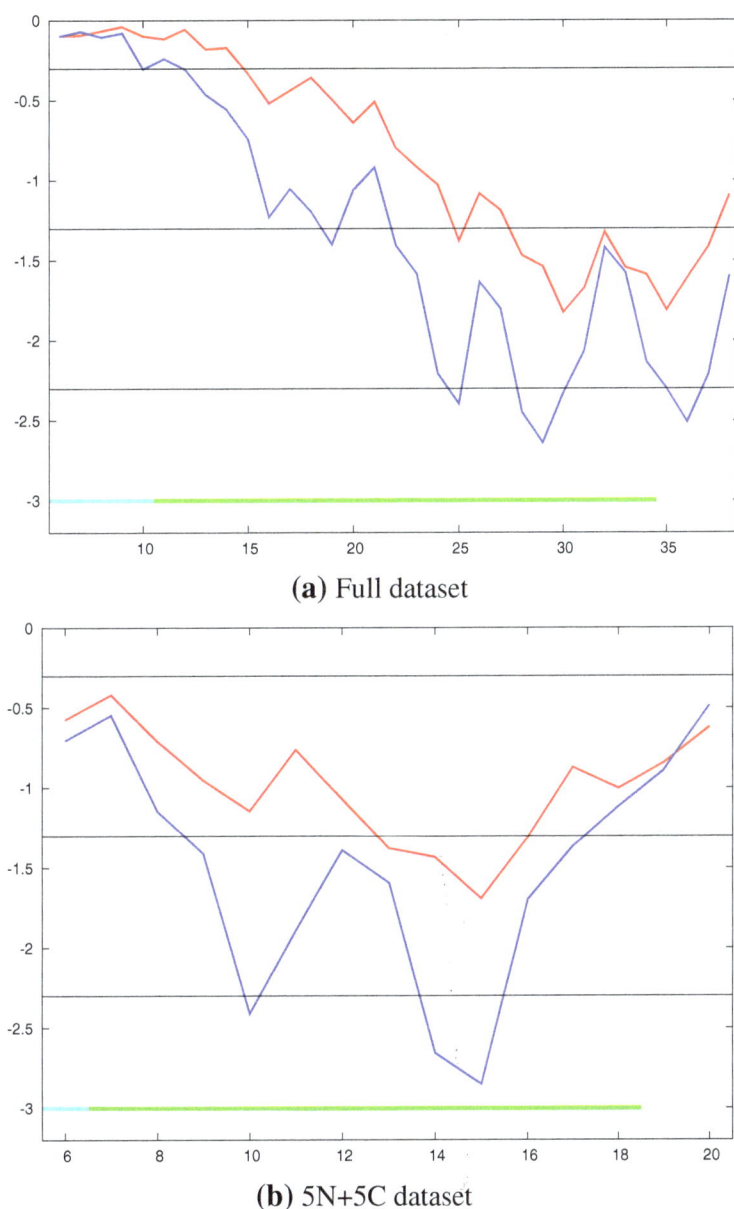

Fig. 10 Asymmetry statistics for like/mixed domain pairs. Given the ranked list of domain pairings, the chance for more domain pairs of like-type to be found higher than the observed order was evaluated from empirical distributions measured by two statistics: the second moment of the rank value (*red*) and the number of like-type pairs in the top half (*blue*). These statistics were calculated for all subsets from the 6 top pairs up to the full set of comparisons (*X*-axis) and for each, the chance of a better score is plotted as the $\log_{10}$ of the probability (*Y*-axis). The *horizontal lines* mark the 0.5, 0.05 and 0.005 levels. The line at the 0.001 level is coloured by the Z-score for each pair as: *green* = over 3 and *cyan* = over 4 sigma. Part (**a**) shows the probabilities calculated from the full set of 7 carboxy and 12 amino domains and part (**b**) shows the same values calculated on a more balanced set of 5 non-redundant carboxy domains and their matching amino domains

the ortho virus domain are related to the foamy virus domains as a result of genetic divergence from a common, double domain ancestor.

Internal duplication
The transposed pairings of N/C and C/N (ortho/foamy) domains still retain a high structural significance and this suggests that the two domains are derived from a common ancestral structure, probably as the result of a prior gene-duplication event that has been retained more clearly in the less embellished foamy virus structures. Comparing the two foamy domains gives a Z-score of 2.077 sigma which, although of marginal significance, supports this model. (Fig. 11a, b).

Table 2 ortho and foamy capsid domain comparison T-test significance

	orthoN	orthoC
foamyN	Avg: 6.67e-01 < 1.32e+00 Tprob = 4.62e-21 **	Avg: 6.51e-01 < 1.25e+00 Tprob = 2.35e-16 **
	StD: 1.61e-01 = 2.12e-01 Fprob = 1.84e-01	StD: 1.17e-01 = 1.89e-01 Fprob = 1.12e-01
foamyC	Avg: 4.92e-01 < 1.29e+00 Tprob = 4.09e-10 **	Avg: 6.22e-01 < 1.30e+00 Tprob = 3.81e-23 **
	StD: 1.02e-01 < 2.21e-01 Fprob = 7.37e-03 **	StD: 1.12e-01 = 1.77e-01 Fprob = 1.20e-01

For each combination of domains between the ortho and foamy viruses, the probability is given that the two means from each distribution (Avg values) were sampled from the same distribution. (i.e., that the native and decoy comparisons are not distinct). All domain pairings are extremely significant. An F-test was used to test if the standard deviations (Std) of each sample were distinct and if not, the a T-test was made on the assumption of equal standard deviations

Such a relationship between the foamy domains implies an equivalent relationship in the ortho viruses and a similar comparison in structures of their N and C domains finds matches with Z-scores ranging from 2 to 4. As with the comparison of the ortho and foamy structures, these can be pooled to allow a joint T-test to be applied. This gave a probability of 10^{-8} that the true N/C domain comparisons were drawn from the decoy distribution, adding strong support to the hypothesis of an ancient gene duplication occurring before the split of the ortho and foamy

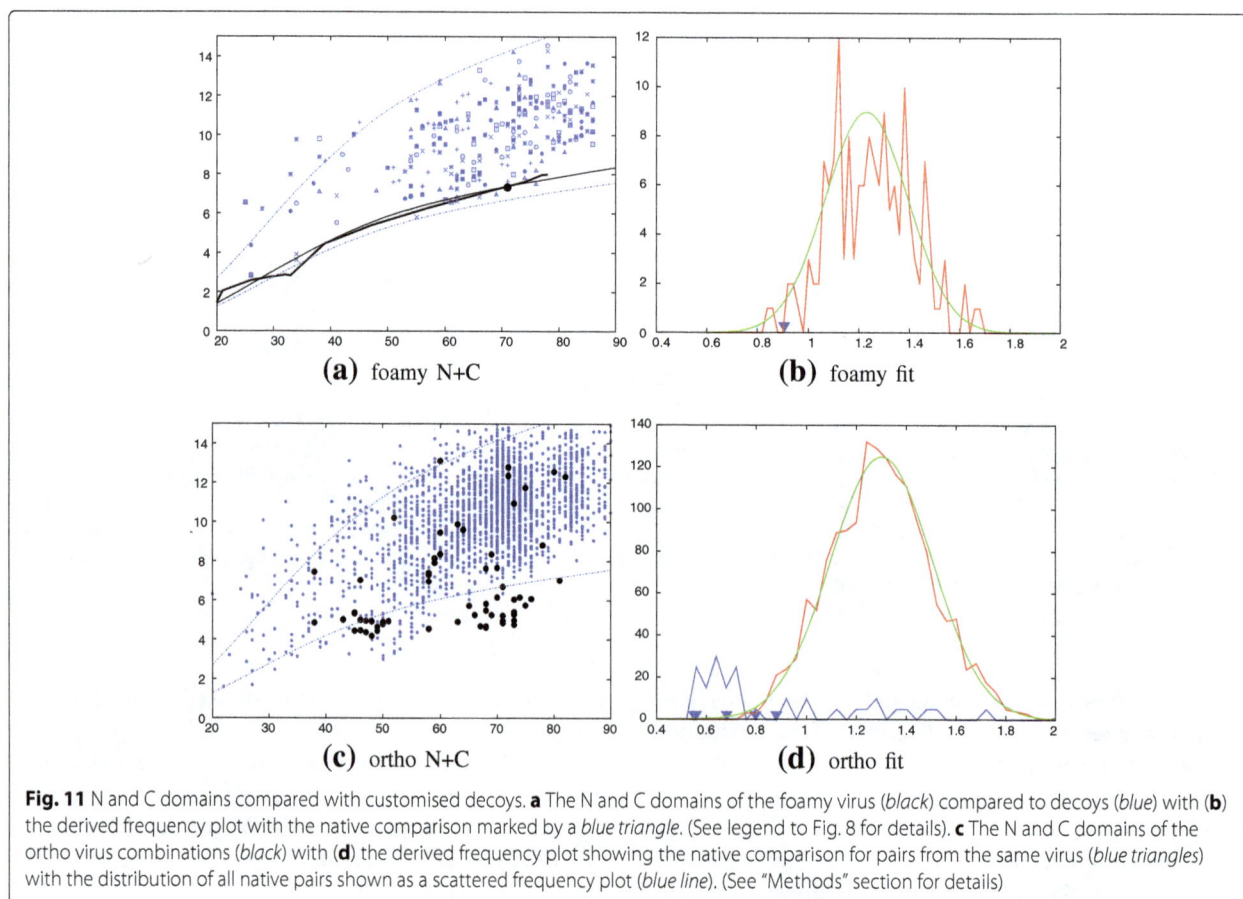

Fig. 11 N and C domains compared with customised decoys. **a** The N and C domains of the foamy virus (*black*) compared to decoys (*blue*) with (**b**) the derived frequency plot with the native comparison marked by a *blue triangle*. (See legend to Fig. 8 for details). **c** The N and C domains of the ortho virus combinations (*black*) with (**d**) the derived frequency plot showing the native comparison for pairs from the same virus (*blue triangles*) with the distribution of all native pairs shown as a scattered frequency plot (*blue line*). (See "Methods" section for details)

virus families. (Fig. 11a, b, *blue triangles*). Supporting this relationship, earlier studies also suggested an internal duplication in the ortho virsuses but were based largely on very distant sequence similarity [12].

This test was applied only to the comparison of domains between viruses with known structures for both domains, however, it is not unreasonable to compare amino and carboxy domains across all viruses. The longer loops in the ortho virus domains gives greater scope of structural variation and a wide range of variation was seen ranging from RMSD values under 4 to over 12. When normalised for length (*a*-value from Equ^n. 1) and partial matches under 60 positions excluded, a distinct cluster remains between $a = 0.5 \ldots 0.8$ (4...6Å RMSD) but still with a long tail to higher values. Despite this tail, the T-test on the distributions is highly significant at 2.7×10^{-17}.

One of the better N/C ortho similarities is shown in Fig.12a, along with the N/C ortho domain superposition in Fig.12b.

Fold-space representation

To summarise the structural relationships among the ortho and foamy domains, the matrix of pairwise comparisons was projected into a three-dimensional fold-space. (See "Methods" for details). This produces a best visual representation of the RMSD values between domains.

As can be seen from Fig. 13, the N and C domains of the ortho viruses form distinct clusters with the foamy C domain lying closer to the ortho C-domain cluster. The foamy N-domain, however, maintains a fairly equal distance from both ortho domain clusters but lies closer to its C-terminal partner.

Discussion and conclusions
Structure comparison
Pairwise significance

The comparison of small domains that are largely composed of α-helices presents a challenging problem in how to interpret the significance of the RMSD values. As the individual helical secondary structure elements (SSEs) constitute a sizeable fraction of the domain, it takes only the chance alignment of a few helices to result in a low RMSD over a large proportion of the structure, giving an apparently meaingful result.

The use of the customised decoy-model sets, as illustrated here, attempts to avoid this problem by recreating a large number of possible folds that were generated using the same (reconnected) SSEs. Moreover, to avoid any chance recreation of native fragments, each comparison always involved the comparison of a native (forward) chain direction with a reversed chain. Using these models, a background distribution of decoy/decoy comparisons allowed us to calculate Z-scores for each native/native comparison between the different Gag proteins. This has

the advantage that every comparison in the background distribution involved two models with the same length, residue packing density and secondary composition as the native pair. These values indicated a clearly significant relationship between the foamy and ortho CA structures.

Direct or transposed domain order?

Although the decoy model alignment strategy did confirm the relationship between the foamy and ortho CA structures, the Z-scores did not point to a clear resolution of whether the domains should have a direct correspondence (NN and CC match) or a transposed relationship (NC and CN) as significant individual matches were found across all pairings. Testing for a bias towards more significant like-domain pairings (NN, CC) in the list of similarities ranked by Z-score confirmed the visual bias towards a natural correspondence but only at a marginal level of significance (around 0.05). By contrast, the application of a T-test on the combined raw comparison data returned a very clear distinction between the direct and the transposed relationships, clearly favouring the more natural forward order.

However, although the "astronomic" probabilities calculated by the T-test seem very convincing, they must be viewed in the light of the much lower probabilities calculated from the asymmetry statistics. Both calculations involve assumptions and are limited by the small number of known structures so neither can be taken as definitive. Nevertheless it would seem likely that the "true" level of significance may lie somewhere between the two results and as both of these objective assessments point in the direction of the NN and CC domain order, there is no reason to adopt the more unexpected transposed domain order.

Evolutionary implications

On the basis of these structural comparisons, and a variety of recently described functional assays [5], we can conclude that the central region of the spumavirus *gag* gene encodes a polypeptide sequence related to that of the corresponding region of orthoretroviral, CA. It therefore seems reasonable to suppose that the last common ancestor of orthoretroviruses and spumaviruses possessed such a sequence. Moreover this region appears to be made up from two related all helical subdomains suggesting a gene duplication event in a common precursor.

In our initial search employing foamy virus CA using the DALI program, we made the observation that the strongest similarity of the foamy virus CA domains was actually with a cellular protein, Arc (Activity-Regulated Cytoskeleton-associated protein). Arc is required for neural synaptic growth and activity [13–16] and misregulation and/or deletion contributes to diseases of cognition [14, 16, 17]. Arc has widespread and clear sequence

(a) ortho

(b) foamy

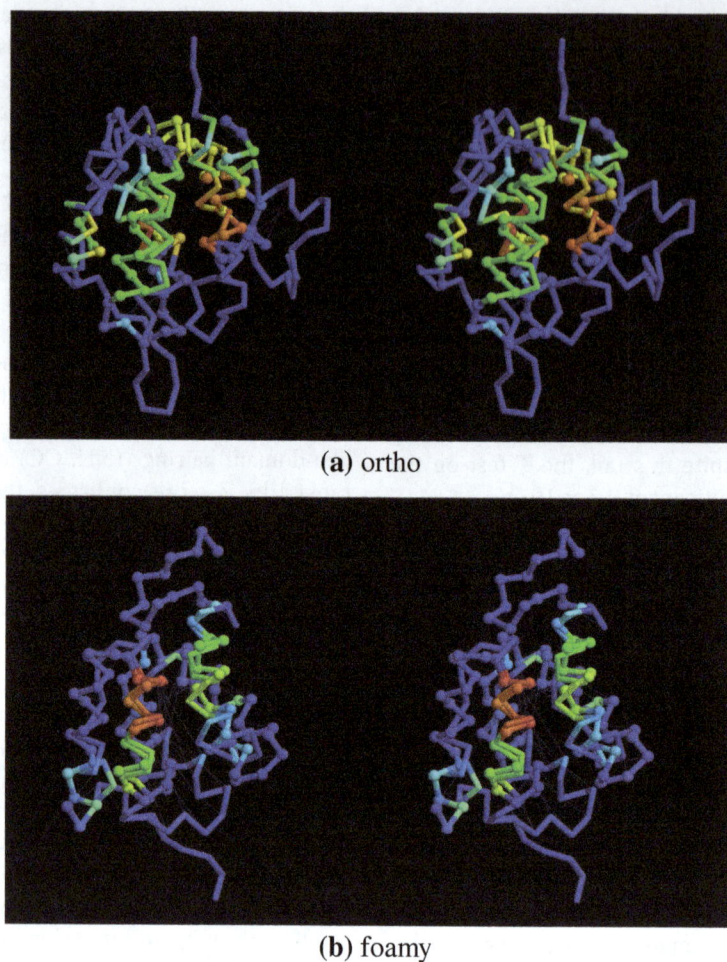

Fig. 12 Amino and carboxy domains superposed. **a** ortho virus domains and (**b**) foamy virus domains are shown as a stereo pair with their α-carbon backbones coloured by the residue similarity score calculated by SAP. (*red* = strong similarity, *blue* = none). The amino terminal domain is distinguished by small balls on its α-carbon positions and the amino terminus lies to the top in both panels

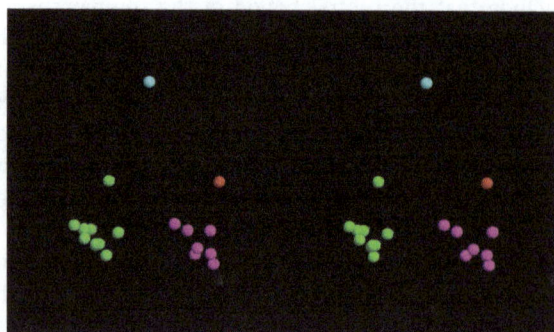

Fig. 13 Fold-space representation of all domains. All the viral domains considered in the paper were projected into a 3D fold-space representing the relationship of their SAP weighted RMSD values. The domains are coloured as: foamyN = *cyan*, foamyC = *red*, orthoN = *green* and ortho C = *magenta*

homologues as far back as insects and probably deeper, giving it a very ancient origin somewhere close to the metazoan root [12, 18] and based on sequence homology Arc is considered to be a relic of an ancient Ty3/Gypsy retrotransposon [8], preserved as a 'living fossil' in metazoan genomes. Given the structural relatedness of foamy virus CA and Arc, this might suggest an equally ancient origin for foamy virus CA. As it is believed that the Ty3/Gypsy family of retrotransposons gave rise to retroviruses [19], it will therefore be of considerable interest to determine whether the Gag of Ty elements also comprise CA proteins with a two-domain structure.

It is also noteworthy that Ty3 Gag is significantly smaller than that of the foamy and orthoretroviruses and although it contains CA related sequences there is no equivalent of either orthoretroviral MA or PFV Gag-NtD, regions of Gag necessary for membrane targeting, budding and extracellular release of virions. Therefore, given the very different structures of MA [20–23] and Gag-NtD [4], this raises the possibility that the MA and Gag-NtD domains of the orthoretroviruses and foamy viruses were co-opted by independent events that has resulted in the viruses employing different mechanisms to facilitate budding from the cell. Notably, Gag from Gypsy, an Errantivirus capable of extracellular replication [24] and Arc contain additional N-terminal domains. In Gypsy-Gag this domain is distantly sequence-related to orthoretroviral MA [12]. By contrast, in Arc it contains a coiled coil region [8] reminiscent of spumavirus Gag-NtD [4, 25] further supporting the notion of a shared origin for Arc and foamy virus Gag that is distinguishable from an alternative acquisition pathway giving rise to Gypsy and the orthoretroviruses.

Methods
Structural data
The foamy virus structures were obtained from the Protein Structure Databank (PDB code:5M1G) [5].

The ortho virus structures used, with their shorthand code in bold and PDB code in teletype, were:

- **BLV**: bovine leukemia virus (deltaretrovirus) 4PH1 (N-ter.dom) and 4PH2 (C-ter.dom) [26],
- **BLV6**: bovine leukemia virus (hexameric) 4PH0 (both dom.s) [26],
- **HIV1**: human immunodeficiency virus 1 (lentivirus) 1AK4 (N-ter.dom) [27] and 1A43 (C-ter.dom) [28],
- **HIV6**: human immunodeficiency virus 1 3H47 (both dom.s) [29],
- **HML2**: human endogenous retrovirus type-K (betaretrovirus) [30],
- **HTLV**: human T-cell leukemia virus (deltaretrovirus) 1QRJ (both dom.s) [31],
- **JSRV**: jaagsiekte sheep Retrovirus (betaretrovirus) 2V4X (N-ter.dom) [32],
- **MLV**: murine leukemia virus (gammaretrovirus) 1U7K (N-ter.dom) [33],
- **MPMV**: Mason-Pfizer monkey virus (betaretrovirus) 2KGF (N-ter.dom) [34],
- **PSIV**: prosimian immunodefficiency virus (ancient lentivirus) 2XGV (N-ter.dom) [35],
- **RELIK**: rabbit endogenous lentivirus type-K (ancient lentivirus) 2XGU (N-ter.dom) [35],
- **RSV**: Rous sarcoma virus (alpharetrovirus) 3G1I (both dom.s) [36].

Structure comparison
DALI
The DALI method for searching the PDB with a structural query [7] was accesed via the server at: http://ekhidna.biocenter.helsinki.fi/ dali_server. The DALI method reports the signifi-

cance of each match with an estimated Z-score which is the raw comparison score, normalised by the combined length of the proteins. Z-scores down to a value of 2 are reported by the program.

The list of DALI hits (ranked by Z-score) were assessed by how many high-scoring capsid structures had been identified. These true/false (T/F) hits were defined simply by protein descriptions that contained the words "CAPSID", "GAG" or "P24". This may have misclassified a few (low scoring) hits to the matrix protein and missed some hits where the primary description refers to a cyclophilin structure solved in complex with the capsid.

DALI reports structural hits in both the full PDB and a reduced collection of structures that have no pair of proteins with over 90% sequence identity, referred to as the 90% non-redundant or PDB-90 collection. It was found, however, that some hits, seen in the full PDB were not found in the PDB-90, for example in Fig. 6, all of the top 31 hits of the N-domain against the full PDB are missing in the PDB-90 hits. The most likely explanation is that the PDB-90 secection has not been updated at the same time as the full collection. For this reason, hits to both databases were monitored.

SAP
The SAP method for structure comparison [6] was run as a local copy which can be accessed at: https:// github.com/WillieTaylor/util. As part of determining the alignment between two structures, the SAP program calculates a similarity score for each pair of matched positions which is how similar the rest of the structure looks from the viewing-frame of the superposed residues. This value can be used both to weight the importance of positions when calculating the (rigid-body) RMSD superposition and to colour positions in the superposed structures [37]. (As in Fig. 3).

If the matched positions are ranked by this value, then RMSD values can be calculated over increasingly larger subsets to high-light the extent of a well matched core before the contribution of variable loops, or domain shifts, leads to higher RMSD values. (As in Fig. 1b).

Decoy structure construction
Reversed structure decoys
Simple structural decoys were generated from native PDB structures by reversing the order of the α-carbon atoms in the PDB file using the Unix command line:

```
cat native.pdb | grep ' CA ' | sort -nr -k2
     > reverse.pdb
```

The reversal of a protein chain does not alter the chirality of the alpha helix and these decoys can be used directly in SAP. However, DALI requires all main-chain atoms and these must be regenerated for the reversed decoys. This was done using the simple `ca2main` program which can also be found at: `https://github.com/WillieTaylor/util`. The method is based on the geometry of the α-carbon-virtual chain using relationships described in ref. [38].

Customised decoys

Customised structural decoys were generated for each comparison using each of the pair of structures being compared to create two pools of decoys then comparing all decoys in the first pool against all decoys from the second but with their chain reversed as described in the previous section.

The decoys were created as described in Ref. [10]: starting by cyclising the chain then introducing new termini in each surface loop to create cyclic permutations. In addition, when three loop regions lie in close proximity, their ends are also reconnected in such a way that if a chain, comprising four segments $(1...4)$ runs from amino (N) to carboxy (C) termini through three adjacent loop regions `a-b`, `c-d` and `e-f` (i.e.: `N,1,a-b,2,c-d,3,e-f,4,C`) then the reconnected chain runs: `N,1,a-d,3,e-b,2,c-f,4,C` with each switch being made at the least disruptive point between a pair of loops. This chain switching does not create any reversed segments which would otherwise form regions of local matching when the whole chain is reversed.

In a pair of structures, if each have four surface loops where breaks can be made, then including the native termini, this gives five cyclic permutations and if two groups of loops can be reconnected then a total of 15 distinct decoys can be made from each native starting structure. As these can be compared pairwise, a pool of 225 decoy derived data points is generated that constitutes the random background against which the native/native comparison can be assessed.

For example, in Fig. 8, the 36 data points marked by a solid circle come from the comparison of six cyclic permutations of a native ortho domain compared with six permutations of a reversed foamy domain that includes a single loop reconnection.

Every pair drawn from this pool will have the same lengths as the two native structures as well as the same secondary structure composition, surface exposure, residue packing density and inertial properties but each decoy will have a different chain fold.

Statistical tests
RMSD length normalisation

The quality of structure comparisons can be characterised by a combination of their RMSD value and the number of matched (superposed) positions. How to combine these values has been the subject of much discussion

over the years and central to this is the expected random RMSD value for two proteins of a given length [39–41]. However, when reviewed [10], all these measures were approximations of a simple square-root function of the protein length (as originally proposed by McLachlan on theoretical grounds [39]) but with an added term to depress the RMSD values obtained with small units or structure that are dominated by secondary structure elements (and super-secondary structure motifs) giving a lower than expected RMSD value. The formula that best captures this is: $R = \sqrt{N(1 - \exp(-N^2/s^2))}$, where, R is the expected random RMSD for N matched positions and s is the damping factor in the inverted Gaussian term (equivalent to the standard deviation in the Normal distribution).

Any point that lies on this line can be considered "exactly" random with those above it being "more" random and those below it "less" random. This can be quantified as a single number which is the value of a scaling factor (a), which when applied to the curve, makes it pass through any given point. If a comparison has an RMSD of R over N positions, then $R = a\sqrt{N(1 - \exp(-N^2/s^2))}$ and when

$$a = R/(\sqrt{N(1 - \exp(-N^2/s^2))}), \quad (1)$$

the line will pass through the data point. This reduces the pair of values (R, N) to a single value a that is a simpler quantity for statistical analysis.

The best value for s is slightly dependent on the nature of the proteins being compared. For artifical (random-walk) models with no secondary structure, no modification will be needed but the proteins considered here have segments of packed alpha helices that can be locally similar over two to three helices. To correct for this, a value of $s = 30$ was used (or $1/s^2 = 0.11$) which is higher than the value of $1/s^2 = 0.03$ used previously. That this is a reasonable fit to the data can be seen in the way the dashed blue lines in Fig. 8 track the upper and lower boundary of the decoy comparison results.

When $a = 1$, the point lies on the random line and when $a = 0$, the RMSD is zero, so values of a that approach this lower bound will be of interest when evaluating similarity.

Frequency plots

The a-values obtained using Equn. 1 were plotted as frequency histograms using using only data points that had a length of $N \pm 10$, where N is the maximum number of matched positions in the comparison of the two native structures. As the sample size is small (typically, 100–300), these plots are quite noisy but their overall distribution does not deviate too greatly from a Gaussian distribution. This was tested on the difference between the observed and ideal cummulative distribution functions (CDFs) using the Kolmogorov-Smirnov test in the statistical package "R". Of the 38 samples from each domain pairing, the null hypothes "that the sample was drawn from a Normal distribution" could be rejected in only two cases with a confidence below the 0.01 significance level or three below the 0.05 level. (See the Additional file 1 for details). The underlying distribution becomes more apparent when the data sets are combined in Fig. 11d.

Previously, a cumulative plot of RMSD was used to select an optimal value for N (giving the minimum a-value). This can be important if the full set of matched positions is dominated by a high deviations from variable loop regions. However, in the current application, the small length of the foamy virus loops meant that this was not an important aspect and the full number of matched positions was taken. Otherwise, the same correction would have to be applied to all decoy comparisons to maintain a fair comparison. (See Fig. 8, where the black dot marks the minimum a-value length).

The mean and standard deviation of the a-values in the $N \pm 10$ region were calculated and the corresponding Normal distribution used to calculate Z-scores for the associated native comparison. (See Fig. 9a, for an example).

T-tests

Data from separate native/native comparisons, with their customised decoy data, were combined giving not only a much larger background population of decoy derived scores but also a small population of native comparison scores that can be tested to calculate the probability that they were drawn from the same population as the decoy data. To do this, a T-test was used which takes the size, mean, and standard deviation of each distribution and calculates a probability. The implementaion of this test was taken from the Numerical Reicpies collection [42] which implements one of two variants of the test depending on whether the distributions have statistically distinct standard deviations. (Routines `ttest()` and `tutest()`). The choice of routine is based on a preapplication of an F-test on the standard-deviations. (Using the routine `ftest()`).

The values quoted in the Results section are for a two-tailed T-test, however, as it is expected that the native comparisons should always be more similar than comparisons between random models, then a one-tailed T-test would be valid, which gives half the probability. As the values in the Tables are so significant and only the relative relationships are of interest, then the choice is unimportant.

Fold-space clustering

The results of the pairwise similarity within a set of structures can be visualised by treating the RMSD values

as Euclidean distances[5] and reducing their dimensionality to sufficiently few dimensions to be visualised: usually 2D or, better 3D, to visualise the space with less distortion. Rather than use a simple multi-dimensional scaling (MDS) method ([43]), the more complicated method of multi-dimensional projection was used ([44], see [45] for a simpler exposition).

This method reduces the dimensionality of the projection in gradual stages with each step employing triangle-inequality balancing and hyper-dimensional real-space refinement. In the real-space refinement stages, a weight can be applied to pairwise distances. (This cannot be done in direct MDS projection, which can only assign a mass to each point). Weights were assigned to distances as a function of their inverse RMSD, up to a maximum value of 1.

The method is robust and has been widely applied to rough models ([46]) and predicted inter-residue distances that constitute highly non-metric data sets ([47]).

Endnotes

[1] This class is also commonly referred to as the Foamy viruses (after the morphological effect they have on infected cells) and will be referred by this name frequently below, with the term orthoretroviruses also contracted to "Ortho viruses".

[2] `http://ekhidna.biocenter.helsinki.fi/ dali_server`, see "Methods" section for details.

[3] True/false hits were defined by protein descriptions with the words "CAPSID", "GAG" or "P24".

[4] Note that reversing the α-carbon backbone does not change the chirality of the αhelices but as DALI requires a full atomic backbone, this must be restored on the reversed chain.

[5] In theory, pairwise RMSD values are guaranteed to constitute a consistent Euclidean metric, but only in N-1 dimensions (where N is the number of structures compared).

Acknowledgements
None.

Funding
The work was supported by the Francis Crick Institute under awards: FC001179 (WRT), FC001162 (JPS) and FC001178 (IAT). The Crick receives its core funding from Cancer Research UK, the UK Medical Research Council, and the Wellcome Trust.

Authors' contributions
WRT, JPS and IAT conceived the work and evaluated the results. WRT executed the computational work. All authors contributed to the manuscript. All authors read and approved the final manuscript.

Competing interests
The authors declare that they have no competing interests.

Author details
[1]Computational Cell and Molecular Biology Laboratory, Francis Crick Institute, Midland Road, NW1 1AT London, UK. [2]Retrovirus-Host Interactions Laboratory, Francis Crick Institute, Midland Road, NW1 1AT London, UK. [3]Macromolecular Structure Laboratory, Francis Crick Institute, Midland Road, NW1 1AT London, UK.

References
1. Lindemann D, Rethwilm A. Foamy virus biology and its application for vector development. Viruses. 2011;3:561–85.
2. Müllers E. The foamy virus Gag proteins: what makes them different?. Viruses. 2013;5:1023–1041.
3. Flügel RM, Pfrepper KI. Proteolytic processing of foamy virus Gag and Pol proteins. Curr Top Microbiol Immunol. 2003;277:63–88.
4. Goldstone DC, Flower TG, Ball NJ, Sanz-Ramos M, Yap MW, Ogrodowicz RW, Stanke N, Reh J, Lindemann D, Stoye JP, Taylor IA. A unique spumavirus Gag N-terminal domain with functional properties of orthoretroviral matrix and capsid. PLoS pathogens. 2013;9:1003376.
5. Ball NJ, Nicastro G, Dutta M, Pollard D, Goldstone DC, Sanz-Ramos M, Ramos A, Müllers E, Stirnnagel K, Stanke N, Lindemann D, Stoye JP, Taylor WR, Rosenthal PB, Taylor IA. Structure of a spumaretrovirus gag central domain reveals an ancient retroviral capsid. PLoS Path. 2016;12: 1005981. doi:10.1371/journal.ppat.1005981.
6. Taylor WR. Protein structure alignment using iterated double dynamic programming. Prot Sci. 1999;8:654–65.
7. Holm L, Sander C. Protein-structure comparison by alignment of distance matrices. J Molec Biol. 1993;233:123–38.
8. Zhang W, Wu J, Ward MD, Yang S, Chuang YA, Xiao M, Li R, Leahy DJ, Worley PF. Structural basis of arc binding to synaptic proteins: implications for cognitive disease. Neuron. 2015;86:490–500.
9. Taylor WR. Protein structure domain identification. Prot Engng. 1999;12: 203–16.
10. Taylor WR. Decoy models for protein structure score normalisation. J. Molec. Biol. 2006;357:676–99.
11. Levitt M, Gerstein M. A unified statistical framework for sequence comparison and structure comparison. Proc Natl Acad Sci USA. 1998;95: 5913–920.
12. Campillos M, Doerks T, Shah PK, Bork P. Computational characterization of multiple gag-like human proteins. Trends in Genetics. 2006;22:285–589.
13. Chowdhury S, Shepherd JD, Okuno H, Lyford G, Petralia RS, Plath N, Kuhl D, Huganir RL, Worley PF. Arc/Arg3.1 interacts with the endocytic machinery to regulate AMPA receptor trafficking. Neuron. 2006;52:445–59.
14. Park S, Park JM, Kim S, Kim JA, Shepherd JD, Smith-Hicks CL, Chowdhury S, Kaufmann W, Kuhl D, Ryazanov AG, et al. Elongation factor 2 and fragile X mental retardation protein control the dynamic translation of Arc/Arg3.1 essential for mGluR-LTD. Neuron. 2008;59:70–83.
15. Shepherd JD, Rumbaugh G, Wu J, Chowdhury S, Plath N, Kuhl D, Huganir RL, Worley PF. Arc/arg3.1 mediates homeostatic synaptic scaling of AMPA receptors. Neuron. 2006;52:475–484.
16. Waung MW, Pfeiffer BE, Nosyreva ED, Ronesi JA, Huber KM. Rapid translation of Arc/Arg3.1 selectively mediates mGluR-dependent LTD through persistent increases in AMPAR endocytosis rate. Neuron. 2008;59: 84–97.
17. Niere F, Wilkerson JR, Huber KM. Evidence for a fragile x mental retardation protein-mediated translational switch in metabotropic glutamate receptor-triggered arc translation and long-term depression. J Neurosci. 2012;32:5924–936.
18. Volff JN. Cellular genes derived from Gypsy/Ty3 retrotransposons in mammalian genomes. Annals New York Acad Sci. 2009;1178:233–43.
19. Llorens C, Fares MA, Moya A. Relationships of gag-pol diversity between Ty3/Gypsy and retroviridae LTR retroelements and the three kings hypothesis. BMC Evol Biol. 2008;8:276.
20. Hill CP, Worthylake D, Bancroft DP, Christensen AM, Sundquist WI. Crystal structures of the trimeric human immunodeficiency virus type 1 matrix protein: implications for membrane association and assembly. Proc Natl Acad Sci USA. 1996;93:3099–104.
21. Prchal J, Srb P, Hunter E, Ruml T, Hrabal R. The structure of myristoylated mason-pfizer monkey virus matrix protein and the role of phosphatidylinositol-(4,5)-bisphosphate in its membrane binding. J Mol Biol. 2012;423:427–38.
22. Rao Z, Belyaev AS, Fry E, Roy P, Jones IM, Stuart DI. Crystal structure of SIV matrix antigen and implications for virus assembly. Nature. 1995;378: 743–7.
23. Riffel N, Harlos K, Iourin O, Rao Z, Kingsman A, Stuart DI, Fry E. Atomic resolution structure of moloney murine leukemia virus matrix protein and its relationship to other retroviral matrix proteins. Structure. 2002;10: 1627–1636.
24. Song SU, Gerasimova T, Kurkulos M, Boeke JD, Corces VG. An env-like protein encoded by a drosophila retroelement: evidence that gypsy is an infectious retrovirus. Gene Dev. 1994;8:2046–2057.
25. Tobaly-Tapiero J, Bittoun P, Giron ML, Neves M, Koken M, Saib A, de The H. Human foamy virus capsid formation requires an interaction domain in the N-terminus of Gag. J Virol. 2001;75:4367–4375.
26. Obal G, Trajtenberg F, Carrion F, Tome L, Larrieux N, Zhang X, Pritsch O, Buschiazzo A. Conformational plasticity of a native retroviral capsid revealed by X-ray crystallography. Science. 2015;349:95–8. doi: 10.1126/science.aaa5182.
27. Gamble TR, Vajdos FF, Yoo S, Worthylake DK, Houseweart M, Sundquist WI, Hill CP. Crystal structure of human cyclophilin A bound to the amino-terminal domain of HIV-1 capsid. Cell. 87:1285–1294.
28. Worthylake DK, Wang H, Yoo S, Sundquist WI, Hill CP. Structures of the HIV-1 capsid protein dimerization domain at 2.6å resolution. Acta Crystallogr., Sect. D. 1999;55:85–92. doi:10.1107/S0907444998007689.
29. Pornillos O, Ganser-Pornillos BK, Kelly BN, Hua Y, Whitby FG, Stout CD, Sundquist WI, Hill CP, Yeager M. X-ray structures of the hexameric building block of the HIV capsid. Cell. 2009;137:1282–1292. doi: 10.1016/j.cell.2009.04.063.
30. Mortuza GB, Dodding MP, Goldstone DC, Haire LF, Stoye JP, Taylor IA. Structure of B-tropic MLV capsid N-terminal domain. J Mol Biol. 376: 1493–1508.
31. Khorasanizadeh S, Campos-Olivas R, Clark CA, Summers MF. Sequence-specific 1H, 13C and 15N chemical shift assignment and secondary structure of the HTLV-I capsid protein. J Biomol NMR. 1999;14:199–200.
32. Mortuza GB, Goldstone DC, Pashley C, Haire LF, Palmarini M, Taylor WR, Stoye JP, Taylor IA. Structure of the capsid amino terminal domain from the betaretrovirus, Jaagsiekte sheep retrovirus. J Molec Biol. 2009;386: 1179–1192.
33. Mortuza GB, Haire LF, Stevens A, Smerdon SJ, Stoye JP, Taylor IA. High-resolution structure of a retroviral capsid hexameric amino-terminal domain. Nature. 2004;431:481–5.
34. Macek P, Chmelik J, Krizova I, Kaderavek P, Padrta P, Zidek L, Wildova M, Hadravova R, Chaloupkova R, Pichova I, Ruml T, Rumlova M, Sklenar V. NMR structure of the N-terminal domain of capsid protein from the mason-pfizer monkey virus. J. Mol. Biol. 2009;392:100–14. doi: 10.1016/j.jmb.2009.06.029.
35. Goldstone DC, Yap MW, Robertson LE, Haire LF, Taylor WR, Katzourakis A, Stoye JP, Taylor IA. Structural and functional analysis of prehistoric lentiviruses uncovers an ancient molecular interface. Cell Host Microbe. 2010;8:248–59.
36. Bailey GD, Hyun JK, Mitra AK, Kingston RL. Proton-linked dimerization of a retroviral capsid protein initiates capsid assembly. Structure. 2009;17: 737–48. doi:10.1016/j.str.2009.03.010.
37. Rippmann F, Taylor WR. Visualization of structural similarity in proteins. J Molec Graph. 1991;9:3–16.
38. Levitt M, Greer J. Automatic identification of secondary structure in globular proteins. J Molec Biol. 1977;114:181–293.
39. McLachlan AD. How alike are the shapes of two random chains?. Biopolymers. 1984;23:1325–1331.
40. Cohen FE, Sternberg MJE. On the prediction of protein structure: the significance of the root-mean-square deviation. J Molec Biol. 1980;138: 321–33.
41. Maiorov VN, Crippen GM. Significance of root-mean-square deviation in comparing three-dimensional structures of globular proteins. J Mol Biol. 1994;235:625–34.
42. Press WH, Flannery BP, Teukolsky SA, Vetterling WT. Numerical Recipes: The Art of Scientific Computing. Cambridge: Cambridge Univ. Press; 1986.

43. Brown NP, Orengo CA, Taylor WR. A protein structure comparison methodology. Computers Chem. 1996;20:359–80.
44. Aszódi A, Taylor WR. Hierarchical inertial projection: a fast distance matrix embedding algorithm. Computers Chem. 1997;21:13–23.
45. Taylor WR, May ACW, Brown NP, Aszódi A. Protein structure: Geometry, topology and classification. Rep Prog Phys. 2001;64:517–90.
46. Taylor WR, Chelliah V, Hollup SM, MacDonald JT, Jonassen I. Probing the "dark matter" of protein fold-space. Structure. 2009;17:1244–1252.
47. Aszódi A, Taylor WR. Folding polypeptide α-carbon backbones by distance geometry methods. Biopolymers. 1994;34:489–506.

Three-dimensional structure model and predicted ATP interaction rewiring of a deviant RNA ligase 2

Sandrine Moreira[1*], Emmanuel Noutahi[2], Guillaume Lamoureux[3] and Gertraud Burger[1]

Abstract

Background: RNA ligases 2 are scarce and scattered across the tree of life. Two members of this family are well studied: the mitochondrial RNA editing ligase from the parasitic trypanosomes (Kinetoplastea), a promising drug target, and bacteriophage T4 RNA ligase 2, a workhorse in molecular biology. Here we report the identification of a divergent RNA ligase 2 (DpRNL) from *Diplonema papillatum* (Diplonemea), a member of the kinetoplastids' sister group.

Methods: We identified DpRNL with methods based on sensitive hidden Markov Model. Then, using homology modeling and molecular dynamics simulations, we established a three dimensional structure model of DpRNL complexed with ATP and Mg2+.

Results: The 3D model of *Diplonema* was compared with available crystal structures from *Trypanosoma brucei*, bacteriophage T4, and two archaeans. Interaction of DpRNL with ATP is predicted to involve double π-stacking, which has not been reported before in RNA ligases. This particular contact would shift the orientation of ATP and have considerable consequences on the interaction network of amino acids in the catalytic pocket. We postulate that certain canonical amino acids assume different functional roles in DpRNL compared to structurally homologous residues in other RNA ligases 2, a reassignment indicative of constructive neutral evolution. Finally, both structure comparison and phylogenetic analysis show that DpRNL is not specifically related to RNA ligases from trypanosomes, suggesting a unique adaptation of the latter for RNA editing, after the split of diplonemids and kinetoplastids.

Conclusion: Homology modeling and molecular dynamics simulations strongly suggest that DpRNL is an RNA ligase 2. The predicted innovative reshaping of DpRNL's catalytic pocket is worthwhile to be tested experimentally.

Keywords: Protein structure, Molecular dynamics simulation, Protein evolution

Background

RNA ligase from phage T4, the work horse of molecular biology research, is the best known member of a large protein family encompassing RNA and DNA ligation enzymes [1]. RNA ligases fall into three classes: (i) RNA ligases type 1, (ii) RNA ligases type 2, and (iii) capping enzymes. All nucleic acid ligases share a characteristic nucleotidyltransferase domain in their N-terminal part with five conserved motifs (I, III, IIIa, IV and V) [2].

Two other classes of enzymes that have RNA ligase activity but lack the above structural features are the LigT phosphoesterases involved in RNA splicing [3–5] and the recently identified RtcB proteins [6, 7]. In the following, the term "RNA ligase family" will refer to the two former classes that contain a nucleotidyltransferase domain.

RNA ligase 1 enzymes are mainly present in viruses, mammals and fungi [8]. This enzyme class is typically involved in defense as exemplified by its founding member, the phage T4 RNL1, which is deployed in the counter-attack against antiviral strategies of bacteria [9], but is also involved in tRNA intron splicing [10] and in

* Correspondence: sandrine.moreira@umontreal.ca
[1]Department of Biochemistry and Robert-Cedergren Centre for Bioinformatics and Genomics, Université de Montréal, Montreal, QC, Canada
Full list of author information is available at the end of the article

the unconventional splicing initiating the unfolded protein response of the endoplasmatic reticulum. RNA ligases 2 have a broad but punctuated distribution across the tree of life [8]: they are found mainly in viruses -with the archetypical example of T4 RNA ligase 2 [11]- and bacteria, while only a few examples are known in archaea and eukaryotes. The biological role of RNA ligases 2 is unknown, except for the members of kinetoplastids [12].

Kinetoplastids (Euglenozoa) are a group of protozoans, some members of which are causing life-threatening human diseases (leishmaniasis, Chagas disease, sleeping sickness) [13]. These species also display a unique mitochondrial genome structure composed of an intricate network of large and small circular chromosomes [14]. Large chromosomes encode typical mitochondrial protein-coding genes. Small circles specify guide RNAs that serve as proofreading templates for editing pre-mRNAs of mitochondrial genes [15, 16]. Editing proceeds by cutting the pre-mRNA molecule at the place of the mismatch, then adding or removing uridines, and finally religating the two parts of the RNA molecule. It is this last step that is performed by RNA ligase 2. Specifically, two different RNA ligases 2 are involved, one dedicated to adding and the second to deleting uridines as exemplified by the ligases TbREL1 and TbREL2 respectively for *Trypanosoma brucei* [17].

Here we report the identification of a putative new member of the RNA ligase 2 family in *Diplonema papillatum*, a member of diplonemids (Euglenozoa), which are the sister group of kinetoplastids. The corresponding gene was discovered in our search of a candidate enzyme involved in the eccentric post-transcriptional processing in *Diplonema* mitochondria [18, 19]. This protist harbors a highly complex mitochondrial genome sharing certain similarities with that of kinetoplastids. First, the *Diplonema* mitochondrial DNA (mtDNA) is also multipartite, as it is composed of hundreds of circular chromosomes of two size classes. The difference and uniqueness of the diplonemid mtDNA is that each chromosome contains one short coding region specifying a fragment of a gene. Each gene module is transcribed separately and then trans-spliced to form full-length mRNAs or structural RNAs. The second resemblance with kinetoplastid mitochondria is RNA editing [18, 20]. Uridine insertion and deletion editing in kinetoplastids involves an RNA ligase 2 to reseal the transcript. In *Diplonema*, RNA editing proceeds by uridine appendage at certain module ends, prior to transsplicing. We hypothesize that an ancestral molecular machinery containing RNA ligase 2 has led to the editosome in kinetoplastids, while it has evolved to perform trans-splicing in the diplonemid branch.

RNA ligases 2 consist of two discrete portions: the N-terminal nucleotidyltransferase domain (amino acids 1–234 in T4) and a C-terminal domain (amino acids 244–329 in T4) responsible for substrate specificity. The ligation reaction of RNA ligase 2 is ATP and Mg^{2+} dependent [10, 21, 22] and proceeds, like all members of the DNA/RNA ligase family, in three steps. During the first step, ATP adenylates the enzyme on the lysine residue of the conserved KxxG tetramer in motif I of the nucleotidyltransferase domain. In step 2, the covalently linked AMP is transferred to the 5′P of the 'downstream' RNA molecule to be ligated. Finally, the 3′OH of the 'upstream' RNA molecule attacks the 5′P of the 'downstream' RNA by releasing AMP and joining the two RNA molecules (Additional file 1: Figure S1). The crystal structure has been determined for only a few family members, notably T4 RNA ligase 2 [23, 24] and one of the two paralogous mitochondrial RNA ligases 2 from *Trypanosoma brucei*, notably in apo form as well as complexed with a magnesium ion and ATP [25].

In this study, we devise a strategy based on hidden Markov models (HMMs) and structural comparisons to identify proteins of large evolutionary distance to wellstudied counterparts in model organisms. Comparative analysis of highly diverged homologs is particularly informative for identifying functionally and structurally important residues that are under elevated selective pressure. Employing this analytic strategy, we identify the gene and model the structure and ligand interactions of a putative RNA ligase 2 from *Diplonema*. The model predicts intriguing innovations in the interaction network between ATP and the residues of the catalytic pocket, which are worthwhile to be tested experimentally by resolving the crystal structure. We discuss possible evolutionary scenarios that led to these innovations.

Results

HMM-based detection of a divergent RNA ligase 2 in *Diplonema*

In general, proteins of *D. papillatum* display a low level of sequence similarity with homologs of other taxa, and are difficult to identify with tools based on sequence similarity such as BLAST [26]. Therefore we employed more sensitive methods based on Hidden Markov Models (HMMs). We used the HMM PF09414.4 from the Protein FAMily database (PFAM) [27], a model that was built based on RNA ligases 2 from all domains of Life including mitochondrial RNA ligases 2 of kinetoplastids. We identified one candidate protein, Dp28902_3, in the conceptual translation of the *Diplonema* draft genome assembly (version no. 2). Expression of this open reading frame was confirmed by RNAseq experiments. The corresponding transcript is poly-adenylated and its steady-state level is about 1/10 compared to the expression of Aspartyl tRNA synthase.

For comparison, we also used HMMs for other RNA and DNA ligase super-families in searches against Dp28902_3 and RNA ligases 2 of *Trypanosoma* (TbREL1, positive control) and the heterolobosean *Naegleria gruberi*. *Naegleria* was chosen because heteroloboseans are the sistergroup of Euglenozoa, and because sequences of this taxon have not been used in building the PFAM HMM. Table 1 summarizes the corresponding *E*-values. Dp28902_3 has the lowest *E*-value with the PF09414 model, a value that is 10^7 times smaller than the second-best match, which was obtained with the HMM of ATP-dependent DNA ligases. Models for proteins that have a different fold (PF02834-LigT, PF01139-RtcB) did not yield significant *E*-values (>0.05) for either Dp28902_3 or the RNA ligases 2 of *Trypanosoma*. Therefore, Dp28902_3 most likely belongs to the RNA ligase 2 family and will be referred to as DpRNL.

DpRNL contains a nucleotidyltransferase domain typical for RNA ligases 2

The RNA/DNA ligase super-family is characterized by a nucleotidyltransferase domain including five subdomains (motifs I, III, IIIa, IV, V) [2] located in the N-terminal portion of the protein. We demonstrate the presence of these motifs in DpRNL by three different methods: sequence alignment against PFAM HMM (Additional file 1: Figure S2); multiple sequence alignment of DpRNL and RNA ligases 2 from kinetoplastids, enterobacteriphage T4, and *Naegleria* (Fig. 1); and structural alignment of DpRNL with RNA ligases 2 for which the three-dimensional (3D) structure has been experimentally determined, notably from *Trypanosoma brucei*, the phage T4, and the archaean *Pyrococcus abyssii* (Fig. 2).

While the five subdomain motifs are well conserved across all RNA ligases 2 and readily recognizable in DpRNL, the rest of the N-terminal portion of the *Diplonema* protein shows only low sequence similarity to established RNA ligases 2 (e.g., ~18 % identity with

TbREL1). DpRNL lacks portions of two loops between domains III and IIIa (TbREL1 amino acid (aa) 163–166 and aa 176–205) that are distinctive for kinetoplastid RNA ligases 2, and that have been shown to interact with RNA [25]. Also missing from DpRNL is the loop between domains IIIa and IV of TbREL1 (aa 262–282), a loop that has been predicted to interact with other proteins of the editosome [25]. Finally, the C-terminal portion of DpRNL (aa 178–203) has no recognizable resemblance with, and its length is also shorter than the corresponding region of other RNA ligases 2.

The 3D model of apo-DpRNL possesses all structural features typical for RNA ligases 2

The global three-dimensional (3D) model of DpRNL was predicted by I-Tasser [28] (Fig. 3, Additional file 2) and validated with SAVes (http://services.mbi.ucla.edu/SAVES/). Nearly all (96.1 %) amino acids have a stereochemical conformation in the "favored" or "allowed" regions of the Ramachandran plot. Only the seven most C-terminal residues are in an unfavorable environment according to the assessment by the tool Verify-3D [29]. While the per-residue analysis of ModFold [30] also found lower quality scores for the C-terminal region, the overall p-value of the model (1.547×10^{-3}) is highly confident. The estimated TM-Score obtained from the standard output of I-Tasser was 0.70 ± 0.12. A TM-score >0.5 usually indicates a model of correct topology, and a TM-score <0.17 means a similarity no better than random. As a whole, the topology of the I-Tasser model of DpRNL is of good quality.

The 3D model of DpRNL is characterized by a core of anti-parallel-twisted β sheets decorated with apical α helices. Two structural sub-domains with similar composition are facing one another. One contains the two extremities of the molecule and consists of an anti-parallel β sheet of four β strands and four α helices. The other sub-domain, corresponding to the middle

Table 1 Identification of the ligase family to which belongs DpRNL[a]

Family	PFAM	*D. papillatum* DpRNL	*N. gruberi* XP_002674912.1	*T. brucei* KREL1	*T. brucei* KREL2
DNA ligase					
[N] ATP dependent	PF01068	3.30×10^{-5}	2.60×10^{-5}	1.00×10^{-3}	1.60×10^{-6}
[N] NAD dependent	PF01653	2.20×10^{-2}	2.90×10^{-2}	–	4.70×10^{-1}
RNA ligase					
[N] Rnl1 defense, splicing	PF09511	2.70×10^{-1}	1.30×10^{-2}	3.40×10^{-1}	4.00×10^{-1}
[N] Rnl2 editing	**PF09414**	**4.90×10^{-12}**	**3.20×10^{-9}**	**7.90×10^{-55}**	**4.30×10^{-53}**
[N] Capping	PF01331	2.70×10^{-1}	1.70×10^{-1}	9.10×10^{-3}	2.10×10^{-1}
LigT	PF02834	–	–	–	–
RtcB splicing	PF01139	–	–	4.80×10^{-1}	–

[a]Family names preceded by an [N] are those containing a Nucleotidyltransferase domain. Each model was searched with HMMer against all the proteins of *Diplonema* papillatum, *Naegleria gruberi* and *Trypanosoma brucei* TREU927. This table presents the *E*-value for the RNA ligases 2 proteins only. The line for the PFAM domain specific for RNA ligases 2 is in bold

Fig. 1 Delineation of the Nucleotidyltransferase domain. Multiple alignment of RNA ligases 2 from *Enterobacteriophage* T4 (T4RNL2), *Diplonema papillatum* (DpRNL), *Naegleria gruberi* (NgRNL), and four kinetoplastids, *Leishmania infantum* JPCM5 (LiREL1, LiREL2), *L. major* Friedlin (LmREL1, LmREL2), *Trypanosoma vivax* Y486 (TvREL1, TvREL2), and *T. brucei* TREU927 (TbREL1, TbREL2). The six sub-domains (I, II, III, IIIa, IV and V) highlighted in orange, cyan, green, blue, yellow and red, respectively are clearly detectable in DpRNL

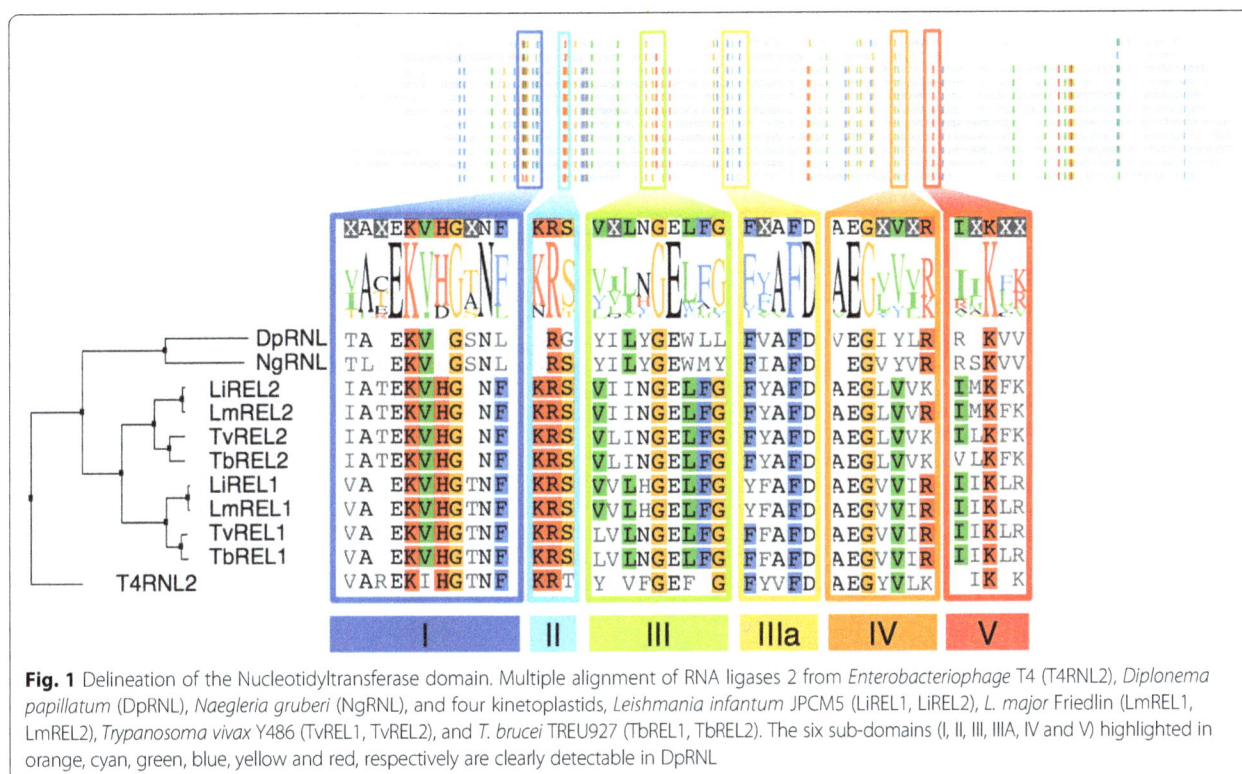

part of the protein, has six β strands and three α helices. The interface between these two sub-domains forms the catalytic pocket of the protein, with the residues of the five nucleotidyltransferase motifs pointing to the pocket's cavity. From the inside to the outside are located motifs I, IV and V on one side, and motifs IIIa, III and II on the other, the two sides facing each other.

Molecular dynamics simulations confirm the stability of the DpRNL 3D model

To assess the stability of the proposed DpRNL model and the relative flexibility of the structural domains, we performed a 50-ns molecular dynamics (MD) simulation. The Root Mean Square Deviation (RMSD) of the backbone α-carbon atoms remained stable after 10 ns of

Fig. 2 Structural alignment with DALI [59] of the *Diplonema* model (first line) and four structures: 1XDN (*Trypanosoma brucei*), 1S68 (*Enterobacteriophage T4*), 2VUG (*Pyrococcus abyssi*), and 3QWU (*Aquifex aeolicus*)

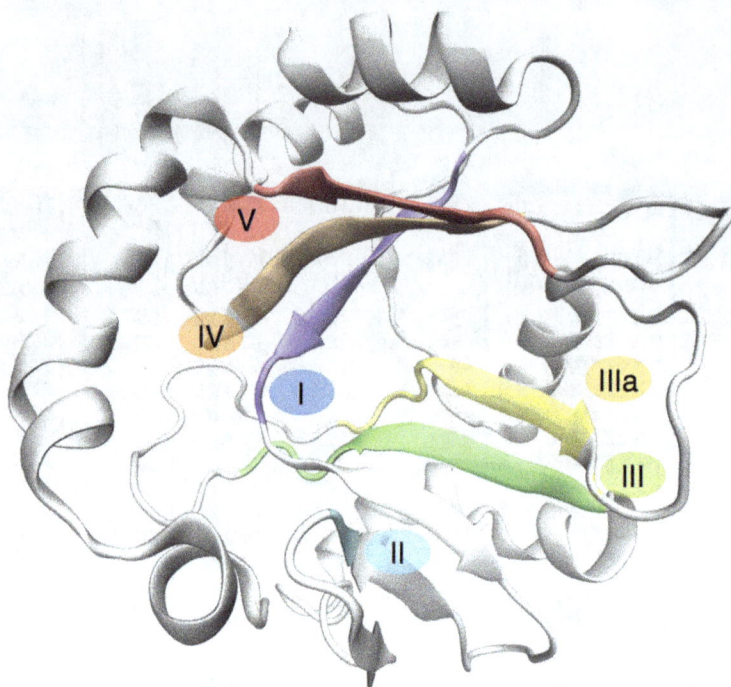

Fig. 3 Three-dimensional model of DpRNL inferred by I-Tasser. The five Nucleotidyltransferase sub-domains are represented in color

simulation with a mean of 4.2 Å (Additional file 1: Figure S3A).

When monitoring the secondary (2D) structure conservation during the simulation (Additional file 1: Figure S4), we observed that the β sheets, which are buried inside the protein, are more stable, whereas the α helices and loops, which are peripheral, are more flexible as reflected by the high Root Mean Square Fluctuation (RMSF) values of the corresponding residues. Specifically, certain residues of the α helices (aa 54–73 and aa 139–154) transiently adopted a 3–10 helix conformation. Flexible α helices and loops are also observed in TbREL1 of Trypanosoma, where the exposed regions of the protein interact with the RNA substrate and with other proteins of the editosome [31]. Therefore, the flexible peripheral regions of DpRNL presumably play a functional role as well.

The C-terminal region of DpRNL is linked to the rest of the molecule by a flexible loop, but this region displays less motion than expected. This is because the C-terminal domain is entangled in a network of hydrogen bonds with more N-terminal amino acids. Most stable are the interactions between the carboxyl group of tyrosine at position 203 (DpRNL_Y203, the last residue in the protein) and the lateral chain of two other residues (DpRNL_R41 with 86 % occupancy and DpRNL_S24 with 46 % occupancy), as well as between the lateral chains of DpRNL_Y203 and DpRNL_Q52. Additional stabilization of this domain comes from a hydrogen bond involving the

carbonyl group of DpRNL_K202 in the main chain and the hydroxyl of DpRNL_S49. In conclusion, the 3D model of DpRNL is stable both at the 2D and 3D level. The observed flexibility parallels that of other RNA ligases 2 [24, 31], providing strong support for DpRNL being a functional member of this protein family.

3D structure comparison of DpRNL with well characterized RNA ligases 2

Compared to recognized RNA ligases 2, DpRNL is more conserved in 3D structure than in sequence. Nevertheless, the β strands of DpRNL are generally shorter than those of its counterparts, resulting in a 15–30 % shorter Nucleotidyltransferase domain compared to the enzymes of *Trypanosoma* or phage T4. Pairwise structural comparison with experimentally confirmed structures (Additional file 1: Table S1) reveals only a moderate fit of DpRNL with TbREL1 (RMSD of 3.4 Å), although kinetoplastids are the sister group of diplonemids. The fit is slightly better with the RNA ligases of T4 (T4RNL2; RMSD of 3.2 Å) and *Pyrococcus abyssii* (PAB1020; PDB id 2VUG; RMSD of 2.3 Å), and the putative DNA ligase from *Aquifex aeolicus* (aq_1106; PDB id 3QWU; RMSD of 2.3 Å; Additional file 3). Note that PAB1020 was initially annotated as DNA ligase, but more recent experimental studies shown that it catalyzes the ligation of RNA [32].

The proteins from *Pyrococcus* and *Aquifex* are both homodimeric with subunits being held together through the interaction of two peripheral α helices [32]. As DpRNL has no region whose sequence resembles that of these interacting helices, we investigated if the two most C-terminal helices of DpRNL allow dimerization through typical hydrophobic interface contacts [33]. The hydrophobicity map of exposed residues (Fig. 4d and Additional file 1: Figure S5) shows that the C-terminal helices of DpRNL do not have the propensity to form an hydrophobic surface comparable to that of the archaean ligases. This suggests that DpRNL is active in a monomeric state as are TbREL1 and T4RNL2.

To determine if the Nucleotidyltransferase domain of DpRNL contains deviant residues otherwise not found in RNA and DNA ligases, we computed a score of «exceptionality» along the structural multiple alignment from selected enzymes including archaeal and kinetoplastid homologs. Each amino acid in *Diplonema* was assigned an exceptionality score based on the proportion of residues in the corresponding alignment column having common physicochemical properties in other ligases (Fig. 4c). The amino acid with the highest score is the tyrosine DpRNL_Y161, a position occupied in all other cases by a different, generally aliphatic residue. The second most deviant amino acid is the valine DpRNL_V177, whose position is generally occupied by a basic residue that non-covalently binds AMP in reaction step 1 [34]. Further exceptional residues in DpRNL are S49, G50, W60, W82, D96, Y104 and R173. The consequences of these substitutions for interactions with RNA and ATP will be discussed in a later section.

Phylogeny of RNA ligases 2
The moderate structural similarity of DpRNL with RNA ligases 2 from the diplonemid sister group raised questions about the phylogenetic relationship of these proteins. We focused our analyses on Excavate taxa, because a broader taxonomic sampling would have resulted in sequences too diverse for meaningful phylogenetic reconstruction. The inferred tree (Additional file 1: Figure S6) shows well supported grouping of kinetoplastid RNA ligases 2, which are split into two subgroups corresponding to the two paralogs (e.g. TbREL1 and TbREL2 in *T. brucei*). The subgroup clustering strongly suggests a duplication of RNA ligases 2 in the kinetoplastid branch prior to the speciation of *Leishmania* and *Trypanosoma*. In contrast, the phylogenetic position of DpRNL in the tree has virtually no support, and the observed affiliation with a homolog from *Naegleria* (heterolobosean) might be an artifact known as long-branch attraction [35, 36]. The phylogenetic reconstruction in this instance suffers from lack of taxa within Euglenozoa (only one diplonemid, no euglenid, and no

basal kinetoplastids), and from low sequence conservation. Nevertheless, the tree indicates that DpRNL diverged prior to the gene duplication event seen in kinetoplastids, and that this protein has no specific relationship to the kinetoplastid RNA ligases 2 that take part in mitochondrial RNA editing.

DpRNL is predicted to interact with RNA in a T4-like fashion
RNA ligases 2 interact with their substrate via two regions of the protein, the C-terminal domain and regions of the N-terminal nucleotidyltransferase domain that have a positive electrostatic potential. Substrate interaction of the C-terminal domain in kinetoplastid RNA ligases 2 is indirect: the four helices bind a protein partner carrying an OB-fold that, in turn, interacts with the substrate. For example, TbREL1 recruits KREPA2, and TbREL2 associates with KREPA1 [37]. In contrast, the C-terminal domain of T4RNL2 alone suffices for efficiently binding the substrate. In DpRNL, the C-terminal domain carries only two short helices making a TbREL_KREPA-like interaction unlikely. In having a positive electrostatic potential and being rich in residues able to interact with RNA, the C-terminal domain of DpRNL resembles that of T4RNL2 [24] (Fig. 4e), and probably also interacts directly with the RNA substrate.

We mentioned earlier that the Nucleotidyltransferase domain of DpRNL lacks the two substrate-binding loops of kinetoplastid RNA ligases 2. RNA interaction of loop 1 (TbREL1 aa 167–177) and loop 2 (TbREL1 aa 190–200) had been predicted based on the crystal structure [25] and the calculation of the ensemble averaged electrostatic potential [31], and has been confirmed by an RNA ligation assay with an N-terminal fragment of TbREL1 containing these two loops [25]. The same study also shows that the equivalent N-terminal portion of T4-RNL2, which lacks these loops, does not display this activity. Again, substrate interaction in the *Diplonema* protein must be different from that in kinetoplastid RNA ligases 2 and rather similar to that of T4RNL2.

In the Nucleotidyltransferase domain of the phage T4RNA ligase 2, RNA interaction is achieved by a patch of positively charged residues located in the exposed region of central beta sheets, as revealed by the crystal structure of T4RNL2 bound to a nicked nucleic acid duplex (PDB id 2HVR). To identify such regions in DpRNL, we computed the electrostatic potential at the solvent-accessible surface of the protein (see Methods). We found a large region in DpRNL's Nucleotidyltransferase domain with strong positive potential [23] (Fig. 4e). Superposition of the DpRNL 3D model onto the T4RNL2 structure with bound RNA duplex shows that the potential is distributed in a pattern similar to that in T4RNL2, and in addition, that the duplex broadly overlaps the positively charged

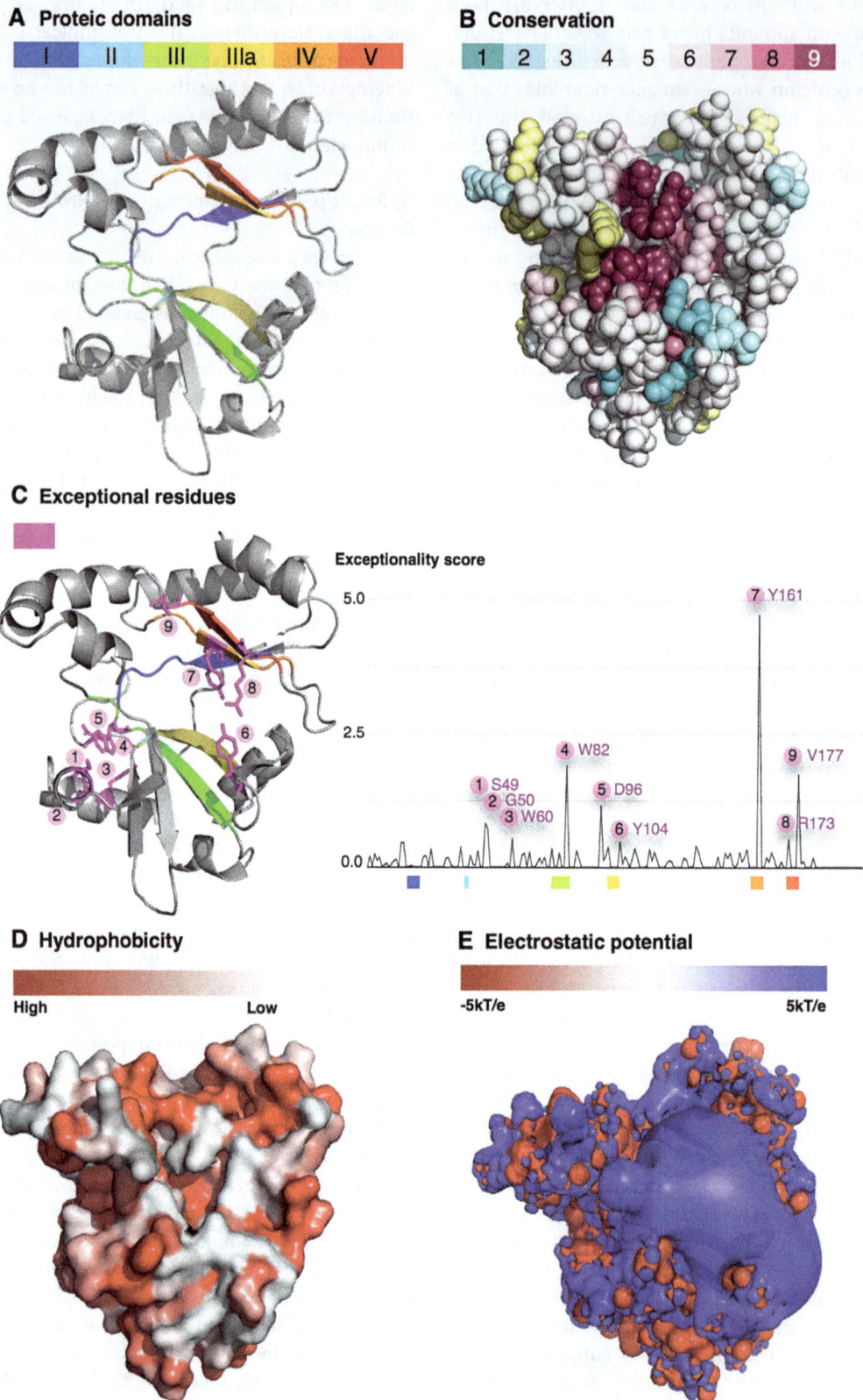

Fig. 4 Protein properties mapped onto DpRNL. **a** Localisation of the five Nucleotidyltransferase sub-domains. **b** Amino acids conserved across the RNA ligase 2 family. The value 9 (dark purple) represents highest conservation. **c** Exceptional residues as determined in this work. **d** Hydrophobicity. **e** Electrostatic potential

regions of DpRNL (Fig. 4e). However, this region in DpRNL is not completely covered by the duplex. Either the substrate is slightly shifted and|or the unoccupied region interacts with another partner. Still, in this superposition, the two C-terminal helices of the *Diplonema* protein wrap themselves around the nucleic acid like a hook, corroborating the predicted position of the RNA substrate in the DpRNL model.

Refinement of the DpRNL structural model by molecular dynamics simulations

RNA ligases 2 typically bind ATP in a covalent fashion during the first step of the catalysis resulting in a ligase-AMP complex (Additional file 1: Figure S1). In a previous section we reported that certain conserved residues otherwise involved in the covalent attachment of AMP, are substituted by different amino acids in DpRNL. To investigate how DpRNL might interact with ATP, we performed an MD simulation after introducing an ATP molecule together with a magnesium ion into the catalytic pocket of the 3D model to mimic the situation at the beginning of the first step of the enzymatic reaction. Our approach has been validated by a control simulation with TbREL1, where ATP and Mg^{2+} assumed stable positions in the catalytic pocket that correspond to those in the crystal structure [25].

MD simulations were performed for 50 and 45 ns. We restrained the position of ATP in the catalytic pocket during the first 15 ns (thereafter called the ATP-restrained production phase) followed by four replicates of unrestrained MD simulation during 35 ns. Second, we conducted three independent ATP-restrained productions of 15 ns, each followed by 30 ns unrestrained MD simulation in order to test whether ATP adopts each time the same position (see Additional file 1: Figure S9). We observed that the most important fluctuations during the entire simulation period took place in peripheral helices and loops, while the core β strands stabilized already during the first 10 ns (see lower RMSF values, Additional file 1: Figure S7). However, the conformation of the catalytic pocket was primarily influenced by the subtle motion of lateral chains in the core β strands that took place during the first 10-ns pre-production phase. In particular, the motion of the residues DpRNL_F101 and DpRNL_Y161, which are among the five residues with the lowest RMSF, had the strongest impact, reshaping the whole interaction network with ATP. Interestingly, DpRNL_Y161, which in the initial structure was perpendicular to ATP, turned around to face both the adenine ring and DpRNL_F101. This rotation occurred already during the MD equilibration phase, and the new position of this residue was retained for the rest of the simulation time in six of the seven replicates. A distinct conformation was adopted by the last replicate for which

the number of distance violations during the ATP-restrained production phase was much higher (18 %), and ATP is more distant from both aromatic residues (5.54 Å from DpRNL_Y161 and 5.79 Å from DpRNL_F101) with a mean angle of 52° (SD = 8.8 Å) with DpRNL_F101 (Additional file 1: Figures S8, S11, Table S2). Such a conformation is incompatible with π-stacking. The conformation obtained by the six consistent simulations will be referred to as the predominant conformation and analyzed in the following sections, while the deviant conformation will be addressed in the Discussion. To summarize, in the predominant 3D model of DpRNL, the pre-production phase locked the catalytic core of the protein in a stable conformation that favors interaction with ATP.

Predicted interactions of DpRNL with adenine and ribose of ATP

We compared the predicted interaction network of ATP in the DpRNL model with that inTbREL1, which is the only enzyme for which both the crystal structure of the protein bound to ATP (1XDN), and detailed molecular dynamics simulations are available [31]. ATP interactions of T4RNL2 are similar to those of TbREL1 (homologous residues are listed in Fig. 6) [23, 24, 31].

The phenylalanine (DpRNL_F101) and tyrosine (DpRNL_Y161), which together sequester the adenine base of ATP in the DpRNL model, establish a π-π stacking interaction with the substrate. This contrasts with the TbREL1 structure, where the base is enclosed by a sandwich composed of the aromatic ring of a phenylalanine (TbREL1_F209, motif IIIa), and a valine (TbREL1_V286). In the *Diplonema* protein the valine is replaced by a tyrosine (DpRNL_Y161), a residue determined as highly exceptional by comparative analysis (see above). This stabilizing interaction reduces greatly the degrees of freedom of the ATP molecule, and gives a significant turn to the interactions in the catalytic pocket by shifting the position of the ligand in DpRNL compared to well characterized RNA ligases. Additional ATP stabilization in DpRNL comes from two hydrogen bonds implicating the amine group of ATP. One hydrogen contacts the carbonyl group of DpRNL_E19 (equivalent to TbREL1_E86) and the other the lateral chain of DpRNL_E18 (which has no equivalent in TbREL1).

In TbREL1, the ribose of the ATP is bound by five residues (TbREL1_I59, TbREL1_K87, TbREL1_N92, TbREL1_R111, TbREL1_E159) allowing the sugar moiety only little mobility. Four out of these five residues (except TbREL1_I59) are conserved in the *Diplonema* protein (Fig. 3), but only two of the counterparts (DpRNL_N25 and DpRNL_E81) interact with the ribose of ATP (Fig. 5). Interactions in the DpRNL model take place indirectly through water molecules, and are weaker than the direct salt bridges in TbREL1, thus allowing the

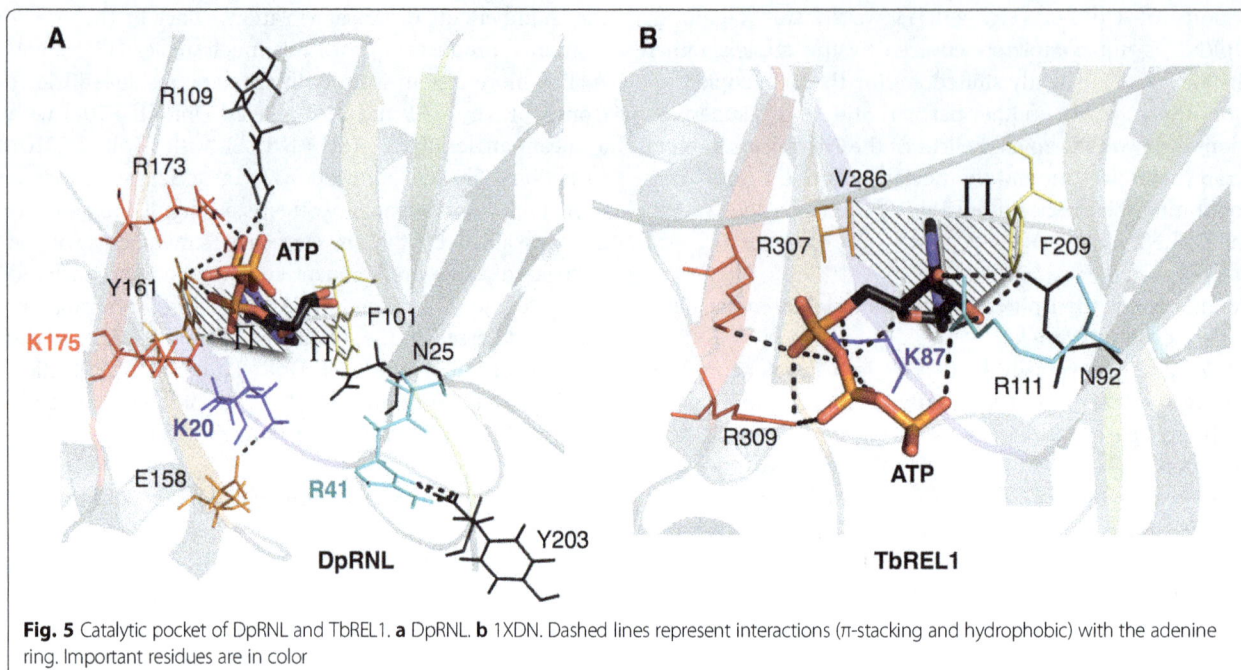

Fig. 5 Catalytic pocket of DpRNL and TbREL1. **a** DpRNL. **b** 1XDN. Dashed lines represent interactions (π-stacking and hydrophobic) with the adenine ring. Important residues are in color

larger motions of the sugar that we observed. The two conserved residues that are not involved in stabilizing the sugar (DpRNL_R41 and DpRNL_K20) play an equally important role as detailed in the following.

The triphosphate tail of ATP engages in a rich network of stabilizing interactions

In the predicted predominant conformation of DpRNL, the triphosphate tail of ATP is stabilized by a network of interactions with three basic residues (DpRNL_R109, DpRNL_R173, and DpRNL_K175). In TbREL1, the triphosphate tail is held in place by five residues, TbREL1_I61, TbREL1_K87, TbREL1_R111, TbREL1_K307 and TbREL1_R309 (see Fig. 5). Among these latter residues, only TbREL1_K307 has the same 3D position and plays the same role as predicted for DpRNL-K175, while TbREL1_I61 has no positional counterpart in DpRNL. The remaining three amino acids have a positional homolog in the DpRNL model, but apparently a different function compared to the Trypanosome protein (Fig. 6).

TbREL1_K87 is the catalytic lysine that in reaction step 1 will covalently bind ATP. This reaction is favored by strong salt bridges between ATP and several other amino acids. DpRNL_K20, the structural equivalent to TbREL1_K87, forms several salt bridges with residues DpRNL_E158, DpRNL_G159 and DpRNL_V21. But instead of promoting the covalent attachment of ATP, the interactions of DpRNL_K20 appear to rather pull this residue away from ATP, the computed distance between DpRNL_K20-Nz and Pα being on average 7.7 Å (Additional file 1: Table S2). A candidate residue for covalently binding ATP could be DpRNL_K175, owing to

its position apical to the Pα at an average distance of 4.3 Å. This distance is comparable to that observed in TbREL1 between K87 and Pα. We propose that the unusual position of ATP in the DpRNL model, as well as the posited substitution of the catalytic lysine, are due to DpRNL_Y161, which, by transforming a simple to a double π-stacking interaction, shifts the position of the ligand.

TbREL1_R111 interacts with the triphosphate tail of ATP, and therefore, the functional homolog of this residue is thought to be DpRNL_R109. However, the positional counterpart of the former residue in our model (DpRNL_R41) plays a radically different role, rather forming hydrogen bridges with residues in the C-terminal region of the protein (maintained for 75.3 % of the frames). It should be stressed that all simulations with TbREL1 have been performed with a sequence lacking the C-terminal domain (because the crystal structure was determined with the N-terminal fragment of the protein), so that interactions with the C-terminal domain are not known. In T4, the crystal structure of the adenylated full-length enzyme revealed a salt bridge between two residues of the C domain, R266 and D292, probably reinforcing its structural integrity [24].

Finally, TbREL1_R309 as well interacts with the triphosphate tail of ATP, and in the homologous position of this residue, we find in the DpRNL model a valine (DpRNL_V177). However, this valine seems not to interact with ATP or any amino acid of the catalytic pocket. The functional homolog of TbREL1_R309 is rather DpRNL_R173. Note that both DpRNL_V177 and DpRNL_R173, are "exceptional" residues, and that a

Nt	TbREL1 and T4RNL2 Residues			DpRNL equivalent		
	TbREL1	T4RNL2	Interaction	Functional	Structural	Interaction
	I59		ATP-ribose	X	-/-	-/-
	I61		ATP-PA	X	-/-	-/-
I	C85	R33	-/-	X →	E18	R117, ATP-A
I	*E86	*E34	ATP-A	→	E19	R90, ATP-A
I	*K87	*K35	ATP-ribose ATP-PA		K20	E158, V21
I	V88	I36	ATP-A	X	V21	K20
	*N92	*N40	ATP-ribose		N25	H_2O --- ATP-ribose
II	R111	*R55	ATP-PB,PG ATP-ribose		R41	Y203
III	*E159	*E99	ATP-ribose		E81	H_2O --- ATP-ribose
IIIa	*F209	*F119	ATP-A		F101	ATP-A
	F222	V220	-/-	X	Dp-R109	ATP-PG,PB, Mg2+
IV	*E283	*E204	Tb-K87?		Dp-E158	K20
IV	V286	V207	ATP-A		Dp-Y161**	ATP-A, ATP-PA
V	I305	A223	-/-	X	Dp-R173**	ATP-PB,PG,PA Y161
V	*K307	*K225	ATP-PA		Dp-K175	ATP-PA
V	*R309	*K227	ATP-PB ATP-PG		Dp-V177**	-/-

Fig. 6 Structurally and functionally equivalent residues in DpRNL, TbREL1 and T4RNL2. Residues on the same line are structural equivalents (at the same position in a structural alignment). Residues having the same functional role are connected with an arrow. Dotted arrows indicate partial functional equivalence. X, no functional equivalent was identified. Residues in *grey* seem not to play a functional role. ATP-A: adenine of ATP; ATP-ribose: ribose of ATP; ATP-PA, PB, PG: phosphate alpha, beta, and gamma, respectively of ATP; *: essential residue; **: exceptional residue; −/−, no structural equivalent identified

non-basic residue at the position corresponding to V177 in T4RNL2 was demonstrated to prevent ligation of ATP [34]). The implications of these findings will be considered in the Discussion section.

Discussion

In the search of an enzyme responsible for the unique trans-splicing in mitochondria of diplonemids, we identified a candidate RNA ligase 2 in the *D. papillatum* genome sequence. Detection of this candidate required the most sensitive HMM search method, because molecular sequences of diplonemids are in general highly divergent [38].

To confirm the sequence-based gene assignment, we constructed a preliminary 3D model of DpRNL that we aligned with RNA ligase 2 family members. Based on the structural sequence alignment, we delineated the boundaries of the predicted functional domains of the *Diplonema* protein. To pinpoint deviant amino acids in the 3D model of DpRNL, we computed a score of exceptionality for each residue. The preliminary structural model was refined by first, eliminating structural inconsistencies and second, performing molecular dynamics simulation. The final model was compared with well-characterized RNA ligases 2.

Available information on how RNA ligases 2 interact with their substrate and ATP comes from crystal structure analysis and enzymatic assays of trypanosome TbREL1 and bacteriophage T4RNL2. In contrast, the presented ligand-binding mode of DpRNL was inferred from molecular dynamics simulations that were based on an *in-silico* modeled 3D structure of the protein.

Homology models built from a template that is very distant in sequence space are usually less reliable and tend to be biased toward the template. Even if the main chains of residues interacting with ATP are correctly placed in the DpRNL model, misplacement of their side chains may influence the simulation of ligand binding. To alleviate these difficulties, we have refined the homology model using extensive MD simulation, and have tested the resulting structure using several metrics (e.g. SAVES, ModFold). The predicted unusual ATP-binding mode in the *Diplonema* protein must be considered with this precautionary note in mind.

How the postulated rewiring of ATP interactions in DpRNL may have evolved

The present model of DpRNL indicates a reorganization of residue-residue and residue-ATP interactions in the catalytic pocket compared to other ligases, entailing that (i) the ribose is less firmly stabilized than in TbTEL1 and T4RNL2, (ii) the conserved lysine DpRNL_K20 in motif I is pulled away from ATP, and (iii) ATP is now contacted by the conserved lysine DpRNL_K175 in motif V. Such a reshaping would most likely impact steps 1 and 2 of the catalysis (Additional file 1: Figure S1; Additional file 1: Figure S10, see legend for detailed description of the hypothesis).

Evolution of such reorganization in the catalytic pocket of DpRNL would require at least two consecutive steps. We speculate that initially, the nearly neutral mutation of a valine to tyrosine DpRNL_Y161 (at the position corresponding to residue 207 in T4RNL2) was made possible by the subsidiary presence of the lysine DpRNL_K175, which incidentally replaced the original catalytic lysine (DpRNL_K20). In this intermediary step, the system could have reverted back to its previous organization. Yet, the accumulation of mutations in a second step (DpRNL_V177, DpRNL_R173 by genetic drift) led to a state with no way back, in the manner of a ratchet [39]. Such a two-step scenario is archetypal of the constructive neutral evolutionary process [40].

As mentioned before, two residues highly conserved at the structural level are predicted to have a different function in DpRNL compared to orthodox RNA ligases 2. These are the ubiquitous lysine (TbREL_K87|T4RNL2_K35) and arginine (TbREL_R111|T4RNL2_R55), which correspond in the structure alignment to DpRNL_K20 and DpRNL_R41, respectively (Fig. 6). Conservation of the residues in *Diplonema* but not their predicted function raises the question about the underlying selection pressure. Interestingly, the catalytic lysine of proven RNA ligases 2 (e.g., TbREL_K87), has been suggested to also interact with the RNA substrate, notably in the reaction step 3 [41] (Additional file 1: Figure S1). Therefore, we speculate that both DpRNL_K20 and DpRNL_R41, may be subject to a negative selection in favor of conserving a second yet unrecognized role. The key message is that the observation of constant sites across an otherwise diverse family is not necessarily indicative of an identical molecular function of the corresponding residues, as residues can play multiple (structural and catalytic) roles in the corresponding protein [42].

The biological process involving DpRNL

We found that sequence- and structure-wise, mitochondrial RNA ligases 2 of kinetoplastids are not the closest homologs of DpRNL. Specifically, the 3D-structure model of DpRNL does not fit better the structure of TbREL compared to that of RNA ligases 2 from a bacteriophage or an archaean. Further, phylogenetic analysis of RNA ligases 2 did not group together the kinetoplastid and diplonemid proteins, but placed DpRNL without support next to a member of the heteroloboseans, a group that emerged prior to Euglenozoa. The large distance between kinetoplastid RNA ligases and DpRNL is probably due to a divergent, accelerated evolution and hyper-specialization of both the kinetoplastid and *Diplonema* proteins. Therefore, we cannot extrapolate from TbREL the biological process in which DpRNL may be involved.

At present it is unknown whether or not DpRNL acts inside mitochondria. There is no recognizable signal in the inferred protein sequence indicative for import into mitochondria or any other subcellular localisation. After translation, DpRNL may either remain in the cytoplasm or be imported into mitochondria by one of the cryptic signals reported for proteins of several other eukaryotes [43]. If DpRNL indeed ends up in mitochondria, then its interaction partner must be fundamentally different to those of the kinetoplastid TbREL, because of significant structural differences between the two proteins (e.g. characteristics of the C-terminal domain, the pattern of electrostatic surface potential, and the absence of interacting loops). Our in silico analyses have prepared the ground for determining experimentally the location of DpRNL in the cell, the protein and RNA partners with which it may interact, and ultimately, via 'guilt by association', the biological process in which it participates.

Conclusion

RNA ligase 2 from bacteriophage T4 is widely used as a tool in molecular biology, in particular for massively parallelized RNA sequencing technologies. Enzyme versions have been engineered with higher efficiency and fidelity than the natural protein. Specifically, the truncated version of the RNA ligase from phage T4 produces less concatemer side products and is 10 times more active than the natural enzyme [23]. Further, attempts have been undertaken to abolish concatemer formation of T4 RNA ligase by directed mutation of specific amino acids (substitution of T4RNL2-K227 by glutamine abolishes reversibility of the second step of the reaction) [34]. Comparative analysis with divergent RNA ligases such as DpRNL are bound to reveal unrecognized evolution-born innovations and to pinpoint residues otherwise not expected to be relevant enzymatically. Our *in-silico* analysis suggests that DpRNL activity relies on structure-function innovations not present in the commonly used RNA ligases, which might reveal suitable for future applications in biotechnology.

Methods

Identification of RNA ligase 2

We identified RNA ligase 2 in the draft version of the *D. papillatum* nuclear genome obtained from a Mira V3.4.1.1 [44] assembly of 7.5 million 454 reads at a coverage of ~ 10×. The search was performed with PFAM [27] domain PF09414 present in kinetoplastid RNA ligase employing HMMer 3 [45, 46] using the maximum sensitivity option (parameter −max). We found a single significant hit (E-value = 1.3e-06) in the *Diplonema* sequence matching a hypothetical protein (DpRNL). The identification of the domains characteristic for the RNA ligase 2 family was first performed by analysing the alignment of DpRNL with the PF09414 HMM domain in the HMMer result file, then by a multiple alignment of the two ligase paralogs from four Leishmania species (*L. braziliensis*: LbrM.20.5890 and LbrM.01.0620; *L.*

mexicana: LmxM.01.0590 and LmxM.20.1730; *L. major* Friedlin: LmjF.20.1730 and LmjF.01.0590; *L. infantum*: LinJ.01.0610 and LinJ.20.1700) and six Trypanosoma species (*T. brucei* TREU927: Tb09.160.2970 and Tb927.1.3030; *T. brucei* Lister strain 427: Tb427.01.3030 and Tb427tmp.160.2970; *T. brucei* gambiense: Tbg972.1.1840 and Tbg972.9.2300; *T. cruzi* CL Brener Esmeraldo-like: Tc00.1047053506363.110 and Tc00.1047053511585.20; *T. cruzi* CL Brener Non-Esmeraldo-like: Tc00.1047053506975.9 and Tc00.1047053510155.20; *T. congolense*: TcIL3000.1.1450 and TcIL3000.9.1420; *T. vivax*: TvY486_0101350 and TvY486_0901490).

The specificity of PF09414 in detecting RNA ligases 2 was evaluated by comparing the score of all PFAM domains of DNA and RNA ligases against (i) the *Diplonema* candidate RNA ligase, (ii) the two well characterized RNA ligases from *Trypanosoma brucei* TREU927 (TbREL1, Gene ID = Tb927.9.4360 and TbREL2, Gene ID = Tb927.1.3030) downloaded from TriTrypDB [47], and (iii) the RNA ligase from *Naegleria gruberi* (XP_002674912.1), a protist diverging basally to Euglenozoa.

Three-dimensional structure modeling

The three-dimensional model of DpRNL has been determined by I-Tasser (the Iterative Threading Assembly Refinement program) web server (http://zhanglab.ccmb.med.umich.edu/I-TASSER/) [48] using default parameters (no restraints, no guide or exclusion template). I-Tasser selected the structure of the DNA ligase of Aquifex aeolicus (PDB ID = 3QWU) as the closest structural homolog of DpRNL and proposed five candidate models. Then, we refined the models with ModRefiner [49], and evaluated the quality of the models with tools available from the SAVeS Web server (Structural Analysis and Verification Server http://services.mbi.ucla.edu/SAVES/) and ModFold [30]. From the five models proposed by I-Tasser, we selected the one having the lowest structural variations compared to the template, and the best structural qualities according to SAVeS.

System preparation for molecular dynamics simulations

Two different molecular dynamics simulation protocol were used for DpRNL. To investigate the stability of our model, we used the apo form of the protein (apo-DpRNL). To examine the interactions between the ligand and the protein, we used DpRNL with bound ATP and Mg^{2+} (DpRNL_ATP+Mg^{2+}). In this experiment, we superimposed DpRNL onto TbREL1, the Trypanosoma homolog of DpRNL crystallised with ATP (PDB ID = 1XDN), and manually copied the ATP and Mg^{2+} residues from 1XDN to the corresponding position in DpRNL. We added hydrogens when needed with WHATIF [50] and rendered

the file CHARMM compatible by employing the PDB Reader of CHARMM-GUI [51].

Molecular dynamics simulations

All molecular dynamics (MD) simulations were performed with the Gromacs 4.0.5, 4.6.5, 5.0.1 and 5.0.2 software [52] and CHARMM27 force field [53]. We modified the charmm27.ff force field [54] files in Gromacs to add topology and parameter information for ATP from toppar_all36_na_nad_ppi.str by following the procedure specified in the Gromacs manual (http://www.gromacs.org/Documentation/How-tos/Adding_a_Residue_to_a_Force_Field). Proteins and ligands were solvated in a cubic box of TIP3P water molecules at a distance of 3 nm (30 Å) from the solute. The net charge of the system was neutralized by addition of six chloride ions for the DpRNL apo system, four chloride ions for DpRNL+ATP+Mg^{2+} and five sodium ions for TbKREL1+ATP+Mg^{2+}. The cut-off for short-range van der Waals and electrostatic interactions was 1.0 nm (default values), and PME (Particle Mesh Ewald) was used for long-range interactions in all simulations. First, we performed an energy minimisation by steepest descent to remove possible spurious contacts until convergence to a maximum force of 1000 kJ/mol/nm on any atom of the system (850 steps). For all MD simulations, the leap-frog formula was used to integrate the equations of motion. Then two MD equilibrations of 100 ps each (25,000 steps with 2 fs timesteps) were performed with restrained positions of protein and ligand. For the first NVT (constant number of particles, volume, and temperature) equilibration, the temperature was set to 300K using the V-rescale thermostat [55] with separate baths for protein and non-protein atoms. Then, for the subsequent NPT (constant number of particles, pressure, and temperature) equilibration, the Parrinello-Rahman barostat [56, 57] was used in addition to the V-rescale thermostat in order to couple the pressure to 1 bar. Following these pre-production steps, MD simulation productions were performed on apo-DpRNL and on holo-DpRNL loaded with ATP and Mg^{2+}.

For DpRNL apo, we performed a 50 ns simulation with 2 fs timesteps. The 2D structure conservation during the simulation period was measured using the timeline plugin of VMD [58]. For DpRNL loaded with ATP and Mg^{2+}, we performed MD simulations of 50 and 45 ns in total. During a preliminary simulation, ATP escaped from the catalytic pocket. Therefore, as a precaution, we restrained its position during the initial 15 ns of the production simulation (referred to as ATP-restrained phase), to let the protein equilibrate around the ligand, and after lifting the restriction, the simulation was continued. First, we used the same 15 ns ATP-restrained simulation (15R0) that we extended by four independent 35-ns

MD simulations (replicates 15R0 + 35_1 to 15R0 + 35_4). Second, we ran three independent 15-ns restrained simulations (15RI, 15RII and 15RIII) followed by 30 ns MD simulations. When measuring the distance between the two molecules during the initial time interval, we noted that the restraint was used in less than 1 % of the frames for all the predominant replicates (15R0, 15RI, 15RII) and in 18 % of the frames for the deviant replicate (15RIII) (Additional file 1: Figure S8). As an anchor of the restraint, we chose DpRNL_F101 because first, this residue is highly conserved among ligases; second, it is positioned deeply inside the catalytic pocket; and third, in *Trypanosoma* TbREL1, the adenine of ATP has been shown to make π-stacking interactions with the homologous position, TbREL1-F209 [25]. We set a distance restraint of 0.3 nm around the initial distance ri between each pair of atoms from the phenyl group of DpRNL_F101 and the pyrimidine ring of the adenine, meaning that there is a component for the restraint added to the potential energy function for $r_i > r_i + 0.30$ nm and $r_i < r_i - 0.30$ nm. Three distances are set: $r_0 = r_i - 0.30$ nm, $r_1 = r_i + 0.30$ nm and $r_2 = r_1 + 1$ nm. The potential for the distance restraints is quadratic below r_0 and between r_1 and r_2, and linear above r_2.

To test whether inserting ATP + Mg^{2+} in the catalytic pocket of DpRNL leads to a realistic positioning of the ligands, we performed a control experiment on TbREL1. To prepare the system, we replaced the selenomethionine used for crystallisation with methionine, then we ran an MD simulation first on the apo protein for 15 ns. Then, we used the structure from the last frame of the previous simulation as a starting point, inserted ATP + Mg^{2+} into the molecule, and ran a simulation for 30 ns.

Exceptional residues
In order to identify exceptional residues in the candidate RNA ligase of *Diplonema*, we computed a score measuring how unexpected each residue of the protein is. Using the I-Tasser model of DpRNL as the query structure, we searched for structural "neighbors" with DALI (http://ekhidna.biocenter.helsinki.fi/dali_server/, [59]): we selected 23 unique RNA and DNA ligases whose structure have the highest percentage of identity and the lowest Root Mean Square Deviation (RMSD), and performed a multiple structural alignment including DpRNL. For subsequent computations, we used the alignment without expanding the gaps, meaning that inserted segments relative to DpRNL are hidden. For each position in the 23 proteins, we computed the entropy s as given by [60] which represents the diversity of amino acids for a given position. The entropy s at position l is $s(l) = - \sum_{i=1}^{6} P_i(l) \log P_i(l)$ where i is the category of amino acid (1: aliphatic, {AVLIMC} 2: aromatic {FWYH}, 3: polar {STNQ}, 4: positive {KR}, 5: negative {DE}, 6: special

{GP}), and $P_i(l)$ is the proportion of amino acids belonging to category i at position l. At a given position, amino-acid categories for which $P_i(l)$ is null are ignored. If the entropy is low, then the position is conserved among family members. The entropy is set arbitrarily to 0 when the position in the multiple alignment contains more than 50 % gaps. We designed an exceptionality score S at position l for amino acids of DpRNL as $Sl = (P_{max}(l) - P_i(l)) / s(l)$ where $P_i(l)$ is the proportion in the previously computed multiple alignment of the amino acid observed at position l for DpRNL, and $P_{max}(l)$ is the proportion of the most abundant amino-acid category (the category that we expect).

3D model analyses
Trajectory analyses were performed with R [61], VMD [62] and PyMOL [63]. Hydrogen bonds were computed using VMD with a distance cutoff of 3.0 Å and an angle cutoff of 30°. The evolution of the secondary structure [58] was computed via the timeline plugin of VMD based on the STRIDE algorithm [64]. The conservation surface was colored with the web server ConSurf [65] using the structural multiple alignment performed by DALI as input and with the Bayesian method for computing the evolutionary rate [66].

The electrostatic potential of the molecule was computed by the classical calculation using the last frame of the simulation, employing the APBS web server (http://www.poissonboltzmann.org/) [67–69] and visualized using the dedicated APBS plugin of PyMOL. The isovalue cut-off for the analyses was set to $+5k_BT/e$ (blue) and $+5k_BT/e$ (red). For DpRNL, this procedure was sufficient to reveal a large region with positive potential, having the propensity to bind RNA. In contrast, for TbREL1, the classical potential calculation (using Delphi [25]) identified only small positive patches. To find a positive region sufficiently large for RNA binding in TbREL1, the authors had to calculate an ensemble average on their 70-ns simulation [31].

Expression
The expression of the gene coding for DpRNL was assessed by mapping RNA-seq reads from a total-RNA library of *D. papillatum* onto the contig carrying the gene. Library construction and read processing have been described earlier [19]. Cutadapt version 1.2.1 [70] was used to remove adapters at 5′ and 3′ termini of reads with an error rate of 0.1 and to clip low-quality sequences with a threshold of 20. Reads <20 nt were discarded, leaving 29 million paired reads, which were mapped with Bowtie2 [71] onto the 1314-nt long contig containing the DpRNL reading frame. Output files in sam format were subsequently transformed into 'bam'

files with SAMtools version 1.4 [72]. Alignments were visualized with tablet version 1.13.05.17 [73].

Phylogenetic reconstruction of RNA ligases 2 from Excavata

We identified RNA ligase 2 proteins in Excavata species by searching with the same PFAM HMM PF09414 as used for *Diplonema*. Sequences were aligned using MAFFT with option "–localpair" (for distantly related species with a single alignable domain). The multiple alignment was refined by successive re-alignment of the sequences on a guiding hmm model built from the alignment with HMMer 3 [45, 46]. The best scoring alignment according to HMMer was selected and filtered with an in-house script to retain positions with less than 30 % gaps and a conservation score greater than 8 as given in the stockholm format. We reconstructed the phylogeny with RaXMLHPC v.7.2.6, a maximum likelihood method, using a gamma distribution to model the heterogeneity of substitution rate over sites and the WAG substitution matrix. A Bootstrap analysis of 100 runs was performed to assess the significance of each node.

Additional files

Additional file 1: Supplementary data. Supplementary result, figures and tables for DpRNL identification, phylogenetic study, 3D model properties, and complementary analyses of MD simulations

Additional file 2: DpRNL model. Atomic coordinates of DpRNL model.

Additional file 3: ITasser threading of DpRNL on 3QWU. Atomic coordinates of DpRNL and 3QWU model.

Abbreviations

2D: Secondary structure; 3D: Tertiary structure; ATP: Adenine triphosphate; DpRNL: RNA ligase 2 from *Diplonema papillatum*; HMM: Hidden Markov model; MD: Molecular dynamics; PFAM: Protein FAMily database; T4RNL2: RNA ligase 2 from bacteriophage T4; TbREL1: RNA ligase 2 from *Trypanosoma brucei*, paralog of KREL1, Tb927.9.4360.

Competing interests

The authors declare that they have no competing interests.

Authors' contributions

SM conceived the project, and all authors participated in the design of the study. EN performed the 3D model determination and refinement. SM performed the comparative analysis, phylogeny, molecular dynamics simulation, and electrostatic calculation, and evaluated the results. GL provided guidance on structural biology and molecular dynamics simulations. GB was involved in the design of the phylogenetic analysis and the interpretation of results. SM and GB wrote the manuscript. All authors read and approved the final manuscript.

Authors' information

Not applicable.

Acknowledgments

The authors acknowledge B.F. Lang (Université de Montréal) for advice in reconstructing the phylogeny. The *Diplonema* nuclear genome is being sequenced in collaboration with Cestmir Vclek (Institute of Molecular Genetics, Prague) and Julius Lukeš (Institute of Parasitology, University of South Bohemia). This work was supported by the Canadian Institute for Health Research [CIHR, grant MOP-79309; to G.B.]. Funding for open access charge: Canadian Institute for Health Research.

Author details

[1]Department of Biochemistry and Robert-Cedergren Centre for Bioinformatics and Genomics, Université de Montréal, Montreal, QC, Canada. [2]Department of Biochemistry, currently Département d'informatique et de recherche opérationnelle (DIRO), Université de Montréal, Montreal, QC, Canada. [3]Department of Chemistry and Biochemistry, Centre for Research in Molecular Modeling (CERMM), Groupe d'étude des protéines membranaires (GÉPROM), Regroupement québécois de recherche sur la fonction, l'ingénierie et les applications des protéines (PROTEO), Concordia University, Montreal, QC, Canada.

References

1. Pascal JM. DNA and RNA ligases: structural variations and shared mechanisms. Curr Opin Struct Biol. 2008;18:96–105.
2. Subramanya HS, Doherty AJ, Ashford SR, Wigley DB. Crystal structure of an ATP-dependent DNA ligase from bacteriophage T7. Cell. 1996;85:607–15.
3. Greer CL, Javor B, Abelson J. RNA ligase in bacteria: formation of a 2″,5″ linkage by an E. coli extract. Cell. 1983;33:899–906.
4. Arn EA, Abelson JN. The 2″–5″ RNA ligase of Escherichia coli. Purification, cloning, and genomic disruption. J Biol Chem. 1996;271:31145–53.
5. Mazumder R, Iyer LM, Vasudevan S, Aravind L. Detection of novel members, structure-function analysis and evolutionary classification of the 2H phosphoesterase superfamily. Nucl Acids Res. 2002;30:5229–43.
6. Popow J, Englert M, Weitzer S, Schleiffer A, Mierzwa B, Mechtler K, et al. HSPC117 is the essential subunit of a human tRNA splicing ligase complex. Science. 2011;331:760–4.
7. Tanaka N, Meineke B, Shuman S. RtcB, a novel RNA ligase, can catalyze tRNA splicing and HAC1 mRNA splicing in vivo. J Biol Chem. 2011;286:30253–7.
8. Popow J, Schleiffer A, Martinez J. Diversity and roles of (t)RNA ligases. Cell Mol Life Sci. 2012;69:2657–70.
9. Amitsur M, Levitz R, Kaufmann G. Bacteriophage T4 anticodon nuclease, polynucleotide kinase and RNA ligase reprocess the host lysine tRNA. EMBO J. 1987;6:2499–503.
10. Greer CL, Peebles CL, Gegenheimer P, Abelson J. Mechanism of action of a yeast RNA ligase in tRNA splicing. Cell. 1983;32:537–46.
11. Ho CK, Shuman S. Bacteriophage T4 RNA ligase 2 (gp24.1) exemplifies a family of RNA ligases found in all phylogenetic domains. Proc Natl Acad Sci U S A. 2002;99:12709–14.
12. Bakalara N, Simpson AM, Simpson L. The Leishmania kinetoplast-mitochondrion contains terminal uridylyltransferase and RNA ligase activities. J Biol Chem. 1989;264:18679–86.
13. Stuart K, Brun R, Croft S, Fairlamb A, Gürtler RE, McKerrow J, et al. Kinetoplastids: related protozoan pathogens, different diseases. J Clin Invest. 2008;118:1301–10.
14. Simpson L, Da Silva A. Isolation and characterization of kinetoplast DNA from Leishmania tarentolae. J Mol Biol. 1971;56:443–73.
15. Blum B, Bakalara N, Simpson L. A model for RNA editing in kinetoplastid mitochondria: RNA molecules transcribed from maxicircle DNA provide the edited information. Cell. 1990;60:189–98.
16. Sturm NR, Simpson L. Kinetoplast DNA minicircles encode guide RNAs for editing of cytochrome oxidase subunit III mRNA. Cell. 1990;61:879–84.
17. Aphasizhev R, Aphasizheva I. Mitochondrial RNA editing in trypanosomes: small RNAs in control. Biochimie. 2014;100:125–31.
18. Marande W, Burger G. Mitochondrial DNA as a genomic jigsaw puzzle. Science. 2007;318:415.
19. Valach M, Moreira S, Kiethega GN, Burger G. Trans-splicing and RNA editing of LSU rRNA in Diplonema mitochondria. Nucl Acids Res. 2014;42:2660–72.
20. Vlcek C, Marande W, Teijeiro S, Lukes J, Burger G. Systematically fragmented genes in a multipartite mitochondrial genome. Nucl Acids Res. 2010;39:979–88.

21. Cranston JW, Silber R, Malathi VG, Hurwitz J. Studies on ribonucleic acid ligase. Characterization of an adenosine triphosphate-inorganic pyrophosphate exchange reaction and demonstration of an enzyme-adenylate complex with T4 bacteriophage-induced enzyme. J Biol Chem. 1974;249:7447–56.

22. Yin S, Ho CK, Shuman S. Structure-function analysis of T4 RNA ligase 2. J Biol Chem. 2003;278:17601–8.

23. Ho CK, Wang LK, Lima CD, Shuman S. Structure and mechanism of RNA ligase. Structure. 2004;12:327–39.

24. Nandakumar J, Shuman S, Lima CD. RNA ligase structures reveal the basis for RNA specificity and conformational changes that drive ligation forward. Cell. 2006;127:71–84.

25. Deng J, Schnaufer A, Salavati R, Stuart KD, Hol WGJ. High resolution crystal structure of a key editosome enzyme from Trypanosoma brucei: RNA editing ligase 1. J Mol Biol. 2004;343:601–13.

26. Altschul SF, Gish W, Miller W, Myers EW, Lipman DJ. Basic local alignment search tool. J Mol Biol. 1990;215:403–10.

27. Punta M, Coggill PC, Eberhardt RY, Mistry J, Tate J, Boursnell C, et al. The Pfam protein families database. Nucl Acids Res. 2012;40(Database issue):D290–301.

28. Zhang Y. I-TASSER server for protein 3D structure prediction. BMC Bioinformatics. 2008;9:40.

29. Bowie JU, Lüthy R, Eisenberg D. A method to identify protein sequences that fold into a known three-dimensional structure. Science. 1991;253:164–70.

30. McGuffin LJ, Buenavista MT, Roche DB. The ModFOLD4 server for the quality assessment of 3D protein models. Nucl Acids Res. 2013;41(Web Server issue):W368–72.

31. Amaro RE, Swift RV, McCammon JA. Functional and structural insights revealed by molecular dynamics simulations of an essential RNA editing ligase in Trypanosoma brucei. PLoS Negl Trop Dis. 2007;1, e68.

32. Brooks MA, Meslet-Cladiére L, Graille M, Kuhn J, Blondeau K, Myllykallio H, et al. The structure of an archaeal homodimeric ligase which has RNA circularization activity. Protein Sci. 2008;17:1336–45.

33. Sheinerman F. Electrostatic aspects of protein–protein interactions. Curr Opin Struct Biol. 2000;10:153–9.

34. Viollet S, Fuchs RT, Munafo DB, Zhuang F, Robb GB. T4 RNA ligase 2 truncated active site mutants: improved tools for RNA analysis. BMC Biotechnol. 2011;11:72.

35. Lartillot N, Brinkmann H, Philippe H. Suppression of long-branch attraction artefacts in the animal phylogeny using a site-heterogeneous model. BMC Evol Biol. 2007;7 Suppl 1:S4.

36. Felsenstein J. Cases in which parsimony or compatibility methods will be positively misleading. Syst Biol. 1978;27:401–10.

37. Park Y-J, Budiarto T, Wu M, Pardon E, Steyaert J, Hol WGJ. The structure of the C-terminal domain of the largest editosome interaction protein and its role in promoting RNA binding by RNA-editing ligase L2. Nucl Acids Res. 2012;40:6966–77.

38. Simpson AGB, Gill EE, Callahan HA, Litaker RW, Roger AJ. Early evolution within kinetoplastids (Euglenozoa), and the late emergence of trypanosomatids. Protist. 2004;155:407–22.

39. Lukeš J, Archibald JM, Keeling PJ, Doolittle WF, Gray MW. How a neutral evolutionary ratchet can build cellular complexity. IUBMB Life. 2011;63:528–37.

40. Stoltzfus A. On the possibility of constructive neutral evolution. J Mol Evol. 1999;49:169–81.

41. Swift RV, Durrant J, Amaro RE, McCammon JA. Toward understanding the conformational dynamics of RNA ligation. Biochemistry. 2009;48:709–19.

42. Todd AE, Orengo CA, Thornton JM. Plasticity of enzyme active sites. Trends Biochem Sci. 2002;27:419–26.

43. Dudek J, Rehling P, van der Laan M. Mitochondrial protein import: common principles and physiological networks. Biochim Biophys Acta. 1833;2013:274–85.

44. Chevreux B, Wetter T, Suhai S: Genome sequence assembly using trace signals and additional sequence information. Comput Sci Biol: Proc. German Conference on Bioinformatics GCB'99 GCB 1999;45–56.

45. Finn RD, Clements J, Eddy SR. HMMER web server: interactive sequence similarity searching. Nucl Acids Res. 2011;39(Web Server issue):W29–37.

46. Eddy SR. Accelerated profile HMM searches. PLoS Comput Biol. 2011;7, e1002195.

47. Aslett M, Aurrecoechea C, Berriman M, Brestelli J, Brunk BP, Carrington M, et al. TriTrypDB: a functional genomic resource for the Trypanosomatidae. Nucl Acids Res. 2010;38(Database issue):D457–62.

48. Roy A, Kucukural A, Zhang Y. I-TASSER: a unified platform for automated protein structure and function prediction. Nat Protoc. 2010;5:725–38.

49. Xu D, Zhang Y. Improving the physical realism and structural accuracy of protein models by a two-step atomic-level energy minimization. Biophys J. 2012;101:2525–34.

50. Vriend G. WHAT IF: a molecular modeling and drug design program. J Mol Graph. 1990;8:52–6.

51. Jo S, Kim T, Iyer VG, Im W. CHARMM-GUI: a web-based graphical user interface for CHARMM. J Comput Chem. 2008;29:1859–65.

52. Pronk S, Páll S, Schulz R, Larsson P, Bjelkmar P, Apostolov R, et al. GROMACS 4.5: a high-throughput and highly parallel open source molecular simulation toolkit. Bioinformatics. 2013;29:845–54.

53. MacKerell AD, Bashford D, Bellott M. All-atom empirical potential for molecular modeling and dynamics studies of proteins. J Phys Chem B. 1998.

54. Bjelkmar P, Larsson P, Cuendet MA, Hess B, Lindahl E. Implementation of the CHARMM Force Field in GROMACS: analysis of protein stability effects from correction maps, virtual interaction sites, and water models. J Chem Theory Comput. 2010;6:459–66.

55. Bussi G, Donadio D, Parrinello M. Canonical sampling through velocity rescaling. J Chem Phys. 2007;126:014101.

56. Parrinello M, Rahman A. Polymorphic transitions in single crystals: a new molecular dynamics method. J Appl Phys. 1981;52:7182–90.

57. Nosé S, Klein ML. Constant pressure molecular dynamics for molecular systems. Mol Phys. 2006;50:1055–76.

58. Kabsch W, Sander C. Dictionary of protein secondary structure: pattern recognition of hydrogen-bonded and geometrical features. Biopolymers. 1983;22:2577–637.

59. Holm L, Rosenström P. Dali server: conservation mapping in 3D. Nucl Acids Res. 2010;38(Web Server issue):W545–9.

60. Mirny LA, Shakhnovich EI. Universally conserved positions in protein folds: reading evolutionary signals about stability, folding kinetics and function. J Mol Biol. 1999;291:177–96.

61. Team RC. R: a language and environment for statistical computing. Vienna: R Foundation for Statistical Computing; 2013.

62. Humphrey W, Dalke A, Schulten K. VMD: visual molecular dynamics. J Mol Graph. 1996;14:33–8– 27–8.

63. Schrödinger L. The PyMOL molecular graphics system, Version 1.3 R1. Py-MOL. 2010.

64. Frishman D, Argos P. Knowledge-based protein secondary structure assignment. Proteins. 1995;23:566–79.

65. Glaser F, Pupko T, Paz I, Bell RE, Bechor-Shental D, Martz E, et al. ConSurf: identification of functional regions in proteins by surface-mapping of phylogenetic information. Bioinformatics. 2003;19:163–4.

66. Mayrose I, Graur D, Ben Tal N, Pupko T. Comparison of site-specific rate-inference methods for protein sequences: empirical Bayesian methods are superior. Mol Biol Evol. 2004;21:1781–91.

67. Holst M, Saied F. Multigrid solution of the Poisson—Boltzmann equation. J Comput Chem. 1993;14:105–13.

68. Holst MJ, Saied F. Numerical solution of the nonlinear Poisson–Boltzmann equation: developing more robust and efficient methods. J Comput Chem. 1995;16:337–64.

69. Dolinsky TJ, Czodrowski P, Li H, Nielsen JE, Jensen JH, Klebe G, et al. PDB2PQR: expanding and upgrading automated preparation of biomolecular structures for molecular simulations. Nucl Acids Res. 2007;35(Web Server issue):W522–5.

70. Martin M. Cutadapt removes adapter sequences from high-throughput sequencing reads. EMBnet J. 2011.

71. Langmead B, Salzberg SL. Fast gapped-read alignment with Bowtie 2. Nat Methods. 2012;9:357–9.

72. Li H, Handsaker B, Wysoker A, Fennell T, Ruan J, Homer N, et al. 1000 genome project data processing subgroup: the sequence Alignment/Map format and SAMtools. Bioinformatics. 2009;25:2078–9.

73. Milne I, Stephen G, Bayer M, Cock PJA, Pritchard L, Cardle L, et al. Using Tablet for visual exploration of second-generation sequencing data. Brief Bioinform. 2013;14:193–202.

4

3D QSAR, pharmacophore and molecular docking studies of known inhibitors and designing of novel inhibitors for M18 aspartyl aminopeptidase of *Plasmodium falciparum*

Madhulata Kumari[1,4], Subhash Chandra[2], Neeraj Tiwari[3] and Naidu Subbarao[4*]

Abstract

Background: The *Plasmodium falciparum* M18 Aspartyl Aminopeptidase (*Pf*M18AAP) is only aspartyl aminopeptidase which is found in the genome of *P. falciparum* and is essential for its survival. The *Pf*M18AAP enzyme performs various functions in the parasite and the erythrocytic host such as hemoglobin digestion, erythrocyte invasion, parasite growth and parasite escape from the host cell. It is a valid target to develop antimalarial drugs. In the present work, we employed 3D QSAR modeling, pharmacophore modeling, and molecular docking to identify novel potent inhibitors that bind with M18AAP of *P. falciparum*.

Results: The PLSR QSAR model showed highest value for correlation coefficient r^2 (88 %) and predictive correlation coefficient (pred_r2) =0.6101 for external test set among all QSAR models. The pharmacophore modeling identified DHRR (one hydrogen donor, one hydrophobic group, and two aromatic rings) as an essential feature of *Pf*M18AAP inhibitors. The combined approach of 3D QSAR, pharmacophore, and structure-based molecular docking yielded 10 novel *Pf*M18AAP inhibitors from ChEMBL antimalarial library, 2 novel inhibitors from each derivative of quinine, chloroquine, 8-aminoquinoline and 10 novel inhibitors from WHO antimalarial drugs. Additionally, high throughput virtual screening identified top 10 compounds as antimalarial leads showing G-scores -12.50 to -10.45 (in kcal/mol), compared with control compounds(G-scores -7.80 to -4.70) which are known antimalarial M18AAP inhibitors (AID743024). This result indicates these novel compounds have the best binding affinity for *Pf*M18AAP.

Conclusion: The 3D QSAR models of *Pf*M18AAP inhibitors provided useful information about the structural characteristics of inhibitors which are contributors of the inhibitory potency. Interestingly, In this studies, we extrapolate that the derivatives of quinine, chloroquine, and 8-aminoquinoline, for which there is no specific target has been identified till date, might show the antimalarial effect by interacting with *Pf*M18AAP.

Keywords: *Plasmodium falciparum*, M18 aspartyl aminopeptidase, 3D QSAR, PLSR, PCR, kNN-MFA, Molecular docking, HTVS, Pharmacophore modeling

Abbreviations: 3D QSAR, 3-dimensional quantitative structure activity relationship; *Pf*M18AAP, *Plasmodium falciparum* M18 Aspartyl Aminopeptidase; PLSR, Partial least square regression; PCR, Principal component regression; kNN-MFA, k-nearest neighbor-molecular field analysis; QN, Quinine; CQ, Chloroquine; 8-AmQN, 8-aminoquinoline; HTVS, High Throughput Virtual Screening; *P. falciparum*, *Plasmodium falciparum*

* Correspondence: nsrao@mail.jnu.ac.in
[4]School of Computational and Integrative Sciences, Jawaharlal Nehru University, New Delhi 110067, India
Full list of author information is available at the end of the article

Background

Malaria, a mosquito-borne disease, kills roughly 627000 people every year, mostly infants in Africa. It affects about 198 million patients annually (World Health Organization, 2013, http://www.who.int/malaria/media/en/). It is caused by parasites which are clubbed under genus *Plasmodium*. Among them, *P. falciparum* is encountered most commonly and is deadliest [1]. Though there are myriad drugs to treat the menace but the increasing instances of resistance against antimalarial drugs are becoming a deepening concern day by day. In recent years, several cases of resistance have been detected across the globe against artemisinin drugs [2]. This underscores the need to discover resilient drugs to combat malaria in future. Therefore, in this effort, several molecular drug targets have been identified to develop new drug candidates. An important drug target is M18 aspartyl aminopeptidase (M18AAP) which is expressed in the cytoplasm of *P. falciparum* by a single copy of *Pf*M18AAP gene. M18AAP interacts with the human erythrocyte membrane protein Spectrin and other proteins during disease kicking off erythrocytic life cycle, and it is essential for the survival of this parasite in Blood cells. It has been reported that the malaria parasites mutated with M18AAP enzyme are not able to survive, which proves that this plays a critical role in the survival of *P. falciparum* and could serve as an important molecular target to develop potential therapeutic agents to control malaria infection [3]. In modern times, virtual screening methods like QSAR, pharmacophore modeling, molecular docking have been proved a valuable tool for rapid discovery of novel drug candidates, e.g., the discovery of O-Acetyl-L-Serine Sulfhydrylase of *Entamoeba histolytica* inhibitors, acetylcholinesterase inhibitors, and antagonists Acetophenazine, fluphenazine and periciazine against Human androgen receptor [4–6]. In the drug development, the study of Quantitative structure-activity relationships (QSAR) plays an important role to analyze the properties of drugs. QSAR is a mathematical model that relates chemical descriptors of compounds to their quantity showing specific biological or chemical activity [7]. The molecular descriptors for the compounds are calculated and used to derive QSAR Model [8]. In the present study, the known bioactive dataset was used to build 3D QSAR models using partial least square regression (PLSR) [9], principal component regression (PCR) [10, 11] and k-nearest neighbor-molecular field analysis (kNN-MFA) methods [12]. After that, pharmacophore mapping was performed to identify the binding modes and structural features of the ligands and followed by molecular docking. The generated models provided a valuable reference which could be applied in the designing of pharmaceuticals with improved antimalarial activity. In the end, virtual screening of antimalarial compounds from ChEMBL Bioassay, and other dataset were also carried out to identify novel potential inhibitors

which could be better as compared to the known inhibitors of *Pf*M18AAP.

Methods

Dataset of experimental *Pf*M18AAP inhibitors

A dataset of 32 compounds known as inhibitors of *Pf*M18AAP was extracted from National Center for Biotechnology Information PubChem bioassay (AID 743024) (https://pubchem.ncbi.nlm.nih.gov/assay/assay.cgi?aid=743024). Another high throughput screened dataset of 3502 known bioactive inhibitors of *Pf*M18AAP was extracted from AID 1822 used for docking studies against *Pf*M18AAP (http://pubchem.ncbi.nlm.nih.gov/assay/assay.cgi?aid=1822). A library of 153,873 compounds was obtained from the ChEMBL antimalarial database used for finding novel inhibitors against *Pf*M18AAP metalloproteinase [https://www.ebi.ac.uk/chembl/]. Additionally, 27 antimalarial drugs described by WHO, 32 analogous of quinine compounds(QN) (AID 660170), 24 analogous of chloroquine (CQ) (AID 404780), and 17 analogous of 8-aminoquinoline(8-AmQN) (AID 554037) were also extracted for molecular docking, 3D QSAR model and pharmacophore similarity search. 2D structures were converted to 3D structures using Corina 2.64v [13] and open babel [14].

Molecular descriptors

The molecular descriptors were calculated by VLifeMDS version 4.3 using Gasteiger-Marsili charge [15, 16]. The *Pf*M18AAP inhibitors compounds along with their activity pIC50 values were given as input for force field calculation. The steric and electrostatic interaction energies are computed using a methyl probe of charge +1.

Development of 3D QSAR models

The biological activity (pIC50) of inhibitors was selected as dependent variables and descriptors as independent variables. The 60 % data for the training set and 40 % for test set were manually selected. The unicolumn statistics were calculated to validate training and test sets. The 3D QSAR models were built using PLSR, PCR, and kNN-MFA by stepwise forward-backward method [17].

3D QSAR Model validation

Internal validation

To perform internal validation (cross validation), a compound is eliminated from the training set and then its biological activity is predicted to validate model accuracy. This step is repeated until the biological activity of every compound in the training set is predicted once. The cross-validated coefficient, q^2 is calculated using the given Eq. (1):

Table 1 Unicolumn statistics for training and test set

DataSet	Column Name	Average	Maximum	Minimum	Standard Deviation	Sum
Training	pIC50	5.6527	6.7200	5.1020	0.4450	90.4430
Test	pIC50	5.6559	6.3400	4.9200	0.4849	62.2146

$$q^2 = 1 - \frac{\sum (y_i - \hat{y}_i)^2}{\sum (y_i - y_{means})^2} \qquad (1)$$

Where, y_i and $\hat{y}_i$ are the actual and predicted activities of the ith molecule in the training set respectively, and y_{means} is the average activity of all the molecules in the training set [18, 19].

External validation

External validation (pred_r^2) is carried out by calculating predicted correlation coefficient (pred_r^2) value using following Eq. (2):

$$pred\_r^2 = 1 - \frac{\sum (y_i - \hat{y}_i)^2}{\sum (y_i - y_{means})^2} \qquad (2)$$

Where, y_i and $\hat{y}_i$ are the actual and predicted activities of the ith molecule in the test set, respectively, and y_{means} is the average activity of all the molecules in the training set.

A Z-score value is calculated by the following Eq. (3):

$$Zscore = \frac{(h - \mu)}{\sigma} \qquad (3)$$

Where, h is the q^2 value calculated for the actual dataset, μ is the average q^2 and σ is the standard deviation calculated for various models built on different random datasets [20].

F-test is Fisher value which indicates statistical significance, a value greater than 30 is considered good, which gives an idea of the chances of failure of the model. On the other hand, q^2_se is the standard deviation of cross validated prediction and r^2_se is standard deviation is a measure of the absolute quality of a model.

Pharmacophore modeling

The pharmacophore model was built using the Phase module of Schrodinger maestro [21]. The same set of inhibitors of *Pf*M18AAP was subjected to LigPrep module which produces high-quality, all-atom 3D structures

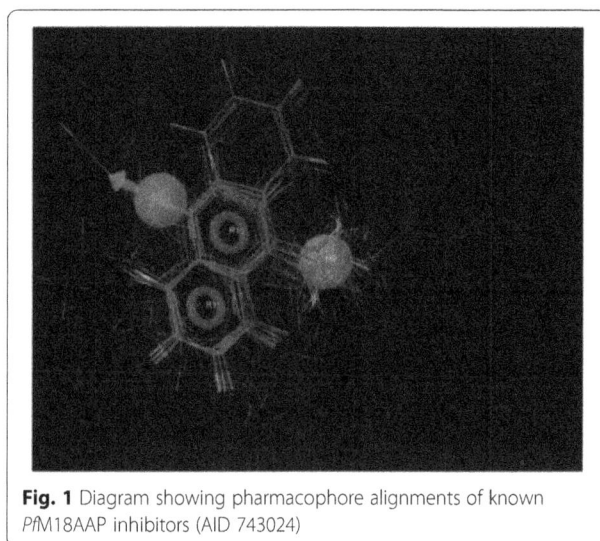

Fig. 1 Diagram showing pharmacophore alignments of known *Pf*M18AAP inhibitors (AID 743024)

with correct chirality. Some pharmacophore hypotheses were generated along with their respective set of aligned conformations. These hypotheses were generated by a systematic variation of many sites and a number of matching active compounds. These selected features were used to build a series of pharmacophore hypotheses by selecting find the common pharmacophore option in phase. The common pharmacophore hypotheses were analyzed using the survival score to yield the best alignment of the active ligands using a maximum overall root mean square deviation (RMSD) value of 2 Å for distance tolerance. Finally, several pharmacophore hypotheses were generated along with their respective set of aligned conformations. All pharmacophore hypotheses were scored for active survival, inactive survival, site, vector, volume, the number of matches, selectivity, energy, active, and inactive terms. Survival score secured by each hypothesis is the measure of the quality of alignment for a particular hypothesis [22].

Docking and scoring

Molecular docking

To understand the nature of the interaction of inhibitors described above [23] with *Pf*M18AAP, molecular docking was performed using GOLD v5.2 (Genetic Optimization for Ligand Docking) [24] and GLIDE module of Schrödinger using [21] against the *Pf*M18AAP. The crystal structure of *Pf*M18AAP (4EME) was obtained from protein data bank (www.rcsb.org/pdb/explore/explore.do?structureId=4eme).

Table 2 The statistical parameters for PLSR, PCR and 3D-QSAR models

Dependent variable	ZScore r^2	ZScore q^2	BestRand r^2	BestRand q^2	Z-Score Pred r^2	Best-Rand Pred r^2
PLSR pIC50	5.96671	2.43240	0.46222	-0.23735	1.64037	0.44031
PCR pIC50	5.11408	2.20918	0.43798	0.09365	1.39477	0.21574

Table 3 The statistical values of top 5 the pharmacophore hypotheses

ID	Survival	Survival inactive	Site	Vector	Volume	Selectivity	Matches	Energy	Activity	Inactive
DHRR.31	11.068	9.052	0.79	0.949	0.527	1.466	14	0.001	6.34	2.016
DHRR.27	11.068	9.052	0.79	0.949	0.527	1.466	14	0.001	6.34	2.016
DHRR.6	10.941	8.863	0.78	0.943	0.548	1.471	14	0.551	6.22	2.079
DHRR.15	10.941	8.863	0.78	0.943	0.548	1.471	14	0.551	6.22	2.079
DHRR.26	10.892	8.81	0.79	0.944	0.535	1.47	14	0	6.15	2.082

Since *Pf*M18AAP requires cofactors for enzymatic activity, Zn was retained during docking analysis [25]. In GOLD docking, the 10 best docked complexes were ranked based on their GOLD fitness score. In GLIDE docking, the top 10 compounds were selected based on G-score. The binding affinity of docked complex was calculated using X-Score v1.2.1 [26]. Protein-ligand interaction was analyzed by using Pymol version 1.1r. www.pymol.org/ and LigPlot + v1.4.5 [27].

Screening of PfM18AAP inhibitors

In this work, High Throughput Virtual Screening (HTVS) used Glide module of the Schrodinger software suite [21]. The ligand libraries were first prepared by adding hydrogen and generating conformations through the LigPrep module. This LigPrep module generated tautomer with the OPLS2005 force field, the total no. of 411,766 output structures were obtained. Then grid on the protein active site was generated. Firstly, HTVS for every ligand library was done and the top 1000 ranked compounds from every library were subjected to Extra-Precision (XP) screening. In both the cases, the structures were flexibly docked on the protein structure. The non-planar conformations were penalized. Structures were having more than 200 atoms or more than 35 rotatable bonds were not docked. Also, the Van Der Waal's radius scaling factor was set to 0.8, and the partial charge cutoff was set to 0.15. From these 1000 compounds, the top 10 compounds from every library were extracted as target-bound complexes. These complexes were re-scored, and their binding affinity was calculated using X-score software.

Results
3D QSAR modeling using PLSR Method

A dataset known as inhibitors of *Pf*M18AAP (AID: 743024) was used for the unicolumn statistics analysis, which showed that the training and test sets were suitable for 3D QSAR model development. The test set is interpolative i.e. derived within the min-max range of the training set. The unicolumn statistics scores were shown in Table 1. The PLSR model demonstrated that descriptors S_356, S_660, E_996, and S_270 are important features to inhibit the activity of *Pf*M18AAP, which represent steric and electrostatic field energy of interactions. The statistical parameters calculated for developed 3D QSAR model for PLSR shown in Table 2. The number suffixed with descriptors represents its position on the 3D spatial grid.

Table 4 Top scoring compounds screened using the selected pharmacophore hypothesis

Compound ID	G-Score (kcal/mol)	Align Score	Vector Score	Volume Score	Fitness	Predicted activity (pIC50)
CHEMBL588000	-10.33	1.4702	0.0537	0.3833	0.2119	5.72
CHEMBL587141	-10.12	0.8484	0.8644	0.4971	1.6545	5.83
CHEMBL529157	-9.81	1.7562	0.3816	0.3651	0.2833	5.85
CHEMBL528484	-9.79	1.5091	0.6425	0.3672	0.7521	5.86
CHEMBL532976	-9.52	1.2208	0.7452	0.2344	0.9623	6.07
CHEMBL2414638	-9.41	0.4596	0.9888	0.3135	1.9194	5.97
CHEMBL601831	-9.37	1.0285	0.6596	0.29897	1.1014	5.85
CHEMBL390368	-9.24	1.0146	0.9530	0.3788	1.4863	5.89
CHEMBL591216	-8.72	0.5189	0.6304	0.3387	1.5367	5.84
CHEMBL465847	-8.08	0.6220	0.7935	0.3477	1.6228	5.87

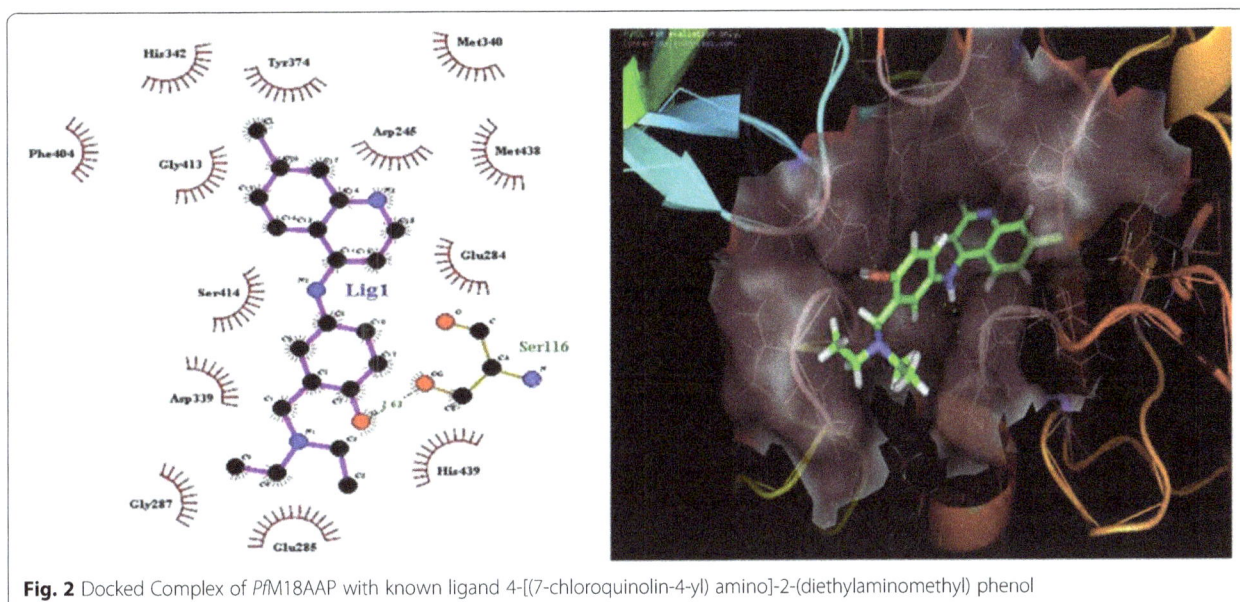

Fig. 2 Docked Complex of *Pf*M18AAP with known ligand 4-[(7-chloroquinolin-4-yl) amino]-2-(diethylaminomethyl) phenol

Equation 4 represents the PLSR 3D QSAR model:

$$pIC50 = -0.0270 \, (S\_356) + 0.0182(S\_660)$$
$$- 0.0905(E\_996) - 0.0125(S\_270) + 6.1966$$

$$(4)$$

Equation 5 represents PCR 3D QSAR model:

$$pIC50 = -0.0321(S\_356) + 0.0147(S\_660) - 0.0886(E\_996)$$
$$- 0.0092(S\_270) + 6.3423$$

$$(5)$$

3D QSAR modeling using PCR

The 3D QSAR Model was developed on the same datasets of molecules by PCR method, and several statistical parameters were calculated which are shown in Table 2. The number suffixed with descriptors represents its position on the 3D spatial grid. This model indicated that descriptors are significant for their biological activities.

3D QSAR Modeling using (kNN-MFA)

The kNN-MFA model shown that the contributing descriptors E_862 (1.0026 1.1562), S_629 (-0.4639 -0.1045) and S_287 (-0.3372, -0.2663) which indicated that degree of amino group shows potent activity. The range at the lattice point E_862 (1.0026, 1.1562) which is positive that means substitution with more electron density could yield more active molecules.

Table 5 Prediction of pIC50 Value of current antimalarial drugs described in the WHO

Compound ID	Generic Name	G-Score (kcal/mol)	Align Score	Vector Score	Volume Score	Fitness	Predicted activity (pIC50)
CHEMBL76	Chloroquine	-3.80	0.1086	0.9996	0.5197	2.4288	6.208
CHEMBL1535	Hydroxychloroquine	-4.53	0.1995	0.9973	0.3341	2.1652	6.207
CHEMBL303933	Piperaquine	-5.30	0.2720	0.9781	0.3049	2.0563	6.19
CHEMBL506	Primaquine	-5.86	0.4463	0.8889	0.4954	2.0124	6.192
CHEMBL2104009	Amquinate	-5.41	0.6142	0.9499	0.355	1.7931	6.205
CHEMBL416956	Mefloquine	-5.28	0.5712	0.7183	0.3248	1.5672	6.20
CHEMBL682	Amodiaquine	-4.48	0.6480	0.7838	0.2687	1.5126	6.385
CHEMBL36	Pyrimethamine	-5.07	0.9521	0.9257	0.3390	1.4712	6.207
CHEMBL339049	Tebuquine	-4.55	0.6093	0.7422	0.2286	1.4630	6.264
CHEMBL35228	Pyronaridine	-5.68	0.6975	0.8185	0.2050	1.4422	6.198

Table 6 Top scoring of QN, CQ and 8 Amino-QN analogous screened using the selected pharmacophore hypothesis

IUPAC Name	G-Score (kcal/mol)	Align Score	Vector Score	Volume Score	Fitness	Predicted activity (pIC50)
(9S)-Cinchonan-9-ol	-4.18	0.8099	0.5259	0.4368	1.2878	5.521
(9S)-6'-Methoxycinchonan-9-ol	-5.47	1.0187	0.7020	0.2737	1.1268	5.85
N-(7-Chloro-4-quinolinyl)-N'-ethyl-1,4-butanediamine	-3.84	0.1101	0.9993	0.5	2.4075	5.98
1,4-Pentanediamine, N4-(7-chloro-4-quinolinyl)-N1, N1-diethyl-Chloroquine	-3.52	0.1053	0.9983	0.4755	2.3861	5.86
PrimaquineN^4-(6-Méthoxy-8-quinoléinyl) -1,4-pentanediamine	-5.14	0.5014	0.9017	0.4005	1.8844	5.75
N^4-{2,6-Diméthoxy-4-méthyl-5-[3-(trifluorométhyl)phénoxy]-8-quinoléinyl} -1,4-pentanediamine	-5.32	0.5274	0.9118	0.2755	1.7478	5.67

Pharmacophore-based screening of *Pf*M18AAP inhibitors

From the Phase Software, ten hypotheses (pharmacophore models) were generated having four features DHRR (one hydrogen bond donor (D), hydrophobic groups (H) and two aromatic rings (R)). These features were common to all of the 15 compounds of the assay. Common pharmacophore hypothesis is shown in Fig. 1. The best model was chosen based on the survival score and pharmacophore based QSAR. The final hypothesis, DHRR.31 model, was selected based on the survival score and pharmacophore based QSAR, which showed the best alignment of the active set along with the site score (0.79), vector score (0.949), and volume score (0.527), top 5 model is shown in Table 3.

Molecular docking

The same data set used for QSAR and Pharmacophore modeling was subjected to the molecular docking analysis. The top 10 compounds showed GOLD fitness score from 60.62 to 39.81 and predicted binding energy from -6.43 to -7.38 kcal/mol (calculated using the X-Score) and G-score from -7.80 to -4.70 kcal/mol (Table 4). The Ligplot + analysis showed that Ser116 and His87 amino acids interact by h-bond interaction, with docked ligands. Since *Pf*M18AAP requires a cofactor for enzymatic activity, docking was performed along with cofactor bound with specific amino acids. A docked complex is depicted in Fig. 2. These results suggest that the novel *Pf*M18AAP inhibitors could be designed considering parameters of docking results leading to new potent drugs against malaria.

Molecular docking analysis was done on another dataset (AID1822:3502 molecules from PubChem Bioassay) known inhibitors of *Pf*M18AAP. The top 10 compounds showed G-score from -7.72 to -6.52 kcal/mol. The G-score indicated that these compounds (Table 5) might bind to *pf*M18AAP

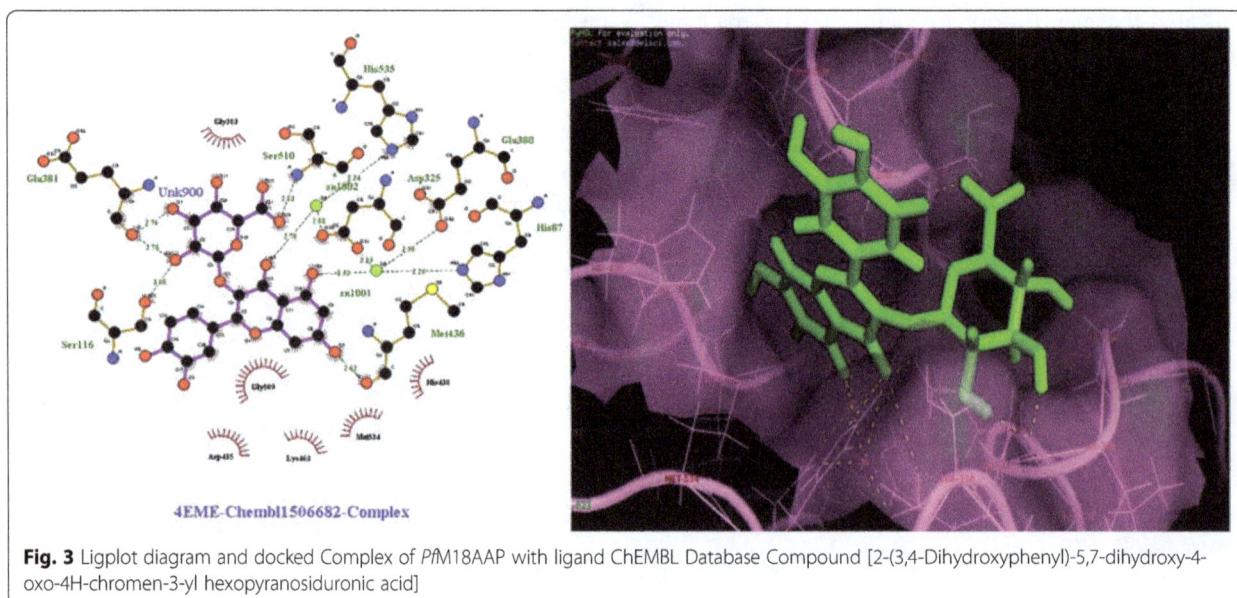

Fig. 3 Ligplot diagram and docked Complex of *Pf*M18AAP with ligand ChEMBL Database Compound [2-(3,4-Dihydroxyphenyl)-5,7-dihydroxy-4-oxo-4H-chromen-3-yl hexopyranosiduronic acid]

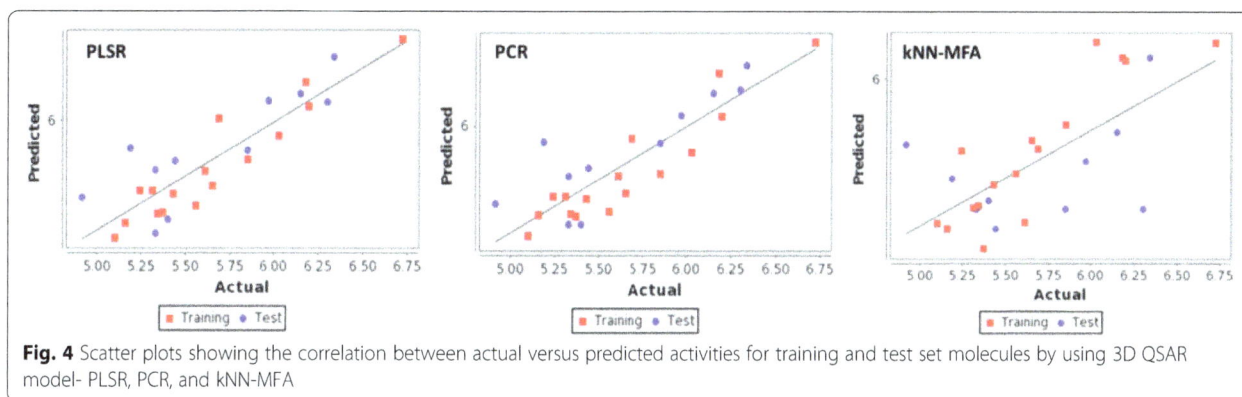

Fig. 4 Scatter plots showing the correlation between actual versus predicted activities for training and test set molecules by using 3D QSAR model- PLSR, PCR, and kNN-MFA

with good binding affinity. Further, predicted binding affinity calculated using X-score for best compounds was found to be in between from -9.54 to -6.51 kcal/mol (Table 5).

HTVS based screening of *Pf*M18AAP inhibitors

ChEMBL antimalarial dataset (153873) was subjected to molecular docking. The top 10 compounds (after docking), based on their G-score are shown in Table 6. The glide score of these compounds varies from -12.50 to -10.45 kcal/mol. The G-score indicated that these compounds (Table 6) have a good binding affinity for *Pf*M18AAP enzyme. Figure 3 shows the docked complex of ligand CHEMBL1506682 (2-(3,4-Dihydroxyphenyl)-5,7-dihydroxy-4-oxo-4H-chromen-3-yl hexopyranosiduronic acid) in the active site of the receptor with best G-score (-12.50 kcal/mol).To further validate *in silico*, predicted binding affinity of the best pose obtained from docking studies for each compound was calculated using

X-score program was found to be in between -8.28 and -6.89 kcal/mol shown in Table 6.

Discussion

The best model was selected through the comparison between fitness plots (Fig. 4) and radar plots for training and test sets (Fig. 5 (a, b)). The linear graphical representation of fitness plots shows the observed and predicted activities of the data set. The radar plots show the training and the test sets separately by the red (actual activity) and blue (predicted activity) lines. The radar plot for training set represents a good r^2 value because the two lines show a good overlap while for the test set a good overlap represents high pred_r^2 value. The PLSR contribution plot for the descriptor is given in Fig. 6 which represents the contribution of various descriptors which are important for the inhibitory activity. In PLSR and PCR models, the negative value in electrostatic field

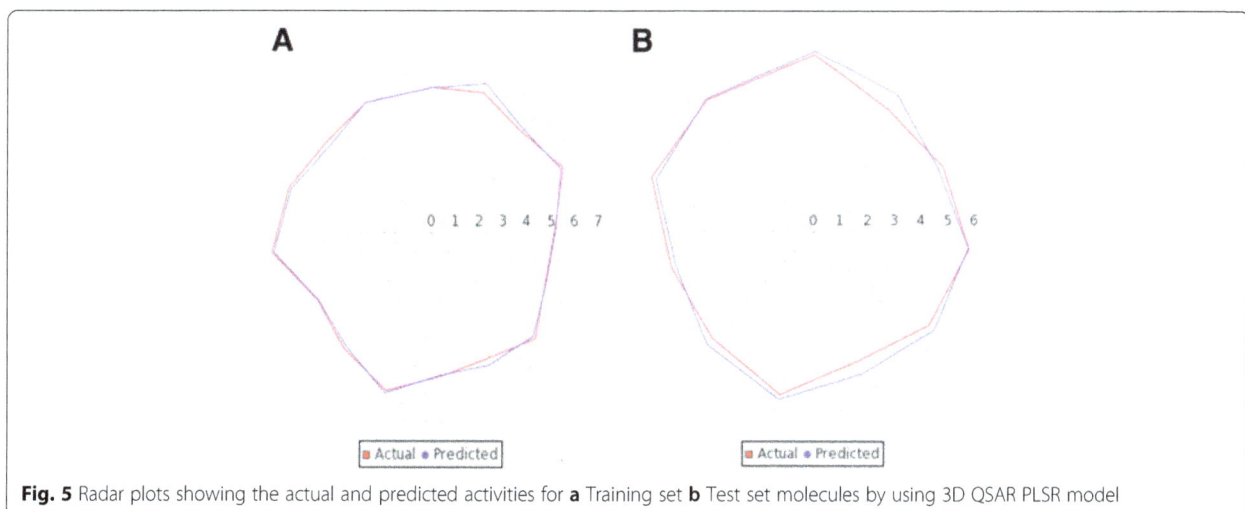

Fig. 5 Radar plots showing the actual and predicted activities for **a** Training set **b** Test set molecules by using 3D QSAR PLSR model

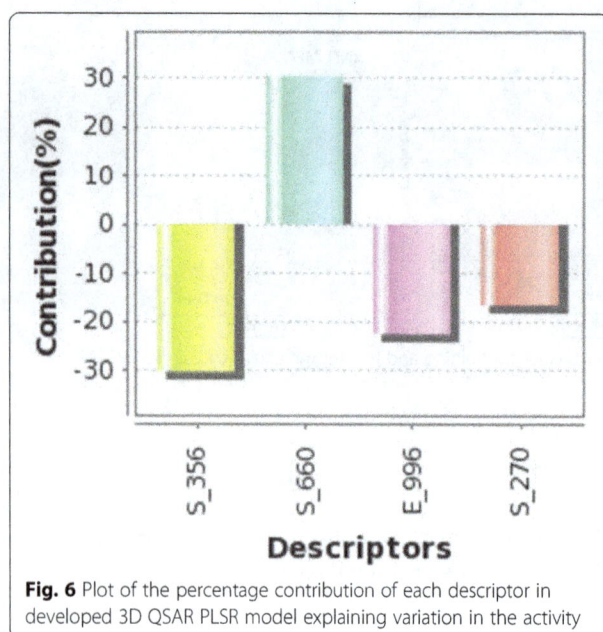

Fig. 6 Plot of the percentage contribution of each descriptor in developed 3D QSAR PLSR model explaining variation in the activity

values in steric descriptors indicate that negative steric potential is favorable for activity, and less lipophilic substitutions or bulky substituents group should be considered in that region, positive value of steric descriptors reveals that positive steric potential is favorable to increase antimalarial activity as in case of 4-[2-(quinolin-4-ylamino)ethyl] benzene-1,2-diol, and more bulky group is advised to prefer in that region. Comparison of statistical parameters of PLSR, PCR, and kNN-MFA, is shown in (Additional file 1) and the predicted pIC50 values in (Additional file 2).

In the present work, we performed screening of CHEMBL antimalarial library to search antimalarial compounds based on the pharmacophoric hypothesis DHRR.31, which resulted in 29,671 compounds. These compounds were subjected to glide docking against *Pf*M18AAP. The top 10 compounds were selected based on the fitness and G-score; predicted activities are shown in Table 7. Further we also carried out screening of 27 WHO antimalarial drugs which resulted in 14 molecules shown in Table 8. Moreover, 17 compounds of 8-aminoquinolines analogous, 24 compounds of CQ analogous and 32 compounds of 8 amino-QN analogous were subjected to screening resulting 17,19, and 22 *Pf*M18AAP inhibitors respectively (Table 8). The resultant top 2 compounds from each analogous were selected based on the fitness and G-score; predicted activities are shown in Table 9. The study found that WHO current antimalarial compound CHEMBL682 (Amodiaquine)

descriptors indicates that negative electronic potential is required to increase antimalarial activity, and more electronegative groups are preferred in that position. Though positive value in kNN-MFA model shows that group that imparting positive electrostatic potential is favorable for antimalarial activity, so less electronegative group should prefer in that region. Similarly, negative

Table 7 Molecular Docking Results for known inhibitors (AID743024) against *Pf*M18AAP

IUPAC Name	Gold Score	G-Score (kcal/mol)	X-Score (kcal/mol)	H Bond	No. of Hydrophobic Interaction	No. of NB Interactions	pIC50 Value
4-[(7-chloroquinolin-4-yl)amino]-2-(diethylamino methyl)phenol	36.57	-5.35	-8.09	Ser116	13	33	6.72
7-chloro-N-[2-(3,4-dimethoxyphenyl)ethyl]quinolin-4-amine	35.17	-5.40	-7.48	-	11	60	6.18
N-[2-(3,4-dimethoxyphenyl)ethyl]-6-ethoxyquinolin-4-amine	33.65	-6.43	-7.08	His342	9	72	5.85
N-[2-(3,4-dimethoxyphenyl)ethyl]isoquinolin-4-amine	33.45	-4.97	-7.17	-	11	59	5.34
4-[2-[(7-chloroquinolin-4-yl)amino]ethyl]benzene-1,2-diol	32.56	-7.80	-7.38	Ser414	12	70	6.2
3-[2-(quinolin-4-ylamino)ethyl]benzene-1,2-diol	32.41	-5.67	-7.36	Glu284 Ser414	10	77	5.56
N-[2-(2-bromo-4,5-dimethoxyphenyl)ethyl]quinolin-4-amine	32.31	-4.85	-7.35	-	11	61	6.34
1-benzyl-N-[2-(3,4-dimethoxyphenyl)ethyl]piperidin-4-amine	32.11	-4.70	-7.10	Ser116	11	61	5.16
4-[2-(quinolin-4-ylamino)ethyl]benzene-1,2-diol	31.89	-5.25	-7.19	Glu284 Ser414	10	62	5.4
4-[3-(acridin-9-ylamino)propyl]benzene-1,2-diol	30.58	-5.65	-7.63	His87 Asp89	6	46	5.43

H Bond Hydrogen-Bond, *NB* Non Bonded

Table 8 Molecular Docking Results for known inhibitors (AID1822) against *Pf*M18AAP

S. No.	Chemical Substance ID	G-Score (kcal/mol)	X-Score (kcal/mol)	HBond	No. of Hydro-phobic Interactions	No. of NB Interactions	% Inhibition
C1	49644635	-7.72	-8.42	Gly509	8	68	32.65
C2	24707924	-7.71	-9.54	Ser116, Asp325, Met436, Lys463	8	76	75.26
C3	26665815	-7.48	-6.51	Ser116, Cys508	4	32	31.93
C4	50086555	-7.36	-7.66	Ser116, Glu380, His438, Ser510	7	35	55.6
C5	49647140	-7.143	-7.04	Ser116, Met436, His438, Lys463	7	29	55.21
C6	47195345	-7.11	-8.14	His438, Asp325, Glu380, His87, His535	7	37	28.24
C7	49644096	-7.07	-7.37	Asp325, Glu380, Ser510, His 535	4	39	37.43
C8	24779308	-6.88	-7.29	His 87,Asp325, Glu380,His535	6	38	37.29
C9	17504161	-6.57	-7.92	Ser116, Asp435, Met436, Lys463	9	32	53.68
C10	11532952	-6.52	-7.57	His438	9	36	36.43

has highest predicted value of pIC50 6.38 which is also present in the known dataset of *Pf*M18AAP with pIC50 value 6.72.

We analyzed the types of interactions of each top ranked compound for known inhibitors (AID1822) against *Pf*M18AAP; 2D plots were generated using Ligplot + software and ligand-protein complex. The number of hydrogen bonded interactions, lipophilic interactions and the number of non-bonded interactions was counted and tabulated in Table 5. It is observed that overall all compounds from C1 to C10 have formed at least 1 (C1 and C10), mostly 4 (C3, C4, C7, C8, and C9), and at most 5 (C6) hydrogen bonds. The total number of lipophilic interactions for each compound varies in between 9 (for C9, C10) and 4 (for C3 and C7). Also, the total number of non-bonded interactions for each compound varies from 29 (for C5) to 76 (for C2). These observations suggest that the compounds C3, C4, C6, C7, C8, and C9 have better specificity as they have more hydrogen bonds and compounds C1, C2, C9, and C10 have good binding affinity due to a high number of hydrophobic contacts. The Compound C1 showed interaction with Glide score -7.72 kcal/mol. The docking poses analysis of C1shows one hydrogen bond (Gly509) interaction with amino acid residues of the protein. The next favorable interaction is shown by C2 with G-score of -7.71 kcal/mol and four hydrogen bond interactions with the active site residues Ser116, Asp325, Met436 and Lys463, 76 nonbonded interactions and inhibition (75.26 %) and eight hydrophobic interactions. The Compound C6 showed highest five hydrogen bond interaction (His438, Asp325, Glu380, His87, and His535). **Asp325** is found to be the most conserved residues, which is present in 6 out of 10 compounds and **Ser116** is found to be the most conserved residues, which is present in 5 out of 10 compounds. Hence, based on the Docking analysis against antimalarial *Pf*M18AAP inhibitors, we conclude that these compounds have a better

affinity with *Pf*M18AAP enzyme, thus are novel potential candidate to develop drugs against malaria.

Further, we also analyzed the interactions of CHEMBL antimalarial library's top ranked inhibitors against *Pf*M18AAP (Table 6). The highest X score of - 11.6 kcal/mol was obtained with the ligand (CHEMBL1506682) having three hydrogen bond (**Ser116, Glu381, and Met436**) interaction with amino acid residues of the protein. The total number of lipophilic interaction for each compound varies in between 9 (CHEMBL602830 and CHEMBL429) and 4 (for CHEMBL511171). This observation suggests that CHEMBL1506682 have better specificity and CHEMBL602830 have a good binding affinity. **Ser510** and **Glu38**0 are found to be the most conserved residues, which is present in 5 out of 10 compounds. Hence, based on the comparison between known bioactive antimalarial M18AAP inhibitors (as control) and top ten novel ChEMBL compounds, we conclude that these compounds could bind to *Pf*M18AAP with better affinity, thus are the potential candidate to develop drugs against malaria.

Conclusions

The present study was aimed at generating the predictive 3D QSAR models capable of revealing the structural requirements for antimalarial inhibitors of *Pf*M18AAP. The comparison of the different statistical parameters of the three models suggests that PLSR model is best due to better internal validation $q^2_=$ 0.6128 and an external test of pred_$r^2_=$ 0.6101. Model 3 (kNN-MFA) also had a good internal validation showing $q^2_=$0.7641, but the external validation had a bad pred_$r^2_=$ 0.0366. Therefore both PLSR and PCR models show potential predictive ability as determined by testing the external test set. Thus, 3D QSAR modeling provided a better understanding of the structural requirements of antimalarial compounds, which could help design potent *Pf*M18AAP inhibitors. Also, pharmacophore mapping was applied to identify the binding modes and structural

Table 9 Top scoring 10 potential inhibitors from CHEMBL antimalarial Library against *Pf*M18AAP

Compound ID and Structure	G Score (in kcal/mol)	X-Score (kcal/mol)	H Bond	No. of Hydrophobic Interactions	No. of NB Interactions
CHEMBL1506682	-12.50	-8.28	Ser116 Glu381 Met436	6 (Gly383, Asp435, His438, Lys463, Gly509, Met534)	46
CHEMBL525132	-11.70	-7.96	Ser510 Glu380	6 (Glu381, Asp435, Met436, His438, Met534, Gly509)	61
CHEMBL245416	-11.63	-6.89	Glu381	7 (Asp435, His438, Tyr470, Gly509, Ser510, Thr511, Met534)	42
CHEMBL66953	-11.57	-7.15	-	6 (Asp435, Met436, His438, Gly509, Met534, Ser510)	33
CHEMBL585601	-11.17	-7.71	Ser510 Glu380	6 (Glu381, Ile382, Gly383, His438, Tyr470, Met534)	54
CHEMBL602830	-11.02	-7.42	Glu380	9 (Ser116, Glu381, Ile382, Gly383, His438, Cys508, Gly509, Ser510, Met534)	62
CHEMBL429	-11.01	-8.10	His438	9 (Ser116, Glu380, Glu381, Asp435, Met436, Tyr470, Gly509, Ser510, Met534)	51
CHEMBL585028	-10.74	-8.26	Ser510	6 (Glu381, His438, Gly509, Met515, Met534, His535)	51
CHEMBL511171	-10.64	-7.51	Ser510 Glu380	4 (Glu381, His438,Gly509, Met534)	35
CHEMBL586200	-10.45	-7.51	Ser510 Glu380	7 (Glu381, Ile382, Gly383 His438, Tyr470, Gly509, Met534)	57

features of the ligands which are important for the biological activity of the inhibitors. The pharmacophore modeling showed that hypothesis DHRR.31 represented the best pharmacophore model for determining *Pf*M18AAP inhibitory activity. Results suggested that the proposed DHRR.31 model can be used to identify the new M18AAP inhibitor and to design a drug rationally for *p. falciparum* from the extensive 3D database of molecules. Further, HTVS using Glide resulted in several potent *Pf*M18AAP inhibitors from ChEMBL antimalarial data set of 153,873 compounds. These novel compounds having an excellent binding affinity with *Pf*M18AAP are better candidates to design the drug in future. Finally, the 3D QSAR model was deployed on different data set to prioritize *Pf*M18AAP inhibitors and predict new inhibitors. Thus, our study advocates the use of combined approaches of 3D QSAR, pharmacophore modeling, and molecular docking to search for novel potential inhibitors unique to *Pf*M18AAP, which is essential and validated drug target involved in performing various enzymatic functions such as hemoglobin digestion, erythrocyte invasion, and parasite growth in the host cell.

Additional files

Additional file 1: The statistical parameters of 3D QSAR models of known bioactive Inhibitors (AID 743024) dataset of *Pf*M18AAP using PLSR, PCR and kNN-MFA methods.

Additional file 2: Comparison between different 3D QSAR models using PLS, PCR and KNN methods for predicting pIC50 values of train set and test set of known bioactive Inhibitors (AID 743024) of *Pf*M18AAP.

Acknowledgment
We would like to thank Dr. Andrew M Lynn, School of Computational and Integrative Sciences, Jawaharlal Nehru University, New Delhi, 110067, India, for proving invaluable suggestions.

Funding
This work was funded by University Performance excellence-II funds (from University Grants Commission), Infrastructure support by Department of Biotechnology through Centre of Excellence in Bioinformatics and also financially supported by Purse funds of Department of Science and Technology, Govt of India.

Authors' contributions
All authors participated in the design of the study. MK and NT performed the comparative analysis of 3D QSAR model, pharmacophore model, and molecular docking. MK, SC, SN and NT wrote the manuscript. All authors read and approved the final manuscript.

Authors' information
Not applicable.

Competing interests
The authors declare that they have no competing interests.

Author details

[1]Department of Information Technology, Kumaun University, SSJ Campus, Almora, Uttarakhand 263601, India. [2]Department of Botany, Kumaun University, SSJ Campus, Almora, Uttarakhand 263601, India. [3]Department of Statistics, Kumaun University, SSJ Campus, Almora, Uttarakhand 263601, India. [4]School of Computational and Integrative Sciences, Jawaharlal Nehru University, New Delhi 110067, India.

References

1. Newton CR, Krishna S. Severe falciparum malaria in children: current understanding of pathophysiology and supportive treatment. Pharmacol Ther. 1998;79(1):1–53.
2. Basco LK, Le Bras J. In vitro activity of artemisinin derivatives against African isolates and clones of Plasmodium falciparum. Am J Trop Med Hyg. 1993; 49(3):301–7.
3. Lauterbach SB, Coetzer TL. The M18 aspartyl aminopeptidase of Plasmodium falciparum binds to human erythrocyte spectrin in vitro. Malar J. 2008;7:161.
4. Nagpal I, Raj I, Subbarao N, Gourinath S. Virtual screening, identification and in vitro testing of novel inhibitors of O-acetyl-L-serine sulfhydrylase of Entamoeba histolytica. PLoS One. 2012;7(2):e30305.
5. Mizutani MY, Itai A. Efficient method for high-throughput virtual screening based on flexible docking: discovery of novel acetylcholinesterase inhibitors. J Med Chem. 2004;47(20):4818–28.
6. Bisson WH, Cheltsov AV, Bruey-Sedano N, Lin B, Chen J, Goldberger N, May LT, Christopoulos A, Dalton JT, Sexton PM, Zhang XK, Abagyan R. Discovery of antiandrogen activity of nonsteroidal scaffolds of marketed drugs. Proc Natl Acad Sci U S A. 2007;104(29):11927–32.
7. Esposito EX, Hopfinger AJ, Madura JD. Methods for applying the quantitative structure-activity relationship paradigm. Methods Mol Biol. 2004;275:131–214.
8. Xue L, Bajorath J. Molecular descriptors in chemoinformatics, computational combinatorial chemistry, and virtual screening. Comb Chem High Throughput Screen. 2000;3(5):363–72.
9. S. Wold AR, Wold H, Dunn WJ. The collinearity problem in linear regression. The partial least squares (PLS) approach to generalized inverses. SIAM J Sci Stat Comp. 1984;5:735.
10. Jolliffe IT. A note on the use of principal components in regression. Appl Stat. 1982;31:300–3.
11. Malashenko Iu R, Romanovskaia VA, Sokolov IG, Kryshtab TP, Liudvichenko ES. Theoretical evaluation of necessity of carbon dioxide assimilation by microorganisms during growth on various substrates. Ukr Biokhim Zh (1978). 1980;52(2):159–63.
12. Ajmani S, Jadhav K, Kulkarni SA. Three-dimensional QSAR using the k-nearest neighbor method and its interpretation. J Chem Inf Model. 2006; 46(1):24–31.
13. Molecular Networks GmbH Computerchemie Erlangen, Germany, 1996.
14. O'Boyle NM, Banck M, James CA, Morley C, Vandermeersch T, Hutchison GR. Open Babel: An open chemical toolbox. J Cheminform. 2011;3:33. doi:10. 1186/1758-2946-3-33.
15. VLifeMDS. Molecular Design Suite Pune: VLife Sciences Technologies Pvt Ltd 4, vol. 3. 2010.
16. Gasteiger J, Marsili M. Iterative partial equalization of orbital electronegativity—a rapid access to atomic charges. Tetrahedron. 1980;36:3219–28.
17. Derksen S, Keselman H. Backward, forward and stepwise automated subset selection algorithms: Frequency of obtaining authentic and noise variables. Brit J Math Stat Psy. 1992;45:265–82.
18. Kohavi R. A study of cross-validation and bootstrap for accuracy estimation and model selection. International joint Conference on artificial intelligence: Lawrence Erlbaum Associates Ltd 1995, 1137-1145.
19. Schuurmann G, Ebert RU, Chen J, Wang B, Kuhne R. External validation and prediction employing the predictive squared correlation coefficient test set activity mean vs training set activity mean. J Chem Inf Model. 2008;48(11):2140–5.
20. Rucker C, Rucker G, Meringer M. y-Randomization and its variants in QSPR/QSAR. J Chem Inf Model. 2007;47(6):2345–57.
21. Maestro, Version 9.1, Schrodinger LLC, NY2008.
22. Kumar V, Kumar S, Rani P. Pharmacophore modeling and 3DQSAR studies on flavonoids as a-glucosidase inhibitors. Der PharmaChemica. 2010;2:324–35.
23. Schoenen FJ, Weiner WS, Baillargeon P, Brown CL, Chase P, Ferguson J, Fernandez-Vega V, Ghosh P, Hodder P, Krise JP, et al. Inhibitors of the Plasmodium falciparum M18 Aspartyl Aminopeptidase, Probe Reports from the NIH Molecular Libraries Program. 2013.
24. Cole JC, Nissink JWM, Taylor R. Protein ligand docking and virtual screening with GOLD, Virtual Screening in Drug Discovery. 2005.
25. Sivaraman KK, Oellig CA, Huynh K, Atkinson SC, Poreba M, Perugini MA, Trenholme KR, Gardiner DL, Salvesen G, Drag M et al. X-ray crystal structure and specificity of the Plasmodium falciparum malaria aminopeptidase PfM18AAP. J Mol Biol. 422(4):495-507.
26. Jones G, Willett P, Glen RC, Leach AR, Taylor R. Development and validation of a genetic algorithm for flexible docking. J Mol Biol. 1997;267(3):727–48.
27. Wang R, Lai L, Wang S. Further development and validation of empirical scoring functions for structure-based binding affinity prediction. J Comput Aided Mol Des. 2002;16(1):11–26.

5

Combined small angle X-ray solution scattering with atomic force microscopy for characterizing radiation damage on biological macromolecules

Luca Costa[1,9†], Alexander Andriatis[1,2†], Martha Brennich[1], Jean-Marie Teulon[3,4,5], Shu-wen W. Chen[3,4,5], Jean-Luc Pellequer[3,4,5]* ⓘ and Adam Round[6,7,8,10]

Abstract

Background: Synchrotron radiation facilities are pillars of modern structural biology. Small-Angle X-ray scattering performed at synchrotron sources is often used to characterize the shape of biological macromolecules. A major challenge with high-energy X-ray beam on such macromolecules is the perturbation of sample due to radiation damage.

Results: By employing atomic force microscopy, another common technique to determine the shape of biological macromolecules when deposited on flat substrates, we present a protocol to evaluate and characterize consequences of radiation damage. It requires the acquisition of images of irradiated samples at the single molecule level in a timely manner while using minimal amounts of protein. The protocol has been tested on two different molecular systems: a large globular tetrameric enzyme (β-Amylase) and a rod-shape plant virus (tobacco mosaic virus). Radiation damage on the globular enzyme leads to an apparent increase in molecular sizes whereas the effect on the long virus is a breakage into smaller pieces resulting in a decrease of the average long-axis radius.

Conclusions: These results show that radiation damage can appear in different forms and strongly support the need to check the effect of radiation damage at synchrotron sources using the presented protocol.

Keywords: β-Amylase, Tobacco mosaic virus, Small angle x-ray scattering (SAXS), Atomic force microscopy (AFM), Radiation damage

Background

The most recent step forward in structural biology for characterizing large molecular assemblies is the integration of several complementary techniques to reach the goal of determining structures at atomic level. In this frame, it is essential to combine information from a variety of physical and chemical origins thus providing a solid basis to understanding molecular function.

This recent development is known as integrative structural biology [1].

Among the different available techniques, small-angle X-ray scattering (SAXS) presents unique advantages. SAXS applied to dilute solutions of proteins is a long established technique in structural biology. It gives ensemble reciprocal space information on the size and shape of macromolecules [2–5]. While the reconstruction of 3D models of proteins from solution scattering data is common, it is an ill-posed problem and typically requires additional constraints such as the maximal distance between two points in a sample D_{max} [6–9]. SAXS data is sensitive to oligomerization or aggregation of biological samples. For example, radiation-induced aggregation has been observed with SAXS data for lysozyme, but

*Correspondence: jean-luc.pellequer@ibs.fr
Contact for BioSAXS: martha.brennich@esrf.fr
†Equal contributors
[3]Univ. Grenoble Alpes, 71 Avenue des Martyrs, 38044 Grenoble, France
[4]CNRS, IBS, 71 Avenue des Martyrs, 38044 Grenoble, France
Full list of author information is available at the end of the article

without any change in folding topology [10]. Irreversible X-ray induced damage, essentially due to free radical formation in the sample at synchrotron sources, are a current limitation of SAXS experiments and often increase the amount of material needed [11] or require radiation protectant such as glycerol or cryo-cooling [12]. However, evaluation of post-SAXS experiment radiation damage on proteins is rarely performed because the allowable doses are highly sample-dependent, and must be determined on a case-by-case basis. A protocol to investigate such radiation damage at SAXS beamline is suggested in this work and makes use of the imaging capability of the atomic force microscope at nanometer resolution.

Atomic force microscopy (AFM) is beginning to make a large impact in the field of structural biology [13–17]. AFM uses a sharp tip located beneath a micro-cantilever that scans across sample molecules usually deposited on flat mineral substrates (e.g. Mica). It can give real space information about the size and shape of particles as well as their physical properties and behaviour in the measurement conditions. Typically, topographical resolution can reach sub-nanometer range when characterizing flat samples [15, 18–20] but it rises up to the nanometer range when measuring isolated biomolecules. The main advantage of AFM resides in its exceptional signal-to-noise ratio where the imaging of a single isolated particle is enough to determine its particular dimensions using standard or improved image processing methods [21]. Moreover, one of the main advantages of an AFM over scanning electron microscopy or transmission electron microscopy is that the sample can be kept in physiological conditions while imaging, such as a liquid buffer for proteins [22], a shared advantage with the SAXS technique. Complementarity between SAXS and AFM techniques allows cross validation thereby increasing the reliability and confidence in the results, and to obtain additional information such as electrochemical properties of a macromolecule based on its binding with surfaces for AFM images. However, to date there are only a few studies which combine these two techniques [23–27].

Here, we describe a protocol for the combined acquisition of bioSAXS and AFM data from the same sample with minimum delay taking advantage of the ESRF user facilities for both techniques. By using remaining (unexposed) sample from the bioSAXS acquisition and diluting it to the required concentration, depositing it onto an atomically flat surface for AFM imaging, the AFM data collection is achieved with no additional sample required over that needed for bioSAXS. Indeed, as little as 1 μL of the sample solution left in the sample changer was diluted 1000 times to a concentration suitable for AFM. With this method, the amount of solution leftover from SAXS is sufficient for many AFM images, which can be very useful in cases where each μL of solution takes large amounts of time and

resources to produce. It is shown that AFM is a useful tool to evaluate the effects of radiation damage by evaluating changes to physical characteristics and electrochemical behaviour of irradiated samples. Such effects result in an increase of the apparent size or in a decrease of the average particle radius due to breakage. It is also shown that the AFM output can be employed as a constraint to interpret SAXS data, reducing ambiguity in the SAXS output. To evaluate the performance of the AFM-SAXS combination, two standard systems have been used: β-Amylase from sweet potato and the tobacco mosaic virus (TMV), a long rod shape plant virus.

Methods
Sample preparation
β-Amylase is a tetrameric enzyme of $\approx$ 200 kDa which catalyzes the hydrolysis of maltose units (two glucose units) for starch. The known crystal structure of β-Amylase (PDB code 1FA2) [28] was used for comparison with experimental data. According to the reconstituted tetrameric structure of β-Amylase, the computed maximum bounding box of Cα atoms has a size of 12.4 $\times$ 12.4 $\times$ 7.5 nm^3 and a radius of gyration of 4.14 nm (all atoms). A 5.5 mg/mL solution of β-Amylase protein from sweet potato (Product Number A8781, Sigma-Aldrich, St. Louis, MO, USA) was prepared by adding 3.75 mL of Tris equilibration buffer (50 mM TRIS-HCl, 100 mM KCl, pH 7.5, 5 % v/v glycerol, Sigma-Aldrich) and filtering with a 0.2 μm filter. The final concentration was verified using a NanoDrop spectrophotometer at OD$_{280}$ ($\epsilon_{1\%}$ = 17.7) [29]. For AFM, the solution was further diluted in Tris equilibration buffer without glycerol.

TMV forms a hollow rod-like structure of about 300 nm length and 19 nm diameter. The crystal structure of TMV was determined and refined by X-ray fiber diffraction [30]. The organization of TMV assembly has been widely studied using imaging techniques such as electron microscopy, AFM, and X-ray microdiffraction [31–34], and it is a common model system for image processing [35]. The regularity of the TMV structure simplifies the comparison of results out of single-image analysis on AFM images with those of other techniques. TMV particles were prepared as previously described [36]. The concentration was determined by spectrophotometric measurement at OD$_{260}$ ($\epsilon_{0.1\%}$ = 3.1) at a value of 26 mg/mL. TMV dilution for AFM and SAXS was performed in deionized water.

Small-angle x-ray scattering (SAXS)
SAXS data collection was performed on BM29 at the ESRF [37]. 50 μL of sample solution was loaded into the automatic sample changer, with 40 μL used for the actual experiment. Scattering data of samples and buffers was acquired at one frame per second for ten seconds while sample was flowing through the capillary using the flow

mode of the automated sample changer [38]. For AFM imaging of radiation damaged samples, the exposed samples were recuperated after exposure and followed the same protocol as the non exposed samples to facilitate comparison. The total flow time was 18 seconds. The X-ray beam energy was 12.5 keV and the beam size was 700×700 μm. The detector distance was 2.864 m. Data was collected at 20 °C. Intensity was normalized to absolute units using background-corrected water. The available q-range ($q = \frac{4\pi}{\lambda} \sin \theta$) was 0.025 nm^{-1} to 4.8 nm^{-1}. Data reduction was done using the standard tools at BM29 yielding the 1D subtracted curves and initial processing to give feedback to the experiment [39, 40]. Extrapolation to zero concentration and determination of the model independent parameters (R_g (radius of gyration), I_0 (Intensity at $q = 0$), molecular mass, etc.) and cross-sectional Guinier analyis done using PRIMUS [41, 42]. Comparison of known structures to the experimental data (β-Amylase) was done using OLIGOMER [41]. Fitting of geometric

models (TMV) was done with Genfit [43]. Computation of model intensities was done using CRYSOL [44] while p(r) (pair distribution) analysis and cross-section pair-distribution using GNOM [42].

Radiation damage investigations were performed testing several irradiation times: 1) one second exposure, 2) standard exposure which corresponds to a total exposure of 10 s while flowing, 3) 5 min exposure, 4) 30 min over-exposure. For both latter exposures, the sample was flowing continuously back and forth through the beam. We have collected a total of 10 patterns for standard exposure and 6 patterns for 5 min and 30 min exposures. The comparison between SAXS and AFM data has been limited to standard, 5 min and 30 min exposures.

Atomic force microscopy (AFM)

AFM images were recorded in amplitude modulation (tapping) mode in liquid [45] with photothermal excitation on Cypher and piezo-dither excitation mode on a

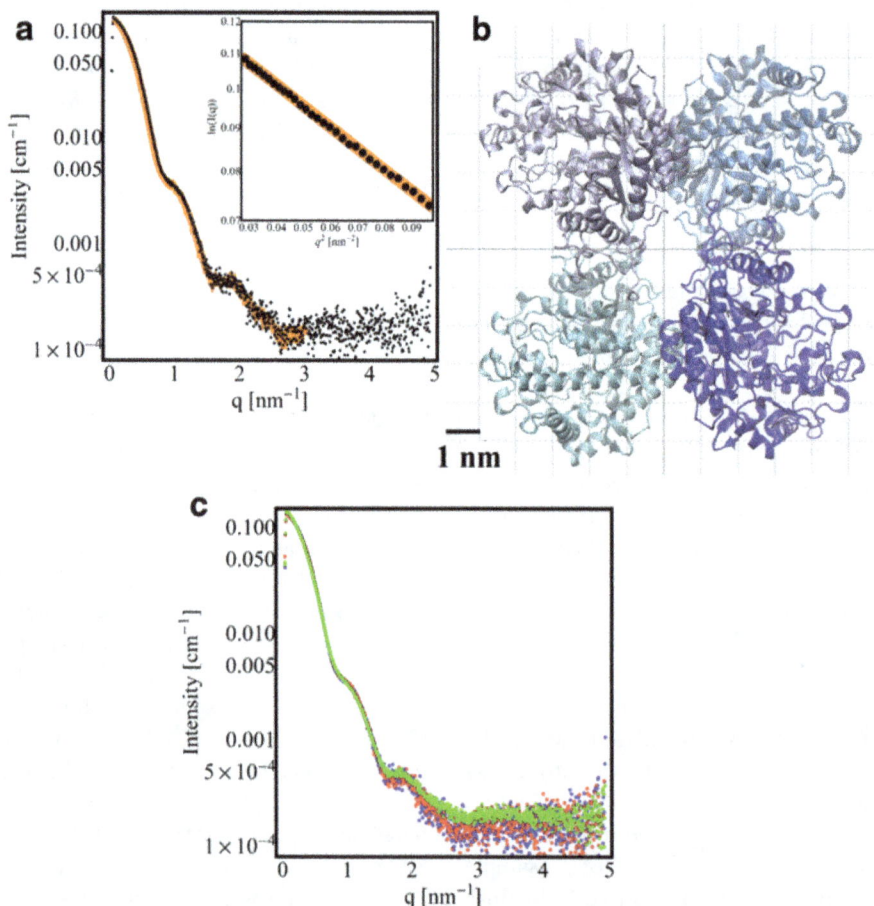

Fig. 1 β-Amylase SAXS data. **a** SAXS curve of β-Amylase (*black dots*) obtained after a standard exposure with fit to a mixture of monomers and tetramers (*orange line*). Inlay: Guinier plot and associated linear fit at low-q^2 ($qR_g < 1.3$). **b** Model of tetrameric β-Amylase (pdb entry 1FA2) [28]. The bounding *box size* of tetrameric β-Amylase is $12.4 \times 12.4 \times 7.5$ nm^3. **c** SAXS data for different exposures time: standard, 5 min and 30 min exposures are plotted in *blue*, *red* and *green*, respectively; very small variations with the exposure time can be observed

MFP3D (both Asylum Research, Santa Barbara, USA). Cantilevers used are the MSNL (triangular lever F, k = 0.6 N/m, F_q = 30 kHz in liquid, Bruker) and Olympus AC40 (k ≈ 0.1 N/m, F_q = 25 kHz in liquid, Olympus). Scan sizes were 10 μm ×10 μm, 5 μm × 5 μm, or 1 μm × 1 μm.

The scan rate was 2 lines per second, the image size was either 512 lines and 512 points per line or 256 lines and 256 points. The atomically flat surfaces were cleaved mica, cleaved mica ion-exchanged with Nickel, and cleaved highly ordered pyrolytic graphite (HOPG). Pre-treated mica was prepared by placing 40 μL of 10 mM $NiCl_2$ on a cleaved mica disk, incubating for 10 minutes, then rinsing with water and drying with nitrogen [46].

To deposit β-Amylase protein onto a surface, 40 μL of 1/1000 diluted solution (4 μg/mL) were placed on the surface and incubated for 15 min. The surface was then rinsed with 1 mL of buffer and covered with 40 μL of buffer for imaging in liquid.

To deposit tobacco mosaic virus (TMV), 40 μL of 1/1000 diluted TMV solution (26 μg/mL) was placed on the surface and incubated for a full hour before rinsing. The surface was then rinsed with 1 mL of water and covered with 40 μL of water for imaging in liquid.

To obtain a representative collection of AFM images, three different 10 μm^2 areas were imaged randomly on a given sample. Within each area, three 5 μm^2 random areas were chosen from which to characterize proteins. If necessary, 1 μm^2 scans were made for areas of particular interest to provide higher resolution images. Image data treatment, such as flattening, was performed using Gwyddion [47] and/or DeStripe [48].

Statistical evaluation of isolated particles in AFM images
Previously flattened AFM height images were used in Gwyddion. When necessary additional flattening was performed to further reduce stripe noise. A grain particle analysis was performed with Gwyddion using the automated thresholding method of Otsu for all images except images of β-Amylase when deposited on bare mica in which case a classical thresholding method of 11–32 % was used instead. Grain size distribution was recorded using the major semiaxis of equivalent ellipse (called here long-axis radius). Ellipse was chosen instead of circle due to the elongated shape of TMV. Consequently, the same numerical measure has been used for β-Amylase and TMV. However, control data did not reach exactly a long-axis radius of 150 nm for TMV (as should be expected from its structure). This is likely due to the fact that the automated ellipsoid fit performed in Gwyddion does not distinguish TMV particles that are touching with each other. Average values were obtained from at least two AFM images for control and over-exposed conditions using the top 2 peaks of the grain particle distribution. Observations made with this criteria would also be valid if we took the top peak of the distribution. The reason to combine the top 2 peaks is that in several distributions, there is not a single major top peak and thus the "correct" value corresponding to such a distribution ought to be the average of nearby peaks. Values reported on figures are average +/− standard deviations.

Protocol for combining SAXS with AFM
When planning combined experiments, it is important to consider the effects of X-ray exposure, both short and

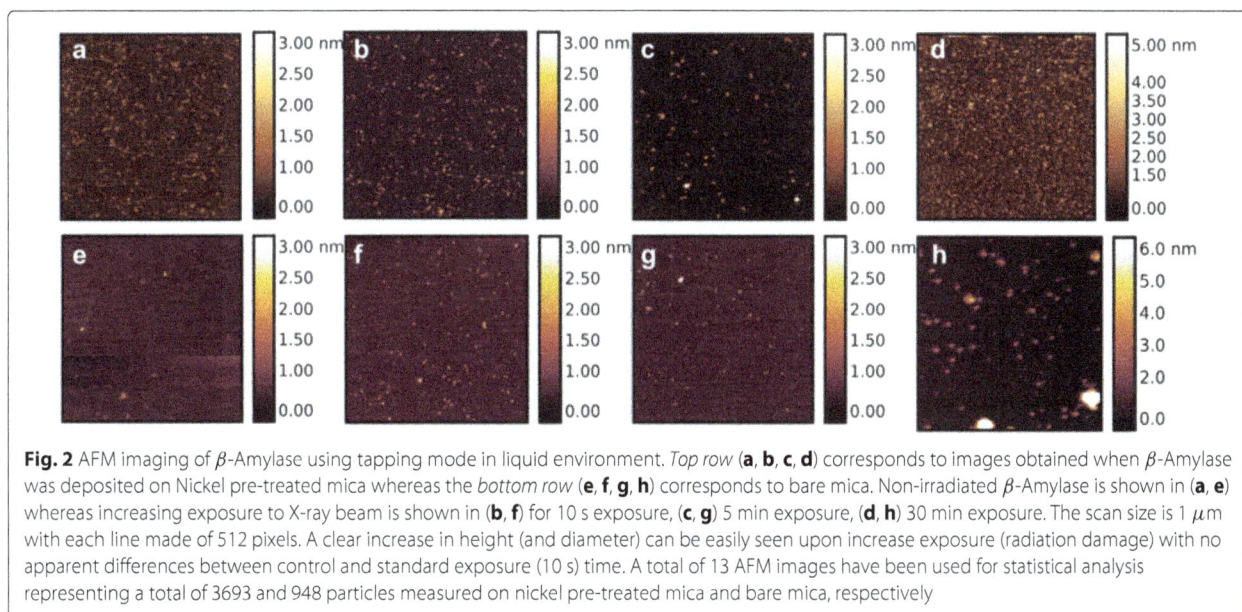

Fig. 2 AFM imaging of β-Amylase using tapping mode in liquid environment. *Top row* (**a**, **b**, **c**, **d**) corresponds to images obtained when β-Amylase was deposited on Nickel pre-treated mica whereas the *bottom row* (**e**, **f**, **g**, **h**) corresponds to bare mica. Non-irradiated β-Amylase is shown in (**a**, **e**) whereas increasing exposure to X-ray beam is shown in (**b**, **f**) for 10 s exposure, (**c**, **g**) 5 min exposure, (**d**, **h**) 30 min exposure. The scan size is 1 μm with each line made of 512 pixels. A clear increase in height (and diameter) can be easily seen upon increase exposure (radiation damage) with no apparent differences between control and standard exposure (10 s) time. A total of 13 AFM images have been used for statistical analysis representing a total of 3693 and 948 particles measured on nickel pre-treated mica and bare mica, respectively

long term, related to the bioSAXS experiment [11, 12]. BioSAXS data is routinely checked for short term variations during exposure and shows there is no variation on the length scales (low resolution size and shape) and time scales (1–10 s at dedicated facilities). Remeasured samples (hours/days after initial exposure) can show significant differences. It is for this reason and the favorable highly dilute state (1000× dilution from bioSAXS) required for AFM that samples were measured by SAXS and only the remaining sample volume (not aspirated by the sample changer) was used for AFM. For studying the effect of radiation damage, samples exposed to the X-ray beam for measurements were recuperated and analyzed by AFM alongside the non-exposed controls. Due to the time needed to prepare the sample for AFM, the time between the two experiments was 20 minutes for β-Amylase and an hour for TMV.

Results

β-Amylase

The background corrected SAXS curve of β-Amylase is shown in Fig. 1a. The mean radius of gyration of the solution is 4.2 nm. To verify that the β-Amylase solution consisted primarily of tetrameric β-Amylase complexes, the scattering of the β-Amylase in solution was compared with the theoretical scattering of monomeric, dimeric, and tetrameric β-Amylase proteins using the program OLIGOMER [41]. Using atomistic models of the protein monomer, dimer, and tetramer, the program fits the theoretical scattering of all three to the observed solution scattering and determines their ratio. β-Amylase was determined to consist of about 94 % tetramers and 6 % monomers (Fig. 1a) without any contribution from dimeric forms ($\chi^2 = 4.19$).

AFM images were obtained with β-Amylase adsorbed on hydrophilic mica in liquid environment using the amplitude modulation (tapping) mode. No significant adsorption of β-Amylase has been observed on hydrophobic graphite (Data not shown). When β-Amylase was deposited onto mica, a uniform distribution of particles was observed, with an average height of 2 nm and an average long-axis radius of 1.87 nm which is smaller than expected from the diameter of β-Amylase crystal structure (Fig. 1b). According to molecular sizes observed with AFM, it is likely that only monomers of β-Amylase are imaged by AFM.

Because the surface of mica is negatively charged, two experimental conditions were used to image irradiated β-Amylase : bare mica or mica pre-treated with $NiCl_2$ solution. Except for native β-Amylase, samples were collected after SAXS beamline exposure. Figure 1c reports SAXS data at different exposure times for β-Amylase. Results of AFM imaging are found in Fig. 2.

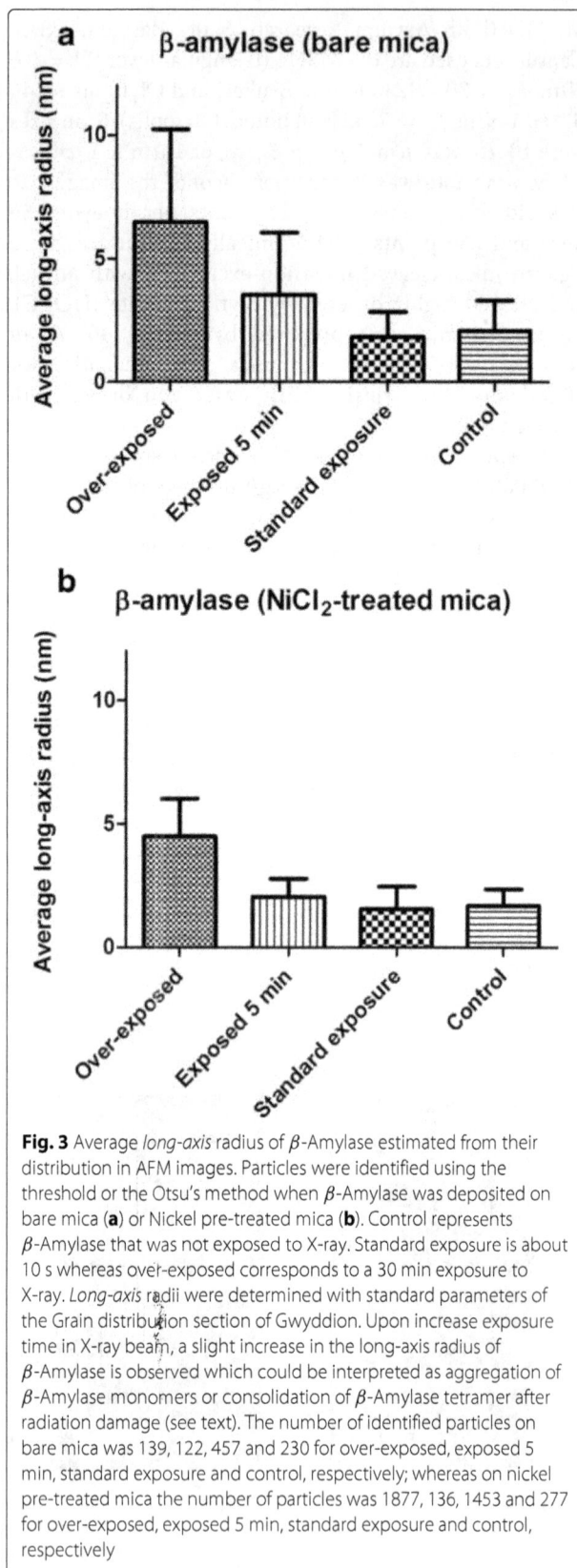

Fig. 3 Average *long-axis* radius of β-Amylase estimated from their distribution in AFM images. Particles were identified using the threshold or the Otsu's method when β-Amylase was deposited on bare mica (**a**) or Nickel pre-treated mica (**b**). Control represents β-Amylase that was not exposed to X-ray. Standard exposure is about 10 s whereas over-exposed corresponds to a 30 min exposure to X-ray. *Long-axis* radii were determined with standard parameters of the Grain distribution section of Gwyddion. Upon increase exposure time in X-ray beam, a slight increase in the long-axis radius of β-Amylase is observed which could be interpreted as aggregation of β-Amylase monomers or consolidation of β-Amylase tetramer after radiation damage (see text). The number of identified particles on bare mica was 139, 122, 457 and 230 for over-exposed, exposed 5 min, standard exposure and control, respectively; whereas on nickel pre-treated mica the number of particles was 1877, 136, 1453 and 277 for over-exposed, exposed 5 min, standard exposure and control, respectively

Table 1 SAXS data-collection and scattering-derived parameters

Data-collection parameters				
Instrument	ESRF BM29			
Beam Geometry	0.7 mm×0.7 mm at sample			
Wavelength (nm)	0.099			
q range (nm^{-1})	0.05-4.95			
Exposure time	1 s per frame			
Concentration, β-amylase	5.5 mg/mL			
Concentration, Tobacco Mosaic Virus	26 mg/mL			
Temperature (K)	293			
Structural Parameters, β-amylase				
Exposure time	1 s	10 s (standard)	5 min	30 min
I_0 (cm^{-1}) [from Guinier]	0.135 ± 0.001	0.135 ± 0.001	0.133 ± 0.1	0.142 ± 0.001
R_g (nm) [from Guinier]	4.14 ± 0.06	4.14 ± 0.06	4.16 ± 0.06	4.35 ± 0.13
I_0 (cm^{-1}) [from $P(r)$]	0.134	0.135	0.135	0.132
R_g (nm) [from $P(r)$]	4.09	4.10	4.12	4.22
D_{max} (nm)	1.10	1.13	1.20	1.30
Porod volume estimate (nm^3)	267	267	266	283
Structural Parameters, Tobacco Mosaic Virus				
Exposure time	1 s	10 s (standard)	5 min	30 min
I_0 (cm^{-1} x nm^{-1}) [from cross-sectional Guinier]	0.312 ± 0.003	0.325 ± 0.003	0.330 ± 0.003	0.317 ± 0.003
cross-sectional R_g (nm) [from cross-sectional Guinier]	6.34	6.40	6.51	6.49
I_0 (cm^{-1} x nm^{-1}) [from cross-sectional $P(r)$]	0.278	0.284	0.289	0.279
cross-sectional R_g (nm) [from cross-sectional $P(r)$]	5.89	5.89	5.88	5.87
cross-sectional D_{max} (nm)	18.0	18.0	18.0	18.0
Molecular-mass determination, β-amylase				
Exposure time	1 s	10 s (standard)	5 min	30 min
Partial specific volume (cm^{3}g^{-1})		0.724		
Contrast ($\Delta\rho \times 10^{10}$cm^{-2})		3.047		
Molecular mass M_r (kDa) [from I_0]	168 ± 1	170 ± 1	167 ± 1	178 ± 1
Molecular mass M_r (kDa) [from Porod Volume] [64]	157	157	156	166
Calculated molecular weight according to sequence (kDa)		224		
Software employed				
Primary data reduction	BM29 online data analysis [65], pyFAI [66], Primus [64]			
1D data processing	PRIMUS			
p(r) analysis	GNOM [42]			
Form factor fitting	GENFIT [43]			
Computation of model intensities	CRYSOL [44]			
Computation of volume fractions of mixtures	OLIGOMER [64]			

Figure 3 shows that increasing X-ray beam exposure provokes an enlargement of the long-axis radius of isolated particles (see Methods for definition of long-axis radius). It can be easily seen that standard exposure time for β-Amylase does not modify significantly the shape of β-Amylase as imaged by AFM. However, a continuous increase of the long-axis radius can be observed when β-Amylase was exposed during 5 min to the X-ray beam. Over-exposure of β-Amylase in the X-ray beam provokes a dramatic increase in the long-axis radius of AFM imaged β-Amylase. The increase in the mean size upon X-ray beam exposure is systematically observed in all experiments both on bare and Nickel pre-treated mica.

Table 1 reports all structural parameters evaluated at different exposures time. Interpretation of SAXS data on various X-ray beam exposure time of β-Amylase indicates that there is no significant change in size, especially the radius of gyration, of irradiated β-Amylase (Fig. 4 obtained treating data presented in Fig. 1c). This is in contrast to AFM measurements (Figs. 2 and 3) for which the magnitude of change of the long-axis radius of β-Amylase due to X-ray beam exposure is large. In AFM, the β-Amylase long-axis radius upon irradiation is close to the

expected size of the β-Amylase tetramer according to its X-ray structure.

TMV

Crystal structure of TMV, as well as electron micrographs, indicate that TMV is about 300 nm long with a diameter of about 19 nm [35]. The background corrected SAXS cross sectional Guinier plot of TMV is shown in the inset of Fig. 5a. The cross-sectional radius of gyration is found to be 6.34 nm. The low q-region of the curve can be fitted with a three shell cylinder model, using parameters comparable to those reported in the literature (Fig. 5c) [49]. The peaks in the region of 3 nm^{-1} can be attributed to fibre diffraction from the helical repeat of TMV (2.9 nm) [50, 51]. The cross-sectional pair distance distribution function shown in Fig. 5b was calculated using $D_{max} = 18$ nm based on the virus height determined by AFM. It is rather symmetric, with its maximum at 7.5 nm. It is noteworthy that the AFM output has been used here as constraint for SAXS data treatment as it is conventionally done using NMR or electron microscopy data. Figure 5e reports SAXS data at different exposure times. TMV SAXS patterns present very small variations in the q-range of 0.05 to 2 nm^{-1}. However, the structural parameters reported in Table 1 for TMV do not show any consistent variation with the exposure time.

TMV particles were imaged with AFM after SAXS measurements both on freshly cleaved HOPG or pre-treated mica (Fig. 6a–f respectively). On freshly cleaved mica no TMV could be detected (data not shown) whereas on Nickel-coated mica, TMV particles of about 17 nm height were detected (Fig. 6d, e, f). The coverage was estimated at 16.9 % using a mask selecting all points higher than 5 nm. When deposited on HOPG (hydrophobic surface), TMV particles of similar height (17 – 18 nm) were observed with a surface coverage of 11.2 % (Fig. 6a, b, c). The observed heights of TMV are similar to values reported in the literature when imaging TMV on mica [52].

Effect of radiation damage is clearly seen in AFM images between control data (Fig. 6a, d) and over-exposed data (Fig. 6c, f) as well as standard exposure (Fig. 6d, e) when TMV was deposited on the hydrophobic HOPG surface. Quantification of such effect is obtained by estimating long-axis radius measurements of TMV particles upon different X-ray beam exposure times (Fig. 7). By combining all AFM data, long-axis radius of TMV in over-exposed data is about 22 nm whereas the value for control data is about 93 nm. It can be seen that, to the contrary of β-Amylase, increasing exposure time on TMV lead to a reduction in its long-axis radius from 4 to 5 fold. A gradual and consistent decrease in TMV long-axis radius can be seen from control, to standard, and over-exposure. It can also be observed that the long-axis radius of control TMV is not 150 nm (perfect fit) but about 2/3 of this

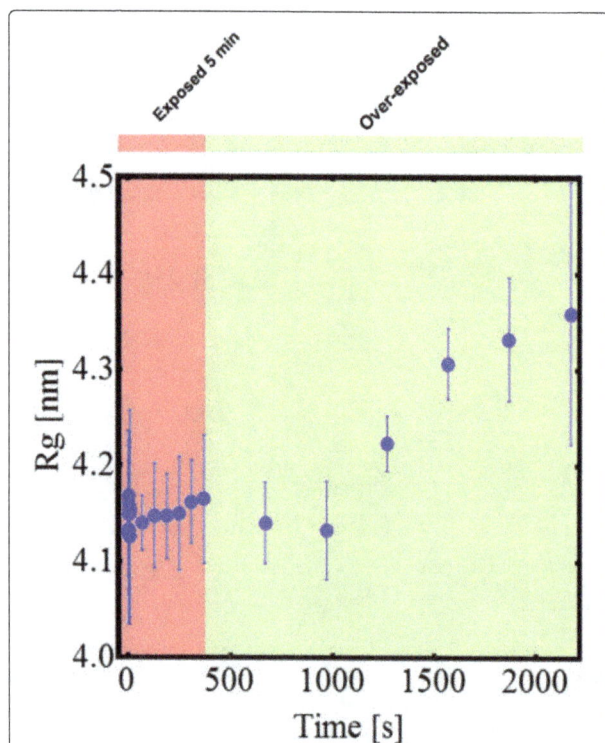

Fig. 4 Evolution of the radius of gyration for the β-Amylase upon X-ray exposure obtained from SAXS data. While there is no significant increase for low exposure time, R_g increases once proteins are over-exposed: this is consistent with the AFM results presented in Fig. 3

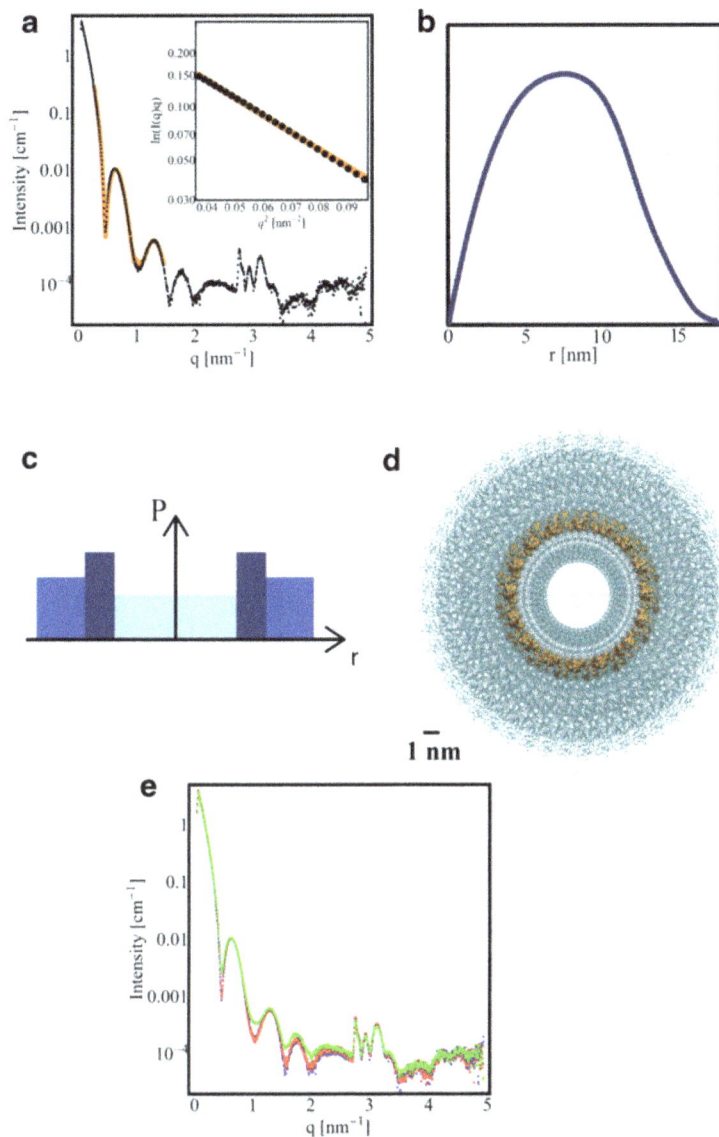

Fig. 5 TMV SAXS data. **a** SAXS curve of TMV (*black dots*) obtained during a standard exposure with fit to a three shell cylinder model (*orange line*). Inlay: Cross-sectional Guinier plot and fit. **b** Radial pair distance distribution function of TMV using the virus diameter found by AFM as constrained for D_{max}. **c** Schematic of the three-shell electron density distribution used as a model in **a**). **d** Atomistic model of the TMV cross-section with the RNA in *orange*. Based on pdb entry 1VTM. **e** SAXS data for different exposures time: standard, 5 min and 30 min exposures are plotted in *blue*, *red* and *green*, respectively

value. It is likely that the 2D image fitting algorithm performs poorly when TMV particle appears in bunch rather than well isolated (Fig. 6). The clear difference observed in AFM images of TMV on mica and HOPG suggests a clear physico-chemical change occurring for TMV upon radiation damage. Whereas native TMV tends to aggregate on HOPG, a more uniform adsorption on HOPG is observed upon X-ray beam exposure concomitantly with a reduction in TMV particle average length.

At the ESRF beamline, it is not possible to perform ultra-small angle X-ray scattering experiments. Consequently, changes in length of TMV could not be obtained using regular SAXS data due to large size of TMV ($\approx$300 nm long).

Discussion

SAXS scattering provides reliable characterization of the average structural properties of biological macromolecules by measuring the scattering curve and interpreting it to determine model-independent structural parameters of molecules. Although 3D reconstructions of shape of macromolecules from scattering curves are possible, they are often not unique. Moreover, SAXS as a technique is able to visualize a wide range of dimensions

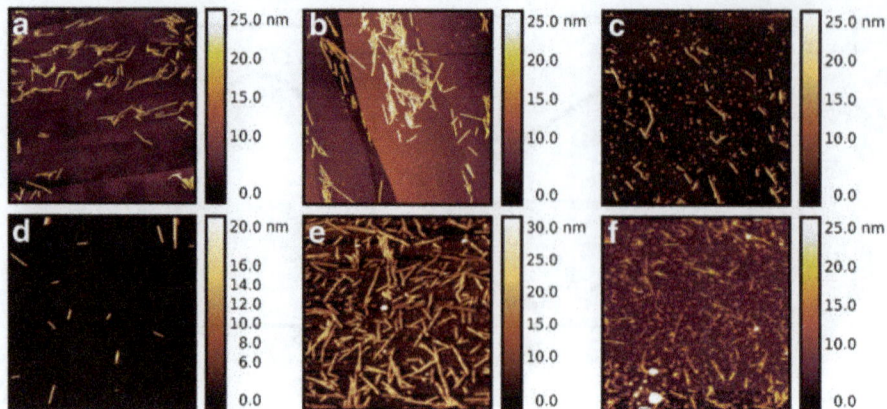

Fig. 6 AFM imaging of TMV using tapping mode in liquid environment. *Top row* (**a**, **b**, **c**) is TMV deposited on HOPG whereas *bottom row* (**d**, **e**, **f**) corresponds to TMV deposited on Nickel pre-treated mica. Non-irradiated TMV is shown in (**a**, **d**). Increasing exposure to X-ray beam is shown in (**b**, **e**) for 10 s exposure, and (**c**, **f**) 30 min exposure. The scan size is 5 μm with each line made of 512 pixels. A clear fragmentation of the 300 nm-long TMV can be observed upon increase in exposure time (radiation damage). To the contrary of β-Amylase (Fig. 2), even at standard exposure time, a beginning of fragmentation is observed for TMV. A total of 14 AFM images has been used for statistical analysis representing 6539 particles on HOPG at 10 μm scan size, 2007 particles on HOPG at 5 μm scan size, and 808 particles on nickel pre-treated mica

depending on the X-ray energy and the angular range observed, but the maximum size observable for any given experiment is limited. For the standard setup at BM29 a q-range of 0.025 to 5 nm^{-1} can be observed which corresponds to a longest particle dimension of approximately 250 nm. As the long axis of tobacco mosaic virus (TMV) is in the order of 300 nm, this dimension could not be measured directly in the standard configuration of BM29. Fortunately, the observed scattering of the TMV is dominated by the circular cross-section of the cylinder, featuring the coat protein and RNA strand allowing direct comparison between SAXS and AFM. The AFM experiments confirmed the rod-like structure of the TMV sample with a cross-section of 17–18 nm. The cross-sectional radius of gyration was calculated as 6.3 nm assuming a rod-like structure. For a homogeneous disc of radius $R, R = \sqrt{2}R_g = 8.9$ nm, which is in direct agreement with the diameter of 18 nm reported in literature [53]. It should be emphasized that the calculation of (cross-sectional) pair distance distribution function requires to find the correct maximum distance D_{max}. In this work, this value was obtained from real space measurements with the AFM allowing a direct access to this parameter, thereby greatly reducing the ambiguity of the analysis. In the case of TMV, using the 18 nm diameter determined by AFM indeed allowed us to calculate the cross-sectional pair distance distribution function without bias.

Radiation damage [54] have been mostly investigated in X-ray crystallography where it was observed that radiation damage on proteins starts with the reduction of metal centers followed by elongation/scission of disulfide bonds and then decarboxylation of Asp and Glu side chains [55]. Moreover, such decarboxylation of acidic amino acids is also observed due to radiation damage with electron microscopy [56].

AFM imaging of single molecules has already been used to observe protective effect of ascorbic acid against double-strand breaks in DNA generated by reactive oxygen species [57]. In addition, AFM has also been coupled with Dynamic Light Scattering technique to help in understanding consequences of radiation-induced conformational change in chromatin structure. It was found that even at low dose (< 0.5 Gy) chromatin shows radiation damage as evidence by a change in hydrodynamic size that was likely due to single-strand breaks in DNA [58]. In SAXS, radiation damage most often present itself as aggregation [59]. Even with sample flow enable, radiation damage in lysozyme, evidenced by an increase in radius of gyration (R_g), still occurs as early as 250 ms exposure time [11]. At increasing dose on lysozyme, an increase in R_g has been observed in relation to radiation damage [60]. Combining SAXS with UV/Visible absorption spectra revealed change in protein solution due to X-ray radiation on bovine serum albumine (BSA) as shown by an increase of R_g from 3.3 to 5 nm [61]. However, it was also found that early effect of radiation damage was an increase of molecular size without any significant unfolding suggesting that radiation damage observed on BSA was compatible with the presence of radical activities [61]. Reduction in radiation damage has been obtained using Cryo-SAXS [12] or using time-resolved SAXS [62]. Besides, fast detection readout allows collection of SAXS before radiation damage occur [60].

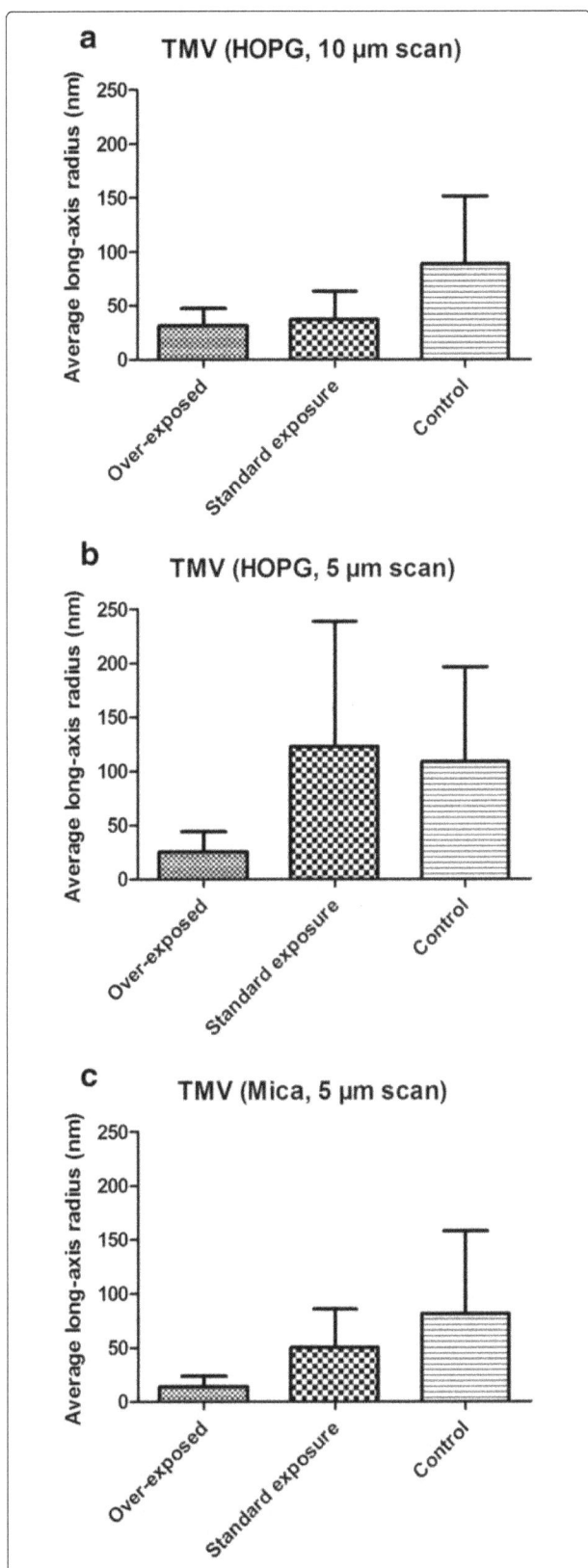

a TMV (HOPG, 10 μm scan)

b TMV (HOPG, 5 μm scan)

c TMV (Mica, 5 μm scan)

Fig. 7 Average long-axis radius of TMV estimated from their distribution in AFM images. Particles were identified using the Otsu's method when TMV was deposited on HOPG (**a**, **b**) or mica (**c**). AFM data have been acquired on unexposed samples (control), after a standard exposure as well as after 30 min exposure (over-exposed). Upon increasing exposure time in X-ray beam, a consistent decrease in the long-axis radius of TMV is observed which corresponds to a fragmentation of TMV particles upon radiation damage. The number of identified particles on HOPG 10 μm was 3912, 2273 and 354 for over-exposed, standard exposure and control, respectively; on HOPG 5 μm the number of identified particles was 1906, 68 and 33 for over-exposed, standard exposure and control, respectively; on mica the number of identified particles was 347, 157 and 304 for over-exposed, standard exposure and control, respectively

In our study, AFM imaging of isolated molecules of β-Amylase revealed a tripling in size upon over-exposure to X-ray beam. From this result, two hypotheses are possible: agglomeration of several β-Amylase monomers or a tightening of β-Amylase tetramer upon X-ray exposure so that when imaged on mica the tetrameric form of β-Amylase is now stable and better preserved than β-Amylase without X-ray exposure. The second hypothesis appears more likely due to SAXS observation that no significant change in R_g was observed upon over X-ray exposition of β-Amylase (Fig. 4 and Table 1), and that β-Amylase remains mostly tetrameric. Besides, it has been shown that one consequence of irradiation damage in synchrotron SAXS experiment was a change of the protein surface due to radical attacks leading to a greater attraction between lysozyme molecules and causing aggregation: a mechanism that could also be envisaged to multimeric proteins such as β-Amylase [10]. The current resolution of AFM does not allow imaging at the atomic scale on isolated proteins to identify more precisely what is the mechanism of such increase in size. In particular, knowing the convolution effect due to tip broadening in AFM images, it is not possible to attribute β-Amylase native tetramers on AFM controlled images (long-axis radius of about 3 nm). The only possible explanation is that only smaller structures are observed, mostly monomers whose presence is also detected by SAXS, for non irradiated β-Amylase while, upon X-ray exposure, AFM images show an expected size of tetrameric β-Amylase. At the moment, it is not possible to speculate about the presence of crosslink in β-Amylase upon X-ray exposure, as the resolution of AFM imaging does not allow such level of details.

Finally, it is striking that AFM imaging can indirectly distinguish between two conformations of tetrameric β-Amylase: native and X-ray over-exposed while SAXS data does not make a significant distinction.

Radiation damage in TMV particle is different from β-Amylase essentially due to the high aspect/ratio of TMV which is a rod-like of 300 nm long. TMV, as most plant viruses, are very stable molecular constructs that can resist harsh storage condition (dessicated) for several years. TMV is consequently a perfect sample for studying radiation damage as no degradation is expected to occur when deposited on mica [36]. Upon increasing exposure time in X-ray beam, a breaking of TMV is consistently seen. In this case, radiation damage on TMV resemble closely that obtained on DNA, i.e. breaking into smaller parts.

It is noteworthy that the imaging substrate surface has a significant importance in AFM. This is for instance brought forward by the comparison between bare mica and nickel pre-treated mica in this study (AFM data not shown). Indeed, TMV adsorption is more efficient on nickel pre-treated mica than on simple mica. However, no substrate is ideal due to the apparent contradictory requirement of strong fixation of biomolecules on a surface with low deformation of adsorbed molecules.

Surface charges of mica may affect the shape of deposited single molecules. For instance, a height reduction of 2 nm is observed when TMV is deposited on mica and imaged in air, while the height difference of TMV when imaged in liquid is close to 0.7 nm (manuscript submitted). However, such reduction in height has never been observed to be concomitant with a reduction of TMV length as observed in this study when TMV is exposed to X-ray beam.

Another difference between TMV and β-Amylase, is that with TMV radiation damage are detected with short exposure time whereas in β-Amylase radiation damage are mostly visible upon over-exposure time. However, a common behavior between β-Amylase and TMV upon X-ray beam exposure is their apparent change in molecular surface properties. Although it is only suggested for β-Amylase, it is clearly observed for TMV. Indeed, when native TMV is deposited on hydrophobic surface (HOPG) there is non uniform binding of TMV on HOPG whereas, upon X-ray beam exposure, TMV displays an increased uniformity in adsorption with HOPG. Because breakage of long TMV particles into smaller pieces, damaged TMV now exposes hidden buried surfaces. From the TMV X-ray structure, such hidden surfaces are known to be rather hydrohobic explaining the sudden increased affinity of irradiated TMV on HOPG. A clear benefit of AFM imaging is observed, first by looking at individual molecules, and second at global properties when changing imaging substrates. Consequently, if reasonable protein binding is observed on HOPG, it could be recommended to use hydrophobic surfaces for imaging X-ray exposed molecules, and thus detect easily the presence of radiation damage by looking at variation of protein binding on HOPG. Such apparent change in molecular surface properties has been already observed in case of lipid model membranes deposited onto silicon substrates in an *in-situ* X-ray - AFM combined experiment [63].

Both X-ray Reflectometry and AFM showed a deacrease of the membranes surface coverage after exposure to X-ray.

Conclusion

The combination of SAXS and AFM can be applied to a variety of different macromolecules and sample surfaces depending on characterization needs and sample properties. Taking advantage of the flexible user access to both the dedicated bioSAXS beamline (BM29) and ESRF surface science laboratory, these experiments can be undertaken on the same visit to the ESRF. AFM imaging requires around two hours and owing to the high dilution factor from SAXS to AFM no additional sample is needed for AFM in addition to SAXS. While SAXS provides rapid characterization of the average properties of a sample, AFM can be used to verify the homogeneity of the sample and provide measurements at the single particle level. As AFM gives direct measurement of single particles it is possible to use the AFM results as additional constraints for modeling purpose thereby extending the possibilities to interpret the SAXS data and reduce ambiguity in the results. The use of combined SAXS-AFM in one experimental visit is facilitated by the presented protocol which enables cross validation, and increased confidence in the conclusions which can be drawn from the experiments. Furthermore, combination of SAXS-AFM is well adapted to study effect of radiation damage on various type of biological samples. Radiation damage is a very complex process and can produce either a change of the protein surface or a breakage of long biological particles, as it has been shown in this work for β-Amylase and TMV, respectively.

Abbreviations
AFM: Atomic force microscopy; EM: Electron microscopy; ESRF: European synchrotron radiation facility; HOPG: Highly ordered pyrolitic graphite; NMR: Nuclear magnetic resonance; SAXS: Small angle x-ray scattering; TMV: Tobacco mosaic virus

Acknowledgments
The authors acknowledge the 2015 GIANT international Internship and the MIT International Science and Technology Initiatives programs. The authors warmly thank Chloe Zubieta for fruitful discussions and Francesco Spinozzi for advice on using Genfit.

Funding
Not applicable.

Authors' contributions
AA and LC acquired AFM data. AA, MB and AR acquired SAXS data. JMT, AA and JLP developed the AFM sample preparation protocols. AA, LC, MB, SWC and JLP treated AFM and SAXS data. AA, AR, JLP, MB and LC prepared this manuscript. AR, LC and JLP conceived this research. All authors read and approved the final manuscript.

Competing interests

The authors declare that they have no competing interests.

Author details

[1]ESRF, The European Synchrotron, 71 Avenue des Martyrs, 38000 Grenoble, France. [2]MIT, 77 Massachusetts Ave., 02139 Cambridge, MA, USA. [3]Univ. Grenoble Alpes, 71 Avenue des Martyrs, 38044 Grenoble, France. [4]CNRS, IBS, 71 Avenue des Martyrs, 38044 Grenoble, France. [5]CEA, IBS, 71 Avenue des Martyrs, 38044 Grenoble, France. [6]European Molecular Biology Laboratory, 71 Avenue des Martyrs, 38000 Grenoble, France. [7]Unit for Virus Host-Cell Interactions, Univ. Grenoble Alpes-EMBL-CNRS, 71 Avenue des Martyrs, 38000 Grenoble, France. [8]Faculty of Natural Sciences, Keele University, Keele, Staffordshire, UK. [9]Present Address: CBS, Centre de Biochimie Structurale, CNRS UMR 5048-INSERM UMR 1054, 29, Rue de Navacelles, 34090 Montpellier, France. [10]Present Address: European XFEL GmbH, Holzkoppel 4, 22869 Schenefeld, Germany.

References

1. Ward AB, Sali A, Wilson IA. Integrative structural biology. Science. 2013;339(6122):913–5.
2. Graewert MA, Svergun DI. Impact and progress in small and wide angle x-ray scattering (saxs and waxs). Curr Opin Struct Biol. 2013;23(5):748–54.
3. Kikhney AG, Svergun DI. A practical guide to small angle x-ray scattering (saxs) of flexible and intrinsically disordered proteins. FEBS Lett. 2015;589(19):2570–7.
4. Putnam CD, Hammel M, Hura GL, Tainer JA. X-ray solution scattering (saxs) combined with crystallography and computation: defining accurate macromolecular structures, conformations and assemblies in solution. Q Rev Biophys. 2007;40(03):191–285.
5. Jacques DA, Trewhella J. Small angle scattering for structural biology expanding the frontier while avoiding the pitfalls. Protein Sci. 2010;19(4):642–57.
6. Konarev PV, Svergun DI. A posteriori determination of the useful data range for small-angle scattering experiments on dilute monodisperse systems. IUCrJ. 2015;2(3):352–60.
7. Semenyuk A, Svergun D. Gnom–a program package for small-angle scattering data processing. J Appl Crystallogr. 1991;24(5):537–40.
8. Bergmann A, Fritz G, Glatter O. Solving the generalized indirect fourier transformation (gift) by boltzmann simplex simulated annealing (bssa). J Appl Crystallogr. 2000;33(5):1212–6.
9. Petoukhov MV, Svergun DI. Ambiguity assessment of small-angle scattering curves from monodisperse systems. Biol Crystallogr. 2015;71(5):1051–8.
10. Kuwamoto S, Akiyama S, Fujisawa T. Radiation damage to a protein solution, detected by synchrotron X-ray small-angle scattering: dose-related considerations and suppression by cryoprotectants. J Synchrotron Radiat. 2004;11(6):462–8.
11. Jeffries CM, Graewert MA, Svergun DI, Blanchet CE. Limiting radiation damage for high-brilliance biological solution scattering: practical experience at the embl p12 beamline petraiii. J Synchrotron Radiat. 2015;22(2):273–9.
12. Meisburger SP, Warkentin M, Chen H, Hopkins JB, Gillilan RE, Pollack L, Thorne RE. Breaking the radiation damage limit with cryo-saxs. Biophys J. 2013;104(1):227–36.
13. Binnig G, Quate CF, Gerber C. Atomic force microscope. Phys Rev Lett. 1986;56(9):930.
14. Parot P, Dufrêne YF, Hinterdorfer P, Le Grimellec C, Navajas D, Pellequer JL, Scheuring S. Past, present and future of atomic force microscopy in life sciences and medicine. J Mol Recognit. 2007;20(6):418–31.
15. Engel A, Müller DJ. Observing single biomolecules at work with the atomic force microscope. Nat Struct MolBiol. 2000;7(9):715–8.
16. Fotiadis D, Scheuring S, Müller SA, Engel A, Müller DJ. Imaging and manipulation of biological structures with the afm. Micron. 2002;33(4):385–97.
17. Costa L, Rodrigues MS, Newman E, Zubieta C, Chevrier J, Comin F. Imaging material properties of biological samples with a force feedback microscope. J Mol Recognit. 2013;26(12):689–93.
18. Schabert FA, Engel A. Reproducible acquisition of escherichia coli porin surface topographs by atomic force microscopy. Biophys J. 1994;67(6):2394.
19. Scheuring S, Ringler P, Borgnia M, Stahlberg H, Müller DJ, Agre P, Engel A. High resolution afm topographs of the escherichia coli water channel aquaporin z. EMBO J. 1999;18(18):4981–7.
20. Chaves RC, Dahmane S, Odorico M, Nicolaes GA, Pellequer JL. Factor va alternative conformation reconstruction using atomic force microscopy. Thromb Haemost. 2014;112(6):1167–73.
21. Chen S-WW, Teulon JM, Godon C, Pellequer JL. Atomic force microscope, molecular imaging, and analysis. J Mol Recognit. 2016;29(1):51–5.
22. Hansma H, Vesenka J, Siegerist C, Kelderman G, Morrett H, Sinsheimer R, Elings V, Bustamante C, Hansma P. Reproducible imaging and dissection of plasmid dna under liquid with the atomic force microscope. Science. 1992;256(5060):1180–4.
23. Hong DP, Fink AL, Uversky VN. Structural characteristics of α-synuclein oligomers stabilized by the flavonoid baicalein. J Mol Biol. 2008;383(1):214–23.
24. Baldock C, Oberhauser AF, Ma L, Lammie D, Siegler V, Mithieux SM, Tu Y, Chow JYH, Suleman F, Malfois M, et al. Shape of tropoelastin, the highly extensible protein that controls human tissue elasticity. Proc Natl Acad Sci USA. 2011;108(11):4322–7.
25. Sander B, Tria G, Shkumatov AV, Kim EY, Grossmann JG, Tessmer I, Svergun DI, Schindelin H. Structural characterization of gephyrin by afm and saxs reveals a mixture of compact and extended states. Acta Crystallogr D Biol Crystallogr. 2013;69(10):2050–60.
26. Wu W, Huang J, Jia S, Kowalewski T, Matyjaszewski K, Pakula T, Gitsas A, Floudas G. Self-assembly of podma-b-pt ba-b-podma triblock copolymers in bulk and on surfaces. a quantitative saxs/afm comparison. Langmuir. 2005;21(21):9721–7.
27. Brus J, Špírková M, Hlavatá D, Strachota A. Self-organization, structure, dynamic properties, and surface morphology of silicaepoxy films as seen by solid-state nmr, saxs, and afm. Macromolecules. 2004;37(4):1346–57.
28. Cheong CG, Eom SH, Chang C, Shin DH, Song HK, Min K, Moon JH, Kim KK, Hwang KY, Suh SW. Crystallization, molecular replacement solution, and refinement of tetrameric β-amylase from sweet potato. Proteins Struct Funct Bioinforma. 1995;21(2):105–17.
29. Takeda Y, Hizukuri S. Improved method for crystallization of sweet potato beta-amylase. Biochim Biophys Acta. 1969;185(2):469–71.
30. Namba K, Stubbs G. Structure of tobacco mosaic virus at 3.6 a resolution: implications for assembly. Science. 1986;231(4744):1401–6.
31. Zenhausern F, Adrian M, Emch R, Taborelli M, Jobin M, Descouts P. Scanning force microscopy and cryo-electron microscopy of tobacco mosaic virus as a test specimen. Ultramicroscopy. 1992;42:1168–72.
32. Drygin YF, Bordunova OA, Gallyamov MO, Yaminsky IV. Atomic force microscopy examination of tobacco mosaic virus and virion rna. FEBS Lett. 1998;425(2):217–21.
33. Gebhardt R, Teulon JM, Pellequer JL, Burghammer M, Colletier JP, Riekel C. Virus particle assembly into crystalline domains enabled by the coffee ring effect. Soft Matter. 2014;10(30):5458–62.
34. Marinaro G, Burghammer M, Costa L, Dane T, De Angelis F, Di Fabrizio E, Riekel C. Directed growth of virus nanofilaments on a superhydrophobic surface. ACS Appl Mater Interfaces. 2015;7(23):12373–9.
35. Chen S-WW, Odorico M, Meillan M, Vellutini L, Teulon JM, Parot P, Bennetau B, Pellequer JL. Nanoscale structural features determined by afm for single virus particles. Nanoscale. 2013;5:10877–86.
36. Trinh MH, Odorico M, Bellanger L, Jacquemond M, Parot P, Pellequer JL. Tobacco mosaic virus as an afm tip calibrator. J Mol Recognit. 2011;24(3):503–10.
37. Pernot P, Round A, Barrett R, De Maria Antolinos A, Gobbo A, Gordon E, Huet J, Kieffer J, Lentini M, Mattenet M, Morawe C, Mueller-Dieckmann C, Ohlsson S, Schmid W, Surr J, Theveneau P, Zerrad L, McSweeney S. Upgraded ESRF BM29 beamline for SAXS on macromolecules in solution. J Synchrotron Radiat. 2013;20(4):660–4.
38. Round A, Felisaz F, Fodinger L, Gobbo A, Huet J, Villard C, Blanchet CE, Pernot P, McSweeney S, Roessle M, Svergun DI, Cipriani F. BioSAXS Sample Changer: a robotic sample changer for rapid and reliable high-throughput X-ray solution scattering experiments. Acta Crystallogr D. 2015;71(1):67–75.
39. Brennich ME, Kieffer J, Bonamis G, De Maria Antolinos A, Hutin S, Pernot P, Round A. Online data analysis at the ESRF bioSAXS beamline, BM29. J Appl Crystallogr. 2016;49(1):203–12.
40. De Maria Antolinos A, Pernot P, Brennich ME, Kieffer J, Bowler MW, Delageniere S, Ohlsson S, Malbet Monaco S, Ashton A, Franke D,

Svergun D, McSweeney S, Gordon E, Round A. ISPyB for BioSAXS, the gateway to user autonomy in solution scattering experiments. Acta Crystallogr. D. 2015;71(1):76–85.

41. Konarev PV, Volkov VV, Sokolova AV, Koch MH, Svergun DI. Primus: a windows pc-based system for small-angle scattering data analysis. J Appl Crystallogr. 2003;36(5):1277–82.

42. Svergun DI. Determination of the regularization parameter in indirect-transform methods using perceptual criteria. J Appl Crystallogr. 1992;25(4):495–503.

43. Spinozzi F, Ferrero C, Ortore MG, De Maria Antolinos A, Mariani P. *GENFIT*: software for the analysis of small-angle X-ray and neutron scattering data of macromolecules in solution. J Appl Crystallogr. 2014;47(3):1132–9.

44. Svergun D, Barberato C, Koch MHJ. *CRYSOL* – a Program to Evaluate X-ray Solution Scattering of Biological Macromolecules from Atomic Coordinates. J Appl Crystallogr. 1995;28(6):768–73.

45. Hansma PK, Cleveland JP, Radmacher M, Walters DA, Hillner PE, Bezanilla M, Fritz M, Vie D, Hansma HG, Prater CB, Massie J, Fukunaga L, Gurley J, Elings V. Tapping mode atomic force microscopy in liquids. Appl Phys Lett. 1994;64(13):1738–40.

46. El Kirat K, Burton I, Dupres V, Dufrene YF. Sample preparation procedures for biological atomic force microscopy. J Microsc. 2005;218(3):199–207.

47. Nečas D, Klapetek P. Gwyddion: an open-source software for spm data analysis. Cent Eur J Phys. 2012;10(1):181–8.

48. Chen S-WW, Pellequer JL. Destripe: frequency-based algorithm for removing stripe noises from afm images. BMC Struct Biol. 2011;11(1):7.

49. Lee B, Lo CT, Thiyagarajan P, Winans RE, Li X, Niu Z, Wang Q. Effect of interfacial interaction on the cross-sectional morphology of tobacco mosaic virus using gisaxs. Langmuir. 2007;23(22):11157–63.

50. Kendall A, McDonald M, Stubbs G. Precise determination of the helical repeat of tobacco mosaic virus. Virology. 2007;369(1):226–7.

51. Franklin RE, Klug A. The splitting of layer lines in x-ray fibre diagrams of helical structures: application to tobacco mosaic virus. Acta Crystallogr. 1955;8(12):777–80.

52. Meillan M, Ramin MA, Buffeteau T, Marsaudon S, Odorico M, Chen S-WW, Pellequer JL, Degueil M, Heuzé K, Vellutini L, et al. Self-assembled monolayer for afm measurements of tobacco mosaic virus (tmv) at the atomic level. RSC Adv. 2014;4(23):11927–30.

53. Klug A. The tobacco mosaic virus particle: structure and assembly. Philos Trans R Soc Lond B Biol Sci. 1999;354:531–5.

54. Garman EF, Weik M. Radiation damage in macromolecular crystallography. Synchrotron Radiat News. 2015;28(6):15–9.

55. Carpentier P, Royant A, Weik M, Bourgeois D. Raman-assisted crystallography suggests a mechanism of x-ray-induced disulfide radical formation and reparation. Structure. 2010;18(11):1410–9.

56. Bartesaghi A, Matthies D, Banerjee S, Merk A, Subramaniam S. Structure of -galactosidase at 3.2-Å resolution obtained by cryo-electron microscopy. Proc Natl Acad Sci USA. 2014;111(32):11709–14.

57. Yoshikawa Y, Hizume K, Oda Y, Takeyasu K, Araki S, Yoshikawa K. Protective effect of vitamin c against double-strand breaks in reconstituted chromatin visualized by single-molecule observation. Biophys J. 2006;90(3):993–9.

58. Jain V, Hassan PA, Das B. Radiation-induced conformational changes in chromatin structure in resting human peripheral blood mononuclear cells. Int J Radiat Biol. 2014;90(12):1143–51.

59. Skou S, Gillilan RE, Ando N. Synchrotron-based small-angle x-ray scattering (saxs) of proteins in solution. Nat Protoc. 2014;9(7):1727.

60. Philipp HT, Koerner LJ, Hromalik MS, Tate MW, Gruner SM. Femtosecond radiation experiment detector for x-ray free-electron laser (xfel) coherent x-ray imaging. 2008 IEEE Nucl Sci Symp Conf Rec. 2010;57(6):3795–9.

61. Haas S, Plivelic TS, Dicko C. Combined saxs/uv–vis/raman as a diagnostic and structure resolving tool in materials and life sciences applications. J Phys Chem B. 2014;118(8):2264–73.

62. Lamb J, Kwok L, Qiu X, Andresen K, Park HY, Pollack L. Reconstructing three-dimensional shape envelopes from time-resolved small-angle X-ray scattering data. J Appl Crystallogr. 2008;41(6):1046–52.

63. Gumí-Audenis B, Carlà F, Vitorino M, Panzarella A, Porcar L, Boilot M, Guerber S, Bernard P, Rodrigues M, Sanz F, et al. Custom afm for x-ray beamlines: in situ biological investigations under physiological conditions. J Synchrotron Radiat. 2015;22(6):1364–71.

64. Petoukhov MV, Franke D, Shkumatov AV, Tria G, Kikhney AG, Gajda M, Gorba C, Mertens HDT, Konarev PV, Svergun DI. New developments in the *ATSAS* program package for small-angle scattering data analysis. J Appl Crystallogr. 2012;45(2):342–50.

65. Brennich ME, Kieffer J, Bonamis G, De Maria Antolinos A, Hutin S, Pernot P, Round A. Online data analysis at the ESRF bioSAXS beamline, BM29. J Appl Crystallogr. 2016;49(1):203–12.

66. Ashiotis G, Deschildre A, Nawaz Z, Wright JP, Karkoulis D, Picca FE, Kieffer J. The fast azimuthal integration Python library: *pyFAI*. J Appl Crystallogr. 2015;48(2):510–9.

6

The observation of evolutionary interaction pattern pairs in membrane proteins

Steffen Grunert[*][†] and Dirk Labudde[†]

Abstract

Background: Over the last two decades, many approaches have been developed in bioinformatics that aim at one of the most promising, yet unsolved problems in modern life sciences - prediction of structural features of a protein. Such tasks addressed to transmembrane protein structures provide valuable knowledge about their three-dimensional structure. For this reason, the analysis of membrane proteins is essential in genomic and proteomic-wide investigations. Thus, many *in-silico* approaches have been utilized extensively to gain crucial advances in understanding membrane protein structures and functions.

Results: It turned out that amino acid covariation within interacting sequence parts, extracted from a evolutionary sequence record of α-helical membrane proteins, can be used for structure prediction. In a recent study we discussed the significance of short membrane sequence motifs widely present in nature that act as stabilizing 'building blocks' during protein folding and in retaining the three-dimensional fold. In this work, we used motif data to define evolutionary interaction pattern pairs. These were obtained from different pattern alignments and were used to evaluate which coupling mechanisms the evolution provides. It can be shown that short interaction patterns of homologous sequence records are membrane protein family-specific signatures. These signatures can provide valuable information for structure prediction and protein classification. The results indicate a good agreement with recent studies.

Conclusions: Generally, it can be shown how the evolution contributes to realize covariation within discriminative interaction patterns to maintain structure and function. This points to their general importance for α-helical membrane protein structure formation and interaction mediation. In the process, no fundamentally energetic approaches of previous published works are considered. The low-cost rapid computational methods postulated in this work provides valuable information to classify unknown α-helical transmembrane proteins and to determine their structural similarity.

Keywords: Membrane proteins, Motif, Evolutionary interaction pattern pair, EIPP, Structural similarity, Protein family affiliation

Background

Membrane proteins shape a special kind of proteins. They feature vital necessary functions in cellular processes of organisms. Fore more essential biological functions such as: photosynthesis, transport of ions and small molecules, signal transduction and light harvesting this are examples of processes which are realised by membrane proteins. The analysis of membrane proteins was shown to be an important part in the comprehension of complex biological processes in the context of proteomics and genomics [1]. Generally, membrane proteins are poorly soluble and cover a wide intra-cellular concentration range. The inaccessibility of many proteomics methods makes membrane protein analyses still an experimentally challenging field [2]. Hence, the number of known three-dimensional structures is relatively small, with 437 non-redundant membrane protein chains currently available [3-5]. Consequently, there is a necessity for approaches that allow to predict structural and functional features of unknown membrane proteins. A variety of methods have been developed to predict structural features from sequence, such as α-helical membrane-spanning helices and extra/intra-cellular domains (i.e. TMHMM [6,7], PHDhtm [8], MEMSAT3 [9]) as well as membrane-spanning β-strands of transmembrane β-barrel proteins

*Correspondence: sgrunert@hs-mittweida.de
[†]Equal contributors
Hochschule Mittweida, University of Applied Sciences, Technikumplatz 17, 09648 Mittweida, Germany

(i.e. BOCTOPUS [10]). Furthermore, a major step toward *ab initio* protein structure prediction has been made through the development of new techniques for mapping energetic interactions in proteins. Here, Lockless and Ranganathan demonstrated [11] a statistical energy function as a good indicator of thermodynamic coupling in proteins. They also showed how sets of interacting residues form connected pathways in the protein fold. An existing basis for efficient energy conduction within proteins has been shown. They called their approach statistical coupling analysis (SCA) that provides the basis for further works in this area. Other approaches dealing in turn with key information to predict protein structures, which can be obtained from homologous sequences and their evolutionary variation because: "The diversity of biologic phenomena arises from the complexity and specificity of biomolecular interactions. Nucleic acid and protein polymers encode and express biologic information through the specific sequence of polymer units (residues). The sequences and corresponding molecular structures are under selective constraints in evolution [12]".

Due to the growth of available protein sequences, many statistical methods have been developed, to compute protein three-dimensional structures from evolutionary context. Diverse contributions were involved to develop sophisticated methods to identify additional key residues that are involved in protein structure and function, especially residues that are strongly conserved within each subfamily but differ between subfamilies [13]. Previous works of Marks et al. [14,15] indicate that rich evolutionary information from genomic sequences can be efficiently mined, leading to information on evolutionary couplings between residues. Morcos et al. [16] have used information about strong constraints on their sequence variability, induced by the three-dimensional structures of homologous proteins. They developed an efficient direct-coupling analysis (DCA) [17,18] implementation to evaluate the accuracy of contact prediction for a large number of protein domains. Later on, Hopf et al. [19] presented a maximum entropy approach to infer evolutionary covariation in pairs of sequence positions of a given protein family. Generated atom models from derived pairwise distance constraints were finally used to predict the full spectrum of protein structures, functional interactions and evolutionary dynamics of unknown three-dimensional structures for 11 transmembrane proteins. A novel approach by Kamisetty et al. [20] utilizes an approximation method to obtain more accurate contact predictions for estimating residue-residue contacts in protein structures. Compared to previous methods, higher accuracy was achieved by integrating structural context and sequence co-evolution information. Hence, their method allow more accurate contact predictions from fewer homologous sequences.

Furthermore, in genome-wide membrane protein sequence analyses, numerous short conserved sequence motifs were identified [21]. These motifs support the understanding of the features that are important for establishing stability and functionality of the folded membrane protein in the membrane environment. Additionally, as addressed in [22], the analysis of sequence motifs in proteins with similar function or structure might help to identify essential functional sites and locations, which contribute to structural stability. Thus, sequence motif analysis can be helpful for numerous applications, e.g. the investigation of mutant proteins, the understanding of protein dynamics and potential effects of mutagens. During evolutionary progress the spatial structure of proteins is generally stronger conserved than the sequential amino acid composition. Adapted to the field of sequence motif analysis, structure-forming motifs point to their general importance in α-helical membrane protein structure formation and interaction mediation [1]. Moreover, hubs and consecutive motifs with high occurrence in certain membrane protein families can be classified as important for family-specific functional characteristics [23]. Finally, the combination of interaction information and sequence motifs with evolutionary variation can be used for three-dimensional structure prediction.

In our work we obtained key information from homologous sequences to separate and predict membrane protein structures in the context of interacting patterns and their evolutionary variation. Patterns as motif representatives are investigated regarding evolutionary covariation. Interaction information contributes to detect interacting patterns with evolutionary background. Here, we report the development of an algorithm that is involved in the extraction of interaction pattern pairs that are evolutionarily influenced. These were used for the investigation of different mutation types, which are provided by evolution to maintain structure and function. Agreeing with previous works we can state that the evolution provides basic building blocks to maintain structure and function. Related to this, family-specific interaction pattern information were used to predict unknown α-helical transmembrane protein structures. We have also tested our method at an already predicted structure of previous work of Hopf et al. [19]. Finally, our approach is not based on recently developed methods like SCA or DCA, but the processing of interaction and secondary structure data for predicting rich helical structure parts leads to the attachment to previous works.

Methods

In the first step, known crystal structures of α-helical membrane proteins were investigated. Structural information were derived from PDBTM [24]. Currently available known α-helical membrane proteins were assigned to

Table 1 The analysed dataset

Protein Family[a]	PDBTM[b]	TMPad[c]	Contacts[d]
PF01036 (Bac_rhodopsin)[e]	130	102	6417
PF00230 (MIP)[f]	44	40	2814

[a] Analysed proteins to corresponding protein family. [b] Number of known structures available from PDBTM [24]. [c] Number of proteins with interaction information available from TMPad [29]. [d] Number of helix-helix contact information available for PDBTM assigned TMPad proteins. [e] Bacteriorhodopsin-like proteins. [f] Major Intrinsic Proteins.

their protein families [25] using Pfam mappings. We have tested our method at two selected families with homologous sequences that contribute to generate coupling statistics (Table 1).

Evolutionary co-variations from pattern alignments (PAs)

Hopf et al. hypothesized and confirmed in their work [19], that the evolution conserves interactions between residues that are important to maintain structure and function. This is done by constraining the sets of mutations that are accepted at interacting sites. To find these constraint interactions within different sequence patterns, we generated PAs using a novel algorithm that detects evolutionary covariation. Aspects of this algorithm are given in this section. However, before elucidating the application of our algorithm, we want to give a short summary on the general definition of short sequence motifs, as well as the aspects of motif detection and information extraction. Consequently, the next steps are involved in motif extraction out of α-helical structures. Like described in previous work of [26] a motif can be written in a generalized, regular expression-like form of XYn, where X and Y correspond to amino acids separated by $n-1$ highly variable positions. For the general purpose, short sequence motifs have been extracted that contribute to build the α-helical structure in the transmembrane environment. Here, a naive text search algorithm was applied for motif extraction. More precisely, the algorithm mainly utilises a sliding sequence frame strategy. Beginning from the start position of the sequence, different window sizes are used to extract the underlying subsequence. Each subsequence is transcribed into its regular expression XYn. More specifically, at each sequence position i and $i + n$ the algorithm returns the N-terminal residue X and the C-terminal residue Y. Note, that X and Y denote any of the 20 canonical amino acids. Redundant duplications were removed. It is known that amino acids are positioned with an average of 3.6 residues per turn in TM-helices [27] and it is also known that motifs with different length are favoured for TM-helix packing [1,28]. Based on this, the number of $n - 1$ variable positions ranges within $2 \leq (n - 1) \leq max$, where max is the maximum helix length of a protein family. Along, for a given protein each motif representative pattern was

searched in all helices. If a pattern was found, the initial pattern (IP) is stored. Here, the IP represents the pattern according to which all others are aligned. To detect evolutionary covariation and to minimize the statistical noise, we have aligned patterns from other structures of the same protein-family. We ensured that these patterns, called subwords (SWs), have up to one mutated variable position and a length of $n_{SW} \leq n_{IP}$. To avoid redundancy and to minimize computational processing time, already aligned SWs were ignored. Each PA returns possible evolutionary covariation at the variable position of the aligned IP. A representative PA example is shown in Figure 1/Pattern Alignment.

Specific evolutionary interaction pattern pairs (EIPPs)

To close the information gap when individual patterns interact with each other, we have decided to derive interaction data information from a known database. Generally, such databases allow a rapid and simple access to the required data. Helix-helix interaction information were derived from TMPad, the TransMembrane Protein Helix-Packing Database [29]. TMPad is an integrated repository of experimentally determined structural folds derived from helix-helix interactions in α-helical membrane proteins. Here, geometric descriptors of helix-helix interactions, topology, lipid accessibility, ligand and binding sites information are provided by TMPad. Currently, 1,107 protein entries, 4,061 protein chains and 17,413 helix-helix interactions are available. Contact information were enriched by Contacts of Structural Units (CSU) [30] derived from Weizmann Institute of Science, which provides different experimental data after the analysis of inter-atomic contacts of structural units of the protein data base (PDB) [31] entries. Now it is able to create a context between structure and helix-helix interaction information adapted to representative patterns of discriminative sequence motifs. After successfully integration of the TMPad-information to find EIPPs, helix-helix interactions were registered. An Interaction pattern pair was extracted when a contact is given only at a variable pattern position. We have ensured that at least one pattern of a given pair has mutations at the variable position. To obtain a statistical overview about the most occurring interacting motif pairs, the corresponding occurrence was recorded for each $XYn - XYn$. EIPPs are specific within the investigated membrane protein family. Such pairs can be considered as family-specific signatures due to their responsibility to build and stabilize the proteins structure by taking into account of the evolutionary space. Each EIPP was labelled with the corresponding protein in which the EIPP was found. Pattern interaction networks were created for final visualization and to support the understanding, how the evolution maintains attractive interaction within an EIPP. Furthermore, the existence of family-specific EIPPs

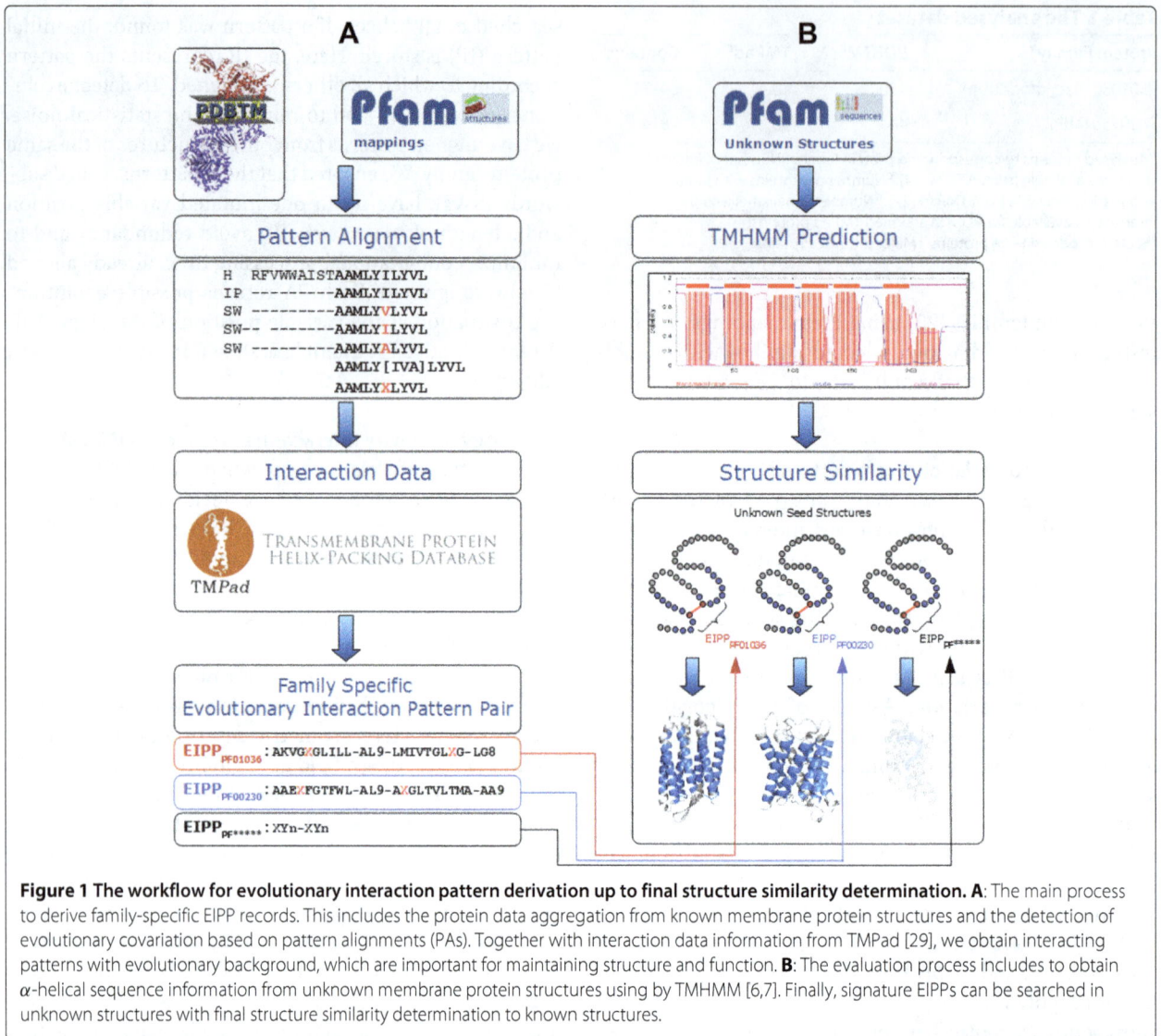

Figure 1 The workflow for evolutionary interaction pattern derivation up to final structure similarity determination. A: The main process to derive family-specific EIPP records. This includes the protein data aggregation from known membrane protein structures and the detection of evolutionary covariation based on pattern alignments (PAs). Together with interaction data information from TMPad [29], we obtain interacting patterns with evolutionary background, which are important for maintaining structure and function. **B**: The evaluation process includes to obtain α-helical sequence information from unknown membrane protein structures using by TMHMM [6,7]. Finally, signature EIPPs can be searched in unknown structures with final structure similarity determination to known structures.

was evaluated by a protein separation task. An evaluation dataset of the investigated Pfam-families PF01036 and PF00230 was derived (Table 2). Redundancy reduction was performed by assuring the family-specific number of transmembrane helices. Transmembrane helical information were obtained using TMHMM Server v. 2.0 [6,7]. Basically, TMHMM performs a prediction of intra/extracellular regions and integral membrane helices based on sequence. Beside per-residue predictions TMHMM also lists underlying per-residue assignment probabilities as an indicator of prediction uncertainty. TMHMM results do not always exhibit the expected typical number of 7 TM-helices (Bacteriorhodopsin-like protein) and 6 TM-helices (Major Intrinsic Proteins) in the evaluation dataset, which leads to the reduction of the evaluation dataset. Eventually, not all sequences of the evaluation dataset were included in the process. Known structure representatives were also removed.

For the further step, protein clusters consisting of all family representative unknown structures were merged, to form a cloud and subsequently sampled. For each cloud member, family-specific EIPPs were applied on TMHMM predicted helices disregarded by mutations and under consideration of different degrees of freedom. Here, a threshold determines the number of approved variable positions within EIPPs. Matches were registered and marked in the respective helices and sequence

Table 2 The evaluation dataset

Protein Family[a]	Proteins[b]	Helices[c]
PF01036 (Bac_rhodopsin)[d]	438	3066
PF00230 (MIP)[e]	6420	38520

This dataset consists of protein family-specific representatives with unknown structures. [a]Analysed proteins to corresponding protein family. [b]Number of proteins available from evaluation dataset. [c]Number of investigated membrane helices. [d]Bacteriorhodopsin-like proteins. [e]Major Intrinsic Proteins.

similarity of the incurred interacting ranges compared to known structures was calculated. In addition, the family-specificity of EIPPs leads to family-specific classifiers and thus to the ability to detect an family affiliation of unknown structures that contain mutation affected homologous sequence parts. Here, it is important to mention that this task is not aimed at developing a new and better approach to classify proteins like Pfam does it with their Hidden Markov models. We will only demonstrate the specificity of mutation affected interacting sequence parts of a given protein family.

Results and discussion

EIPPs were derived from known crystal structures of different membrane protein families. PAs provide evolutionarily induced variable positions within EIPPs. Like previously described, evolutionary covariation have been detected in EIPPs. In some cases, aligned SWs with up to

one mutated position are responsible for multiple covariation within an EIPP member. One could have given the evolution more leeway and aligned SWs could have been designed with more than one mutated position, because it is a fact that the evolution allows more variance at the variable pattern positions to maintain structure and function. Our results show that the evolution provides basic building blocks, which are significant for the transmembrane environment like described in previous works [1,21,23]. The evolution itself determines the sequence variability and thus the variance of the variable pattern positions. If we consider each EIPP member as a basic building block we obtain a global view for this interacting sequence part in relation to a single residue. Thereby, we bypass the analysis of each residue to obtain structurally interacting units. The visualization of generated pattern interaction networks (Figure 2) supports the understanding, which pattern pairs of different length are generally

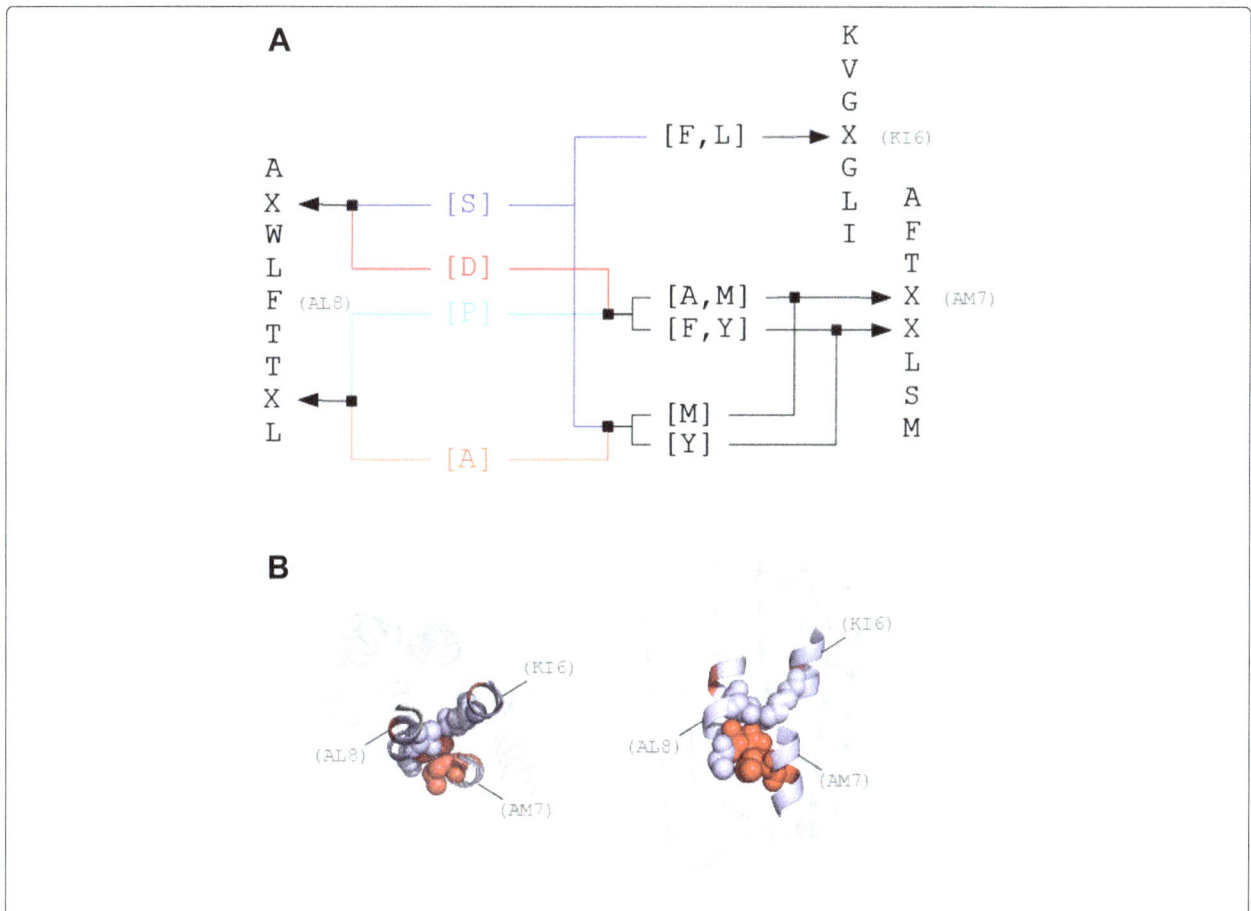

Figure 2 Examples of spatially interacting sub-sequences with respect to their corresponding pattern interaction network. A: More specifically, KI6 and AM7 representative patterns (right) interact with AL8 (left). All patterns have mutations. Mutational positions are marked with X. Possible amino acid replacements for AL8 representative (left) are coloured and arrows point to the respective X position. Black arrows point to the respective X position of the KI6 and AM7 representatives. With this interaction network we can track, which substitutions occur during the evolution, without influence on the interaction. **B**: The top and side view illustrations of Bacteriorhodopsin trimer (PDB-Id:1brr) are indicating where the interacting patterns are present in the helices of chain A. Generally, spheres illustrate residue-residue contacts. Red coloured spheres illustrate variable positions (X).

involved in spatial interaction by taking into account the evolutionary background. We obtain important information about variable pattern positions that are subjected to a mutation without influencing attractive pattern interactions. The application of interaction tree schemes can lead to better indicators in laboratory mutagen investigations. More specifically, this supports the investigation of mutational variants causing different diseases like e.g. Nephrogenic diabetes insipidus.

Incidentally, for reasons of incomplete TMPad information not all position specific mutations are an integral part of our EIPPs. Only EIPP related mutations were collected if any contact could be detected from TMPad. Regarding this tree information, different known

structures of PF01036 were analysed for EIPPs. The investigation of Rhodopsin-like proteins represents a major subject of research. Here different structure-function studies were performed [32,33]. Further, the investigation of active core fluctuations, the folding core and kinetics and the involved residues have been treated extensively in previous studies [34-36]. In this work, Bacteriorhodopsin-like protein structures were used to evaluate the derived EIPPs. Representatives of the statistically most interacting motifs were searched. Furthermore, long motif XYn ($n = 9$) representative patterns show a greater tendency to interact more frequently than short ones, because of the larger number of possible residue-residue interaction combinations. The examples given in Figure 3 show,

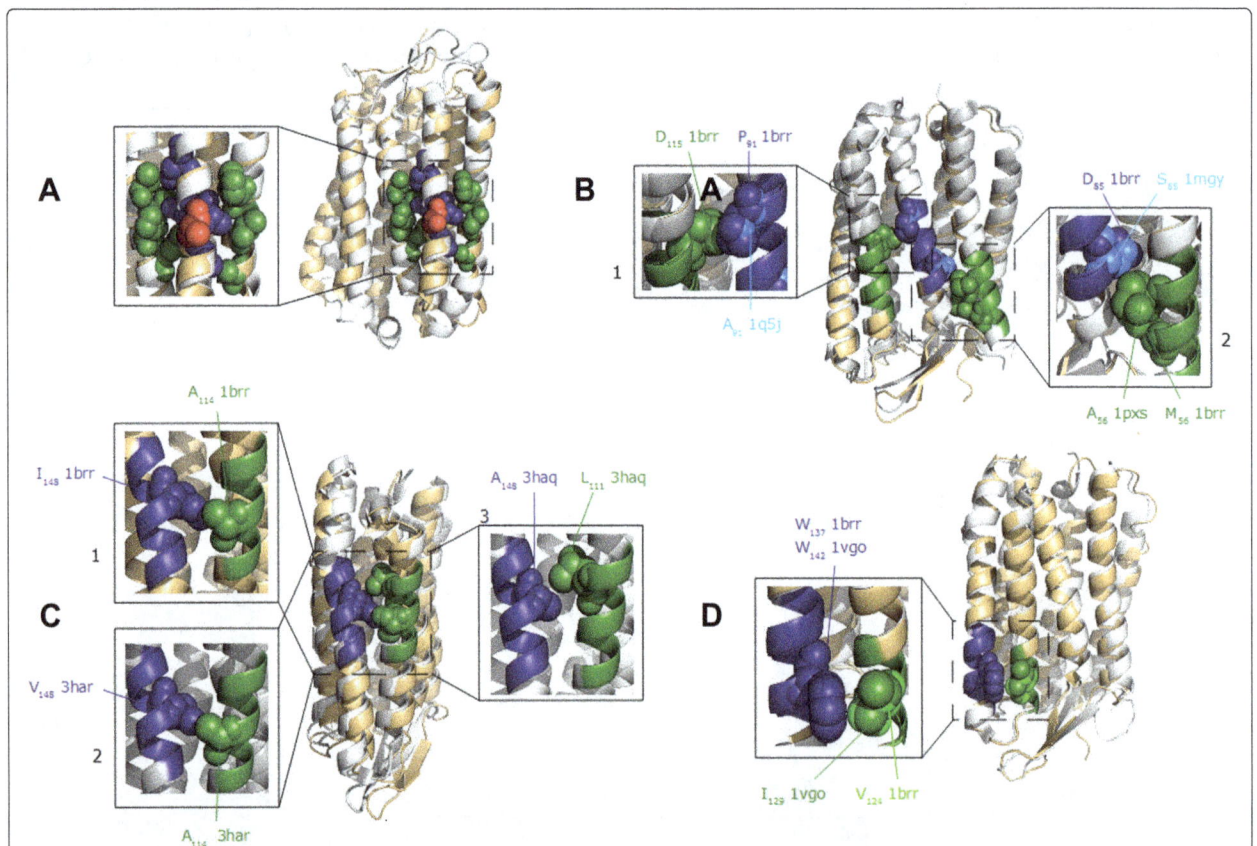

Figure 3 Mutation interaction types. Four mutation interaction types are present. Labelled spheres indicate which amino acid at specified position is present related to PDB-Id. **A**: Simple evolutionary replacements (red) around the blue and green interacting residue spheres. **B**: Interacting AL9 motifs (blue and green) with evolutionary residue substitution without loss of interaction. Mutations at one or at both interaction partner are possible. **B1**: Asp$_{115}$ at the second position of AL9-motif pattern representative AD$_{115}$GlMIGTL interacts with Ala$_{91}$ or Pro$_{91}$ of AL9-motif pattern representative A[SD]$_{85}$WLFTT[AP]$_{91}$LL . This is made possible by the same orientation of Ala$_{91}$ and Pro$_{91}$ towards its interacting counterpart. **B2**: Analogously, fourth position of AL9-motif pattern representative AFT[MA]$_{56}$YLSMLL is designed variable with Ala$_{56}$ or Met$_{56}$ and interacts with Asp$_{85}$ or Ser$_{85}$ reason by same orientation in space. **C**: If contact information will be lost by mutation, the responsible destabilizing amino acid will be compensated by another position, in order to maintain attractive residue pair interaction [16]. **C1/C2**: Ile$_{148}$ and Val$_{148}$ at fifth position of AL9-motif pattern representative AMLY[VIA]$_{148}$LYVL (blue) are able to interact with Ala$_{114}$ at sixth position of LI8-motif representative LAL$_{111}$VGA$_{114}$DGI (green). **C3**: Mutation with Ala$_{148}$ causes that contact will be lost reason by to short distance to Ala$_{114}$ counterpart. Here, Leu$_{111}$ at third position of LI8-motif compensates the destabilizing amino acid. Evolution aims at maintaining stabilizing interactions. **D**: Trp$_{137/142}$ is an evolutionary coupling residue which interacts with Ile$_{129}$ or Val$_{124}$ by full changeable residue environment around Trp$_{137/142}$. This means that the evolutionary degree of freedom allows it to change all variable positions of an interacting pattern by keeping the conserved interaction residue.

how different EIPPs comprise structural tasks and spatial interactions. Specifically, the evolution presents how EIPPs contribute to emerge different evolutionary mutation types. These types describe the sequence variability on a closer way, which has no significant influence on the protein structure and function.

These are described in more detail below:

1. Simple residue replacements that are not involved in any interaction. Tend to be an important block within an EIPP member, thus the structure can be folded without any task to build important spatial contacts (Figure 3A).

2. Contact specific mutations within evolutionary patterns. An amino acid with the responsibility to build a spatial contact to another helix will be replaced by an amino acid without modifications of the residue-residue interaction network. This can only be realized using amino acids with similar properties of the replaced residues. Here, the length and the spatial orientation play a major role to be a suitable replacement. As injunctive contact example shown in Figure 3B1: The replacement of Pro_{91} (PDB-Id: 1brr) with Ala_{91} (PDB-Id: 1q5j) within the AL9-motif representative $A[DS]_{85}WLFTT[PA]_{91}LL$ has no influence to maintain the injunctive contact to their counterpart D_{115} within the AL9-motif representative $AD_{115}GIMIGTGL$. The extended contact (Figure 3B2) between helix-helix interaction at positions 85 and 56 shows how evolutionary sequence variability contributes in such a manner that both interaction residues can be replaced by another without loosing the family-specific important contact. Here, Asp_{85} (PDB-Id: 1brr) is replaced by Ser_{85} (PDB-Id: 1mgy) within the AL9-motif representative $A[DS]_{85}WLFTT[PA]_{91}LL$. It has no influence to maintain the injunctive contacts to their counterparts Met_{56} (PDB-Id: 1brr) and Ala_{56} (PDB-Id: 1pxs) within the AL9-motif representative $AFT[MA]_{56}YLSMLL$.

3. Morcos et al. [16] explained the simplicity between evolutionary substitutions and residue-residue contacts. "If two residues of a protein or a pair of interacting proteins form a contact, a destabilizing amino acid substitution at one position is expected to be compensated by a substitution of the other position over the evolutionary time-scale, in order for the residue pair to maintain attractive interaction". For in-depth discussions and evaluations see [16]. These results can be seen in our frequently interacting motif pair AL8-LI8. shown in Figure 3C. C1/C2: Here, the fifth variable position of AL9-motif representative $AAMLY[VAI]_{148}LYVL$. Val_{148} and Ile_{148} have a coupling with Ala_{114} of the LI8 representative $LAL_{111}VGA_{114}DGI$. C3: Mutation at position 148 with tiny Ala_{148} leads to the loss of contact to Ala_{114}. Here, Leu_{111} compensates the loss of contact by interacting with tiny Ala_{148}.

4. A fundamental change of variable motif positions right down to contact specific position. Thereby, common amino acids take place to cope the complete change. Such amino acids are e.g. tryptophane (Trp) with the important role in membrane proteins as described in previous work [37].

In the following, a summary on how to use EIPP data for structure prediction is given. As a proof of concept, 116,810 EIPPs (PF01036) and 63,283 EIPPs (PF00230) (Table 3) were extracted from known structures of the corresponding protein families (see Additional file 1). Here, the number of EIPPs is given by interacting patterns with different lengths. These include interaction members with permanently assigned positions and members that are

Table 3 Number of EIPPs derived from 130 Bacteriorhodopsin-like and 44 Major Intrinsic Protein structures

Variable positions	EIPPs	EIPPs
	PF01036	PF00230
2	5754	4988
3	7656	5930
4	8784	6326
5	9864	6398
6	10382	6594
7	10302	6087
8	10529	5470
9	9692	4936
10	8727	4196
11	7797	3428
12	6545	2748
13	5538	2129
14	4569	1533
15	3498	1031
16	2530	645
17	1867	375
18	1278	218
19	801	131
20	437	68
21	187	35
22	64	12
23	8	4
24	1	1
Σ	116810	63283

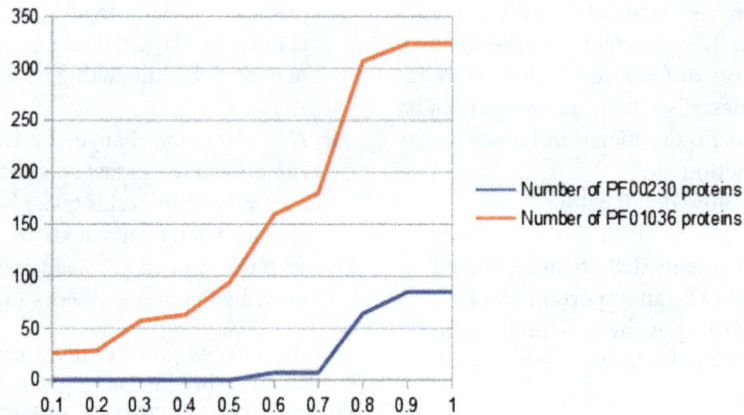

Figure 4 Classification result for Bacteriorhodopsin-like (PF01036) representative unknown structures. 372 of 438 representative proteins have been correctly assigned to PF01036. The greater the evolutionary degree of freedom (x-axis), the more variability occurs within PF01036-EIPPs. This leads to more classified proteins. On the other side, EIPPs become more unspecific for a membrane protein family which leads to wrong classified. In this case, PF01036-EIPPs were covered in 85 PF00230-proteins.

evolutionarily influenced. The rediscovery of EIPPs in unknown membrane protein structures of different families leads to the separation and finally to the determination of a membrane protein family affiliation. However, this is influenced by the evolutionary degree of freedom within EIPPs. With increasing variability of the variable position and under considering of the number of amino acids of a given interacting pattern, EIPPs can be recovered in other membrane protein families. That means, the greater the number of amino acids of a EIPP and the lower the evolutionary degree of freedom, the more specific is a EIPP for a membrane protein family. This has a significant impact on correctly classified proteins. In this context, the recovery of EIPPs in unknown membrane protein structures leads to the following classification results as shown in Figures 4 and 5.

Here, 372 of 438 (PF01036) and 5,993 of 6,420 (PF00230) representative proteins have been correctly assigned to their families under the consideration of the evolutionary degree of freedom. Caused by the increase of variable positions, EIPPs became more non-specific for a membrane protein family and more proteins are incorrectly assigned. Misclassified indicate no EIPPs in the investigated membrane helices and thus no sequence similarity due to heterologous sequence parts. The reason is the restriction to allow only single mutations within aligned SWs. This leads to the fact that not all positions are considered by our algorithm. Sequence homology causes generated EIPPs to be a part of current unknown structures of the investigated protein family. Generally, our classification result shows that unknown structures can be assigned to a membrane protein family

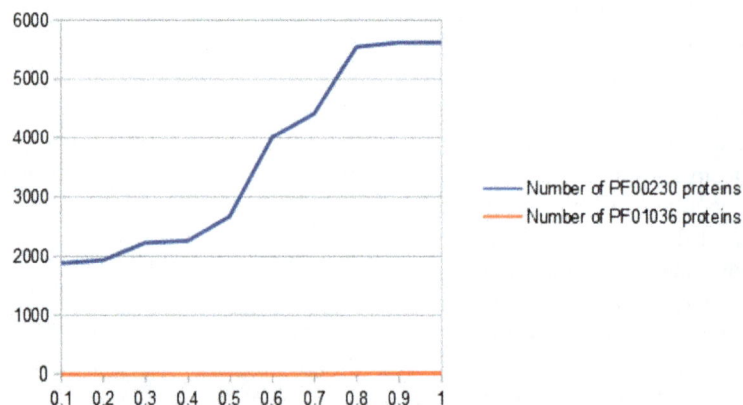

Figure 5 Classification result for major intrinsic protein (PF00230) representative unknown structures. 5,993 of 6,420 representative proteins have been correctly assigned to PF00230. The greater the evolutionary degree of freedom (x-axis), the more variability occurs within PF00230-EIPPs. This leads to more classified proteins. On the other side, EIPPs become more unspecific for a membrane protein family which leads to wrong classified. In this case, PF00230-EIPPs were covered in 14 PF01036-proteins.

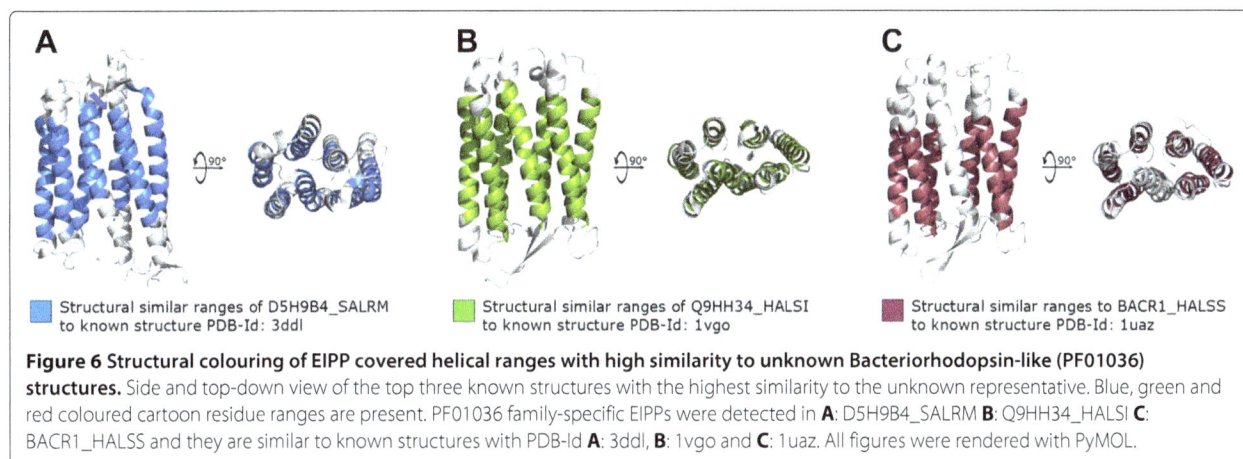

Figure 6 Structural colouring of EIPP covered helical ranges with high similarity to unknown Bacteriorhodopsin-like (PF01036) structures. Side and top-down view of the top three known structures with the highest similarity to the unknown representative. Blue, green and red coloured cartoon residue ranges are present. PF01036 family-specific EIPPs were detected in A: D5H9B4_SALRM B: Q9HH34_HALSI C: BACR1_HALSS and they are similar to known structures with PDB-Id A: 3ddl, B: 1vgo and C: 1uaz. All figures were rendered with PyMOL.

by our described method. Furthermore, registered EIPPs were marked and compared to known structures. As shown in Figures 6 and 7, the three representatives are present. These have a high structural similarity to known protein structures of the families (PF01036, PF00230). D5H9B4_SALRM, Q9HH34_HALSI and BACR1_HALSS are the top three representatives, where the most PF01036-EIPPs have been detected in TM-helices. G7RII8_ECOC1, AQP5_MOUSE and PIP27_MAIZE are three freely selected PF00230-structures with high similarity. Further similarity results are given in the attached Additional file 2.

The appropriate statistic is present in Tables 4 and 5. Considered as a whole, predicted helical ranges and finally the whole unknown structure can be compared structurally to similar known structures. For D5H9B4_SALRM this means, that 91.2% of the helical ranges be covered by PF01036-EIPPs. Followed by Q9HH34_HALSI with 90.5% and BACR1_HALSS with 85.2% structural similarity. Analogously, G7RII8_ECOC1 with 90.2%, AQP5_MOUSE with 85.2% and PIP27_MAIZE with 83.8% are covered by PF00230-EIPPs. A further evaluation has been performed. Hopf et al. have predicted [19] the unknown structures of ADR1_HUMAN with structural similarity to Bacteriorhodopsin (Pfam: PF01036, PDB-Id: 3hao) and LIVH_ECOLI with structural similarity to permease protein BtuC (Pfam: PF01032, PF00005, PDB-Id: 1l7v) in their work. We have used both structures and considered these as unknown structures. Transmembrane α-helical information predicted by TMHMM were applied to the classification task. ADR1_HUMAN could successfully be assigned to PF01036 and LIVH_ECOLI to PF00005. For ADR1_HUMAN this means that six of seven helices were structurally predicted with 100% similarity. The helical range of helix number 6 (H6) was covered by EIPPs with 86.4%. Besides, helical ranges of LIVH_ECOLI have high similarity to known structures of PF00005 (H1: 72.7%, H2: 50.0%, H3: 100%, H4: 90.9%, H5: 72.7%, H6: 94.1%, H7: 100%). This confirms the structure prediction result of Hopf et al. addressed to the structural similarity of ADR1_HUMAN to Bacteriorhodopsin and LIVH_ECOLI to permease protein BtuC.

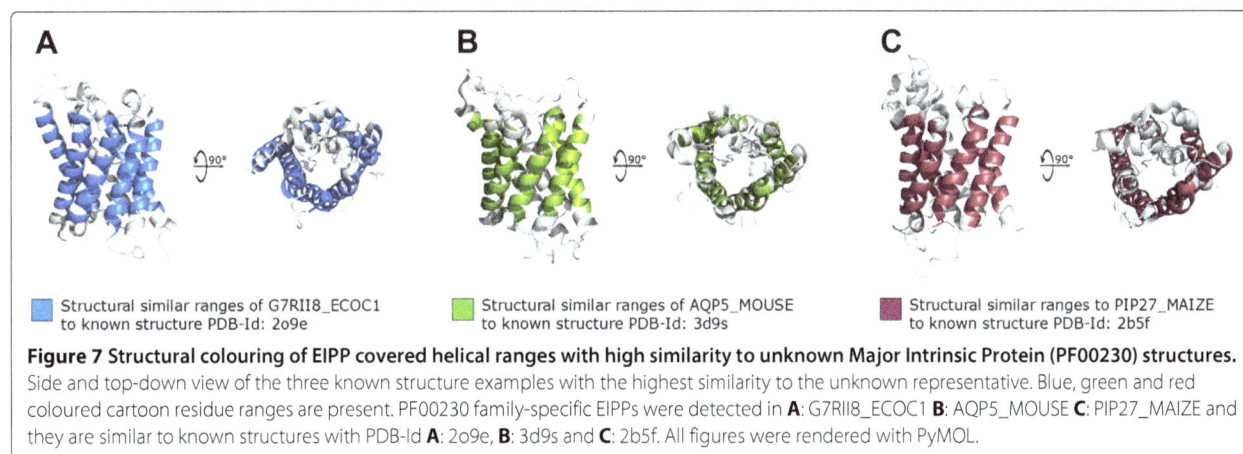

Figure 7 Structural colouring of EIPP covered helical ranges with high similarity to unknown Major Intrinsic Protein (PF00230) structures. Side and top-down view of the three known structure examples with the highest similarity to the unknown representative. Blue, green and red coloured cartoon residue ranges are present. PF00230 family-specific EIPPs were detected in A: G7RII8_ECOC1 B: AQP5_MOUSE C: PIP27_MAIZE and they are similar to known structures with PDB-Id A: 2o9e, B: 3d9s and C: 2b5f. All figures were rendered with PyMOL.

Table 4 Structurally similar helical ranges of unknown PF01036-structures

	D5H9B4_SALRM		Q9HH34_HALSI			BACR1_HALSS		
Helix	Amino acids	Similarity	Helix	Amino acids	Similarity	Helix	Amino acids	Similarity
1	23	95.6%	1	23	82.6%	1	23	73.9%
2	20	95%	2	23	91.3%	2	23	78.2%
3	18	88%	3	23	95.6%	3	23	0%
4	18	100%	4	23	86.9%	4	20	90%
5	23	91.3%	5	20	100%	5	20	70%
6	23	95.6%	6	23	91.3%	6	23	69.5%
7	23	73.9%	7	23	86.9%	7	23	82.6%

For each Bacteriorhodopsin-like protein, the number of amino acids of individual TMHMM predicted helices are given. Similarity values describe consistent helical ranges, which are covered by EIPPs.

Table 5 Structurally similar helical ranges of unknown PF00230-structures

	G7RII8_ECOC1		AQP5_MOUSE			PIP27_MAIZE		
Helix	Amino acids	Similarity	Helix	Amino acids	Similarity	Helix	Amino acids	Similarity
1	23	100%	1	23	73.9%	1	23	86.9%
2	23	86.9%	2	23	73.9%	2	23	82.6%
3	23	100%	3	23	73.9%	3	23	78.2%
4	23	78.2%	4	18	94.4%	4	20	100%
5	23	78.2%	5	23	73.9%	5	23	78.2%
6	18	100%	6	18	94.4%	6	18	77.7%

For each Major Intrinsic Protein, the number amino acids of individual TMHMM predicted helices are given. Similarity values describe consistent helical ranges, which are covered by EIPPs.

Moving forward, we discuss the structural similarity results. EIPPs as interacting structural blocks are specific within a membrane protein family and for the folding of each TM-helix within a membrane protein. To recover EIPPs on a unknown structure sequence, EIPPs must occur in the helix that reflects the known structure. In this case, we had to fall back on TMHMM, a known secondary prediction tool. This dependence means that the discussed approach does not perform better than the best secondary prediction tool. On the other side, EIPPs provide TM-helical information from known structures. This leads to the possibility chance to refine secondary structure prediction tools and can be discussed in further works. Finally, our method can be used to improve sequence-based methods for classification and protein homology detection.

Conclusion

In this work, we have demonstrated an approach for extracting short, spatially interacting amino acid sub-sequences - so called evolutionary interaction pattern pairs (EIPPs) - from known crystal structures of α-helical membrane protein families and underlying sequence data of protein family members. Finally, it is outlined how EIPPs can be utilized to predict protein structure. Here, covariation within motif representative homologous sequence patterns have been detected using a pattern alignment algorithm. In combination with interaction information from TMPad [29], EIPPs were obtained and employed to generate interaction trees. Thereby, we are able to show how different interacting patterns differ evolutionarily. Moreover, they have been evaluated using known structures of Bacteriorhodopsin-like proteins and discussed in detail. Here, different mutation types emerge to create an evolutionary instrument to realise sequence variability within a protein family. Furthermore, EIPPs have been used to generate family-specific classifiers. Representative proteins with unknown secondary structure have been used to predict α-helical sequence information using TMHMM [6,7]. Finally, family-specific protein separation has been performed and the structural similarity to known structures of the related protein family has been calculated. Addressed to structure similarity, our method describes how different interacting patterns with evolutionary background contribute to register a protein family affiliation. We are also able to determine the most similar unknown to known structures of a given α-helical membrane protein family. We also produced a good agreement with recently published studies that the evolution provides basic building and interacting blocks for maintaining structure and function. Due to sequence homology such blocks are repeated and we have proven structural conservation. The contemplation of a sequence from the perspective of such blocks facilitates the understanding how membrane protein structures of a family are constructed. Last but not least, low-cost rapid computational methods can be developed to support, extend or refine classification and prediction methods for membrane proteins.

Additional files

Additional file 1: EIPP data. Includes derived EIPP information from families (PF00230, PF01036) with tab separated values. Can be viewed with a simple text editor. Each line consists of 7 columns: source pattern, source RegEx, destination pattern, destination RegEx, source helix, destination helix, corresponding PDB-Ids.

Additional file 2: Similarity results. Includes two text files for each protein family (PF00230, PF01036). Each file shows prediction results in the context of the evolutionary degree of freedom (EDF). For each protein, original and predicted helical range information are given. The end of a file shows the prediction winners.

Abbreviations
TM: Transmembrane; SCA: Statistical coupling analysis; DCA: Direct coupling analysis; PDBTM: Protein data bank of transmembrane proteins; Pfam: The protein families database; PA(s): Pattern alignment(s); IP(s): Initial pattern(s); SW(s): Sub-word(s); TMPad: Transmembrane protein helix-packing database; EIPP(s): Evolutionary interaction pattern pair(s); CSU: Contacts of structural units; PDB: Protein data bank; SG: Steffen Grunert; DL: Dirk Labudde.

Competing interests
The authors declare that they have no competing interests.

Authors' contributions
SG and DL participated in the design of this study. SG designed all methods and performed the implementation. SG evaluated the results. DL provided valuable consultation on structural biology and procedural steps. SG and DL wrote the manuscript. Both authors read and approved the final manuscript.

Acknowledgements
The authors would like to thank the Free State of Saxony and the European Social Fond (ESF) for financial support. In addition, SG and DL thank to Rico Beier, Sebastian Bittrich and Florian Kaiser, which have revised the manuscript.

References
1. Grunert S, Florian H, Dirk L. Structure topology prediction of discriminative sequence motifs in membrane proteins with domains of unknown functions. Struct Biol. 2013;2013:10.
2. Sadowski PG, Groen AJ, Dupree P, Lilley KS. Sub-cellular localization of membrane proteins. Proteomics. 2008;8(19):3991–4011. doi:10.1002/pmic.200800217.
3. Bowie JU. Solving the membrane protein folding problem. Nature. 2005;438(7068):581–9. doi:10.1038/nature04395.
4. Tusnady GE, Dosztanyi Z, Simon I. Transmembrane proteins in the protein data bank: identification and classification. Bioinformatics. 2004;20(17):2964–72. doi:10.1093/bioinformatics/bth340.
5. Tusnady GE, Dosztanyi Z, Simon I. Pdbtm: selection and membrane localization of transmembrane proteins in the protein data bank. Nucleic Acids Res. 2005;33(Database issue):275–8. doi:10.1093/nar/gki002.
6. Krogh A, Larsson B, von Heijne G, Sonnhammer EL. Predicting transmembrane protein topology with a hidden Markov model: application to complete genomes. J Mol Biol. 2001;305(3):567–80. doi:10.1006/jmbi.2000.4315.
7. Moeller A, Croning B, Apweiler C. Evaluation of methods for the prediction of membrane spanning regions. Bioinformatics. 2001;17(7):646–653.

8. Rost B, Casadio R, Fariselli P, Sander C. Transmembrane helices predicted at 95% accuracy. Protein Sci. 1995;4(3):521–33. doi:10.1002/pro.5560040318.

9. Jones DT. Improving the accuracy of transmembrane protein topology prediction using evolutionary information. Bioinformatics. 2007;23(5):538–44. doi:10.1093/bioinformatics/btl677.

10. Hayat S, Elofsson A. Boctopus: improved topology prediction of transmembrane β-barrel proteins. Bioinformatics. 2012;28(4):516–22. doi:10.1093/bioinformatics/btr710.

11. Lockless SW, Ranganathan R. Evolutionarily conserved pathways of energetic connectivity in protein families. Science. 1999;286(5438):295–9.

12. Reva B, Antipin Y, Sander C. Determinants of protein function revealed by combinatorial entropy optimization. Genome Biol. 2007;8(11):232. doi:10.1186/gb-2007-8-11-r232.

13. Casari G, Sander C, Valencia A. A method to predict functional residues in proteins. Nat Struct Biol. 1995;2(2):171–8.

14. Marks DS, Colwell LJ, Sheridan R, Hopf TA, Pagnani A, Zecchina R, et al. Protein 3d structure computed from evolutionary sequence variation. PLoS One. 2011;6(12):28766. doi:10.1371/journal.pone.0028766.

15. Marks DS, Hopf TA, Sander C. Protein structure prediction from sequence variation. Nat Biotechnol. 2012;30(11):1072–80. doi:10.1038/nbt.2419.

16. Morcos F, Pagnani A, Lunt B, Bertolino A, Marks DS, Sander C, et al. Direct-coupling analysis of residue coevolution captures native contacts across many protein families. Proc Natl Acad Sci USA. 2011;108(49):1293–301. doi:10.1073/pnas.1111471108.

17. Lunt B, Szurmant H, Procaccini A, Hoch JA, Hwa T, Weigt M. Inference of direct residue contacts in two-component signaling. Methods Enzymol. 2010;471:17–41. doi:10.1016/S0076-6879(10)71002-8.

18. Weigt M, White RA, Szurmant H, Hoch JA, Hwa T. Identification of direct residue contacts in protein-protein interaction by message passing. Proc Natl Acad Sci USA. 2009;106(1):67–72. doi:10.1073/pnas.0805923106.

19. Hopf TA, Colwell LJ, Sheridan R, Rost B, Sander C, Marks DS. Three-dimensional structures of membrane proteins from genomic sequencing. Cell. 2012;149(7):1607–21. doi:10.1016/j.cell.2012.04.012.

20. Kamisetty H, Ovchinnikov S, Baker D. Assessing the utility of coevolution-based residue-residue contact predictions in a sequence- and structure-rich era. Proc Natl Acad Sci USA. 2013;110(39):15674–9. doi:10.1073/pnas.1314045110.

21. Liu Y, Engelman DM, Gerstein M. Genomic analysis of membrane protein families: abundance and conserved motifs. Genome Biol. 2002;3(10):0054.

22. Jackups R Jr, Liang J. Combinatorial model for sequence and spatial motif discovery in short sequence fragments: examples from beta-barrel membrane proteins. Conf Proc IEEE Eng Med Biol Soc. 2006;1:3470–3. doi:10.1109/IEMBS.2006.259727.

23. Grunert S, Labudde D. Graph representation of high-dimensional alpha-helical membrane protein data. BioData Min. 2013;6(1):21. doi:10.1186/1756-0381-6-21.

24. Kozma D, Simon I, Tusnady GE. Pdbtm: Protein data bank of transmembrane proteins after 8 years. Nucleic Acids Res. 2013;41(Database issue):524–529. doi:10.1093/nar/gks1169.

25. Punta M, Coggill PC, Eberhardt RY, Mistry J, Tate J, Boursnell C, et al. The pfam protein families database. Nucleic Acids Res. 2012;40(Database issue):290–301. doi:10.1093/nar/gkr1065.

26. Liu Y, Engelman DM, Gerstein M. Genomic analysis of membrane protein families: abundance and conserved motifs. Genome Biol. 2002;3(10):0054.

27. Branden C, Tooze J. Introduction to protein structure. New York: Garland Publishing; 1991. doi: 10.1016/0307-4412(92)90129-A.

28. Senes A, Gerstein M, Engelman DM. Statistical analysis of amino acid patterns in transmembrane helices: the gxxxg motif occurs frequently and in association with beta-branched residues at neighboring positions. J Mol Biol. 2000;296(3):921–36. doi:10.1006/jmbi.1999.3488.

29. Lo A, Cheng C-W, Chiu Y-Y, Sung T-Y, Hsu W-L. Tmpad: an integrated structural database for helix-packing folds in transmembrane proteins. Nucleic Acids Res. 2011;39(Database issue):347–55. doi:10.1093/nar/gkq1255. http://bio-cluster.iis.sinica.edu.tw/TMPad/.

30. Sobolev V, Sorokine A, Prilusky J, Abola EE, Edelman M. Automated analysis of interatomic contacts in proteins. Bioinformatics. 1999;15(4):327–32. http://ligin.weizmann.ac.il/cgi-bin/lpccsu/LpcCsu.cgi.

31. Berman HM, Westbrook J, Feng Z, Gilliland G, Bhat TN, Weissig H, et al. The protein data bank. Nucleic Acids Res. 2000;28(1):235–42.

32. Stern LJ, Khorana HG. Structure-function studies on bacteriorhodopsin. x. individual substitutions of arginine residues by glutamine affect chromophore formation, photocycle, and proton translocation. J Biol Chem. 1989;264(24):14202–8.

33. Subramaniam S. The structure of bacteriorhodopsin: an emerging consensus. Curr Opin Struct Biol. 1999;9(4):462–8. doi:10.1016/S0959-440X(99)80065-7.

34. Wood K, Lehnert U, Kessler B, Zaccai G, Oesterhelt D. Hydration dependence of active core fluctuations in bacteriorhodopsin. Biophys J. 2008;95(1):194–202. doi:10.1529/biophysj.107.120386.

35. Curnow P, Di Bartolo ND, Moreton KM, Ajoje OO, Saggese NP, Booth PJ. Stable folding core in the folding transition state of an alpha-helical integral membrane protein. Proc Natl Acad Sci USA. 2011;108(34):14133–8. doi:10.1073/pnas.1012594108.

36. Schlebach JP, Cao Z, Bowie JU, Park C. Revisiting the folding kinetics of bacteriorhodopsin. Protein Sci. 2012;21(1):97–106. doi:10.1002/pro.766.

37. Schiffer M, Chang CH, Stevens FJ. The functions of tryptophan residues in membrane proteins. Protein Eng. 1992;5(3):213–4.

7

Molecular dynamics simulation of the opposite-base preference and interactions in the active site of formamidopyrimidine-DNA glycosylase

Alexander V. Popov[1], Anton V. Endutkin[1,2], Yuri N. Vorobjev[1,2]* and Dmitry O. Zharkov[1,2]*

Abstract

Background: Formamidopyrimidine-DNA glycosylase (Fpg) removes abundant pre-mutagenic 8-oxoguanine (oxoG) bases from DNA through nucleophilic attack of its N-terminal proline at C1′ of the damaged nucleotide. Since oxoG efficiently pairs with both C and A, Fpg must excise oxoG from pairs with C but not with A, otherwise a mutation occurs. The crystal structures of several Fpg–DNA complexes have been solved, yet no structure with A opposite the lesion is available.

Results: Here we use molecular dynamic simulation to model interactions in the pre-catalytic complex of *Lactococcus lactis* Fpg with DNA containing oxoG opposite C or A, the latter in either *syn* or *anti* conformation. The catalytic dyad, Pro1–Glu2, was modeled in all four possible protonation states. Only one transition was observed in the experimental reaction rate pH dependence plots, and Glu2 kept the same set of interactions regardless of its protonation state, suggesting that it does not limit the reaction rate. The adenine base opposite oxoG was highly distorting for the adjacent nucleotides: in the more stable *syn* models it formed non-canonical bonds with out-of-register nucleotides in both the damaged and the complementary strand, whereas in the *anti* models the adenine either formed non-canonical bonds or was expelled into the major groove. The side chains of Arg109 and Phe111 that Fpg inserts into DNA to maintain its kinked conformation tended to withdraw from their positions if A was opposite to the lesion. The region showing the largest differences in the dynamics between oxoG:C and oxoG:A substrates was unexpectedly remote from the active site, located near the linker joining the two domains of Fpg. This region was also highly conserved among 124 analyzed Fpg sequences. Three sites trapping water molecules through multiple bonds were identified on the protein–DNA interface, apparently helping to maintain enzyme-induced DNA distortion and participating in oxoG recognition.

Conclusion: Overall, the discrimination against A opposite to the lesion seems to be due to incorrect DNA distortion around the lesion-containing base pair and, possibly, to gross movement of protein domains connected by the linker.

Keywords: DNA glycosylase, 8-oxoguanine, Fpg, Molecular dynamics, Opposite-base specificity, Reaction mechanism

* Correspondence: ynvorob@niboch.nsc.ru; dzharkov@niboch.nsc.ru
[1]SB RAS Institute of Chemical Biology and Fundamental Medicine, 8 Lavrentieva Ave., Novosibirsk 630090, Russia
Full list of author information is available at the end of the article

Background

Formamidopyrimidine-DNA glycosylase (Fpg or MutM) is a bacterial DNA repair enzyme that removes several abundant oxidized bases from DNA. The principal substrate bases of Fpg are 8-oxoguanine (oxoG), 2,6-diamino-4-oxo-5-formamidopyrimidine (fapyG) and 2,4-diamino-5-formamidopyrimidine (fapyA) [1, 2] but the enzyme also can recognize several dozens of other damaged purines and pyrimidines [3–10]. By excision of a damaged base, Fpg initiates base excision repair (BER), which engages AP endonucleases, a DNA polymerase and a DNA ligase to restore the integrity of DNA [11, 12].

The activity of Fpg towards oxoG has attracted much attention due to abundance and biological importance of this lesion, induced in DNA by oxidative metabolism byproducts, oxidative stress, and ionizing radiation [13]. Steric and electrostatic repulsion between the substituent at C8 and the sugar–phosphate atoms effectively pushes oxoG towards the *syn* conformation, in which oxoG forms a Hoogsteen pair with A [14, 15]. Misincorporation of A by DNA polymerases, in the absence of repair, leads to a G → T transversion after the second round of replication.

Systems for repair of oxoG have been found in all cellular organisms. The tendency of oxoG to form pairs with both C and A presents a challenge to its repair: both oxoG:C and oxoG:A pairs must be converted into G:C pairs. This requirement is not trivial since a simple removal of oxoG from an oxoG:A mispair would generate a G → T transversion after the repair. This problem is circumvented by repair of oxoG:A pairs in two sequential rounds of BER [16]. The non-damaged (but inappropriately incorporated) A is removed first and replaced with C, and the resulting oxoG:C pair is then repaired through the excision of oxoG. In *E. coli*, the mutagenic potential of oxoG is counteracted by three enzymes, Fpg, MutT, and MutY, collectively known as a "GO system". Fpg excises oxoG from oxoG:C pairs but has little activity towards oxoG:A substrates to prevent G → T transversions [1, 17]. Another DNA glycosylase, MutY, specifically excises A from A:oxoG mispairs. If G in DNA is oxidized to oxoG, it will inevitably be paired with C and will be removed by Fpg. If, on the other hand, A is incorporated during replication opposite an unrepaired oxoG, the resulting oxoG:A mispair will be a substrate for MutY but not for Fpg. The repair synthesis then has a chance to incorporate C opposite oxoG [16].

The function of the GO system therefore critically depends on the selectivity of Fpg to the base opposite to the damaged one. X-ray structures are available for free Fpg protein from *Thermus thermophilus* (*Tth*-Fpg) and for various types of complexes of DNA with Fpg from *Escherichia coli* (*Eco*-Fpg), *Geobacillus stearothermophilus* (*Bst*-Fpg) and *Lactococcus lactis* (*Lla*-Fpg) [18–33]

(Fig. 1a). Based on these structures, kinetic data, and computational modeling, a reaction mechanism has been suggested that involves a nucleophilic attack at C1′ of oxoG by a lone electron pair of the secondary amino group of the deprotonated N-terminal Pro1 residue, assisted by protonation of O4′ in the deoxyribose moiety [33–36]. As a result, the *N*-glycosidic bond is broken, the deoxyribose ring is opened, and a Schiff base covalent intermediate between Fpg and DNA is formed (Fig. 1b). This series of events is followed by two sequential steps of elimination of the 3′- and 5′-phosphates and hydrolysis of the Schiff base. However, many questions about the initial stages of the reaction still remain. For example, the mechanism of oxoG recognition in the active site of the enzyme is unclear, and the mechanism of proton transfer in the multistep reaction is unknown. Notably, no structural or modeling data is available for Fpg in a complex with oxoG:A-containing DNA, limiting our knowledge of the mechanisms of rejection of this functionally relevant mispair. In this work, we use molecular dynamics approach to analyze the structure of complexes of Fpg with oxoG-containing DNA (either A or C opposite the lesion) to get an insight into the reasons behind the opposite-base selectivity of the enzyme and into the dynamic features of the immediate precatalytic complex involving oxoG.

Methods
Model preparation

The starting model for the MD analysis of Fpg bound to oxoG-containing DNA was the X-ray structure of *Lla*-Fpg in a complex with a 14-mer DNA duplex (Fig. 1a,c) containing a non-hydrolysable carbocyclic analog of fapyG (PDB ID 1XC8) [23]. The lesion was changed to oxoG using the following protocol. The initial structure of the oxoG base was taken from *Bst*-Fpg coordinates (PDB ID 1R2Y) [22]. The base was aligned for the best fit to the fapyG ring and incorporated into the PDB file instead of fapyG. The methylene group in the cyclopentane ring isosteric to O4′ was manually changed to oxygen. Heavy atoms of the side chains lacking in the structure were built using the Missing Heavy Atom Restoration module of BioPASED molecular modeling package [37]. Out of 397 water molecules found in the crystal unit cell, seven that reside in the enzyme's active site or in its immediate vicinity were retained as explicit water, otherwise the modeling was done in implicit water to broaden the sampled conformational space. To analyze the effect of A *vs* C placed opposite oxoG, three sets of simulations were performed: one with oxoG opposite C (henceforth termed C models), another with oxoG opposite A(*anti*) (Aa models), and the third, with oxoG opposite A(*syn*) (As models). All models containing A opposite to the lesion were constructed by base

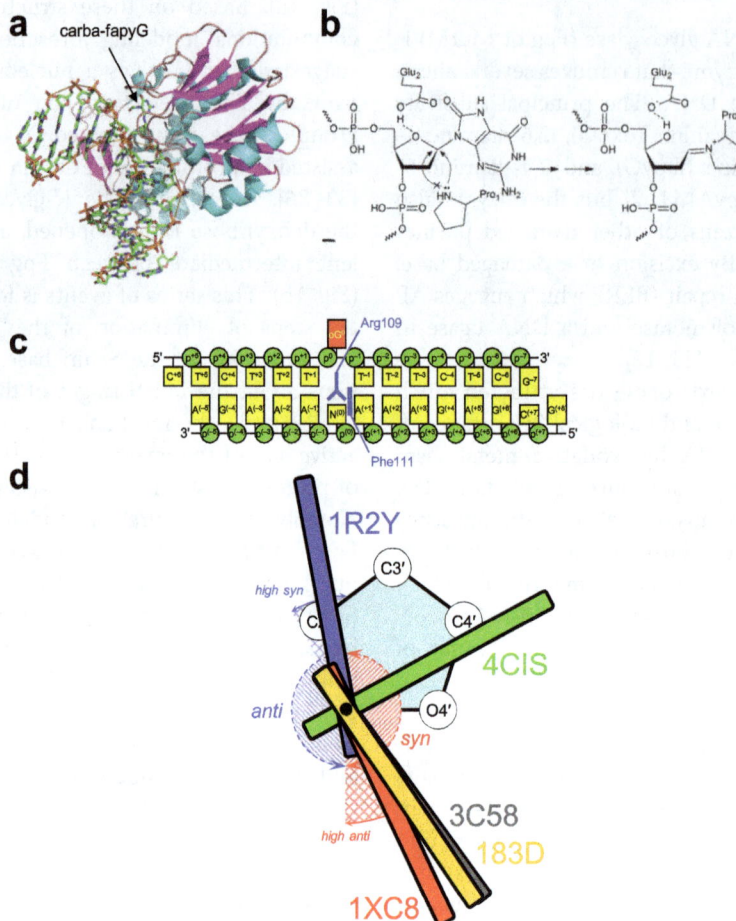

Fig. 1 a, Structure of *Lla*-Fpg (1XC8) used as a starting model. The protein is colored according to its secondary structure (*cyan*, α helices; *magenta*, β sheets; *coral*, loops); the DNA is colored by atom type (*green*, C; *blue*, N; *red*, O; *orange*, P). An *orange* line is drawn through P atoms in DNA to highlight an axial kink induced by Fpg binding. **b,** Mechanism of oxoG excision by Fpg proposed from the structural data [35]. The S_N2 displacement occurs in the C1′ → O4′ direction rather than in the C1′ → N9 direction. **c,** Schematic representation of the modeled DNA duplex and numbering of DNA bases and phosphates (p). $N^{(0)}$ is either C or A. Positions of Arg109 and Phe111 in the complex are indicated. **d,** Schematic position of the damaged base relative to the sugar plane in the structures of free oxoG-containing DNA (183D, [56]) or Fpg–DNA complexes containing various purine-derived lesions everted into the active site (1XC8, 3C58, 4CIS and 1R2Y; see structure details in the text)

replacement. For each group, four simulations were done, with different protonation states of Pro1 and Glu2: deprotonated Pro1 and Glu2 (PRN-GLU models), protonated Pro1 and deprotonated Glu2 (PRO-GLU), deprotonated Pro1 and protonated Glu2 (PRN-GLH), and protonated Pro1 and Glu2 (PRO-GLH models) (Table 1). The starting structures were checked for errors using the PDB Validator tool of BioPASED [37]. All models were energy-minimized in 500 steps of Fletcher energy optimization algorithm and finally refined by simulated annealing MD for 500 ps using the BioPASED package [37]. The AMBER force field parameters for oxoG and neutral Pro1 were from [38]. Force field parameters for neutral glutamate were from AMBER ff99 [39]. The parameters for the rest of the protein, including Zn^{2+}, were taken from the classic Amber ff99 force field. The protonation states of the other residues were

Table 1 Mean r.m.s.d. values of the models and their standard deviations (Å) over the last 8 ns of the runs

	PRN-GLH	PRN-GLU	PRO-GLH	PRO-GLU
	Global			
oxoG:C	1.50 ± 0.06	1.58 ± 0.07	1.69 ± 0.07	1.51 ± 0.05
oxoG:A(*anti*)	1.75 ± 0.05	1.43 ± 0.06	1.66 ± 0.06	1.45 ± 0.06
oxoG:A(*syn*)	1.68 ± 0.06	1.53 ± 0.06	1.41 ± 0.06	1.74 ± 0.06
	Active site			
oxoG:C	1.37 ± 0.10	1.55 ± 0.08	1.60 ± 0.13	1.58 ± 0.11
oxoG:A(*anti*)	1.89 ± 0.06	1.57 ± 0.08	2.16 ± 0.06	1.86 ± 0.07
oxoG:A(*syn*)	1.66 ± 0.12	1.61 ± 0.06	1.44 ± 0.10	1.92 ± 0.13

selected to match physiological pH conditions; e. g., E5 was modeled negatively charged. The Zn^{2+} ion was described as a single atom with four distance-based harmonic restraints to bind it to the coordinating cysteins and to maintain the correct geometry. Implicit counterion correction was applied by scaling charges of phosphate groups by a factor of 0.2 [40].

Molecular dynamics

Molecular dynamics simulations (10 ns) were performed using the BioPASED molecular dynamics modeling software [37] using the AMBER ff99 force field with BioPASED modifications and EEF1 analytical implicit solvent model [41], with an integration time step of 1 fs. The system was gradually heated from 10 K to 300 K during 50 ps and equilibrated at this temperature (the heating time was 150, 200, and 250 ps in the repeat runs of the PRO-GLH models). A classic molecular dynamics trajectory was generated in the NVT ensemble with harmonic restraints of 0.001 $kcal/A^2$ for the protein and water atoms, 0.25 $kcal/A^2$ for the atoms of the terminal nucleotides, and 0.0025 $kcal/A^2$ for the rest of the DNA atoms. Coordinates of each atom of the system were saved each 2 ps, thus producing a trajectory size of 5000 snapshots. The trajectories were analyzed using MDTRA [42], a part of the BISON package [43]. Trajectories were compared using moving MWZ method [44] with bins of 50 snapshots. Statistically significant differences in parameters between different models were estimated using F-test, with false discovery rate method (Benjamini–Hochberg procedure) employed to correct for multiple comparisons [45]. Hydrogen bonds were searched using MDTRA [42]. Structures were visualized and rendered using VMD [46], RasMol [47] and PyMOL (Schrödinger, Portland, OR).

pH dependence of Fpg activity

Eco-Fpg was purified as described [19]. Oligonucleotides were synthesized in-house from commercially available phosphoramidites (Glen Research, Sterling, VA). An oligonucleotide 5′-CTCTCCCTTCXCTCCTTTCCTCT-3′ (X = oxoG) was ^{32}P-labeled using polynucleotide kinase (SibEnzyme, Novosibirsk, Russia) and γ[^{32}P]ATP (PerkinElmer, Waltham, MA) following the manufacturer's protocol and annealed to a complementary strand placing C or A opposite oxoG. The reactions (20 μl) included 25 mM sodium phosphate buffer (H3PO4/NaH2PO4, NaH2PO4/Na2HPO4, and Na2HPO4/Na3PO4 conjugate pairs spanning the range of pH 4.0–9.0), 100 nM duplex oligonucleotide substrate, and either 2 nM (steady-state experiments) or 500 nM Fpg (single-turnover experiments). The reaction was allowed to proceed for 1 min either at 37 °C with 2 nM Fpg or at 0 °C with 500 nM Fpg, and was stopped by adding 20 μl of formamide/EDTA gel loading dye and heating at 95° for 3 min. The products were separated by electrophoresis in 20% polyacrylamide gel/8 M urea and quantified by phosphorimaging using a Molecular Imager FX system (Bio-Rad Laboratories, Hercules, CA). Three independent experiments have been performed. Calculation of pK_a for Pro1 and Glu2 were done using PROPKA v3.1 [48]. Circular dichroism spectra were recorded on a JASCO J-600 CD spectrometer (JASCO Analytical Instruments, Tokyo, Japan) at 25 °C in 30 mM Na phosphate with a 1-nm step.

Evolutionary analysis

A taxonomically balanced sample of 124 bacterial Fpg sequences (limited to two sequences per taxonomic family) was constructed by protein BLAST search [49] in the NCBI non-redundant protein sequences database using *Eco*-Fpg as a query, filtered for conservation of the N-terminal Pro–Glu dipeptide and C-terminal zinc finger, and clipped from the absolutely conserved N-terminal Pro to the absolutely conserved Gln after the fourth Cys of the zinc finger as described [50, 51]. Alignment of multiple sequences and neighbor-joining tree construction was performed using Clustal Omega [52]. Hierarchical analysis of conservation of physicochemical properties in the alignments was done using AMAS [53] with 5% atypical residues allowed; the results are reported as conservation numbers (C_n).

Results and discussion
General model considerations
Selection of the starting structure
Currently, the Protein Data Bank holds 56 released X-ray structures of Fpg, belonging to four bacterial species and sampling several points along the reaction coordinate [18–33]. Our selection of the starting structure for MD was guided by the following considerations. First, it should contain DNA with the damaged base still in place, residing in the enzyme's active site. Second, minimal deviation from the wild-type enzyme recognizing oxoG should be present. Third, the structure should have good resolution (<2.0 Å), with as few as possible residues missing.

Based on these considerations, we have chosen 1XC8, the 1.95-Å structure of wild-type *Lla*-Fpg bound to DNA containing a non-cleavable carbacyclic fapyG analog (carba-fapyG [23]) as a starting model (Fig. 1a). In carba-fapyG, a methylene group substitutes for O4′, and the damaged base, fapyG, is different from oxoG only by the absence of a bond between N9 and C8 of the purine heterocycle O4′ [54]. The *Lla*-Fpg is nearly identical to *Eco*-Fpg in its selectivity for C *vs* A as the opposite base [17].

OxoG glycosidic angle

Besides 1XC8, the structures of Fpg bound to DNA with an extrahelical damaged base include *Lla*-Fpg bound to DNA containing carbacyclic *N*5-benzyl-fapyG (3C58 [26]), carbacyclic oxoG (4CIS [33]) or 5-hydroxy-5-methylhydantoin (2XZF, 2XZU [29]) and *Bst*-Fpg bound to DNA containing oxoG (1R2Y) or 5,6-dihydrouracil (1R2Z) [22]. With the exception of 1R2Y and 4CIS, the damaged bases in the structures are quite different from oxoG. In the 1R2Y structure, oxoG is present in DNA, and the cleavage is prevented by changing the absolutely conserved catalytic Glu2 residue into Gln [22]. In this structure, oxoG is often stated to be in the *syn* conformation in the active site, yet its χ angle (101°) is actually in the *anti* domain (namely, in its border range, so-called "*high syn*") (Fig. 1d). On the contrary, oxoG opposite C in B-DNA is usually stated to be *anti* as it forms regular Watson–Crick bonds [55, 56], yet its χ angle in the crystal structure (−55°, [56]) is actually in the *syn* domain. Carba-fapyG in 1XC8 is in the *high anti* range (χ = −64°), and only in 4CIS, carba-oxoG is unambiguously *syn* (χ = 27°, Fig. 1d). Moreover, oxoG in 4CIS does not form the same set of hydrogen bonds with the active site as in 1R2Y. The possibility of conformation artifacts induced by E2Q mutation has been amply discussed in the literature [23, 38, 57–61]. Therefore, we have chosen 1XC8, which straddles the *syn/anti* border (Fig. 1d) as our starting model and allowed the conformation to drift into the most preferable χ range during MD.

Opposite base glycosidic angle

While the oxoG:A pair in B-DNA exists as oxoG(*syn*):A(*anti*), this does not mean that the same conformations will be observed in the complex with Fpg, since the hydrogen bonds within the mispair are lost upon oxoG eversion, and the conformation of the nucleotides is governed largely by their interaction with the protein residues. The most relevant example is given by another oxoG mispair, oxoG:G, which adopts the conformation oxoG(*syn*):G(*anti*) in the B-DNA duplex [62] but the G opposite to the lesion flips into *syn* and forms two strong hydrogen bond with an Arg residue when this duplex is bound to *Bst*-Fpg [20]. Therefore, in addition to the oxoG:C pair, we have constructed two series of models with oxoG:A, with A in either *anti* or *syn* conformation to fully explore the range of possible dynamics of Fpg–oxoG:A complex.

Solvent

Although MD in explicit solvent is common nowadays, recent advances in implicit solvent models revived the popularity of this alternative [63–65]. The major advantages of implicit solvent over the explicit one are speed, better estimates of solvation and folding energy, wider

coverage of conformational space and more accurate account for pH and residue ionization. The latest versions of Poisson–Boltzmann, generalized Born and hybrid implicit/explicit models are comparable with explicit solvent-based calculations with respect to agreement with experimental free energy data [63–65]. Although the acceleration of conformation sampling afforded by implicit solvent strongly depends on the modeled system, direct comparisons show a 7–10-fold increase for a system with several conformational transitions [66]. Since our primary interest was to sample a wide range of conformations available for the Fpg–substrate complexes, we have chosen a hybrid model combining an implicit solvent with explicit strongly bound water molecules; such approaches retain the advantages of implicit methods but significantly improve quality of protein–DNA interface models [67].

Protonation state of the catalytic dyad

The ionizable groups of Pro1 and Glu2 directly participate in the enzymatic reaction. Mechanistically, the nucleophilic attack by Pro1 at C1'[oxoG] requires N[Pro1] to carry a lone electron pair (Fig. 1b). On the other hand, opening of the deoxyribose ring involves protonation of its O4', which is near Oε2 of Glu2 (Fig. 1b); quantum mechanics/molecular mechanics (QM/MM) simulations show that O4' protonation provides a low-barrier path to glycosidic bond cleavage by Fpg and its eukaryotic functional analog, OGG1 [33, 36, 68]. From several structures Fpg–DNA complexes, it has been suggested that the proton is shuttled from N[Pro1] to Oε2[Glu2], perhaps through a network of crystallographic water molecules present in the active site [19, 26]; this possibility was also favored by QM/MM analysis [69]. However, no attempt to estimate pK_a of Pro1 and Glu2 has been reported in the literature. It is possible that a mixture of Pro1/Glu2 ionization states exists in the active centers of different Fpg molecules at physiological pH; although only one of them (PRN-GLH) is permissive for the reaction chemistry, the path to this state may go through other states. Therefore, we have performed MD of the full system for four ionization states of the Pro1–Glu2 catalytic dyad: PRN-GLH, PRN-GLU, PRO-GLH, and PRO-GLU (see Methods).

Overall model characterization

The root mean square deviation (r.m.s.d.) with respect to the backbone of the starting structure of the complex was calculated every 2 ps. R.m.s.d. values of all the models increased rapidly during the first 500 ps of the dynamics and stabilized at approximately 1.6 Å (see Table 1 and Additional file 1: Fig. S1). Overall r.m.s.d. of all models was similar; however, in the active center (defined as all protein residues with at least one atom within 4 Å of oxoG nucleotide or the opposite C/A nucleotide, plus three nucleotide pairs centered on the oxoG), the r.m.s.d. of the C models was significantly lower than in the A models ($p < 10^{-4}$). The DNA backbone displayed higher mobility than the protein backbone: average r.m.s.d. of the DNA residues was greater by 0.54–0.98 Å depending on the model. The overall complex conformation was stable along the whole trajectory with a radius of gyration ~20 Å for each model (20.17 ± 0.04 to 20.30 ± 0.05 Å). The angle of DNA kink was also stable ($55° \pm 2°$ to $63° \pm 2°$, depending on the model).

Simulation reproducibility

To test the consistency of results between independent runs, we have selected three models (PRO-GLH-C, PRO-GLH-Aa, and PRO-GLH-As) and performed three additional simulations with each one, to the total of nine additional simulations, using different heating times (150, 200, and 250 ps) to provide different conditions for the start of the production run. Then the resulting four trajectories for each model (one original and three new) were compared. The r.m.s.d. values of individual runs were similar (1.0–1.5 Å over the last 8 ns, Additional file 1: Fig. S1). The inter-run r.m.s.d. were expectedly higher (1.9–2.2 Å, Additional file 1: Fig. S1) but still did not show significant divergence of the models. Stable hydrogen bonds, including model-specific ones, were well consistent across the four runs (Additional file 1: Fig. S5B); the 90% cut-off of the mean occurrence identified as stable all Watson–Crick bonds and 79% of the main-chain bonds observed in the 1XC8 crystal structure.

Pro–Glu catalytic dyad
Arrangement of the reacting groups in the models
We have sampled the population of two key distances of the Fpg–DNA complex, N[Pro1]...C1′[oxoG] and Oε2[Glu2]...O4′[oxoG] in all our models (Fig. 2, Table 2). In all C models (Fig. 2a, D, G, J), the distribution of Oε2[Glu2]...O4′[oxoG] distances was unimodal and produced similar central values (Table 2). On the contrary, the N[Pro1]...C1′[oxoG] distance was less stable: in some models, two peaks in the distribution histogram were clearly observed, indicative of stable conformational basins (Table 2).

Another important parameter in the reaction of base excision is the angle of attack by Pro1 at C1′. Two mechanisms for S_N2 displacement initiating Schiff base formation have been considered for bifunctional DNA glycosylases: with the C1′–O4′ bond or C1′–N9 bond breaking first [35, 70]. Enzyme-catalyzed S_N2 reactions require a 10°–20° alignment of the nucleophile lone pair

and carbon antibonding orbital [71, 72]. The ideal attack geometry for Pro1 is thus 107° for the C1′[oxoG]... N[Pro1]...Cδ[Pro1] angle and 180° for the X... C1′[oxoG]...N[Pro1] angle where X is either O4′[oxoG] or N9[oxoG]. As can be seen from Fig. 3 and Table 2, the C1′...N...Cδ angle of all PRO models, as well as PRN-GLH-C, PRN-GLH-As, and PRN-GLU-C, either lied in the acceptable domain or made appreciable excursions to it. All models were incompatible with the C1′–N9 direction of nucleophilic attack (Table 2). The O4′...C1′...N angle for 7 of 12 models lied in the acceptable domain, and was close to this range in other models, consistent with the C1′–O4′ attack. The opposite base had no consistent effect on the Pro1 approach angle.

Defining the "optimal geometry" as N[Pro1]... C1′[oxoG] distance < 4 Å, Oε2[Glu2]...O4′[oxoG⁰] distance < 4.5 Å, and C1′[oxoG⁰]...N[Pro1]...Cδ[Pro1] and O4′[oxoG]...C1′[oxoG]...N[Pro1] within 20° of the ideal values, we have sampled the population of the zone with all four parameters optimal (Table 2). All A(*anti*) models showed the optimal geometry very rarely. For C and A(*syn*) models, PRO-GLH was most populated, followed by PRN-GLH. Interestingly, PRN models were more selective towards C vs A(*syn*). Other sensible definitions of "optimal" Pro1 and Glu2 distances (e. g., the lowest quartile of the respective distance population), also showed C models spending more time in the optimal conformation than A models. The preference of C models for the optimal geometry was also evident in the repeat runs of the PRO-GLH models (Table 2).

pK_a estimate of the catalytic dyad residues
To get an independent estimate of the protonation state of Pro1 and Glu2, we have used PROPKA, an empirical algorithm based on the spatial proximity of charged residues [48]. In addition to our starting structure, we have considered several other PDB structures of Fpg from different species (Additional file 2: Table S1). In all cases, pK_a of Pro1 was notably lowered (by 0.35–2.99 units) compared with the reference pK_a of N-terminal Pro, while pK_a of Glu2 was considerably higher (by 1.54–3.64 units) than the reference pK_a of the internal Glu side chain. Similar pK_a changes were reported for phage T4 endonuclease V, another DNA glycosylase employing the N-terminal amino group and a Glu carboxyl as a catalytic dyad [73]. Interestingly, structures of free Fpg and Fpg bound to undamaged DNA with the sampled base still intrahelical displayed more acidic pK_a for Glu2, suggesting that this group may be specifically activated upon eversion of the damaged nucleotide. Although PROPKA considers the influence of nucleic acid ligands on amino acid ionization potential only approximately, it

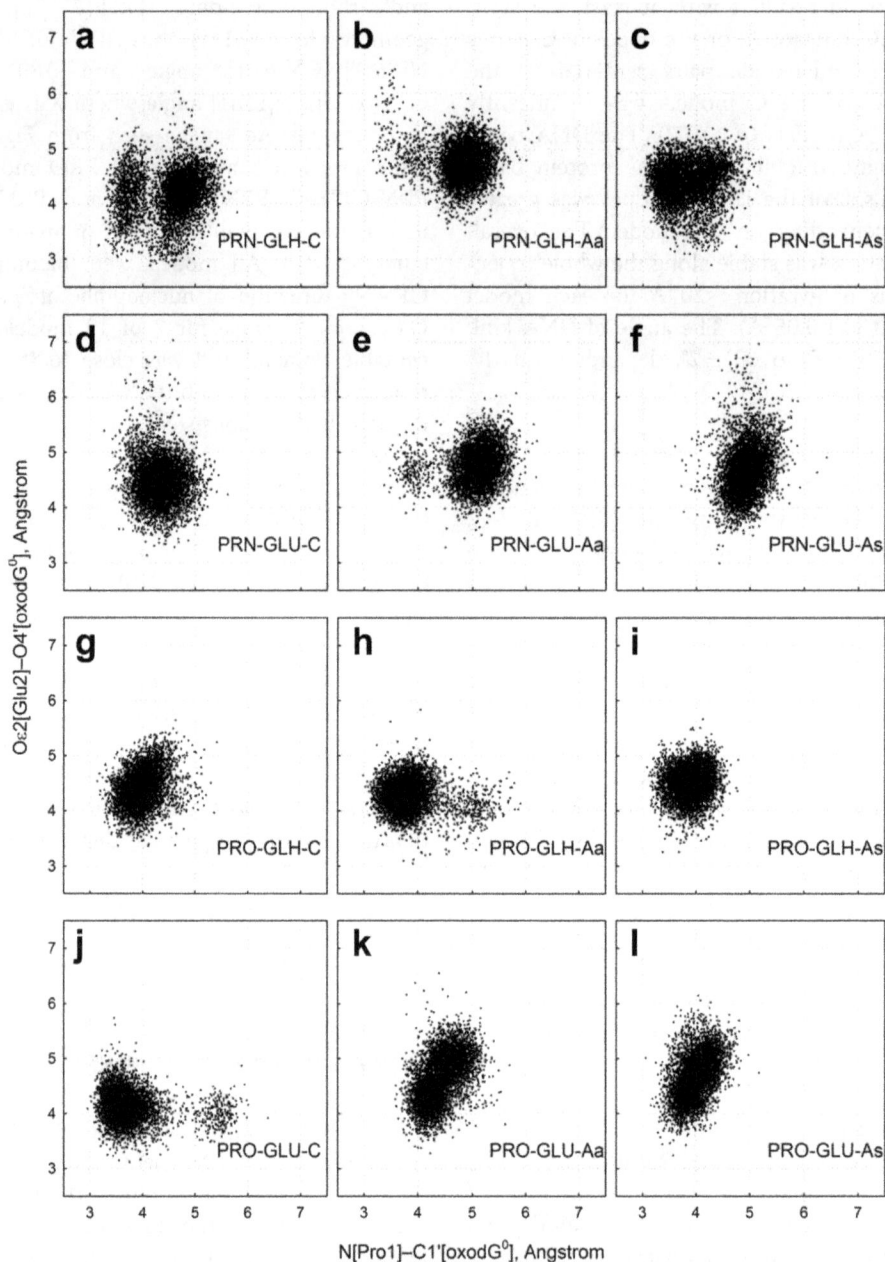

Fig. 2 Distances N[Pro1]...C1'[oxoG^0] and Oε2[Glu2]...O4'[oxoG^0] during the simulation with different protonation states of N[Pro1] and Oε2[Glu2] (**a–l**, the model nature is indicated in the respective panels)

is nevertheless clear that in Fpg, Pro1 is considerably more acidic, and Glu2, more basic than expected.

pH profile of Fpg activity

To assess the functional importance of the catalytic dyad protonation states experimentally, we have analyzed the pH profile of activity for *Eco*-Fpg, assuming that the mechanistic features of base excision will be conserved in *Eco*-Fpg and *Lla*-Fpg. Usually, when an enzyme's active site possesses two functionally important ionizable groups, one of which has to be protonated while the

other has to be deprotonated for activity, the pH dependence is characteristically bell-shaped. For DNA glycosylases, such bell-shaped dependence was shown for human alkyladenine glycosylase, which is monofunctional, structurally different from Fpg, and uses a histidine and a glutamate residue as a general acid and a general base, respectively [74]. On the contrary, Fpg showed a single transition in the activity over a pH range of 5 units (pH 4 to pH 9) (Fig. 4). This was observed both under single-turnover conditions, where the rate is limited by the catalytic step of the reaction (Fig. 4a) and

Table 2 Key distances and angles around the reacting C1′ atom of oxoG

Model		Distance N[Pro1]... C1′[oxoG⁰], Å	Distance Oε2[Glu2]... O4′[oxoG⁰], Å	Angle C1′[oxoG⁰]... N[Pro1]... Cδ[Pro1], degrees	Angle O4′[oxoG⁰]... C1′[oxoG⁰]...N[Pro1], degrees	# of snapshots with optimal geometry
C	PRN-GLH	3.73 (3.41–4.04)ᵃ 4.74 (4.29–5.20)	4.20 (3.41–4.79)	61 (55–95)	150 (125–169)	409
	PRN-GLU	4.31 (3.83–4.88)	4.46 (3.87–5.12)	116 (80–122)	168 (145–171)	293
	PRO-GLH	3.97 (3.55–4.48)	4.42 (3.90–4.96)	119 (113–122)	165 (156–168)	1077 (788)ᵇ
	PRO-GLU	3.59 (3.27–4.17) 5.43 (5.01–5.84)	4.13 (3.68–4.60)	107 (78–122)	145 (135–151)	141
A (anti)	PRN-GLH	4.86 (4.06–5.30)	4.75 (4.24–5.27)	52 (49–77)	153 (136–157)	0
	PRN-GLU	5.12 (4.35–5.57)	4.76 (4.22–5.29)	52 (49–75)	158 (132–165)	0
	PRO-GLH	3.79 (3.34–4.78)	4.27 (3.83–4.74)	100 (89–107)	141 (136–152)	73 (277)
	PRO-GLU	4.39 (3.96–4.97)	4.62 (3.92–5.33)	124 (94–133)	168 (156–172)	78
A (syn)	PRN-GLH	3.82 (3.26–4.54)	4.44 (3.71–4.94)	101 (89–113)	155 (136–167)	225
	PRN-GLU	4.94 (4.52–5.38)	4.65 (4.01–5.48)	55 (48–62)	149 (135–160)	0
	PRO-GLH	3.92 (3.43–4.31)	4.49 (4.02–4.95)	119 (107–122)	167 (157–170)	961 (137)
	PRO-GLU	3.99 (3.57–4.43)	4.65 (4.02–5.34)	118 (93–121)	154 (144–164)	184

ᵃMedian and 90% range in parentheses
ᵇAverage over three repeat runs in parentheses

steady-state conditions, where the reaction rate, in the case of Fpg, is a function of both the catalytic step and product release (Fig. 4b). The pK_a values calculated from a two-state model were 6.8 ± 0.1 and 7.5 ± 0.3 for the single-turnover and steady-state conditions, respectively; the increase in pK_a under the steady-state conditions is likely due to a pH effect on the partially rate-limiting product release step. Circular dichroism spectra showed no considerable change in the Fpg structure at pH 4 (Additional file 1: Fig. S2), so the activity changes most likely can be assigned to the ionization of critical active site residues.

Since Pro1 has to be deprotonated for the reaction, the rising activity *vs* pH plot with a single inflection may be explained by this deprotonation and suggests that the equilibrium ionization state of Glu2 is not rate-limiting. As discussed below, Glu2 may be conveniently protonated by a water molecule trapped in the active site.

Other interactions of the catalytic dyad
Pro1 did not form stable interactions within the active site in the models, consistent with the available structural information on the pre-catalytic Fpg complexes. In contrast, in 1XC8 and other known structures of Fpg, Glu2 accepts two hydrogen bonds from the amides of Ile172 and Tyr173 (or their counterparts in other Fpgs). These bonds were stable in all our simulations independently of the protonation state of Glu2. Several available structures (1K82, 1L1T, 1L1Z [19, 20]) strongly suggest that these two bonds hold the carboxylic group of Glu2 in a position suitable for interaction with a nearby water molecule, and, later in the reaction, with

O4′ of the damaged nucleotide that becomes a hydroxyl after base excision and sugar ring opening. Importantly, a water molecule (see the section "Water molecules in the Fpg–DNA complex") was the only stably interacting group other than Ile172 and Tyr173, and this interaction was not affected by the protonation state of Glu2.

Fpg interactions with oxoG and the opposite base
Stability of oxoG in the base-binding pocket
To inquire what features of Fpg–substrate interactions may explain poor substrate properties of oxoG:A mispairs, we have compared the dynamics of C, Aa and As models. The eversion angle of oxoG [75] was similar in all models (median range 74°–87°), indicative of full insertion of the damaged nucleoside into the enzyme's active site. Starting from *high anti* χ = −64° in 1XC8, the orientation of the oxoG base in all models spontaneously drifted towards the *anti* range, with 9 of 12 models remaining mostly in this range, with brief excursions into the *syn* domain (Fig. 5a). Two models, PRO-GLH-Aa and PRO-GLU-C, had reverted to *syn*, remaining in its *high anti* sub-range (χ = −83° ± 12°), whereas a single model, PRO-GLU-Aa, ventured deeper into the *anti* range (χ = −131° ± 11°; Fig. 5a).

In the *high syn* 1R2Y model of *Bst*-Fpg, four consecutive main chain amide nitrogens belonging to the Thr220–Tyr224 loop form a crown around O⁶ of oxoG, positioned at distances and angles suitable for hydrogen bond formation (Fig. 5c). Surprisingly, even though the oxoG base is rotated nearly 180° from its position in 1R2Y, the same set of contacts is maintained by the homologous loop Ser218–Tyr222 of *Lla*-Fpg (Fig. 5b, d). This observation agrees well with

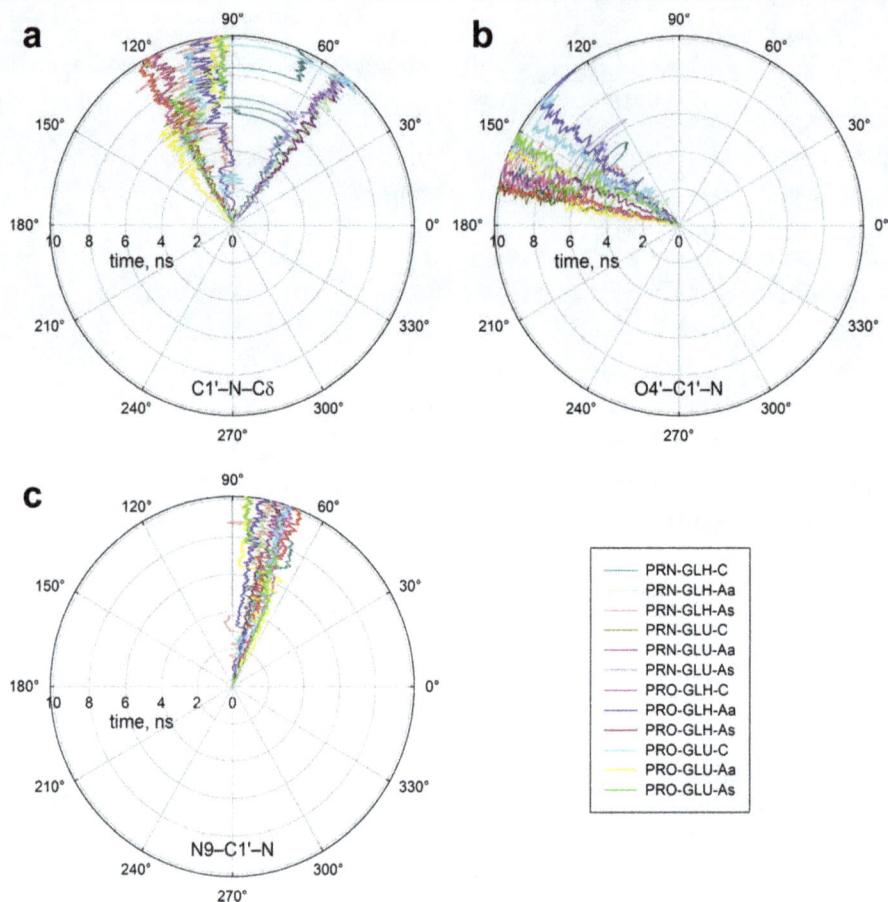

Fig. 3 Angles C1′[oxoG⁰]…N[Pro1]…Cδ[Pro1] **(a)**, O4′[oxoG⁰]…C1′[oxoG⁰]…N[Pro1] **(b)** and N9[oxoG⁰]…C1′[oxoG⁰]…N[Pro1] **(c)** in the models. Moving average of 50 consecutive snapshots is plotted vs time. The traces are color-coded: *dark cyan*, PRN-GLH-C; *light lime*, PRN-GLH-Aa; *coral*, PRN-GLH-As; *olive*, PRN-GLU-C; *dark magenta*, PRN-GLU-Aa; *light blue*, PRN-GLU-As; *magenta*, PRO-GLH-C; *blue*, PRO-GLH-Aa; *red*, PRO-GLH-As; *cyan*, PRO-GLU-C; *yellow*, PRO-GLU-Aa; *green*, PRO-GLU-As

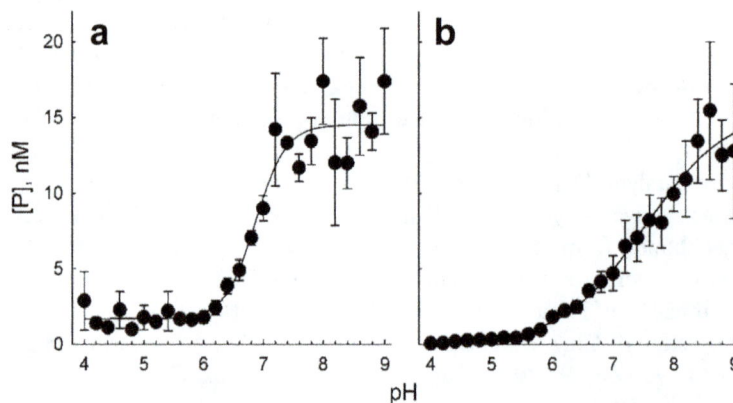

Fig. 4 pH dependence of Fpg activity. **a,** single-turnover conditions (500 nM Fpg, 100 nM substrate, 0 °C). **b,** steady-state conditions (2 nM Fpg, 100 nM substrate, 37 °C). See Methods for details

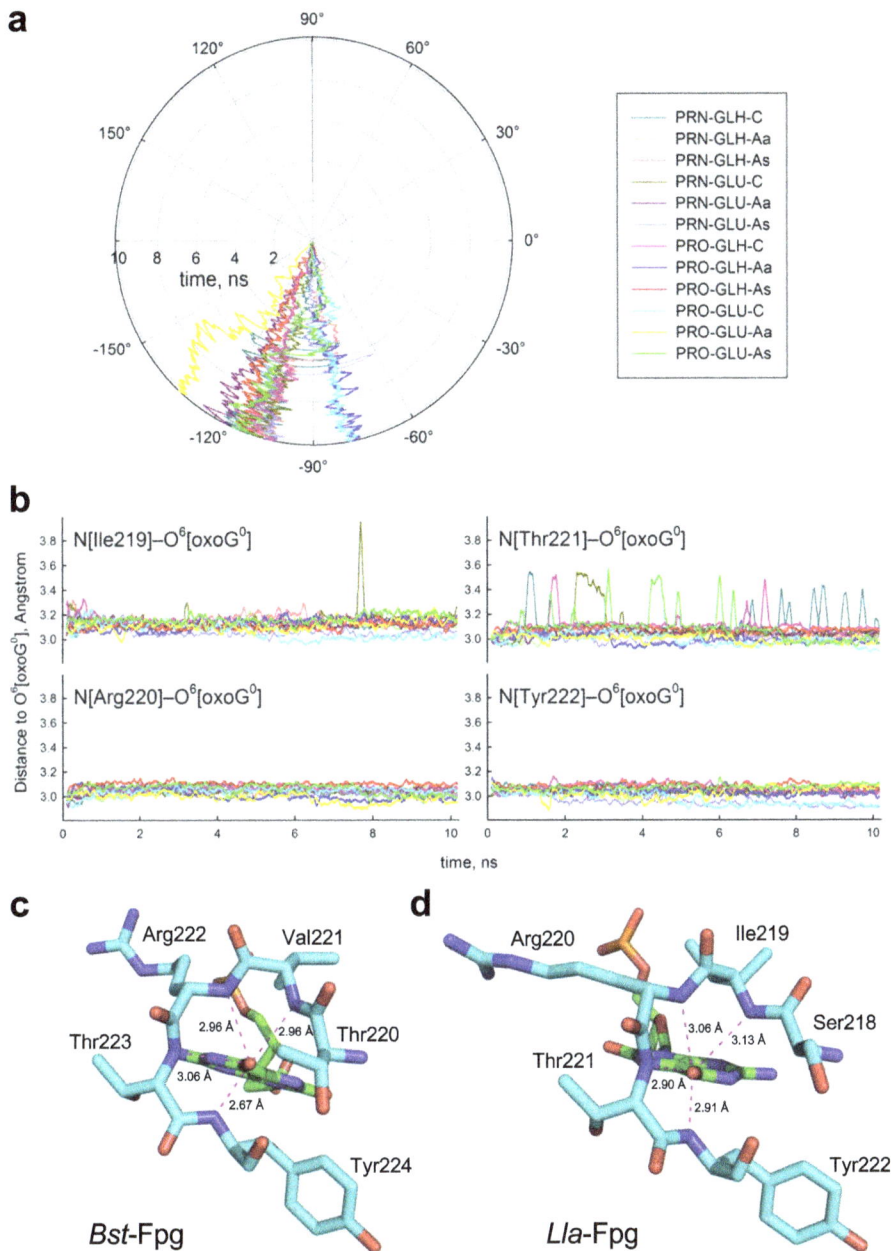

Fig. 5 a, χ angle evolution during the simulation. **b,** distances between O⁶[oxoG⁰] and main chain amide nitrogen atoms of Ile119, Arg220, Thr221, and Tyr222. Moving average of 50 consecutive snapshots is plotted *vs* time. The colors of the traces are the same as in Fig. 3. **c,** loop Thr220–Tyr224 of *Bst*-Fpg forms an extensive set of contacts with O⁶ of oxoG in *high syn* orientation (χ = 101°, 1R2Y). **d,** a homologous loop Ser218–Tyr222 of *Lla*-Fpg forms the same set of contacts with O⁶ when oxoG is flipped around the glycosidic bond (χ = –103°, PRO-GLH-C model, 9 ns)

the literature data on simulation of *Bst*-Fpg with oxoG forced into the *anti* conformation [60] and with the same pattern of contacts to O^6 in the 1XC8 structure of *Lla*-Fpg [23]. Notably, the "distinguishing" bond between the main chain carbonyl of Ser220 and pyrrolic N7 of oxoG, seen in *Bst*-Fpg 1R2Y structure but absent from *Lla*-Fpg 1XC8, was not observed in our simulations.

Interactions and dynamics of the opposite base

In all reported structures of Fpg bound to DNA with the fully everted damaged nucleotide, specific recognition of C opposite to the lesion is governed by two hydrogen bonds from Arg109 after its insertion into the DNA void: Nε[Arg109]–O^2[$C^{(0)}$] and Nη2[Arg109]–N3[$C^{(0)}$]. If G substitutes for C, Nε and Nη2 of the inserted Arg form slightly suboptimal bonds with N7 and O^6, respectively, of the G base in the *syn* orientation, whereas T in place of C retains a bond with Nε[Arg109] but experiences a clash between two hydrogen bond donors, N3[T] and Nη2[Arg109] [20].

In all our C models, $O^2[C^{(0)}]$ and $N3[C^{(0)}]$ maintained stable ~3 Å bonds with their interaction partners throughout the simulation (Additional file 1: Fig. S3A, B). In contrast, $A^{(0)}$ in our *anti* simulations existed in two configurations. In both GLU-Aa models, it remained intrahelical in the *anti* orientation, stabilized by a hydrogen bond between its exocyclic N^6 and the O2P non-bridging oxygen of $A^{(+1)}$ (Fig. 6a, b and Additional file 1: Fig. S3C). In both GLH-Aa models, $A^{(0)}$ is pushed towards the major groove and rotated halfway to the *syn* orientation, so it essentially lies extrahelically in the major groove with the Arg109 guanidine moiety stacked against $A^{(0)}$ base (Fig. 6b). In the *syn* family of models, the $A^{(0)}$ base was more stable. In all *syn* models, Nη2[Arg109] donated a hydrogen bond to $N7[A^{(0)}]$ (Additional file 1: Fig. S3D). The kinked conformation of DNA also allowed the exocyclic amino group of $A^{(0)}$ to form additional non-canonical hydrogen bonds with other nucleotides: $N^6[A^{(0)}]–O^4[T^{+1}]$ and $N^6[A^{(0)}]–O2P[A^{(+2)}]$ for a considerable part of the trajectories (Fig. 6c and Additional file 1: Fig. S3E, F). Considering the tendency of *anti* $A^{(0)}$ to be expelled out of the stack, it is thus likely that in the Fpg-bound oxoG:A mispair (where, in the absence of the protein, A is *anti* [14, 15]), the orphaned A ultimately adopts *syn* conformation.

Despite the geometry of the As models is less disturbed in comparison with the Aa models, oxoG:A experimentally is still a poor substrate for Fpg. A comparison of equilibrium binding, steady-state and pre-steady state kinetics of the *E. coli* enzyme [4, 76] suggests that the Fpg–oxoG:A complex forms quickly but then is much slower to proceed to the catalytically competent conformation than the Fpg–oxoG:C complex, leading to a ~15-fold higher apparent K_M (14 nM for oxoG:C vs 190 nM for oxoG:A [4]; 8.7 nM for oxoG:C vs 150 nM for oxoG:A [76]). At the same time, the observed effect of A on k_{cat} is minor [4, 76]. Since K_M reflects the population of the last pre-catalytic state, it is tempting to suggest that only oxoG:A(*syn*) may be capable of attaining the catalytically competent conformation, thus necessitating the *anti–syn* transition in the course of the productive reaction with oxoG:A and partly explaining its slow progression with this substrate where A is initially *anti* [14, 15].

Aromatic wedge-induced distortion

In all known structures of Fpg bound to DNA, a Phe residue (Phe111 in *Lla*-Fpg) is inserted between the sampled base pair and the base pair 3′ to it (Fig. 1c). The sampled pair is significantly buckled but the strain is relieved upon everting the damaged base, with the Phe wedge remaining to contact the orphaned base and its neighbor in the undamaged strand [25, 77]. Interestingly, in the structures of *Bst*-Fpg containing C, T or G opposite a reduced AP site, the Phe wedges overlap almost perfectly [20]. However, in our models, the Phe residue showed significant mobility: in 8 out of 12 models, Phe111 retreated back into the minor groove. This movement was accompanied with a significant turn of the $A^{(+1)}$ base, which maintained stacking with Phe111: in 9 out of 12 models, the area of contact between the Phe111 side chain and the adenine was larger than in

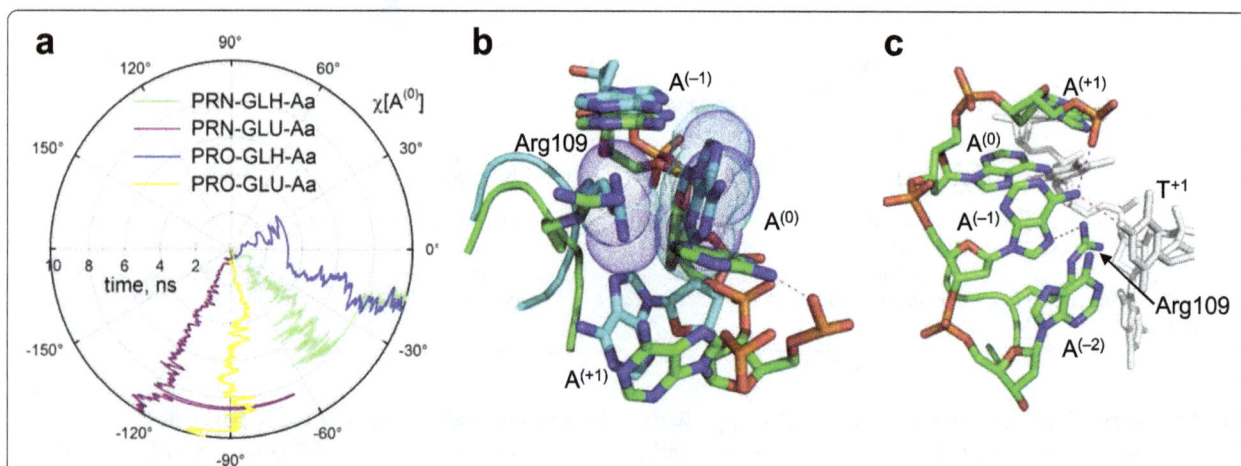

Fig. 6 Conformation of the models around the orphaned nucleotide. **a,** glycosidic angle of the orphaned $A^{(0)}$ nucleotide in the Aa models. **b,** overlay of structures from two snapshots at 6 ns (PRN-GLU-Aa model, carbons colored *green*; PRO-GLH-Aa model, *cyan*, the same structure as in Panel J but slightly turned for a clearer view) showing the central three bases (the non-damaged strand) and Arg109. The hydrogen bond between N6[$A^{(0)}$] and O2P[$A^{(+1)}$] in the PRO-GLH-Aa model and stacking between the partially extrahelical $A^{(0)}$ and Arg109 in the PRN-GLU-Aa model are shown. **c,** structures from an 8-ns snapshot (PRO-GLH-As model) showing the central four base pairs and Arg109. The non-damaged DNA strand is colored. Note the hydrogen bonds formed by the orphaned A base with other nucleotides and Arg109

1XC8 for more than half of the simulation (Fig. 7a and Additional file 1: Fig. S4). As a result, the $T^{-1}:A^{(+1)}$ pair was grossly distorted, mostly by the propeller twist movement (Fig. 7a, b). In the remaining four models, one (PRO-GLH-C) displayed brief aborted attempts to retract Phe in the same manner (with full retraction in one of the repeats), in one (PRO-GLH-Aa), the $A^{(+1)}$ moved by a buckling motion allowing Phe to unstack and adopt an alternative conformation without leaving the double helix, and only in two models (PRN-GLH-C and PRN-GLU-Aa) the initial conformation of the wedge and the adjacent nucleotides was stable.

Specific distant interactions in Fpg–DNA complexes
Model-specific hydrogen bonds
In order to select out inter- and intramolecular interactions specific for oxoG:C, we have searched for hydrogen bonds that existed (i. e., had an energy > 1.2 kcal/mol) in > 1% of the snapshots. Around 600 such hydrogen bonds were found in each model. In all models, less than 50% of the found bonds existed for more than 90% of the snapshots, and less than 25% of the found bonds existed in less than 25% of the snapshots (Additional file 1: Fig. S5A, B). The former class may be considered to represent stable, functionally important hydrogen bonds, whereas the latter one is most likely due to conformational fluctuations. Therefore, all detected hydrogen bonds were first analyzed with respect to their occurrence in these categories (≤90% vs > 90% and ≤ 25% vs > 25%). Pearson's mean square contingency coefficients (φ) for pairwise comparison between different models showed no significant contribution of protonation state or substrate into the overall distribution of bonds in the high- and low-stability categories.

We then searched for the bonds that were consistently different between C, Aa, and As models, selecting those deviating > 3σ from the mean distance between the models (Fig. 8 and Additional file 1: Fig. S6A–C). Only a few bonds consistently showed different stability in all C vs A comparisons irrespective of the *syn* or *anti* conformation of $A^{(0)}$. Unsurprisingly, some of these were formed by the orphaned base itself (Fig. 8). Notably, the oxoG nucleotide, the O^6-binding crown loop, and the Pro1–Glu2 catalytic dyad formed no model-specific bonds. Moreover, a comparison of bonds specific for protonation states (PRO vs PRN, GLU vs GLH) revealed only a few isolated bonds remote from the active site (Additional file 1: Fig. S6D, E).

Fpg regions with C-specific bonds outside the active site
The most prominent opposite-base-specific feature in the protein structure was a cluster at the start of the C-terminal domain immediately next to the interdomain linker (residues Glu134–Phe140). Most of the amino acid residues there engaged in multiple bonds, forming a network, which existed in two stable configurations. In one, which was statistically significantly more often observed in A models, Thr136 formed two bonds with Glu134, one with Asp139, and one with Phe140 (N[Thr136]–O[Glu134], Oγ[Thr136]–O[Glu134], N[Asp139]–Oγ[Thr136], N[Phe140]–Oγ[Thr136]), and a N[Tyr137]–Oε1/Oε2[Glu138] was present. A completely different set of bonds was characteristic of C models (Oγ[Thr136]–Oε1/Oε2[Glu138], N[Asp139]–Oε1/Oε2[Glu138], N[Phe140]–O[Thr136]). As a result, the Glu134–Phe140 loop adopted different conformations in the C and A models (Fig. 8). Importantly, the conservation of Fpg sequence is

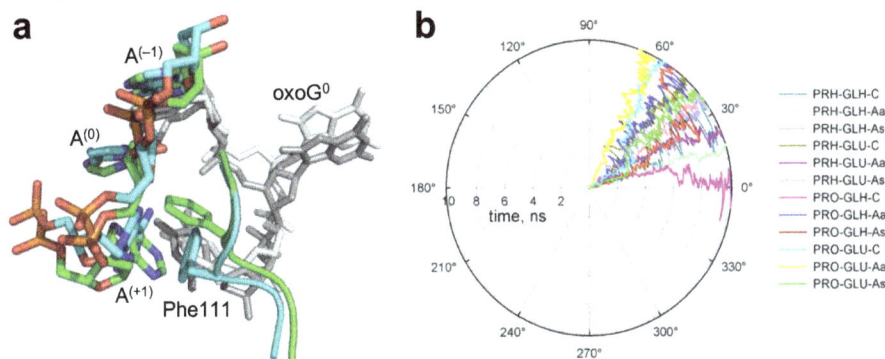

Fig. 7 a, overlay of the structures (PRN-GLU-C model) illustrating the retraction of the intercalating side chain of Phe111. The structure with carbons colored *green* is the starting structure after minimization (0 ns); the structure with carbons colored *cyan* is at 8 ns. The protein backbone (residues 109–113) is shown in the cartoon representation, colored in the same way, with Phe111 presented as a stick model. The N, O, and P atoms are colored *blue, red,* and *orange,* respectively. In DNA, only the non-damaged strand is colored. Note the protein backbone movement, accompanied with ~90° Phe111 ring turn, and the corresponding turn of $A^{(+1)}$ to keep the phenyl ring stacked with the purine heterocycle. **b,** propeller twist angle (ω) of the pair $T^{-1}:A^{(+1)}$. In B-DNA (PDB ID 355D) [80], ω = 13° ± 4°

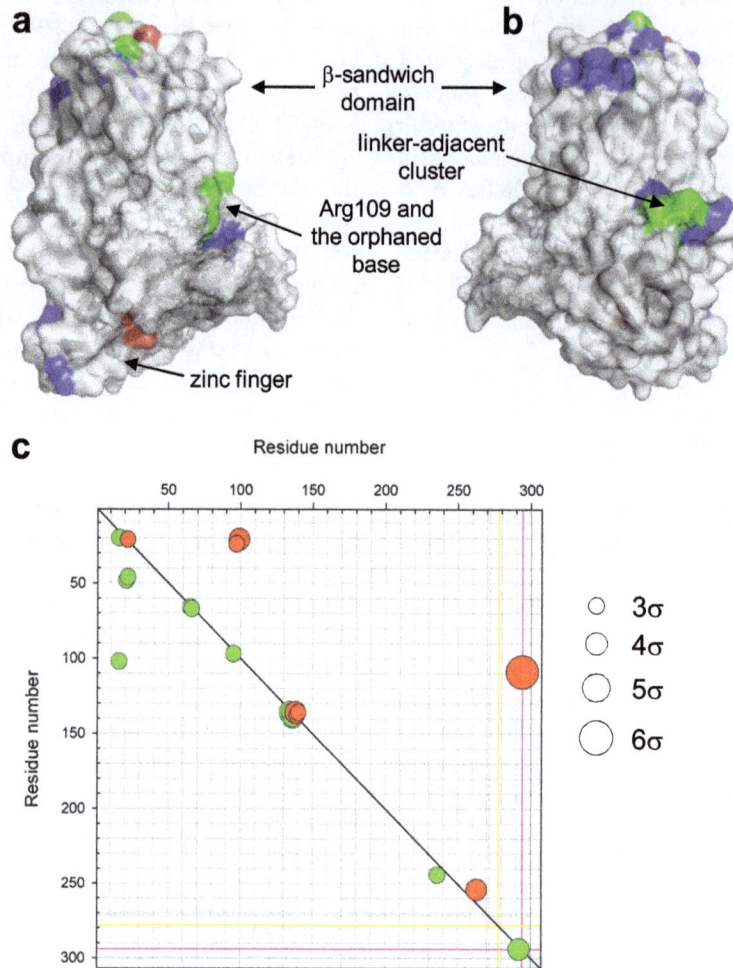

Fig. 8 a, surface representation of the PRO-GLH-C model (8 ns) showing parts of the molecule where C/A-specific hydrogen bonds are found. Residues forming C-specific bonds only are colored *red*, those forming exclusively A-specific bonds are *blue*, and the residues forming alternative bonds in C and A models are *green*. **b,** the same model as in **a** rotated 180° around the vertical axis. **c,** interaction difference map showing pairs of hydrogen bond-forming amino acids specific (>3σ difference in bond occurrence calculated over all pairs of models) for C models (*red*) or A models (*green*). Residues 1–271, protein; 272–285, damaged DNA strand; 286–299, complementary DNA strand; 300, Zn²⁺; 301–307, structural water molecules. The *yellow* line marks the position of oxoG⁰, the *magenta* line, the position of C⁽⁰⁾/A⁽⁰⁾

quite high in this region (Additional file 1: Fig. S7), underlying its functional significance despite its position well away from the active site.

The only other region of known functional importance where consistently different bonds existed was the β-hairpin zinc finger, a structural motif in Fpg involved in major groove tracking and lesion recognition [35] (Fig. 8). Several C/A-specific hydrogen bonds were scattered in the β-sandwich domain around the C-terminal end of the long α-helix αA, which carries the catalytic Pro–Glu dyad at the other end (Fig. 8). The functional significance of this region is not clear; most C/A specific residues here are located in surface loops and are not conserved (Additional file 1: Fig. S7).

Fpg regions with A(syn) and A(anti)-specific bonds outside the active site

In addition, we have searched for bonds specific for A models in different (*anti* or *syn*) conformations of the orphaned A (Additional file 1: Fig. S6C). Most of the differences were encountered between protein and DNA, and within DNA, reflecting the conformational changes inflicted by introducing the disfavored A base. The protein residues affected by the conformation of the orphaned nucleotide showed little overlap with the C/A-specific interactions. The most prominent Aa/As-specific contacts were formed by Tyr29/Arg31 and His91/Lys110, two elements that coordinate the phosphates flanking the orphaned A, and Lys155 that contacts DNA a few nucleotides away from the lesion but is important for Fpg activity [6]. The Glu134–Phe140 C/A-specific linker-adjacent region showed no significant difference between Aa and As models.

Water molecules in the Fpg–DNA complex
Dynamics of structural water in Fpg

The structure of *Lla*-Fpg–DNA complex, 1XC8, contains the total of 397 water molecules. However, only 22 of those reside at the protein–DNA interface and only seven are buried at it (i. e., have < 10% solvent exposure). The structures of Fpg–DNA complexes from different species, as well as the structures of the homolog of Fpg, *Eco*-Nei, in a complex with DNA [78], suggest that several water molecules form a tight network of bonds in the enzyme's active site that may serve to shuttle protons during the concerted cleavage of three bonds catalyzed by Fpg.

We have explicitly modeled the seven water molecules buried at the protein–DNA interface and determined whether they form hydrogen bonds with two or three Fpg or DNA donors or acceptors at the same snapshot. Such water bridges, if persistent, may indicate an important role of water in structure maintenance or reaction mechanism. There were no significant differences between models or between groups of models in the number of water bridges. One particular pair of acceptors, $O\epsilon1/O\epsilon2[Glu76]$ and $O^8[oxoG^0]$, was consistently found bridged by two water molecules in 8 of 12 models. In several models, such multiple water-mediated connections existed between the non-bridging phosphate oxygens of $oxoG^0$ and T^{+1} and between $O2P[oxoG^0]$ and $N\eta1[Arg109]$ but their occurrence was much less common. No donor/acceptor triplets were connected by multiple bridges.

In order to single out the preferred sites of water binding in the Fpg–DNA structure, we have looked in more detail at the water bridges with the occurrence above a threshold of 2000 (for pairs) or 1500 (for triplets). These thresholds cut off the lowest quartile of the cumulative distribution of bridges averaged over all twelve models, i.e., they define the bridges that collectively account for >75% of all occurrences (Additional file 1: Fig. S8). The most frequent triplet was formed by $O\epsilon1/O\epsilon2[Glu2]$, $O\epsilon1[Glu5]$ and $N^2[oxoG^0]$ (Fig. 9a); it was found in the high range in 11 out of 12 models and was not far below the 1500 cut-off (1273) in the remaining one (PRO-GLH-Aa). A cluster of spots habitually occupied by a water molecule was near $N\epsilon[Arg260]$ and $N[Gly261]$ in the protein and non-bridging oxygens of $oxoG^0$ and T^{+1} in DNA (Fig. 9c). Usually, a single water molecule was found in this region at any one snapshot, alternating between different triplets of donors and acceptors. Finally, $O\epsilon1/O\epsilon2[Glu76]$ formed triplets with $N\eta1[Arg109]$ and $O^2[T^{+1}]$ or $O^8[oxoG^0]$ (9 out of 12 models in total) with two water molecules involved (Fig. 9d). Other triplets,

even those passing the threshold of 1500, were found in 1–3 models and are not expected to be significant.

Possible role of structural water molecules in Fpg mechanism

The identified stable triplets are suggestive of an important role of water in the mechanism of action of Fpg. The water molecule trapped between $O\epsilon1/O\epsilon2[Glu2]$, $O\epsilon1[Glu5]$ and $N^2[oxoG^0]$ is located at the position suitable for proton transfer to Glu2, required for the protonation of O4′ of oxoG nucleotide; water-mediated proton transfer to Glu2 was earlier proposed on structural reasons [19, 26]. In the GLH models, this water stably donated a bond only to the unprotonated $O\epsilon1[Glu2]$ (73% bond occurrence averaged over all GLH models, compare with 5% for the bond to $O\epsilon2$) but when Glu2 was charged, $O\epsilon2$ accepted this hydrogen with a higher frequency (64% and 32% bonds to $O\epsilon1$ and $O\epsilon2$, respectively) (Fig. 9b). In the GLH-Aa models, the protonated $O\epsilon2$ showed a tendency to donate a hydrogen bond to the water molecule rather than accept one (23% in PRN-GLH-Aa, 58% in PRO-GLH-Aa, 0–7% in other GLH models), consistent with poor substrate properties of *anti* A. It should be mentioned that in QM/MM analysis of fapyG excision by Fpg this water molecule was inhibitory to the reaction, preventing the protonation of O4′ by neutral Glu2 [36] and should be displaced from its crystallographic position after donating a proton to $O\epsilon2$.

The water molecule bridging the protein residues with the phosphates of T^{+1} and $oxoG^0$ may be important for distorting the DNA duplex. Notably, the distance between the phosphorus atoms $P[T^{+1}]$ and $P[oxoG^0]$ is significantly shorter than in the regular B-DNA in all models. This pinching of the phosphates around T^{+1}, together with wedging of Phe111 and insertion of Met75 and Arg109, assists in kinking the DNA axis by ~60° and eversion of the damaged nucleotide.

The tightly coordinated two-water bridge to $O^8[oxoG^0]$ presents an intriguing conundrum. On one hand, water-mediated recognition of this unique carbonyl would be an attractive mechanism of direct oxoG sensing in the active site pocket. On the other hand, Glu76, which in our models participates in the water coordination, is present only in a small branch of the Fpg family tree consisting of two closely phylogenetically related groups, Bacilli and Mollicutes (which include *L. lactis* and *G. stearothermophilus*), while in all other Fpg sequences this position is occupied by Ser/Thr with very rare exceptions (Additional file 1: Fig. S9, Additional file 3). In the structure of *Bst*-Fpg, the presence of Glu76 stabilizes the everted oxoG in the *high syn* conformation through hydrogen bonding with $N^2[oxoG]$, whereas its *in silico* replacement with Ser reverts the preferred χ angle to the *anti* domain [60]. In *Eco*-Fpg, Ser74 and Lys217 correspond to

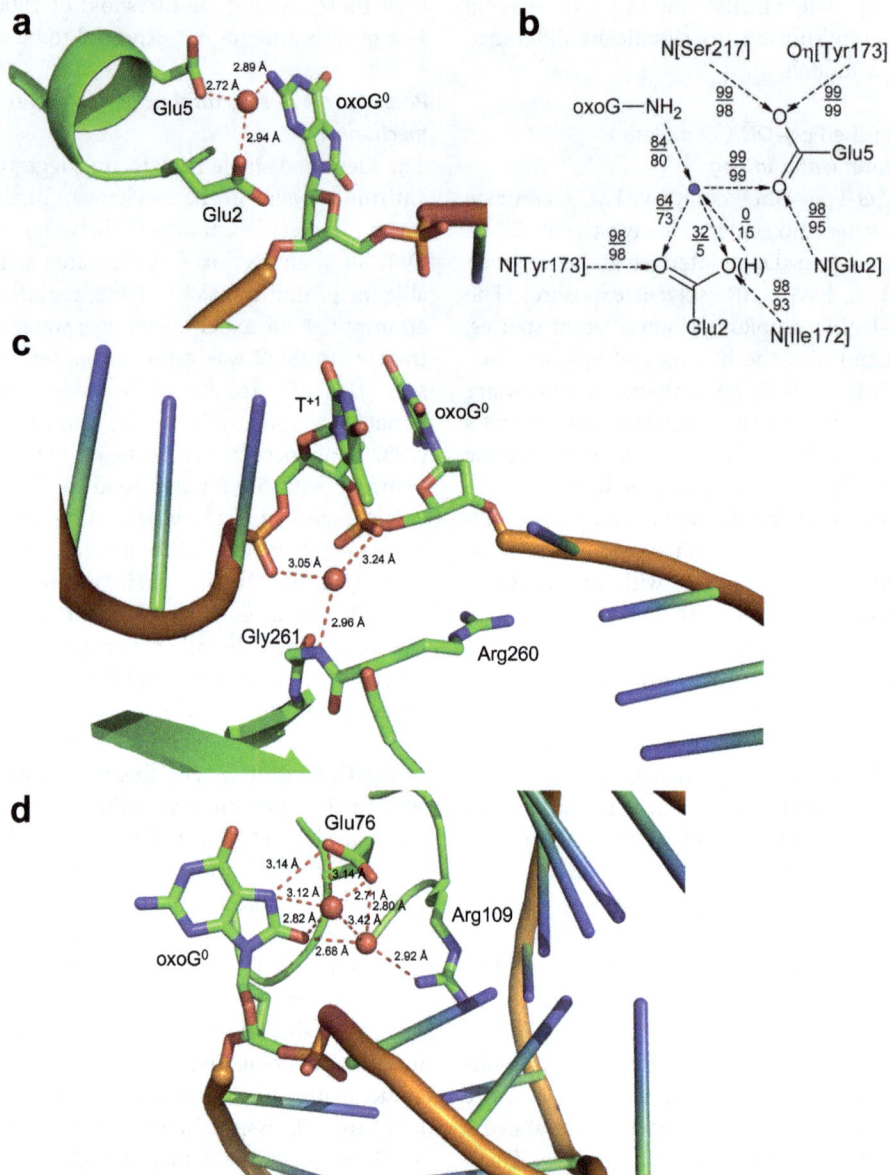

Fig. 9 View of the PRO-GLH-C model (8 ns, the same snapshot as in Fig. 8) showing water traps. Protein and DNA residues coordinating the water molecule (*red ball*) are shown as a stick model and colored according to atom type (*green*, C; *blue*, N; *red*, O; *orange*, P). Other parts of the complex are either shown as a cartoon model or hidden for clarity. Distances between possible hydrogen bond donors and acceptors are indicated by dashed lines. **a,** Glu2, Glu5, and oxoG⁰. **b,** schematic representation of hydrogen bonds formed by the water molecule (*blue dot*) trapped between Glu2, Glu5, and oxoG⁰. The numbers indicate percentage of snapshots in which the bond is observed, averaged over all GLU models (top) or GLH models (bottom). **c,** Arg260, Gly261, oxoG⁰ and T⁺¹. **d,** Glu76, Arg109, and oxoG⁰

Glu76 and Arg220 of *Lla*-Fpg, and Lys217 forms a direct hydrogen bond with O⁸[oxoG] [38]. Obviously, there are several ways by which Fpg enzymes can employ the residues at these positions to the effect of recognizing the exocyclic oxygen at C8 either directly or indirectly.

Conclusion

The opposite-base selectivity of Fpg and some other DNA glycosylases (eukaryotic OGG1 and TDG, bacterial MutY and Mug, etc.) is extremely important for the prevention of mutations in the course of DNA repair. Analysis of the causes of this selectivity is complicated due to the paucity of structures of DNA glycosylases bound to substrates with disfavored opposite bases. In the case of Fpg, no structure containing oxoG:A, the biologically relevant disfavored mispair, is available. Our modeling effort was mostly undertaken to analyze possible structural features of such a complex and reveal those that could explain low activity of Fpg on oxoG:A substrates.

As our models suggest, introduction of A opposite to oxoG indeed distorts the protein–DNA interface within ±2 base pairs around the lesion site, outside of which DNA exists as a normal duplex. Arg-109 and Phe-111, two residues that Fpg inserts into DNA to sharply kink it and maintain oxoG everted from the base stack, tended to withdraw if A was opposite to the lesion, indicating that the pre-catalytic complex of Fpg with oxoG:A is inherently unstable. Interestingly, although the oxoG:A mispair adopts oxoG(*syn*):A(*anti*) conformation in free DNA, our models showed that upon Fpg binding and oxoG eversion, the orphaned A is more stable as a *syn* conformer, engaged both in hydrogen bonding with Arg-109 and in base stacking. We speculate that Fpg binding to oxoG(*syn*):A(*anti*) may be energetically disadvantageous and require rotation of the A base around the glycosidic bond for rare events of base excision; direct test of this hypothesis would require solving the structure of Fpg–DNA(oxoG:A) complex or stopped-flow kinetics with a series of fluorescent reporter bases incorporated next to A, in which case the *anti*–*syn* transition may be expected to be observed in the fluorescent traces.

Analysis of model-specific hydrogen bonds unexpectedly revealed a cluster of highly conserved residues next to the interdomain linker of Fpg, which adopted alternative conformations when C or A was in the opposite strand. This cluster is remote from what is usually considered the active site of Fpg; however, it is packed against a helix–two-turn–helix motif that is present in all Fpg family members and partly forms the DNA-binding groove. Of note, it has been shown that in Nei, a homolog of Fpg specific for oxidized pyrimidines, a structural rearrangement of the linker and the region adjacent to it induces productive DNA binding [79]. Thus, our models add weight to a hypothesis of indirect readout by DNA glycosylases, which states that recognition of damaged bases is not limited to formation of specific bonds but greatly relies on the differences in energetics and dynamics of protein and DNA parts that may be far away from the moiety being recognized.

Structural and kinetic data together with QM/MM modeling of Fpg favor the reaction chemistry that combines a nucleophilic attack at C1′ of oxoG by N[Pro1] residue and protonation of O4′ of oxoG by Oε2[Glu2] [33, 35, 36]. The latter step is important since it affords a ~60 kcal/mol lower barrier to glycosidic bond breakage compared to base protonation as the leaving group activation [33]. Such mechanism requires Pro1 to be in the unprotonated, and Glu2, in the protonated state immediately before the reaction, implying that both these residues should change their preferred protonation state. Our measurements of the pH dependence of Fpg activity suggest that only one group is ionized in a pH-dependent manner, in which case it is consistent with Pro1 N-terminal secondary amine losing a proton at increasing pH. Consequently, the ionization state of Glu2 in the Fpg–DNA complex shows no evidence of being pH-dependent, which means that the assembled active site is capable of protonating Glu2, possibly using a water molecule as a proton shuttle. The arrangement of the reacting atoms is only consistent with the reaction stereochemistry with S_N2 displacement of O4′ as the first step, in agreement with the QM/MM data [33]. Since the substitution of Gln for Glu2 inactivates the enzyme, which rules out simple hydrogen bonding as the primary function of Glu2, the mechanistic implication of our results is that Glu2 has to be deprotonated again later in the reaction, likely by the nascent alkoxide O4′, and contribute its charge to the stabilization of the transition state of the departing oxoG base. In the QM/MM simulation, several consecutive acts of proton transfer between Oε2[Glu2], O4′[oxoG], N[Pro1], and O^8[oxoG] allow the enzyme to lower the highest barrier in the reaction from 71 kcal/mole (as with direct oxoG protonation path) to 13 kcal/mole relative to the lesion recognition complex [33]; a similar energy gain was calculated for fapyG excision [36].

Finally, our modeling concerned only the pre-catalytic complex of Fpg–DNA. It is now clear that the selectivity of DNA glycosylases is not determined exclusively by interactions in their pre-catalytic complexes, the structures of which are relative easy to establish by X-ray crystallography, but also relies on several kinetic gates along the full recognition pathway, including primary encounter and damaged base eversion. Future modeling of the early steps of recognition of oxoG-containing pairs will add clarity to our understanding of the opposite-base discrimination by Fpg.

Additional files

Additional file 1: Figure S1. A, R.m.s.d. of the models over time. The traces are color-coded: dark *cyan*, PRN-GLH-C; *light lime*, PRN-GLH-Aa; *coral*, PRN-GLH-As; *olive*, PRN-GLU-C; dark *magenta*, PRN-GLU-Aa; *light blue*, PRN-GLU-As; *magenta*, PRO-GLH-C; *blue*, PRO-GLH-Aa; *red*, PRO-GLH-As; *cyan*, PRO-GLU-C; *yellow*, PRO-GLU-Aa; *green*, PRO-GLU-As. **B,** Reproducibility of the repeat runs. R.m.s.d. of the initial run (*red*) and three repeat runs (*green, blue,* and *magenta*) of the PRO-GLH-C model are shown together with the cross-run r.m.s.d. between two pairs of the repeat runs (*black* and *green*). Repeat runs of other models produced similar within-run and cross-run r.m.s.d. values and are not shown. **Figure S2.** Circular dichroism spectrum of Fpg at pH 4.0 (*black circles*) and pH 7.6 (*white circles*). **Figure S3.** Conformation of the models around the orphaned nucleotide. **A,** distance Nε[Arg109]…O^2[C$^{(0)}$] in the C models. **B,** distance Nη2[Arg109]…N3[C$^{(0)}$] in the C models. **C,** distance N^6[A$^{(0)}$]…O2P[A$^{(+1)}$] in the Aa models. **D,** distance Nη2[Arg109]…N7[A$^{(0)}$] in the As models. **E,** distance N^6[A$^{(0)}$]…O4[T^{+1}] in the As models. **F,** distance N^6[A$^{(0)}$]…O2P[A$^{(+2)}$] in the As models. Moving average of a 50-snapshot window is shown in all panels. **Figure S4.** Occluded area (inaccessible to a 1.4 Å probe) between Phe111 side chain and A$^{(+1)}$ base. The colors of the traces are the same as in Fig. S1. The dashed line indicates the occluded area in the 1XC8 structure. Moving average of a 50-snapshot window is shown. **Figure S5. A,** Cumulative distribution of the occurrence of hydrogen bonds in the *Lla*-Fpg–DNA complex. **B,** Overall reproducibility of hydrogen bonds in replicate PRO-GLH runs. Dots show the coefficient of variation for the occurrence of a particular hydrogen bond calculated over four replicates plotted against the mean occurrence of the bond. The histograms show the distribution of the mean occurrence. The scale in all panels is the same. Numbers above the graphs indicate the percentage of hydrogen bonds with the mean occurrence >90%. **Figure S6.** Interaction difference maps showing pairs of hydrogen-bond forming amino acids specific (>3σ difference in bond occurrence calculated over all pairs of models) for: **A,** C models (*red*) or Aa models (*blue*); **B,** C models (*red*) or As models (*cyan*); **C,** As models (*cyan*) or Aa models (*blue*); **D,** PRO models (*red*) or PRN models (*blue*); **E,** GLH models (*red*) or GLU models (*blue*). Larger circles correspond to larger deviations from the mean occurrence. Residues 1–271, protein; 272–285, damaged DNA strand; 286–299, complementary DNA strand; 300, Zn^{2+}; 301–307, structural water molecules. The *yellow* line marks the position of oxoG^0, the *magenta* line, the position of C$^{(0)}$/A$^{(0)}$. **Figure S7.** Conservation of Fpg sequence. **A,** plot of conservation number C_n against the residue position. **B,** view of the PRO-GLH-C model (8 ns) colored according to C_n. **C,** the same model as in **B** rotated 180° around the vertical axis. The orientation of the molecule in **B** and **C** is the same as in Fig. 8. **Figure S8.** Rank plot of water-mediated bridges (top 100 occurrences) in the Fpg–DNA structures (**A–L**, the model nature is indicated in the respective panels). *Red*, pairs; *blue*, triplets. Dashed lines indicate cutoffs of 2000 snapshots for pairs and 1500 snapshots for triplets. Insets show cumulative distribution frequencies of pairs and triplets. **Figure S9.** Cladogram of Fpg sequences. The tree was constructed as described in Methods and visualized using TreeDyn [81].

Additional file 2: Table S1. pK_a of Pro1 and Glu2 in selected Fpg structures.

Additional file 3: Alignment of 124 sequences from the Fpg family. See Methods for sequence selection and alignment details.

Acknowledgments

The modeling was performed on an NKS-30T cluster at the SB RAS Supercomputing Center.

Funding

The work was supported by Russian Foundation for Basic Research (grant 17-04-01761-a). The funding body had no role in the design of the study and collection, analysis, and interpretation of data and in writing the manuscript.

Authors' contributions

DOZ and YNV have designed the study. AVP has carried out the MD simulations. AVE has performed biochemical experiments. AVP and YNV have contributed custom software. AVP, YNV and DOZ have participated in the analysis and interpretation of MD trajectories, and writing of the manuscript. All authors read and approved the final manuscript.

Competing interests

The authors declare that they have no competing interests.

Author details

[1]SB RAS Institute of Chemical Biology and Fundamental Medicine, 8 Lavrentieva Ave., Novosibirsk 630090, Russia. [2]Novosibrsk State University, 2 Pirogova St., Novosibirsk 630090, Russia.

References

1. Tchou J, Kasai H, Shibutani S, Chung M-H, Laval J, Grollman AP, Nishimura S. 8-oxoguanine (8-hydroxyguanine) DNA glycosylase and its substrate specificity. Proc Natl Acad Sci U S A. 1991;88:4690–4.
2. Karakaya A, Jaruga P, Bohr VA, Grollman AP, Dizdaroglu M. Kinetics of excision of purine lesions from DNA by *Escherichia coli* Fpg protein. Nucleic Acids Res. 1997;25:474–9.
3. Boiteux S, O'Connor TR, Laval J. Formamidopyrimidine-DNA glycosylase of *Escherichia coli*: cloning and sequencing of the *fpg* structural gene and overproduction of the protein. EMBO J. 1987;6:3177–83.
4. Tchou J, Bodepudi V, Shibutani S, Antoshechkin I, Miller J, Grollman AP, Johnson F. Substrate specificity of Fpg protein: recognition and cleavage of oxidatively damaged DNA. J Biol Chem. 1994;269:15318–24.
5. Hatahet Z, Kow YW, Purmal AA, Cunningham RP, Wallace SS. New substrates for old enzymes: 5-hydroxy-2'-deoxycytidine and 5-hydroxy-2'-deoxyuridine are substrates for *Escherichia coli* endonuclease III and formamidopyrimidine DNA *N*-glycosylase, while 5-hydroxy-2'-deoxyuridine is a substrate for uracil DNA *N*-glycosylase. J Biol Chem. 1994;269:18814–20.
6. Rabow LE, Kow YW. Mechanism of action of base release by *Escherichia coli* Fpg protein: role of lysine 155 in catalysis. Biochemistry. 1997;36:5084–96.
7. Jurado J, Saparbaev M, Matray TJ, Greenberg MM, Laval J. The ring fragmentation product of thymidine C5-hydrate when present in DNA is repaired by the *Escherichia coli* Fpg and Nth proteins. Biochemistry. 1998;37: 7757–63.
8. Gasparutto D, Ait-Abbas M, Jaquinod M, Boiteux S, Cadet J. Repair and coding properties of 5-hydroxy-5-methylhydantoin nucleosides inserted into DNA oligomers. Chem Res Toxicol. 2000;13:575–84.
9. Zhang Q-M, Miyabe I, Matsumoto Y, Kino K, Sugiyama H, Yonei S. Identification of repair enzymes for 5-formyluracil in DNA: Nth, Nei, and MutM proteins of *Escherichia coli*. J Biol Chem. 2000;275:35471–7.
10. Krishnamurthy N, Muller JG, Burrows CJ, David SS. Unusual structural features of hydantoin lesions translate into efficient recognition by *Escherichia coli* Fpg. Biochemistry. 2007;46:9355–65.
11. Friedberg EC, Walker GC, Siede W, Wood RD, Schultz RA, Ellenberger T. DNA repair and mutagenesis. Washington, D.C.: ASM Press; 2006.
12. Zharkov DO. Base excision DNA repair. Cell Mol Life Sci. 2008;65:1544–65.
13. Evans MD, Dizdaroglu M, Cooke MS. Oxidative DNA damage and disease: induction, repair and significance. Mutat Res. 2004;567:1–61.
14. Kouchakdjian M, Bodepudi V, Shibutani S, Eisenberg M, Johnson F, Grollman AP, Patel DJ. NMR structural studies of the ionizing radiation adduct 7-hydro-8-oxodeoxyguanosine (8-oxo-7*H*-dG) opposite deoxyadenosine in a DNA duplex. 8-Oxo-7*H*-dG(*syn*)•dA(*anti*) alignment at lesion site. Biochemistry. 1991;30:1403–12.
15. McAuley-Hecht KE, Leonard GA, Gibson NJ, Thomson JB, Watson WP, Hunter WN, Brown T. Crystal structure of a DNA duplex containing 8-hydroxydeoxyguanine-adenine base pairs. Biochemistry. 1994;33: 10266–70.
16. Grollman AP, Moriya M. Mutagenesis by 8-oxoguanine: an enemy within. Trends Genet. 1993;9:246–9.
17. Duwat P, de Oliveira R, Ehrlich SD, Boiteux S. Repair of oxidative DNA damage in gram-positive bacteria: the *lactococcus lactis* Fpg protein. Microbiology. 1995;141:411–7.
18. Sugahara M, Mikawa T, Kumasaka T, Yamamoto M, Kato R, Fukuyama K, Inoue Y, Kuramitsu S. Crystal structure of a repair enzyme of oxidatively damaged DNA, MutM (Fpg), from an extreme thermophile, *Thermus thermophilus* HB8. EMBO J. 2000;19:3857–69.
19. Gilboa R, Zharkov DO, Golan G, Fernandes AS, Gerchman SE, Matz E, Kycia JH, Grollman AP, Shoham G. Structure of formamidopyrimidine-DNA glycosylase covalently complexed to DNA. J Biol Chem. 2002;277:19811–6.
20. Fromme JC, Verdine GL. Structural insights into lesion recognition and repair by the bacterial 8-oxoguanine DNA glycosylase MutM. Nat Struct Biol. 2002;9:544–52.
21. Serre L, Pereira de jésus K, boiteux S, zelwer C, castaing B. Crystal structure of the *lactococcus lactis* formamidopyrimidine-DNA glycosylase bound to an abasic site analogue-containing DNA. EMBO J. 2002;21:2854–65.
22. Fromme JC, Verdine GL. DNA lesion recognition by the bacterial repair enzyme MutM. J Biol Chem. 2003;278:51543–8.
23. Coste F, Ober M, Carell T, Boiteux S, Zelwer C, Castaing B. Structural basis for the recognition of the FapydG lesion (2,6-diamino-4-hydroxy-5-formamidopyrimidine) by formamidopyrimidine-DNA glycosylase. J Biol Chem. 2004;279:44074–83.

24. Pereira de Jésus K, Serre L, Zelwer C, Castaing B. Structural insights into abasic site for Fpg specific binding and catalysis: comparative high-resolution crystallographic studies of Fpg bound to various models of abasic site analogues-containing DNA. Nucleic Acids Res. 2005;33:5936–44.

25. Banerjee A, Santos WL, Verdine GL. Structure of a DNA glycosylase searching for lesions. Science. 2006;311:1153–7.

26. Coste F, Ober M, Le Bihan Y-V, Izquierdo MA, Hervouet N, Mueller H, Carell T, Castaing B. Bacterial base excision repair enzyme Fpg recognizes bulky N^7-substituted-FapydG lesion via unproductive binding mode. Chem Biol. 2008;15:706–17.

27. Qi Y, Spong MC, Nam K, Banerjee A, Jiralerspong S, Karplus M, Verdine GL. Encounter and extrusion of an intrahelical lesion by a DNA repair enzyme. Nature. 2009;462:762–6.

28. Qi Y, Spong MC, Nam K, Karplus M, Verdine GL. Entrapment and structure of an extrahelical guanine attempting to enter the active site of a bacterial DNA glycosylase, MutM. J Biol Chem. 2010;285:1468–78.

29. Le Bihan Y-V, Izquierdo MA, Coste F, Aller P, Culard F, Gehrke TH, Essalhi K, Carell T, Castaing B. 5-Hydroxy-5-methylhydantoin DNA lesion, a molecular trap for DNA glycosylases. Nucleic Acids Res. 2011;39:6277–90.

30. Qi Y, Nam K, Spong MC, Banerjee A, Sung R-J, Zhang M, Karplus M, Verdine GL. Strandwise translocation of a DNA glycosylase on undamaged DNA. Proc Natl Acad Sci U S A. 2012;109:1086–91.

31. Sung R-J, Zhang M, Qi Y, Verdine GL. Sequence-dependent structural variation in DNA undergoing intrahelical inspection by the DNA glycosylase MutM. J Biol Chem. 2012;287:18044–54.

32. Sung R-J, Zhang M, Qi Y, Verdine GL. Structural and biochemical analysis of DNA helix invasion by the bacterial 8-oxoguanine DNA glycosylase MutM. J Biol Chem. 2013;288:10012–23.

33. Sadeghian K, Flaig D, Blank ID, Schneider S, Strasser R, Stathis D, Winnacker M, Carell T, Ochsenfeld C. Ribose-protonated DNA base excision repair: a combined theoretical and experimental study. Angew Chem Int Ed. 2014;53: 10044–8.

34. Sun B, Latham KA, Dodson ML, Lloyd RS. Studies of the catalytic mechanism of five DNA glycosylases: probing for enzyme-DNA imino intermediates. J Biol Chem. 1995;270:19501–8.

35. Zharkov DO, Shoham G, Grollman AP. Structural characterization of the Fpg family of DNA glycosylases. DNA Repair. 2003;2:839–62.

36. Blank ID, Sadeghian K, Ochsenfeld C. A base-independent repair mechanism for DNA glycosylase—no discrimination within the active site. Sci Rep. 2015; 5:10369.

37. Popov AV, Vorob'ev YN. GUI-BioPASED: a program for molecular dynamics simulations of biopolymers with a graphical user interface. Mol Biol (Mosk). 2010;44:648–54.

38. Perlow-Poehnelt RA, Zharkov DO, Grollman AP, Broyde S. Substrate discrimination by formamidopyrimidine-DNA glycosylase: distinguishing interactions within the active site. Biochemistry. 2004;43:16092–105.

39. Hornak V, Abel R, Okur A, Strockbine B, Roitberg A, Simmerling C. Comparison of multiple Amber force fields and development of improved protein backbone parameters. Proteins. 2006;65:712–25.

40. Ravishanker G, Auffinger P, Langley DR, Jayaram B, Young MA, Beveridge DL. Treatment of counterions in computer simulations of DNA. Rev Comput Chem. 1997;11:317–72.

41. Lazaridis T, Karplus M. Effective energy function for proteins in solution. Proteins. 1999;35:133–52.

42. Popov AV, Vorobjev YN, Zharkov DO. MDTRA: a molecular dynamics trajectory analyzer with a graphical user interface. J Comput Chem. 2013;34:319–25.

43. Vorobjev YN. Study of the mechanism of interaction of oligonucleotides with the 3'-terminal region of tRNA^{Phe} by computer modeling. Mol Biol (Mosk). 2005;39:777–84.

44. Mauget S. Time series analysis based on running Mann-Whitney Z Statistics. J Time Ser Anal. 2011;32:47–53.

45. Benjamini Y, Hochberg Y. Controlling the false discovery rate: a practical and powerful approach to multiple testing. J R Stat Soc Ser B Stat Methodol. 1995;57:289–300.

46. Humphrey W, Dalke A, Schulten K. VMD: Visual molecular dynamics. J Mol Graph. 1996;14:33–8.

47. Sayle RA, Milner-White EJ. RASMOL: biomolecular graphics for all. Trends Biochem Sci. 1995;20:374–6.

48. Søndergaard CR, Olsson MHM, Rostkowski M, Jensen JH. Improved treatment of ligands and coupling effects in empirical calculation and rationalization of pK_a values. J Chem Theory Comput. 2011;7:2284–95.

49. Altschul SF, Madden TL, Schäffer AA, Zhang J, Zhang Z, Miller W, Lipman DJ. Gapped BLAST and PSI-BLAST: a new generation of protein database search programs. Nucleic Acids Res. 1997;25:3389–402.

50. Zharkov DO, Grollman AP. Combining structural and bioinformatics methods for the analysis of functionally important residues in DNA glycosylases. Free Radic Biol Med. 2002;32:1254–63.

51. Zharkov DO. Predicting functional residues in DNA glycosylases by analysis of structure and conservation. In: Practical Bioinformatics. Edited by Bujnicki JN. Berlin–Heidelberg: Springer-Verlag. 2004;15:243-61.

52. Sievers F, Wilm A, Dineen D, Gibson TJ, Karplus K, Li W, Lopez R, McWilliam H, Remmert M, Söding J, et al. Fast, scalable generation of high-quality protein multiple sequence alignments using clustal omega. Mol Syst Biol. 2011;7:539.

53. Livingstone CD, Barton GJ. Protein sequence alignments: a strategy for the hierarchical analysis of residue conservation. Comput Appl Biosci. 1993;9: 745–56.

54. Ober M, Linne U, Gierlich J, Carell T. The two main DNA lesions 8-oxo-7,8-dihydroguanine and 2,6-diamino-5-formamido-4-hydroxypyrimidine exhibit strongly different pairing properties. Angew Chem Int Ed. 2003;42:4947–51.

55. Oda Y, Uesugi S, Ikehara M, Nishimura S, Kawase Y, Ishikawa H, Inoue H, Ohtsuka E. NMR studies of a DNA containing 8-hydroxydeoxyguanosine. Nucleic Acids Res. 1991;19:1407–12.

56. Lipscomb LA, Peek ME, Morningstar ML, Verghis SM, Miller EM, Rich A, Essigmann JM, Williams LD. X-ray structure of a DNA decamer containing 7,8-dihydro-8-oxoguanine. Proc Natl Acad Sci U S A. 1995;92:719–23.

57. Zaika EI, Perlow RA, Matz E, Broyde S, Gilboa R, Grollman AP, Zharkov DO. Substrate discrimination by formamidopyrimidine-DNA glycosylase: a mutational analysis. J Biol Chem. 2004;279:4849–61.

58. Amara P, Serre L, Castaing B, Thomas A. Insights into the DNA repair process by the formamidopyrimidine-DNA glycosylase investigated by molecular dynamics. Protein Sci. 2004;13:2009–21.

59. Amara P, Serre L. Functional flexibility of Bacillus stearothermophilus formamidopyrimidine DNA-glycosylase. DNA Repair. 2006;5:947–58.

60. Song K, Hornak V, de los Santos C, Grollman AP, Simmerling C. Computational analysis of the mode of binding of 8-oxoguanine to formamidopyrimidine-DNA glycosylase. Biochemistry. 2006;45:10886–94.

61. Song K, Kelso C, de los Santos C, Grollman AP, Simmerling C. Molecular simulations reveal a common binding mode for glycosylase binding of oxidatively damaged DNA lesions. J Am Chem Soc. 2007;129:14536–7.

62. Thiviyanathan V, Somasunderam A, Hazra TK, Mitra S, Gorenstein DG. Solution structure of a DNA duplex containing 8-hydroxy-2'-deoxyguanosine opposite deoxyguanosine. J Mol Biol. 2003;325:433–42.

63. Chen J, Brooks III CL, Khandogin J. Recent advances in implicit solvent-based methods for biomolecular simulations. Curr Opin Struct Biol. 2008;18:140–8.

64. Vorobjev YN. Advances in implicit models of water solvent to compute conformational free energy and molecular dynamics of proteins at constant pH. Adv Protein Chem Struct Biol. 2011;85:281–322.

65. Kleinjung J, Fraternali F. Design and application of implicit solvent models in biomolecular simulations. Curr Opin Struct Biol. 2014;25:126–34.

66. Anandakrishnan R, Drozdetski A, Walker RC, Onufriev AV. Speed of conformational change: comparing explicit and implicit solvent molecular dynamics simulations. Biophys J. 2015;108:1153–64.

67. Li S, Bradley P. Probing the role of interfacial waters in protein–DNA recognition using a hybrid implicit/explicit solvation model. Proteins. 2013;81:1318–29.

68. Sadeghian K, Ochsenfeld C. Unraveling the base excision repair mechanism of human DNA glycosylase. J Am Chem Soc. 2015;137:9824–31.

69. Sowlati-Hashjin S, Wetmore SD. Computational investigation of glycosylase and β-lyase activity facilitated by proline: applications to FPG and comparisons to hOgg1. J Phys Chem B. 2014;118:14566–77.

70. Stivers JT, Jiang YL. A mechanistic perspective on the chemistry of DNA repair glycosylases. Chem Rev. 2003;103:2729–60.

71. Storm DR, Koshland Jr DE. A source for the special catalytic power of enzymes: orbital steering. Proc Natl Acad Sci U S A. 1970;66:445–52.

72. Bruice TC, Brown A, Harris DO. On the concept of orbital steering in catalytic reactions. Proc Natl Acad Sci U S A. 1971;68:658–61.

73. Fuxreiter M, Warshel A, Osman R. Role of active site residues in the glycosylase step of T4 endonuclease V. Computer simulation studies on ionization states. Biochemistry. 1999;38:9577–89.

74. O'Brien PJ, Ellenberger T. Human alkyladenine DNA glycosylase uses acid-base catalysis for selective excision of damaged purines. Biochemistry. 2003; 42:12418–29.

75. Song K, Campbell AJ, Bergonzo C, de los Santos C, Grollman AP, Simmerling C. An improved reaction coordinate for nucleic acid base flipping studies. J Chem Theory Comput. 2009;5:3105–13.

76. Kuznetsov NA, Koval W, Zharkov DO, Vorobjev YN, Nevinsky GA, Douglas KT, Fedorova OS. Pre-steady-state kinetic study of substrate specificity of *Escherichia coli* formamidopyrimidine-DNA glycosylase. Biochemistry. 2007; 46:424–35.

77. Kuznetsov NA, Bergonzo C, Campbell AJ, Li H, Mechetin GV, de los Santos C, Grollman AP, Fedorova OS, Zharkov DO, Simmerling C. Active destabilization of base pairs by a DNA glycosylase wedge initiates damage recognition. Nucleic Acids Res. 2015;43:272–81.

78. Zharkov DO, Golan G, Gilboa R, Fernandes AS, Gerchman SE, Kycia JH, Rieger RA, Grollman AP, Shoham G. Structural analysis of an *Escherichia coli* endonuclease VIII covalent reaction intermediate. EMBO J. 2002;21:789–800.

79. Golan G, Zharkov DO, Feinberg H, Fernandes AS, Zaika EI, Kycia JH, Grollman AP, Shoham G. Structure of the uncomplexed DNA repair enzyme endonuclease VIII indicates significant interdomain flexibility. Nucleic Acids Res. 2005;33:5006–16.

80. Shui X, McFail-Isom L, Hu GG, Williams LD. The B-DNA dodecamer at high resolution reveals a spine of water on sodium. Biochemistry. 1998;37:8341–55.

81. Chevenet F, Brun C, Bañuls A-L, Jacq B, Christen R. TreeDyn: towards dynamic graphics and annotations for analyses of trees. BMC Bioinformatics. 2006;7:439.

8

Endocrine disruption: *In silico* perspectives of interactions of di-(2-ethylhexyl)phthalate and its five major metabolites with progesterone receptor

Ishfaq A. Sheikh[1], Muhammad Abu-Elmagd[2], Rola F. Turki[3,4], Ghazi A. Damanhouri[1], Mohd A. Beg[1]* and Mohammed Al-Qahtani[2]

From 3rd International Genomic Medicine Conference
Jeddah, Saudi Arabia. 30 November - 3 December 2015

Abstract

Background: Di-(2-ethylhexyl)phthalate (DEHP) is a common endocrine disrupting compound (EDC) present in the environment as a result of industrial activity and leaching from polyvinyl products. DEHP is used as a plasticizer in medical devices and many commercial and household items. Exposure occurs through inhalation, ingestion, and skin contact. DEHP is metabolized to a primary metabolite mono-(2-ethylhexyl)phthalate (MEHP) in the body, which is further metabolized to four major secondary metabolites, mono(2-ethyl-5-hydroxyhexyl)phthalate (5-OH-MEHP), mono(2-ethyl-5-oxyhexyl)phthalate (5-oxo-MEHP), mono(2-ethyl-5-carboxypentyl)phthalate (5-cx-MEPP) and mono[2-(carboxymethyl)hexyl]phthalate (2-cx-MMHP). DEHP and its metabolites are associated with developmental abnormalities and reproductive dysfunction within the human population. Progesterone receptor (PR) signaling is involved in important reproductive functions and is a potential target for endocrine disrupting activities of DEHP and its metabolites. This study used *in silico* approaches for structural binding analyses of DEHP and its five indicated major metabolites with PR.

Methods: Protein Data bank was searched to retrieve the crystal structure of human PR (Id: 1SQN). PubChem database was used to obtain the structures of DEHP and its five metabolites. Docking was performed using Glide (Schrodinger) Induced Fit Docking module.

Results: DEHP and its metabolites interacted with 19-25 residues of PR with the majority of the interacting residues overlapping (82-95 % commonality) with the native bound ligand norethindrone (NET). DEHP and each of its five metabolites formed a hydrogen bonding interaction with residue Gln-725 of PR. The binding affinity was highest for NET followed by DEHP, 5-OH-MEHP, 5-oxo-MEHP, MEHP, 5-cx-MEPP, and 2-cx-MMHP.

Conclusion: The high binding affinity of DEHP and its five major metabolites with PR as well as a high rate of overlap between PR interacting residues among DEHP and its metabolites and the native ligand, NET, suggested their disrupting potential in normal PR signaling, resulting in adverse reproductive effects.

Keywords: Docking, Progesterone receptor, DEHP, 5-OH-MEHP, 5-oxo-MEHP, MEHP, 5-cx-MEPP, 2-cx-MMHP

* Correspondence: mabeg51@gmail.com
[1]King Fahd Medical Research Center, King Abdulaziz University, PO Box
80216, Jeddah 21589, Kingdom of Saudi Arabia
Full list of author information is available at the end of the article

Background

The chemical industry contributes significantly to the prosperity and economic development of modern society. However, many chemical compounds that are discharged into the environment due to industrial activity and leaching from consumer products interfere with the physiological functions of the exposed human and animal populations and are referred to as endocrine disrupting compounds (EDCs) [1–3].

Di-(2-ethylhexyl)phthalate (DEHP) is a high volume plasticizer used as a softener in polyvinyl chloride industry with a 54 % market share (2010 data) and is considered as one of the most common EDCs present in the environment [4]. DEHP is frequently used in the manufacture of medical devices, blood storage bags, surgical gloves, dialysis equipment, cosmetics, household and personal items such as soap, shampoo, detergents, adhesives, vinyl flooring, shower curtains, plastic bags, garden hoses, children's toys, and many other plastic products [4]. Exposure of human population to DEHP occurs continuously through inhalation, ingestion, and skin contact [5]. Recently [6], DEHP was detected in 74 % of 72 common food items including infant foods, chicken, pork and other food items in a market in Albany, New York. DEHP is metabolized in the body by hydrolysis to a primary metabolite, mono-(2-ethylhexyl)phthalate (MEHP), which is then further metabolized into multiple hydroxylative and oxidative secondary metabolites [7, 8]. The four major secondary metabolites of

DEHP are mono(2-ethyl-5-hydroxyhexyl)phthalate (5-OH-MEHP), mono(2-ethyl-5-oxyhexyl)phthalate (5-oxo-MEHP), mono(2-ethyl-5-carboxypentyl)phthalate (5-cx-MEPP) and mono[2-(carboxymethyl)hexyl]phthalate (2-cx-MMHP) [7, 8]. A simplified metabolic pathway of 5 major metabolites of DEHP is illustrated (Fig. 1).

DEHP and its metabolites have been detected in various human body fluids such as blood and breast milk [9], follicular fluid [10], amniotic fluid [11], cord blood of newborns [12] and urine [5] indicating immense potential for adverse health effects. Monoester metabolites rather than native DEHP are thought to be responsible for toxicity of DEHP [13] with secondary metabolites displaying a 100 fold increase in embryo-toxicity compared to MEHP [14]. In a recent study [15], positive associations were reported between total DEHP metabolites, MEHP, 5-OH-MEHP, and 5-oxo-MEHP levels in urine and plasma estradiol and ratio of estradiol to testosterone. Higher levels of MEHP, 5-OH-MEHP, and 5-oxo-MEHP were associated with lower sperm concentration, lower sperm motility, higher sperm apoptosis, and ROS generation [16]. Prenatal exposure with DEHP and its metabolites has been associated with reduced gestational age for pregnancies bearing male fetus [17], anogenital distance problems in male babies [18–20], cryptorchidism [21], altered reproductive hormone levels [22], hypospadias [23], intellectual and motor development in children [24], and preterm birth [25, 26]. Retrospective analyses of DEHP metabolites in pregnancy

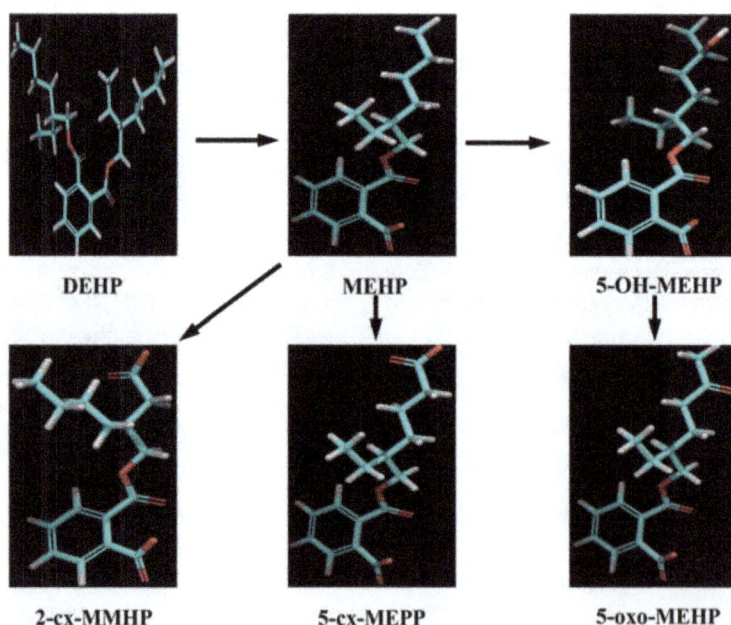

Fig. 1 Two dimensional representation and a simplified pathway of di-(2-ethylhexyl)phthalate (DEHP) and its five major metabolites, mono-(2-ethylhexyl)phthalate (MEHP), mono-(2-ethyl-5-hydroxyhexyl)phthalate (5-OH-MEHP), mono-(2-ethyl-5-oxyhexyl)phthalate (5-oxo-MEHP), mono-(2-ethyl-5-carboxypentyl)phthalate (5-cx-MEPP), and mono-[2-(carboxymethyl)hexyl]phthalate (2-cx-MMHP)

serum of mothers [27] indicated that prenatal exposure of children to DEHP was associated with reproductive problems during adolescence; higher 5-OH-MEHP level in prenatal maternal serum was related with lower semen volume and lower sperm concentrations and higher 5-cx-MEPP was associated with lower free testosterone concentrations.

Studies in rats and mice have also shown that exposure to DEHP can induce deleterious reproductive and endocrine effects [28–31]. In rats, prenatal DEHP treatment was associated with developmental abnormalities in male pups such as cryptorchidism, anogenital problems, and malformations of epididymis, vas deferens, seminal vesicles, prostate, and external genitalia collectively called as the phthalate syndrome, which is similar to effects of DEHP exposure in men [32, 33]. In vitro, MEHP and 5-OH-MEHP decreased gonocyte number and increased gonocyte apoptosis in rat testis organ culture [34].

In general, EDCs have been proposed to exert their toxic effects through interactions with nuclear steroid receptors, sex steroid binding proteins, and steroid enzymatic pathways regulating reproductive and endocrine functions [1]. Progesterone receptor (PR) belongs to the family of nuclear receptors and binds to progesterone, which is an important hormone involved in female reproductive function and maintenance of pregnancy [35, 36] as well as an important modulator of male reproductive function [37]. Interference in PR signaling leads to reproductive dysfunction and pregnancy failure [38]. Recently [39], docking studies of PR with three stereoisomers of DEHP have been reported. Docking of DEHP and its primary metabolite, MEHP, with PR have also been reported [40], however, the important secondary metabolites were not included in the study.

This study aimed at analyzing and comparing the structural binding characteristics of DEHP and its five major metabolites, MEHP, 5-OH-MEHP, 5-oxo-MEHP, 5-cx-MEPP, and 2-cx-MMHP with PR using *in silico* approaches. The study involved the delineation of the binding mechanism of all the six xeno-ligands with PR by molecular docking simulation and comparing the distinctive binding pattern and the interacting residues.

Methods

Data retrieval

The molecular structures of DEHP and its five major metabolites, MEHP, 5-OH-MEHP, 5-oxo-MEHP, 5-cx-MEPP, and 2-cx-MMHP were retrieved from PubChem compound database. The two dimensional structures of the ligands are illustrated (Fig. 1) and their abbreviations and PubChem compound identities (CIDs) are presented (Table 1). Schrodinger 2015 suite with Maestro 10.3 (graphical user interface) software (Schrodinger, LLC, New York, NY, 2015) was used for docking studies of DEHP and its five metabolites [39].

Protein selection and preparation

The Protein Data Bank (PDB; http://www.rcsb.org/) was searched to retrieve the crystal structure of human PR (PDB code: 1SQN) with a resolution of 1.45 Å. The crystal structure was a co-complex with bound ligand, norethindrone (NET). The preparation of the co-complex crystal structure for docking analysis was done using protein preparation wizard workflow of Schrodinger Glide (Schrodinger suite 2015-3; Schrodinger, LLC) and was described in detail [39]. Briefly, the PDB structure was imported to docking software Glide and using protein preparation wizard workflow, OPLS-2005 force field, and Prime 3.0 module software water molecules were removed, hydrogen atoms and charges were added, and loops and missing side chains were built. The hydrogen bonding network was optimized and finally a geometry optimization was performed to a maximum root-mean-square deviation (RMSD) of 0.30 Å. For generating grid boxes, bound ligand (NET) in crystal complex was selected and used for docking of DEHP and its five metabolites.

Ligand preparation, conformational search

The methodology described above [39] was employed to draw ligand structures (Fig. 1) using Maestro 10.3 (Maestro, version 10.3, Schrodinger, LLC). LigPrep module (Schrodinger 2015: LigPrep, version 3.1, Schrodinger, LLC) was used for preparation of ligands and correct molecular

Table 1 Nomenclature, commonly used abbreviations, and PubChem IDs of di-(2-ethylhexyl)phthalate and its five major metabolites for docking study with human progesterone receptor (PR)

S.No.	Name	Abbreviation	PubChem ID
1	Di-(2-ethylhexyl)phthalate	DEHP	8343
2	Mono-(2-ethylhexyl)phthalate	MEHP	20393
3	Mono-(2-ethyl-5-hydroxyhexyl)phthalate	5-OH-MEHP	170295
4	Mono-(2-ethyl-5-oxyhexyl)phthalate	5-oxo-MEHP	119096
5	Mono-(2-ethyl-5-carboxypentyl)phthalate	5-cx-MEPP	149386
6	Mono-[2-(carboxymethyl)hexyl]phthalate	2-cx-MMHP	187353
7	Norethindrone	NET	6230

geometries and ionization at biological pH 7.4 were obtained by using the OPLS-2005 force field software.

Induced fit docking

Schrodinger's Induced Fit Docking (IFD) module was used for docking analyses of the DEHP and its five metabolites MEHP, 5-OH-MEHP, 5-oxo-MEHP, 5-cx-MEPP, and 2-cx-MMHP [39]. The ligands were submitted as starting geometries to IFD which is capable of sampling the minor changes in the backbone structure as well as robust conformational changes in side chains [41]. A softened-potential docking is performed in the first IFD stage where docking of the ligand occurs into an ensemble of the binding protein conformations. Subsequently, complex minimization for highest ranked pose is performed where both ligand and binding sites are free to move.

Binding energy calculations

The ligand binding affinity calculations against the crystal complex was executed using Prime module of Schrodinger 2015 with MMGB-SA function.

Results

Successful execution of IFD for docking simulation of DEHP and its five major metabolites, MEHP, 5-OH-MEHP, 5-oxo-MEHP, 5-cx-MEPP, and 2-cx-MMHP against the ligand binding pocket of PR resulted in multiple docking poses for each ligand. The best pose for each ligand was analyzed further for *in silico* data considerations and the resulting data is presented here (Figs. 2, 3, 4, 5, 6 and 7). Similarly, for the co-complex bound ligand (NET) of PR the data for the best pose after IFD are illustrated (Fig. 8).

Docking complexes of DEHP and its five major metabolites, MEHP, 5-OH-MEHP, 5-oxo-MEHP, 5-cx-MEPP, and 2-cx-MMHP displayed interactions with 19-25

amino acid residues of PR (Figs. 2, 3, 4, 5, 6 and 7, Table 2). The bound ligand, NET, displayed interactions with 22 residues of PR in the NET-PR docking complex (Fig. 8; Table 2). DEHP and its five metabolites shared 18-21 PR interacting residues with the bound native PR ligand, NET, (commonality of 82-95 %; Table 2). For each of the native ligand, NET, and DEHP and its five metabolites, 16 PR interacting residues (Leu-718, Asn-719, Leu-721, Gln-725, Met-756, Met-759, Val-760, Leu-763, Arg-766, Phe-778, Leu-797, Met-801, Leu-887, Tyr-890, Cys-891, Met-909) were common (Table 3). The PR interacting residues, Leu-715 and Thr-894 were also common between bound ligand, NET, and DEHP and all of its metabolites except MEHP, and residue Gly-722 was common between NET and all ligands except 5-OH-MEHP (Table 3). In addition, two residues, Trp-755 and Phe-905 were common among NET and 4 of 6 ligand molecules (not shown). DEHP and each of its five metabolites and bound native ligand, NET, formed a hydrogen bonding interaction against residue Gln-725 of PR. In addition, MEHP, 5-oxo-MEHP, 5-cx-MEPP, and 2-cx-MMHP each formed two hydrogen bonding interactions with residue Arg-766 of PR. The metabolite 5-OH-MEHP formed only one hydrogen bonding interaction with Arg-766 but was also involved in a hydrogen bonding interaction with another residue, Asn-719, of PR. The IFD score, Dock score, and Glide score for all the docked xeno-ligands and bound native ligand, NET, are presented (Table 2). The binding affinity values (MMGB-SA values) were highest for NET followed by DEHP, 5-OH-MEHP, 5-oxo-MEHP, MEHP, 5-cx-MEPP, and 2-cx-MMHP (Table 2).

Discussion

Di-(2-ethylhexyl)phthalate (DEHP) is a widely used phthalate compound representing more than half of all phthalate compounds manufactured worldwide for use in the industry as a plasticizer. Several reviews showed

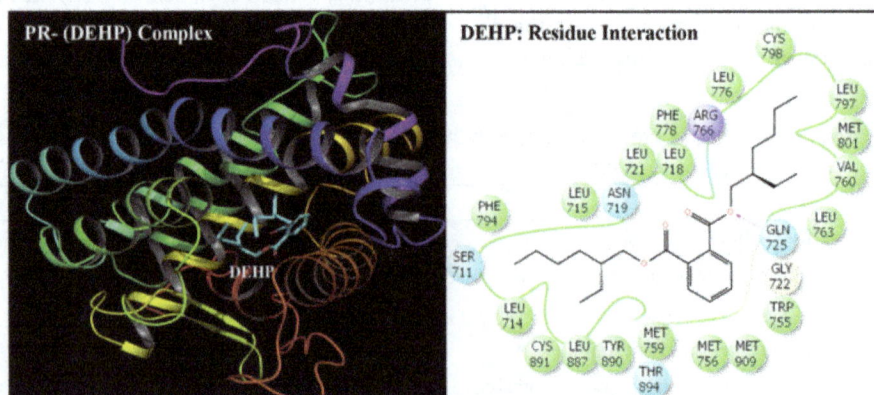

Fig. 2 Ribbon form representation of docking complex of human progesterone receptor (PR) with di-(2-ethylhexyl)phthalate (DEHP) (left panel). Amino-acid residues in the binding pocket of PR involved in interactions with DEHP (right panel)

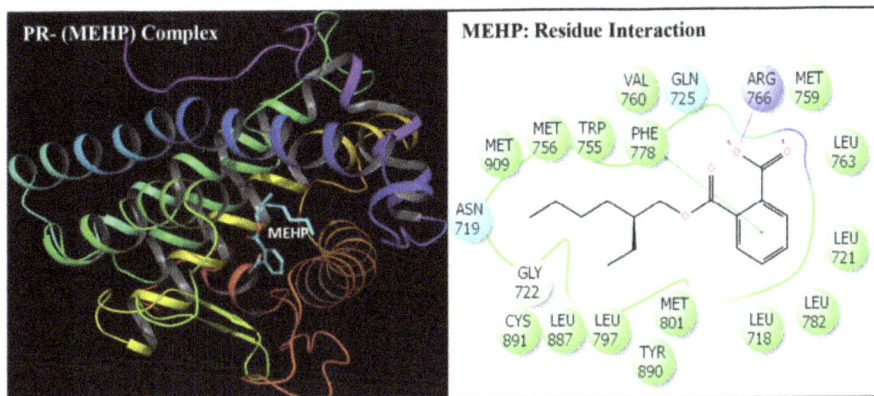

Fig. 3 Ribbon form representation of docking complex of human progesterone receptor (PR) with mono-(2-ethylhexyl)phthalate (MEHP) (left panel). Amino-acid residues in the binding pocket of PR involved in interactions with MEHP (right panel)

that DEHP is a universally prevalent environmental contaminant and behaves as a reproductive and developmental toxin [5, 28, 29, 32, 42]. Several epidemiological reports have identified DEHP and its metabolites as the cause of adverse effects on various systems of the body including endocrine and reproductive system [28, 29, 32]. Many studies have reported developmental problems during prenatal period and postnatal period in unborn and new born children as a result of gestational exposure of mothers to DEHP [18–20, 24, 42, 43]. In women, higher urinary or serum levels of DEHP and its metabolites were associated with problems in conception, endometriosis, and high rates of miscarriage, delayed or preterm gestation, and pregnancy associated toxemia and preeclampsia [25, 26, 28, 29]. In men, higher urinary or serum levels of DEHP and its metabolites were linked with lower semen volume, lower sperm concentrations, lower sperm motility, higher sperm apoptosis, and lower testosterone concentrations [5, 16, 27, 32, 44, 45]. Due to side effects of DEHP mentioned above, it has been banned since 2009 in the United States for use in children's toys and the European Union has also classified DEHP as a reproductive toxicant. However, DEHP continues to be manufactured and used in many countries across the world.

DEHP is metabolized in the body by hydroxylative and oxidative reactions to many metabolic products which include five major metabolites: MEHP, 5-OH-MEHP, 5-oxo-MEHP, 5-cx-MEPP, and 2-cx-MMHP ([7, 8]; see Introduction section). The toxicity of DEHP in the body is attributed mainly to the actions of its secondary metabolites [13, 14]. Progesterone receptor signaling is an essential pathway controlling reproductive function and is involved in reproductive periodicity and establishment and maintenance of pregnancy [35, 36]. DEHP and the indicated five major metabolites can act as potential xenoligands for PR and disrupt the normal progesterone

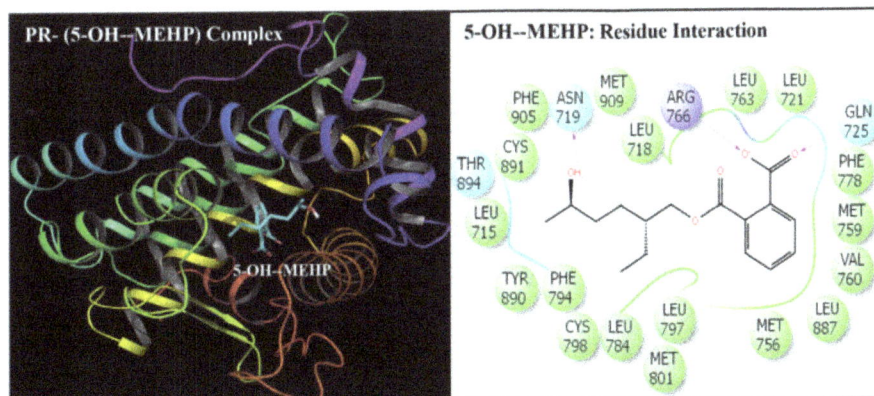

Fig. 4 Ribbon form representation of docking complex of human progesterone receptor (PR) with mono-(2-ethyl-5-hydroxyhexyl)phthalate (5-OH-MEHP) (left panel). Amino-acid residues in the binding pocket of PR involved in interactions with 5-OH-MEHP (right panel)

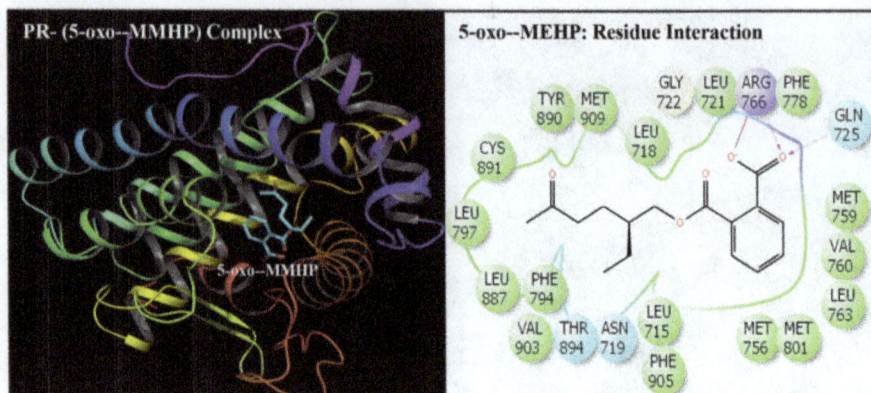

Fig. 5 Ribbon form representation of docking complex of human progesterone receptor (PR) with mono-(2-ethyl-5-oxyhexyl)phthalate (5-oxo-MEHP) (left panel). Amino-acid residues in the binding pocket of PR involved in interactions with 5-oxo-MEHP (right panel)

signaling pathway and this could be one of the important mechanisms which lead to adverse effects in the human population. In the present study, docking simulations of DEHP and its five major metabolites namely, MEHP, 5-OH-MEHP, 5-oxo-MEHP, 5-cx-MEPP, and 2-cx-MMHP were performed with PR and comparison of docking displays and interacting residues was performed among the ligands and the co-complex bound native ligand, norethindrone (NET) of PR crystal structure.

Induced Fit Docking of DEHP and its five metabolites with PR showed that all the six xeno-ligands fitted well into the steroid binding pocket of the receptor. The high binding affinity values, IFD scores, and dock scores indicated that the docking complexes formed by DEHP, MEHP, 5-OH-MEHP, 5-oxo-MEHP, 5-cx-MEPP, and 2-cx-MMHP with PR were in their most favorable conformation. A number of important PR amino acid residues interacted through hydrophobic and hydrogen-bonding

interfaces with each of the six xeno-ligands during docking simulation contributing to the ligand-PR docking complex stability. A consistent and high overlapping (82-95 % commonality) of the interacting residues of PR among native bound ligand, NET, and DEHP and its metabolites suggested a common platform of action. This was further supported by the fact that 16 of the 22 PR residues interacting with bound native ligand, NET, also interacted with DEHP and each of the five metabolites. In addition, DEHP and each of its five metabolites, and bound native ligand, NET, formed a hydrogen bonding interaction against residue Gln-725 of PR altogether pointing to the common structural binding characteristics of the native bound ligand and the six xeno-ligands. Commonality of structural binding characteristics of bound native ligand, NET, and DEHP and its metabolites with PR suggest, on a preliminary basis, potential disruption of PR function by DEHP and its metabolites.

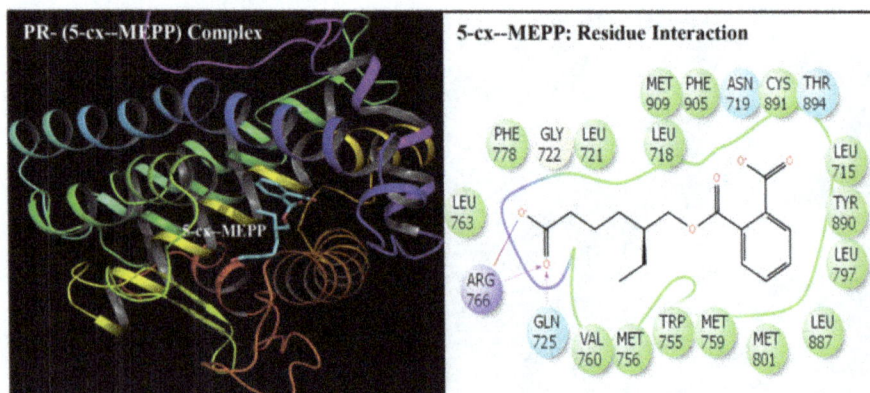

Fig. 6 Ribbon form representation of docking complex of human progesterone receptor (PR) with mono-(2-ethyl-5-carboxypentyl)phthalate (5-cx-MEPP) (left panel). Amino-acid residues in the binding pocket of PR involved in interactions with 5-cx-MEPP (right panel)

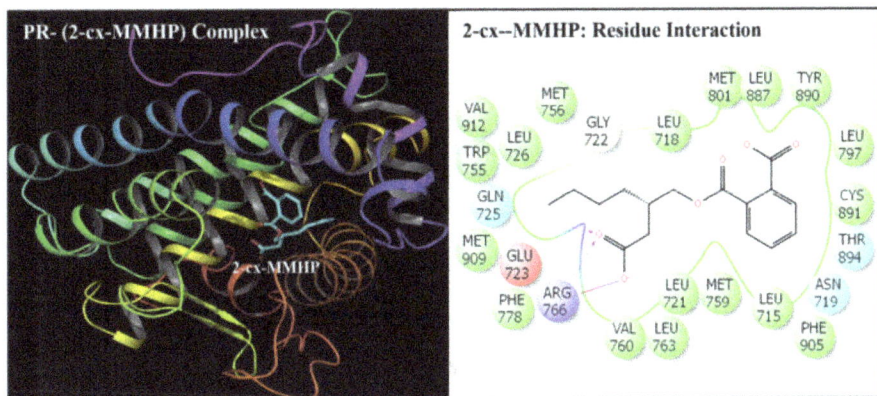

Fig. 7 Ribbon form representation of docking complex of human progesterone receptor (PR) with mono-[2-(carboxymethyl)hexyl]phthalate (2-cx-MMHP)(left panel). Amino-acid residues in the binding pocket of PR involved in interactions with 2-cx-MMHP (right panel)

To the best of our knowledge, the current study is the first structure based report for docking stimulation of secondary metabolites of DEHP with PR. In vitro competitive binding of DEHP and its metabolites with PR are seemingly unavailable. Docking studies of PR with three stereoisomers of DEHP have recently been reported [39]. Docking of DEHP and its primary metabolite, MEHP, with PR have also been reported [40]. The results of the current study with docking of DEHP and PR support the results of the reported study [40] showing residues Gln-725, Arg-766 and Phe-778 as the crucial interacting residues of PR interaction with DEHP. The importance of the current study lies in the fact that the secondary metabolites of DEHP viz. 5-OH-MEHP, 5-oxo-MEHP, 5-cx-MEPP, and 2-cx-MMHP are the best biomonitoring markers of DEHP in the urine or blood and are potentially more potent disruptors because of their long elimination half-life compared to the primary metabolite, MEHP [7, 8]. Approximately 75 % of a single dose of DEHP was excreted in urine within two days; 67 % was excreted within the first 24 h which included 6 % MEHP, 23 % 5-OH-MEHP, 15 % 5-oxo-MEHP, 19 % 5-cx-MEPP, and 4 % 2-cx-MMHP (Koch et al. [8]). Of the 3.8 % excreted in the next 24 h, more than 75 % included 5-cx-MEPP and 2-cx-MMHP and the rest included 5-OH-MEHP and 5-oxo-MEHP indicating long elimination half-lives of the former two secondary metabolites.

Although not related to the progesterone receptor, DEHP treatment inhibited progesterone secretion from human luteal cells in culture [46]. Furthermore, in vivo treatment of DEHP decreased secretion of progesterone in mice [47] and in vitro treatment of MA-10 mouse Leydig cells with MEHP resulted in inhibition of steroidogenesis including progesterone secretion [48]. Interestingly, in sheep, DEHP causes shortening of

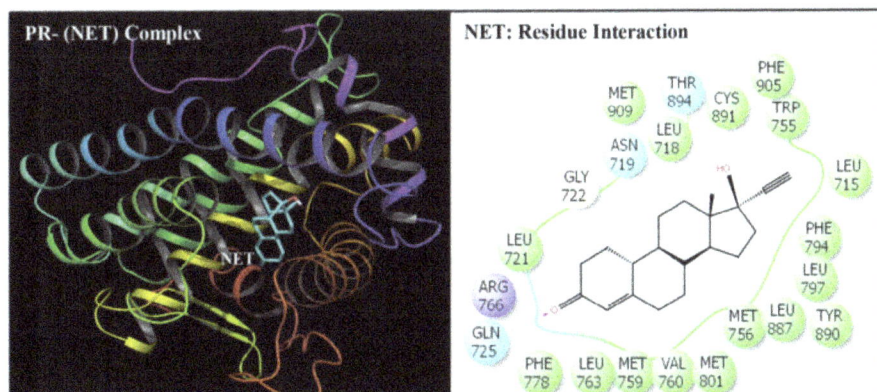

Fig. 8 Ribbon form representation of docking complex of human progesterone receptor (PR) with native co-complex ligand norethindrone (NET) (left panel). Amino-acid residues in the binding pocket of PR involved in interactions with NET (right panel)

Table 2 Number of interacting residues, number and percentage of residues common with native ligand norethindrone (NET), Induced Fit Docking (IFD) Score, Dock score, Glide score and binding affinity values (MMGB-SA values) of di-(2-ethylhexyl)phthalate (DEHP), mono-(2-ethylhexyl)phthalate (MEHP), mono-(2-ethyl-5-hydroxyhexyl)phthalate (5-OH-MEHP), mono-(2-ethyl-5-oxyhexyl)phthalate (5-oxo-MEHP), mono-(2-ethyl-5-carboxypentyl)phthalate (5-cx-MEPP), and mono-[2-(carboxymethyl)hexyl]phthalate (2-cx-MMHP) and native co-complex ligand, NET, after IDF with human progesterone receptor (PR)

S. no.	Ligand	Number of interacting residues	Number of interacting residues common with NET (%)	IFD score	Docking score (Kcal/mol)	Glide score (Kcal/mol)	MMGB-SA (Kcal/mol)
1	DEHP	25	20 (91 %)	-563.15	-9.59	-9.59	-131.26
2	MEHP	19	18 (82 %)	-560.47	-8.40	-8.40	-84.18
3	5-OH-MEHP	22	20 (91 %)	-561.24	-8.95	-8.95	-90.78
4	5-oxo-MEHP	22	20 (91 %)	-561.72	-8.83	-8.83	-87.24
5	5-cx-MEPP	21	21 (95 %)	-562.87	-10.52	-10.52	-80.01
6	2-cx-MMHP	24	21 (95 %)	-562.01	-9.02	-9.02	-68.59
7	NET	22	22 (100 %)	-566.25	-12.13	-12.13	-139.00

estrous cycles due to a reduction in the size and lifespan of CL, however, in contrast to mice, an increase in circulating concentrations of progesterone was noted [49]. Conversely, MEHP treatment was associated with an increase in steroidogenesis including progesterone concentrations in cultured rat ovarian follicles [50]. Apparently, direct studies involving treatments with secondary metabolite compounds namely 5-OH-MEHP, 5-oxo-MEHP, 5-cx-MEPP, and 2-cx-MMHP in laboratory animals or in in vitro cell cultures are not available. It goes without saying that no single mechanism or pathway can explain the endocrine disrupting effects of DEHP and its metabolites

Table 3 Amino-acid residues of human progesterone receptor that were common among co-complex natural ligand, norethindrone (NET), and di-(2-ethylhexyl)phthalate (DEHP) and its five major metabolites, mono-(2-ethylhexyl)phthalate (MEHP), mono-(2-ethyl-5-hydroxyhexyl)phthalate (5-OH-MEHP), mono-(2-ethyl-5-oxyhexyl)phthalate (5-oxo-MEHP), mono-(2-ethyl-5-carboxypentyl)phthalate (5-cx-MEPP), and mono-[2-(carboxymethyl)-hexyl]phthalate (2-cx-MMHP)

S. no	Interacting residue	S. no	Interacting residue
1	Leu-715[a]	12	Phe-778
2	Leu-718	13	Leu-797
3	Asn-719	14	Met-801
4	Leu-721	15	Leu-887
5	Gly-722[b]	16	Tyr-890
6	Gln-725	17	Cys-891
7	Met-756	18	Thr-894[a]
8	Met-759	19	Met-909
9	Val-760		
10	Leu-763		
11	Arg-766		

Amino-acid residues indicated by superscript [a] were not shared by MEHP and the residue indicated by superscript [b] was not shared by 5-OH-MEHP

on reproductive and endocrine systems in the human body. As an example, PPAR alpha was thought to be a possible pathway of adverse effects of DEHP in mice, however, the toxic effects were observed despite the use of PPAR alpha null mice suggesting the involvement of additional pathways [51]. Besides the PR signaling pathway, multiple other pathways could mediate the adverse effects of DEHP and its metabolites in the body. Androgen receptor pathway could also be an important mechanism as agonistic (androgenic) and antagonistic (antiandrogenic) actions of DEHP and other phthalate compounds have been shown at the androgen receptor level [52]. This study showed that DEHP and all its five major metabolites were able to bind to PR with structural binding characteristics that were common with the bound native ligand, NET, of PR. Hence, DEHP and its five metabolites have potential to interfere with the binding of progesterone to its receptor resulting in adverse effects and the dysfunction of progesterone signaling.

Conclusion

This study was undertaken to understand the structural binding mechanisms of DEHP and its five major metabolites (MEHP, 5-OH-MEHP, 5-oxo-MEHP, 5-cx-MEPP, and 2-cx-MMHP) with PR in order to predict their potential adverse effects on progesterone signaling. The results indicated, a high percentage of overlap (82-95 %) among the interacting residues of PR for the native bound ligand, NET, and for DEHP and its metabolites. The structural binding similarities were further supported by a common hydrogen bonding interaction between Gln-725 residue of PR and DEHP, each of its five metabolites, and bound native ligand, NET. Therefore, on a preliminary basis, the six xeno-ligands have potential disruptive activities in the binding of progesterone to its receptor resulting in the dysfunction of progesterone signaling and adverse effects.

Abbreviations

2-cx-MMHP: mono[2-(carboxymethyl)hexyl]phthalate; 5-cx-MEPP: mono(2-ethyl-5-carboxypentyl)phthalate; 5-OH-MEHP: mono(2-ethyl-5-hydroxyhexyl)phthalate; 5-oxo-MEHP: mono(2-ethyl-5-oxyhexyl)phthalate; DEHP: Di-(2-ethylhexyl)phthalate; EDC: Endocrine disrupting compound; IFD: Induced fit docking; MEHP: mono-(2-ethylhexyl)phthalate; PR: Progesterone receptor; RMSD: Root-mean-square deviation; WHO: World Health Organization

Acknowledgements

The authors are thankful to M. S. Gazdar, Head of the library at KFMRC, for help with online journals and providing access to books and journals.

Declaration

This article has been published as part of *BMC Structural Biology* Volume 16 Supplement 1, 2016: Proceedings of the 3rd International Genomic Medicine Conference: structural biology. The full contents of the supplement are available online at http://bmcstructbiol.biomedcentral.com/articles/supplements/volume-16-supplement-1.

Funding

This project was funded by the Deanship of Scientific Research (DSR), King Abdulaziz University, Jeddah, under grant no. (HiCi-1434-117-9). The authors, therefore, acknowledge with thanks the DSR technical and financial support. Publication charges for this article were funded by the Center of Excellence in Genomic Medicine Research, King Abdulaziz University, Jeddah, Kingdom of Saudi Arabia.

Authors' contributions

IAS conducted literature search, conceived and designed the experiments, performed the experiments, analyzed the data, prepared manuscript outline, and revision. MAB conducted literature search, conceived and designed the experiments, analyzed the data, prepared manuscript outline, and revision. MAE, RFT, GAD and MAQ conducted literature search, participated in manuscript preparation, drafting, critical review and revision. All authors have read and approved the final manuscript.

Competing interests

The authors declare that they have no competing interests.

Author details

[1]King Fahd Medical Research Center, King Abdulaziz University, PO Box 80216, Jeddah 21589, Kingdom of Saudi Arabia. [2]Centre of Excellence in Genomic Medicine Research, King Abdulaziz University, Jeddah, Kingdom of Saudi Arabia. [3]KACST Innovation Center in Personalized Medicine, King Abdulaziz University, Jeddah, Kingdom of Saudi Arabia. [4]Department of Obstetrics and Gynecology, King Abdulaziz University Hospital, Jeddah, Kingdom of Saudi Arabia.

References

1. Diamanti-Kandarakis E, Bourguignon JP, Giudice LC, Hauser R, Prins GS, Soto AM, et al. Endocrine-disrupting chemicals: an Endocrine Society scientific statement. Endocr Rev. 2009;30(4):293–342.
2. WHO-UNEP. State of the science of endocrine disrupting chemicals 2012. Edited by Bergman Å, Heindel JJ, Jobling S, Kidd KA, Zoeller RT editors. WHO Press, Geneva, Switzerland, 2013, pp 1–260. http://www.who.int/ceh/publications/endocrine/en/
3. Gore AC, Crews D, Doan LL, La Merrill M, Patisaul H, Zota A. Introduction to endocrine disrupting chemicals (EDCs) — a guide for public interest organizations and policy makers. Endocrine Society reports and white papers, 2014; pp 1–76 (http://ipen.org/sites/default/files/documents/ipen-intro-edc-v1_9a-en-web.pdf).
4. Guo Y, Kannan K. A survey of phthalates and parabens in personal care products from the United States and its implications for human exposure. Environ Sci Technol. 2013;47(24):14442–9.
5. Heudorf U, Mersch-Sundermann V, Angerer J. Phthalates: toxicology and exposure. Int J Hyg Environ Health. 2007;210:623–34.
6. Schecter A, Lorber M, Guo Y, Wu Q, Yun S, Kannan K, et al. Phthalate concentrations and dietary exposure from food purchased in New York State. Environ Health Perspect. 2013;121(4):473–9.
7. Koch HM, Bolt HM, Angerer J. Di(2-ethylhexyl)phthalate (DEHP) metabolites in human urine and serum after a single oral dose of deuterium-labelled DEHP. Arch Toxicol. 2004;78(3):123–30.
8. Koch HM, Bolt HM, Preuss R, Angerer J. New metabolites of di(2-ethylhexyl)phthalate (DEHP) in human urine and serum after single oral doses of deuterium-labelled DEHP. Arch Toxicol. 2005;79(7):367–76.
9. Hogberg J, Hanberg A, Berglund M, Skerfving S, Remberger M, Calafat AM, et al. Phthalate diesters and their metabolites in human breast milk, blood or serum, and urine as biomarkers of exposure in vulnerable populations. Environ Health Perspect. 2008;116(3):334–9.
10. Krotz SP, Carson SA, Tomey C, Buster JE. Phthalates and bisphenol do not accumulate in human follicular fluid. J Assist Reprod Genet. 2012;29:773–7.
11. Huang PC, Kuo PL, Chou YY, Lin SJ, Lee CC. Association between prenatal exposure to phthalates and the health of newborns. Environ Int. 2009;35(1):14–20.
12. Lin L, Zheng LX, Gu YP, Wang JY, Zhang YH, Song WM. Levels of environmental endocrine disruptors in umbilical cord blood and maternal blood of low-birth-weight infants. Zhonghua Yu Fang Yi Xue Za Zhi. 2008;42(3):177–80.
13. Kavlock R, Boeckelheide K, Chapin R, Cunningham M, Faustman E, Foster P, et al. NTP Center for the Evaluation of Risks to Human Reproduction: Phthalates expert panel report on the reproductive and developmental toxicity of di(2-ethylhexyl)phthalate. Reprod Toxicol. 2002;16(5):529–53.
14. Regnier J, Bowden C, Lhuguenot J. Effects on rat embryonic development in vitro of di-(2-ethylhexyl) phthalate (DEHP) and its metabolites. Toxicol CD— Official J Soc Toxicol. 2004;78:187.
15. Fong JP, Lee FJ, Lu IS, Uang SN, Lee CC. Relationship between urinary concentrations of di(2-ethylhexyl) phthalate (DEHP) metabolites and reproductive hormones in polyvinyl chloride production workers. Occup Environ Med. 2015;72(5):346–53.
16. Huang LP, Lee CC, Fan JP, Kuo PH, Shih TS, Hsu PC. Urinary metabolites of di(2-ethylhexyl) phthalate relation to sperm motility, reactive oxygen species generation, and apoptosis in polyvinyl chloride workers. Int Arch Occup Environ Health. 2014;87(6):635–46.
17. Weinberger B, Vetrano AM, Archer FE, Marcella SW, Buckley B, Wartenberg D, et al. Effects of maternal exposure to phthalates and bisphenol A during pregnancy on gestational age. J Matern Fetal Neonatal Med. 2014;27(4):323–27.
18. Bustamante-Montes LP, Hernández-Valero MA, Flores-Pimentel D, García-Fábila M, Amaya-Chávez A, Barr DB, et al. Prenatal exposure to phthalates is associated with decreased anogenital distance and penile size in male newborns. J Dev Orig Health Dis. 2013;4(4):300–6.
19. Bornehag CG, Carlstedt F, Jönsson BA, Lindh CH, Jensen TK, Bodin A, et al. Prenatal phthalate exposures and anogenital distance in Swedish boys. Environ Health Perspect. 2015;123(1):101–7.
20. Swan SH, Sathyanarayana S, Barrett ES, Janssen S, Liu F, Nguyen RH, et al. First trimester phthalate exposure and anogenital distance in newborns. Hum Reprod. 2015;30(4):963–72.
21. Swan SH. Environmental phthalate exposure in relation to reproductive outcomes and other health endpoints in humans. Environ Res. 2008;108(2):177–84.
22. Araki A, Mitsui T, Miyashita C, Nakajima T, Naito H, Ito S, et al. Association between maternal exposure to di(2-ethylhexyl) phthalate and reproductive hormone levels in fetal blood: the Hokkaido study on environment and children's health. PLoS One. 2014;9(10):e109039.
23. Ormond G, Nieuwenhuijsen MJ, Nelson P, Toledano MB, Iszatt N, Geneletti S, et al. Endocrine disruptors in the workplace, hair spray, folate supplementation, and risk of hypospadias: case control study. Environ Health Perspect. 2009;117(2):303–7.
24. Kim Y, Ha EH, Kim EJ, Park H, Ha M, Kim JH, et al. Prenatal exposure to phthalates and infant development at 6 months: Prospective mothers and children's environmental health (moceh) study. Environ Health Perspect. 2011;119(10):1495–500.

25. Ferguson KK, McElrath TF, Ko YA, Mukherjee B, Meeker JD. Variability in urinary phthalate metabolite levels across pregnancy and sensitive windows of exposure for the risk of preterm birth. Environ Int. 2014;70:118–24.

26. Ferguson KK, McElrath TF, Meeker JD. Environmental phthalate exposure and preterm birth. JAMA Pediatr. 2014;168(1):61–7.

27. Axelsson J, Rylander L, Rignell-Hydbom A, Lindh CH, Jönsson BA, Giwercman A. Prenatal phthalate exposure and reproductive function in young men. Environ Res. 2015;138:264–70.

28. Kay VR, Chambers C, Foster WG. Reproductive and developmental effects of phthalate diesters in females. Crit Rev Toxicol. 2013;43(3):200–19.

29. Hannon PR, Flaws JA. The effects of phthalates on the ovary. Front Endocrinol (Lausanne). 2015;6:8.

30. Niermann S, Rattan S, Brehm E, Flaws JA. Prenatal exposure to di-(2-ethylhexyl) phthalate (DEHP) affects reproductive outcomes in female mice. Reprod Toxicol. 2015;53:23–32.

31. Moyer B, Hixon ML. Reproductive effects in F1 adult females exposed in utero to moderate to high doses of mono-2-ethylhexylphthalate(MEHP). Reprod Toxicol. 2012;34(1):43–50.

32. Kay VR, Bloom MS, Foster WG. Reproductive and developmental effects of phthalate diesters in males. Crit Rev Toxicol. 2014;44(6):467–98.

33. Lioy PJ, Hauser R, Gennings C, Koch HM, Mirkes PE, Schwetz BA, et al. Assessment of phthalates/phthalate alternatives in children's toys and childcare articles: Review of the report including conclusions and recommendation of the Chronic Hazard Advisory Panel of the Consumer Product Safety Commission. J Expo Sci Environ Epidemiol. 2015;25(4):343–53.

34. Chauvigne F, Menuet A, Lesne L, Chagnon MC, Chevrier C, Regnier JF, et al. Time- and dose-related effects of di-(2-ethylhexyl) phthalate and its main metabolites on the function of the rat fetal testis in vitro. Environ Health Perspect. 2009;117(4):515–21.

35. Wang H, Dey SK. Roadmap to embryo implantation: clues from mouse models. Nat Rev Genet. 2006;7(3):185–99.

36. Ellmann S, Sticht H, Thiel F, Beckmann MW, Strick R, Strissel PL. Estrogen and progesterone receptors: from molecular structures to clinical targets. Cell Mol Life Sci. 2009;66(15):2405–26.

37. Oettel M, Mukhopadhyay AK. Progesterone: the forgotten hormone in men? Aging Male. 2004;7(3):236–57.

38. Lydon JP, Demayo FJ, Funk CR, Mnai SK, Hughes AR, Montgomery CA, et al. Mice lacking progesterone receptor exhibit pleiotropic reproductive abnormalities. Genes Dev. 1995;9(18):2266–78.

39. Sheikh IA. Stereoselectivity and the potential endocrine disrupting activity of di-(2-ethylhexyl)phthalate (DEHP) against human progesterone receptor: a computational perspective. J Appl Toxicol. 2016;36(5):741–7.

40. Sarath Josha MK, Pradeepa S, Vijayalekshmy Amma KS, Sudha Devi R, Balachandran S, Sreejith MN et al. Human ketosteroid receptors interact with hazardous phthalate plasticizers and their metabolites: an in silico study. J Appl Toxicol. 2015; doi: 10.1002/jat.3221. (Epub ahead of print).

41. Nabuurs SB, Wagener M, de Vlieg J. A flexible approach to induced fit docking. J Med Chem. 2007;50(26):6507–18.

42. Braun JM, Sathyanarayana S, Hauser R. Phthalate exposure and children's health. Curr Opin Pediatr. 2013;25:247–54.

43. Kolarik B, Naydenov K, Larsson M, Bornehag CG, Sundell J. The association between phthalates in dust and allergic diseases among bulgarian children. Environ Health Perspect. 2008;116(1):98–103.

44. Meeker JD, Calafat AM, Hauser R. Urinary metabolites of di(2-ethylhexyl) phthalate are associated with decreased steroid hormone levels in adult men. J Androl. 2009;30(3):287–97.

45. Mendiola J, Meeker JD, Jorgensen N, Andersson AM, Liu F, Calafat AM, et al. Urinary concentrations of di(2-ethylhexyl) phthalate metabolites and serum reproductive hormones: pooled analysis of fertile and infertile men. J Androl. 2012;33(3):488–98.

46. Romani F, Tropea A, Scarinci E, Federico A, Dello Russo C, Lisi L, et al. Endocrine disruptors and human reproductive failure: the in vitro effect of phthalates on human luteal cells. Fertil Steril. 2014;102:831–7.

47. Li N, Liu T, Zhou L, He J, Ye L. Di-(2-ethylhcxyl) phthalate reduces progesterone levels and induces apoptosis of ovarian granulosa cell in adult female ICR mice. Environ Toxicol Pharmacol. 2012;34:869–75.

48. Zhou L, Beattie MC, Lin CY, Liu J, Traore K, Papadopoulos V, et al. Oxidative stress and phthalate-induced down-regulation of steroidogenesis in MA-10 Leydig cells. Reprod Toxicol. 2013;42:95–101.

49. Herreros MA, Gonzalez-Bulnes A, Iñigo-Nunez S, Contreras-Solis I, Ros JM, Encinas T. Toxicokinetics of di(2-ethylhexyl) phthalate (DEHP) and its effects on luteal function in sheep. Reprod Biol. 2013;13:66–74.

50. Inada H, Chihara K, Yamashita A, Miyawaki I, Fukuda C, Tateishi Y, et al. Evaluation of ovarian toxicity of mono-(2-ethylhexyl) phthalate (MEHP) using cultured rat ovarian follicles. J Toxicol Sci. 2012;37:483–90.

51. Zhang Y, Ge R, Hardy MP. Androgen-forming stem leydig cells: identification, function and therapeutic potential. Dis Markers. 2008;24:277–86.

52. Specht IO, Toft G, Hougaard KS, Lindh CH, Lenters V, Jönsson BA, et al. Associations between serum phthalates and biomarkers of reproductive function in 589 adult men. Environ Int. 2014;66:146–56.

9

A computational assessment of pH-dependent differential interaction of T7 lysozyme with T7 RNA polymerase

Subhomoi Borkotoky and Ayaluru Murali*◉

Abstract

Background: T7 lysozyme (T7L), also known as N-acetylmuramoyl-L-alanine amidase, is a T7 bacteriophage gene product. It involves two functions: It can cut amide bonds in the bacterial cell wall and interacts with T7 RNA polymerase (T7RNAP) as a part of transcription inhibition. In this study, with the help of molecular dynamics (MD) calculations and computational interaction studies, we investigated the effect of varying pH conditions on conformational flexibilities of T7L and their influence on T7RNAP -T7L interactions.

Results: From the MD studies of the T7L at three different pH strengths *viz.* 5, neutral and 7.9 it was observed that T7L structure at pH 5 exhibited less stable nature with more residue level fluctuations, decrease of secondary structural elements and less compactness as compared to its counterparts: neutral pH and pH 7.9. The T-pad analysis of the MD trajectories identified local fluctuations in few residues that influenced the conformational differences in three pH strengths. From the docking of the minimum energy representative structures of T7L at different pH strengths (obtained from the free energy landscape analysis) with T7RNAP structures at same pH strengths, we saw strong interaction patterns at pH 7.9 and pH 5. The MD analysis of these complexes also confirmed the observations of docking study. From the combined *in silico* studies, it was observed that there are conformational changes in N-terminal and near helix 1 of T7L at different pH strengths, which are involved in the T7RNAP interaction, thereby varying the interaction pattern.

Conclusion: Since T7L has been used for developing novel therapeutics and T7RNAP one of the most biologically useful protein in both *in-vitro* and *in vivo* experiments, this *in silico* study of pH dependent conformational differences in T7L and the differential interaction with T7RNAP at different pH can provide a significant insight into the structural investigations on T7L and T7RNAP in varying pH environments.

Keywords: T7 lysozyme, T7 RNA polymerase, Molecular dynamics simulation, Docking, Principal component analysis, T-pad analysis

Background

The ~17 kDa lysozyme of bacteriophage T7 (T7L) or simply T7 lysozyme, also known as N-acetylmuramoyl-L-alanine amidase or endolysin, is a product of class II gene of T7 bacteriophage genome. Endolysins have a wide array of usage such as antimicrobial agents [1], food safety [2], against phytopathogenic bacteria [3], enzybiotics [4], disinfectants [5] etc. Endolysins have been categorized into four classes: i) glycosidases (muramidase),

ii) endopeptidases, iii) amidohydrolases (amidase), and iv) lytic transglycosylases. Endolysins infect both Gram-positive and Gram-negative bacteria; while the former type contains multiple domains, the later one generally represents single-domain globular proteins (15–20 kDa). T7 lysozyme (T7L) falls into the later type of the endolysins. The T7L is a bi-functional protein that cuts amide bonds in the bacterial cell wall and also inhibits transcription by T7 RNA polymerase. It lyses a range of Gram-negative bacteria by hydrolyzing the amide bond between N-acetylmuramoyl residues and the L-alanine of the peptidoglycan layer. The zinc amidase, T7L, has a zinc atom located in the cleft bound directly to three amino

* Correspondence: murali@bicpu.edu.in
Centre for Bioinformatics, School of Life Sciences, Pondicherry University,
Puducherry 605014, India

acids and a water molecule; however, the presence of zinc is required for amidase activity but not for inhibition of T7RNAP [6, 7].

As levels of T7L rise, transcriptional-inhibited T7RNAP-T7L complexes form. Though this complex can catalyze the synthesis of short RNA molecules, it fails to clear the abortive initiation phase. It has been found that T7L does not bind to the active site of T7RNAP; alternatively, it binds to a remote site. The RNAP interaction domain (amino acids 2–52) of T7L interacts with T7RNAP at portions of its N-terminal domain (amino acids 307, 309–312), finger sub-domain (amino acids 720, 721, 724, 726, 728 and 736) and palm extended foot module (amino acids 844, 850–853 and 855). The binding of lysozyme may induce two types of conformational changes in these portions either by i) altering the orientation of these portions relative to each other as compared to the apo T7RNAP or ii) by hindering possible conformational changes that may be required during various stages of the transcriptional cycle [8].

T7L has five α-helices and five β-sheets (Fig. 1) and its optimal amidase activity is around pH 7–7.5. This activity decreases to 50% at pH 6.0 and drops significantly with a further decrease in pH with a considerable loss in secondary structural content [7]. Although pH dependent lytic activity of T7L has been studied extensively [7], there is no information on how pH dependence influences the other activity of this bi-functional protein i.e., inhibition of transcription by T7 RNA polymerase. The single subunit polymerase from bacteriophage T7 phage, T7RNAP has a wide array of applications in biological research ranging from over expression of heterologous genes under the control of the T7 promoter [9] to industrial biotechnology [10] and synthetic biology [11]. Due to its advantageous properties, the T7RNAP has been expressed in different environments of prokaryotes to eukaryotes with different

organelle (cell nucleus) [12], organs (human liver cell line) [13] etc. Thus a study of pH dependent interaction analysis will help researchers to regulate this biologically important enzyme in low pH environments with the help of mutational studies. It was revealed from a pH based activity profile of wild type T7RNAP [14] that the polymerase has substantial enzymatic activity in the range 7.9 to 9.5, but at a low pH (pH 5) and beyond pH 11 the enzyme exhibited diminished or no activity. Since in this study, our interest was to find out the pH dependent differential interaction of T7 lysozyme with T7 RNA polymerase, we have selected an acidic pH (pH 5) and a basic pH (pH 7.9).

Methods
Structure of T7L and T7RNAP
Since the current study is focused on T7RNAP inhibition by T7L, we have used the lysozyme structure from the T7RNAP-T7L crystallographic complex (PDB ID: 1ARO) [8]. The structure of T7RNAP at different pH strengths viz. pH 5, neutral and pH 7.9 were taken from our earlier work [15] (Borkotoky S, Meena CK, Bhalerao GM, Murali A.: An in-silico glimpse into the pH dependent structural changes of T7 RNA polymerase: a protein with simplicity, Manuscript Submitted).

Molecular dynamics simulation of T7L at different pH strengths
The structure of T7L was subjected to exhaustive molecular dynamics simulation (MDS) up to 40 ns with GROMACS (Groningen Machine for Chemical Simulations) 4.5 simulation package [16] under three different pHs viz. 5, neutral and 7.9 using Gromos force field [17]. First, the topologies for each pH strength were generated by setting protonation and deprotonation states of K, R, D, E and H residues as identified by the H++ server [18]. Each system was settled in a cubic box where the

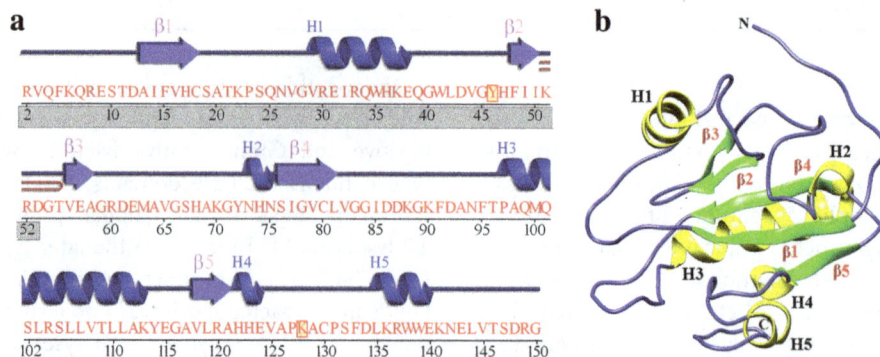

Fig. 1 Structural description of T7 lysozyme (PDB ID: 1ARO_L): (**a**) Secondary structure representation of T7L. Helices labeled as H1, H2 etc. whereas beta sheets and beta hairpin are represented by β and ▭. The T7RNAP binding domain is highlighted in the *grey box*. The diagram was generated from PDBsum server; (**b**) three dimensional representation of T7L. Helices are colored in *yellow* and β sheets are colored in *green* and numbered as per their occurrence

edge of the box from the molecule was set to 1.5 nm in all directions. SPC216 water model was used to solvate the box based on periodic boundary conditions. The net charge of the system was maintained for the pH 5.0 and pH 7.9 structure of T7L after protonation and de-protonation step while the neutral pH structure was neutralized by replacing the water molecules with Cl^- and Na^+ counter ions based on their net charge. Each system was minimized by steepest descent algorithm up to a maximum of 50,000 steps and a convergence tolerance of 1000 kJ $mol^{-1}nm^{-1}$, following which, conjugate gradient algorithm was used with the same steps and convergence tolerance. For long-range interactions, the PME method was used with a 1.0 nm cut-off. Then, equilibrations were carried out for 100 ps for each system with NVT (constant number of particles, volume, and temperature) with modified Berendsen thermostat with velocity rescaling at 310 K and a 0.1 ps time step, Particle Mesh Ewald coulomb type for long-range electrostatics with Fourier spacing 0.16 followed by NPT (constant number of particles, pressure, and temperature) with Parrinello–Rahman pressure coupling at 1 bar, with a compressibility of 4.5×10^{-5} bar^{-1} and a 2 ps time constant. Finally, the equilibrated different pH systems were subjected to 40 ns MD simulation with a time-step of 2 fs. Further analyses of the MD trajectories were carried out using the utilities associated with GROMACS package such as g_rms and g_rmsf to obtain the RMSD (root-mean-square deviation) and the RMSF (root-mean-square fluctuation) values while g_gyrate and do_dssp to calculate the radius of gyration and secondary structure for each time frame.

Principal component and free energy landscape (FEL) analysis

To describe the high amplitude concerted motion in a trajectory, principal component analysis (PCA) or Essential Dynamics [19, 20] was carried out based on their eigenvectors of the mass-weighted covariance matrix of protein atomic fluctuations. The cosine contents (c_i) of each principal component (p_i) of covariance matrix were calculated to generate the free energy landscape defined by PCA analysis. The GROMACS in-built utility "g_covar" was used to generate the covariance matrix using protein backbone as a reference structure for the rotational fit and "g_anaeig" was used to analyze and plot the eigenvectors. Principal components with smaller cosine content values, in general, below 0.2 can yield qualitatively better results with the observation of single basin [21]. Therefore, the first 6 principal components of T7L simulation trajectory at each of different pH strengths were extracted and analyzed based on their cosine values. The cosine content was calculated with "g_analyze" utility.

FEL was constructed using cosine contents lesser than 0.2 of the first two projection eigenvectors (defined as PC1 and PC2 respectively). The minimum energy structure extracted from the minimum energy basins of the FELs were used for further analyses.

Cluster analysis

The RMSD-based structural clustering was performed by the tool g_cluster within GROMACS. The backbone atoms were used in the clustering using GROMOS algorithm [22] to extract clusters of similar conformations. The central structure of the cluster was picked as the representative.

T-pad analysis

T-pad [23] is a tool which can effectively calculate the intrinsic plasticity of protein residues, along with the occurrence of transitions between distinct residue conformations. This information regarding residue-wise flexibility of the protein and backbone conformational transitions is important for its biological functionality. This analysis calculates protein angular dispersion of the angle ω (PADω) to quantitatively analyze local fluctuations and transitions of individual residues. PADω is a function of ω ($= \Phi + \psi$), and hence dependent on both Ramachandran angles Φ and ψ (CSΦ and CSψ). Unlike torsion angle $C\alpha - C - N - C\alpha$ in a peptide bond (CSω), PADω allows quantitative comparisons among residues and among proteins by keeping a narrow range of ω between 0° and 180°. PADω reads as follows,

$$PAD\omega = \frac{180}{\pi} \cos^{-1}\left[\frac{1-CS\omega}{1+CS\omega}\right]$$

T-pad analysis identifies fluctuations (F), long transitions (T) and short transitions (t). A fluctuation (F) is attributed to the fluctuations along a given direction and those along two separate directions are identified as a transition. The difference in long and short transition depends on PAIω (a function of CSω and the Angular Transition Index ATIω). The transitions having PAIω between 30° and 60° are attributed as long transitions and those between 60° and 90° are identified as short transitions. A detailed theory of T-pad tool has been reported by Caliandro et al. [23]. T-pad tool has been successfully used over time to answer various biological questions [24–26].

Here, T-pad analysis was conducted based on molecular dynamics trajectories to understand the structural transitions of the T7L structure at different pH strengths. The MDS trajectories, for T-pad analysis, at pH 5, neutral and pH 7.9 were prepared by extracting the trajectories corresponding to the minimum energy basins of the FEL using the *trjconv* tool of GROMACS.

Docking of T7L andT7RNAP at different pH

To study the effect of variable pH strengths on the interactions between T7L and T7RNAP, we have used HADDOCK (High Ambiguity Driven protein-protein DOCKing) server [27] for data-driven biomolecular docking. HADDOCK web server has correctly solved structures of more than 60 biomolecular complexes available in PDB and besides has outperformed in CAPRI (Critical Assessment of Predicted interactions) blind docking experiment [28]. In HADDOCK, experimental data (e.g., from mutagenesis, mass spectrometry or a variety of NMR experiments, residual dipolar couplings (RDCs) or hydrogen/deuterium exchange) are entered in the form of active and passive residues. HADDOCK then converts them into a series of Ambiguous Interaction Restraints (AIR). Docking in HADDOCK was performed in three major steps that involve a rigid-body energy minimization, a semi-flexible refinement in torsion angle space and a final refinement in explicit solvent. From each step, a defined set of best complexes is progressed to next stage [29]. The docked complexes are ranked on the basis of the sum of electrostatics, van der Waals, and AIR energy terms. The docking was performed using the structures of T7L obtained from the previous step at different pH strengths and T7RNAP from our earlier work (Borkotoky S, Meena CK, Bhalerao GM, Murali A.: An in-silico glimpse into the pH dependent structural changes of T7 RNA polymerase: a protein with simplicity, Manuscript Submitted). The best-docked complexes were selected based on internal energy complex and binding energy from HADDOCK energies and binding affinities ΔG (kcal/mol) and dissociation constants K_d (M) calculated from PRODIGY server [30]. This method uses a simple but robust descriptor of binding affinity based only on structural properties of a protein–protein complex (a combination of the number of contacts at the interface of a protein–protein complex and on properties of the non-interacting surface). The accuracy of this method can be verified by its observed Pearson's Correlation coefficient of 0.73 between the predicted and measured binding affinities on the benchmark described by Kastritis *et al.*[31]. To confirm these results, the docked complexes were further subjected to MD simulation.

Molecular dynamics simulation of T7L and T7RNAP complexes

The complexes of T7L and T7RNAP obtained from the docking step were subjected to molecular dynamics simulation (MDS) up to 60 ns with GROMACS 4.5 simulation package to observe the stability of the docked complexes. Since both the constituents of the complexes were obtained from exhaustive MDS studies and are the minimum energy conformations at respective pH values,

protonation was not considered and total charge was neutralized by adding ions to each system. Rest of the MDS protocol was kept same as that of T7L. The number of hydrogen bonds between the complexes was calculated using the g_hbond utility of GROMACS.

Results

Molecular dynamics simulations of T7L

In order to study the pH effect on T7L, we performed MD simulations in different pH conditions using GROMACS. To measure the conformational stability of the proteins after 40 ns of simulations at pH values 5, neutral and 7.9 various utilities were used.

The RMSD (Root Mean Square Deviation) profile (Fig. 2a) for backbone residues showed that the RMSD, during the 40 ns of simulation period, exhibited stabilization at 0.3 and 0.2 nm for pH strengths of 7.9 and neutral respectively, whereas, at pH 5, it exhibited comparatively increased deviation with RMSD close to 0.7 nm.

The RMSF (Root Mean Square Fluctuation) profiles were also calculated for T7L at different pH strengths and are shown in Fig. 2b. From the RMSF plot, it can be seen that the overall fluctuations of the protein were maximum for pH 5 among the three pH values simulated. At pH 5, it was noticed that the N-terminal (residues 2–7) showed the highest fluctuations in the range 0.6 to 1.2 and other significant fluctuations were observed near helix 1 (residues 38–40). Compared to the fluctuations seen at pH 5, these residues showed considerably lower fluctuations in the range of 0.2 to 0.4 nm at other two pH strengths. However, in the C-terminal region (residues 49–50), T7L showed stronger fluctuations (in the range of 0.6–0.7 nm) at pH 7.9 compared to other two pH strengths.

The plot for Rg (radius of gyration) variation during the simulation time (Fig. 2c) showed that the maximum values for pH strength of 5 were close to 1.7 nm whereas for pH 7.9 and neutral it was close to 1.5 nm. The radius of gyration is a parameter used as an indicator to determine the compactness of the protein and the lowest Rg value corresponds to compactness. Hence, we can say that the compactness of T7L structure decreases at pH 5.

The secondary structure profile (Fig. 2d) of the simulated trajectories showed that there is a decrease of secondary structure at pH 5. This observation is in agreement with the recent circular dichroism (CD) spectroscopy profiles [7], wherein it was reported a decrease in structural content with a decrease in pH strength.

Principal component and free energy landscape analysis

Generally, the first ten eigenvectors represent most of the motions that illustrate the relevant combined motions within an atomic system [24]. We retrieved the

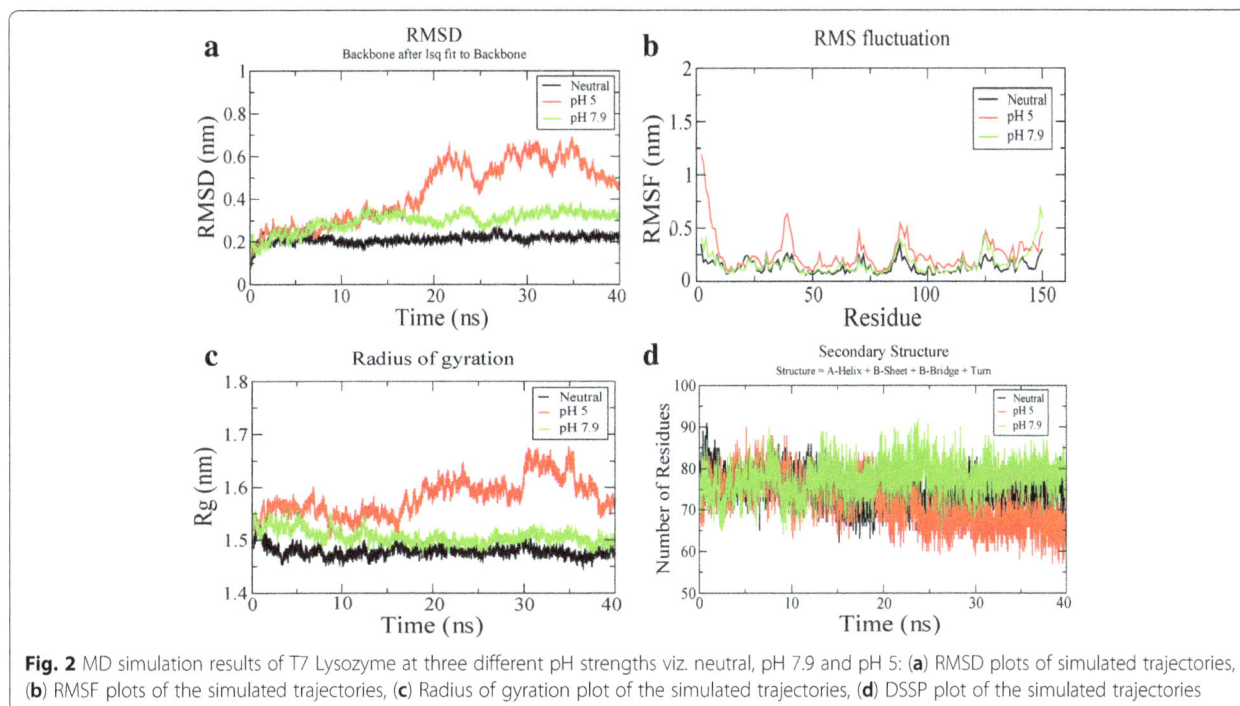

Fig. 2 MD simulation results of T7 Lysozyme at three different pH strengths viz. neutral, pH 7.9 and pH 5: (**a**) RMSD plots of simulated trajectories, (**b**) RMSF plots of the simulated trajectories, (**c**) Radius of gyration plot of the simulated trajectories, (**d**) DSSP plot of the simulated trajectories

first five, the tenth, and the twentieth projections from the protein trajectories at each pH during 40 ns simulation and projected them onto the eigenvectors as obtained from the covariance matrices (Fig. 3). Steep curves of eigenvalues were obtained after plotting eigenvalues against the eigenvectors at each pH (Fig. 3d), and it was observed that 90% of the backbone motion is covered by the first 20 eigenvectors. These results indicate that the motions of the backbone reached their equilibrium fluctuations in the first ten eigenvectors. The trajectories were projected onto the planes defined by two eigenvectors (the tenth and twentieth eigenvectors) from the backbone coordinate covariance matrix for each pH (Fig. 3e). A strong correlation was observed between the projections of these two eigenvectors onto the plane of the backbone motion at each pH, and they filled the expected ranges almost completely which indicates that there is no high projection observed far from the diagonal. This clearly supports that the MD simulation time interval used to extract the trajectories are sufficient for the FEL graph generation and further analysis. The principal components (PCs), extracted with cosine content closer to 0.2, PC1 and PC2, for individual trajectories at different pH strengths, were used to construct the FEL contour maps (Fig. 3). The contour map at pH 5 (Fig. 3a) showed multiple energy clusters depicting the structural transition to distinct active conformational states. These multiple clusters are due to the fluctuation of N-terminal region which is otherwise more stable at other two conditions (also can be seen in RMSF plot). The coordinates from FEL map with minimum energy

cluster (at 26 ns) were used to retrieve the low energy representative structure. On the other hand, the FEL maps at neutral (Fig. 3b) and pH 7.9 (Fig. 3c) showed only one energy minimum at 39 ns and 28 ns, and lowest energy representative structures corresponding to the coordinates at 39 ns and 28 ns were retrieved respectively.

Cluster analysis
Cluster analysis combined with FEL analysis allows us to establish whether a correlation exists between structural similarity and minimum energy basins in the sampled trajectories. As differences appear at RMSD graph at about 15 ns, with clear differences between 20 ns and 30 ns, the first 15 ns of the trajectory were not used to determine the average structures. Different RMSD cutoffs were adopted for cluster analysis at each pH i.e., the average RMSD values derived from each RMSD matrices (Additional file 1) were selected (for pH 5, cutoff: 0.33 nm, pH 7, cutoff: 0.13 nm, pH 7.9, cutoff: 0.16 nm). The central structure of the highly populated cluster was picked at each pH (Fig. 4). The central structures obtained from the trajectories at pH 5 (at 27 ns), neutral (at 24 ns) and pH 7.9 (at 32 ns) were represented by clusters having the highest number of structures and each representative structure was superimposed with the representative structures obtained from FEL analysis (Fig. 4). It was observed that the structures were highly similar and the time frames used to extract the structures from FEL were also represented by the top clusters obtained from the cluster analysis (Additional file 1). Hence we used the FEL derived structures for further analysis.

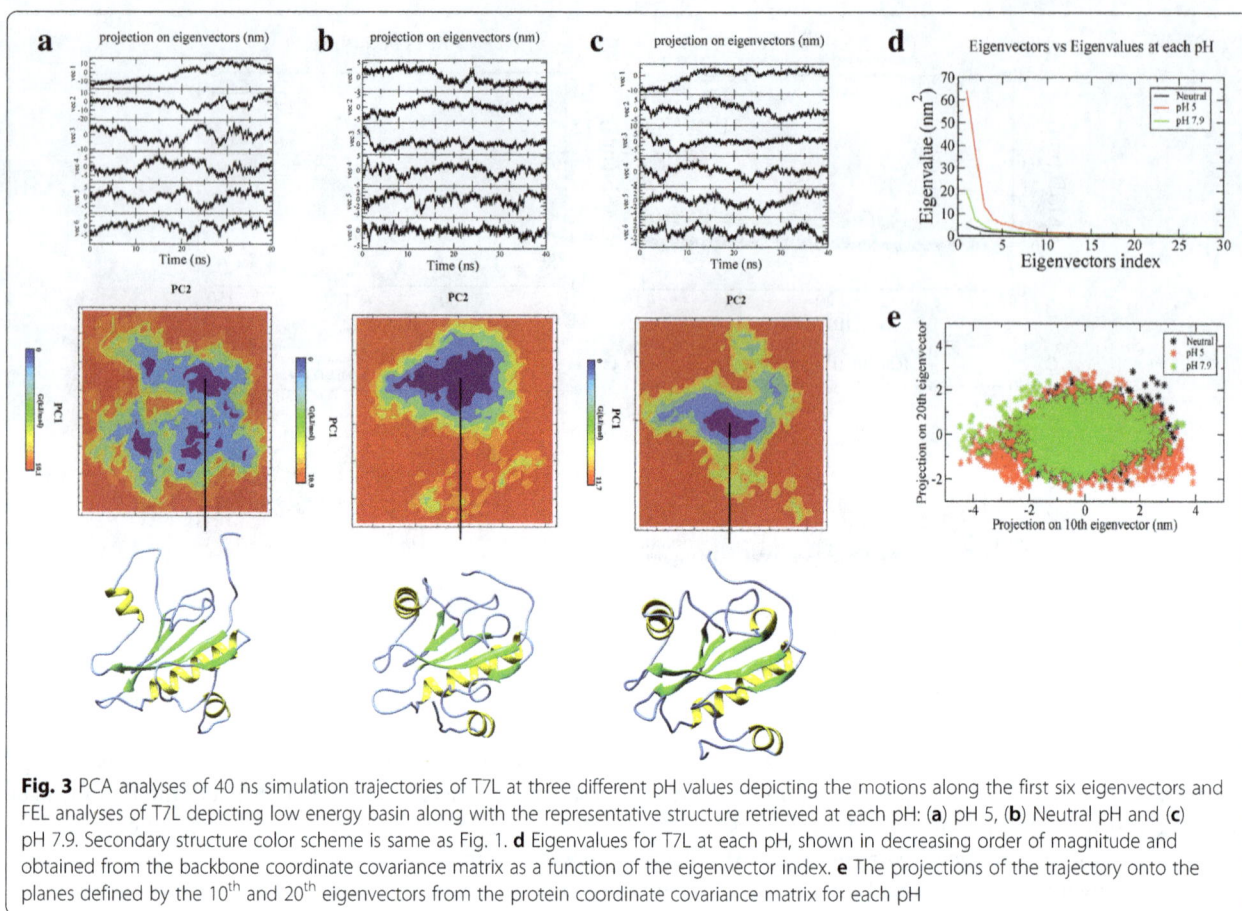

Fig. 3 PCA analyses of 40 ns simulation trajectories of T7L at three different pH values depicting the motions along the first six eigenvectors and FEL analyses of T7L depicting low energy basin along with the representative structure retrieved at each pH: (**a**) pH 5, (**b**) Neutral pH and (**c**) pH 7.9. Secondary structure color scheme is same as Fig. 1. **d** Eigenvalues for T7L at each pH, shown in decreasing order of magnitude and obtained from the backbone coordinate covariance matrix as a function of the eigenvector index. **e** The projections of the trajectory onto the planes defined by the 10th and 20th eigenvectors from the protein coordinate covariance matrix for each pH

Structural transition in T7L

To get into the detail of structural transitions of T7L at different pH strengths, we attempted to understand the local fluctuations and conformational transitions using respective MD trajectories.

At pH 5, the N-terminal residues (2–7) that showed the highest fluctuations in the RMSF plot were checked for their local fluctuations. The residues Arg 2 (33.2°), Gln 4 (58.4°), Gln 7 (67.1°) showed transitions, Val 3 (73.9°) showed short transition and residues Phe 5 and Lys 6 showed fluctuations with PAD degrees 53° and 61° (Fig. 5a). At neutral pH, Arg 2 (36.2°) and Phe 5 (54.3°) showed transitions, Val 3 and Gln 4 showed fluctuations with PAD degrees 39.7° and 47.8° and residue Lys 6 and Gln 7 showed fluctuations with PAD degrees 33.9° and 40.5° (Fig. 4b). At pH 7.9, Arg 2, Phe 5 and Lys 6 showed fluctuations with PAD degrees 41.4°, 44.3° and 35.7° while Val 3 and Gln 4 and Gln 7 showed transitions with PAD degrees 42.9°, 31.15° and 43.5° respectively (Fig. 4c).

Other residues identified with significant transitions at pH 5 are Ala 117 (72.3°), Val 125 (83.7°) and Ala 126 (77°). Whereas, short transitions are found at residues Glu 115 (65.3°) and Gly 116 (81.7°); and highest fluctuations were found in Asp 148 (61.2°) (Fig. 5a).

At neutral pH (Fig. 5b), highest transitions are found at residues Gln 39 (65.5°), Gly 40 (80.4°), Trp 41 (60.3°), Ser 147 (74.2°) and Asp 148 (83.8°). The highest short transition was found at Gly 150 (58.9°) and top fluctuations are found in Gly 116 (60.8°), Ala 117 (60.7°) and Arg149 (64.4°).

At pH 7.9 (Fig. 4c), highest transitions are found at residues Pro 23 (83.8°), Ser 24 (89.5°), Ala (62.9°), Asp 148 (68.3°), and Gly 150 (72.3°). The residues showing highest PAD degrees of short transitions are Glu 115 (97.86°) and Gly116 (80.7°); and the highest fluctuation was found at Asn 26 with PAD degree 58.3°.

T7L and T7RNAP interactions at different pH

After docking the representative structures of T7RNAP at different pH strengths with T7 lysozyme in HAD-DOCK server, the best interaction models were selected based on HADDOCK score and energies (internal energy complex and binding energy). The complex of T7RNAP with T7L was noticed to have fair internal and binding energy in comparison to those of other counterparts (Table 1). These complexes were further submitted for PRODIGY analysis to calculate binding affinity (ΔG) and dissociation constant (K_d). The complex at pH 7.9

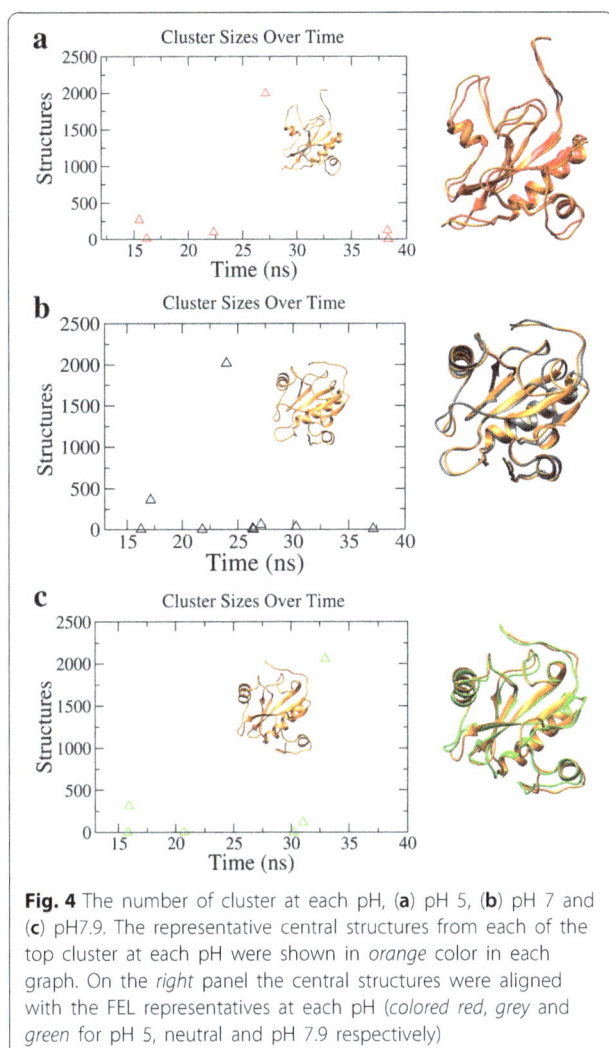

Fig. 4 The number of cluster at each pH, (**a**) pH 5, (**b**) pH 7 and (**c**) pH7.9. The representative central structures from each of the top cluster at each pH were shown in *orange* color in each graph. On the *right* panel the central structures were aligned with the FEL representatives at each pH (*colored red, grey* and *green* for pH 5, neutral and pH 7.9 respectively)

showed strongest binding affinity with ΔG = –12.5 kcal/mol and better dissociation constant 7.1e–10 M. The complexes at pH 5 and neutral pH were observed to have lesser affinity –12.2 kcal/mol and –10.0 kcal/mol respectively, whereas dissociation constants were found to be 1.2e–09 M and 5.0e–08 M. The lysozyme complex with pH 7.9 representative forms 12 H-bonds and 179 non-bonded contacts and 4 salt bridges. While for pH 5 and neutral pH, 12 and 14 H-bonds were observed respectively (Additional files 2, 3 and 4). The number of non-bonded contacts was found to be 149 and 150 for complexes at pH5 and neutral pH respectively, while the number of salt bridges was found to be 1 and 5 respectively. At all three pH values, lysozyme was seen to be interacting with the N-terminal domain, finger sub-domain and extended foot module. These results indicate that the conformational changes occurred at T7 RNAP and T7L at pH 7.9 are suitable for strong interaction; on the other hand, at lower

pH, the attained conformations do not contribute to a strong interaction. Details of the interactions are included in the additional files 1, 2 and 3.

Stability analysis of T7L and T7RNAP complexes
From the 60 ns MDS study of the T7L-T7RNAP complexes obtained from the earlier step, we saw that the complex at pH 7.9 demonstrated a stable dynamics after approximately 40 ns time period than the other two counterparts (Fig. 6a). The complex at pH 5 was found to be more stable than the neutral counterpart after 40 ns. These RMSD patterns confirm the docking results (Table 1) where we saw similar patterns in binding affinities and dissociation constants. The number of hydrogen bonds (H-bond) formed between the complexes (Fig. 6b) showed that after 40 ns both the complexes at pH 7.9 and pH 5 have increased the number than the neutral pH counterpart, thereby contributing to the higher stability as seen in the RMSD plot.

Discussion
We conducted the current study to gain clearer insights into the residue level differences in the T7L structure at different pH strengths and how the changes affect the interaction with T7RNAP. The MD simulation studies at different pH strengths showed that the lysozyme structure at pH 7.9 and neutral pH are stable in comparison to pH 5, with higher residue level fluctuations at low pH prominently at N-terminal region and near α-helix 1 (H1). The number of secondary structure elements of the T7L was observed to be decreased in lower pH (pH 5) in agreement with the experimental results [7]. We have also observed a compactness of the structure of T7L under pH 7.9 and neutral pH compared to pH 5. The multiple low energy basins of the FEL of pH 5 trajectory (Fig. 3) in comparison of the single low energy basins at neutral and pH 7.9 also showed the overall stability of T7L at higher pH. From PCA and FEL analysis, we obtained the minimum energy representative structures at different pH strengths. These structures (Fig. 7) are also in agreement to the radius of gyration plot (Fig. 2c), where an open conformation was observed at pH 5. This observation was linked to the loss of secondary structure content at pH 5 (Fig. 2d). The RMSD based clustering analysis also shown strong agreement to FEL analysis.

As we have seen a flipping out movement in the N-terminal region (Fig. 3a), we investigated the N-terminal end (residues Arg 2-Gln 7) showing high fluctuations in the RMSF plot (Fig. 2b) for local fluctuations and conformational transitions. From the T-pad analysis (Fig. 5) it can be proposed that the short transition at Val 3, transition at the two glutamines at positions 4 and 7 and fluctuation at Lys 6 are responsible for the flipping out nature

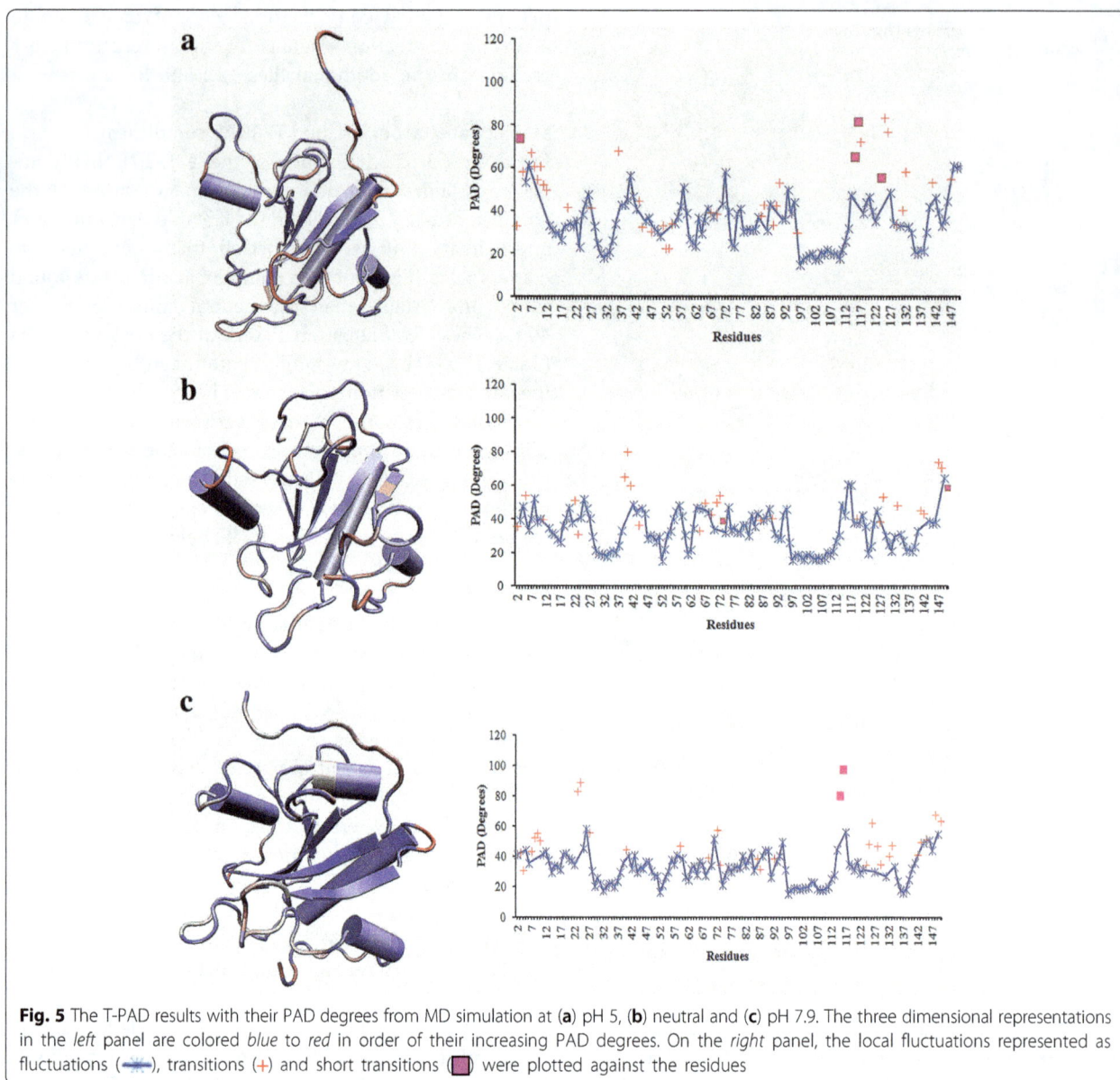

Fig. 5 The T-PAD results with their PAD degrees from MD simulation at (**a**) pH 5, (**b**) neutral and (**c**) pH 7.9. The three dimensional representations in the *left* panel are colored *blue* to *red* in order of their increasing PAD degrees. On the *right* panel, the local fluctuations represented as fluctuations (━✳━), transitions (+) and short transitions (■) were plotted against the residues

as evidenced by the high PAD degrees in the range of 61°–74° (Fig. 7). The possible mechanism for closed conformation of the N-terminal loop at pH 7 is that it was maintained via a series of hydrophobic interactions between the N-terminal loop and the loop connecting β3 and H2 ($L_{3,2}$). Here, the $L_{3,2}$ contains both polar and nonpolar amino acids which are not stimulated by their protonation state; hence the aliphatic side chains of $L_{1,2}$ form a series of hydrophobic interactions (Additional file 5). In case of the free energy representative structure of

Table 1 HAADOCK docking results for the complexes of T7RNAP and T7L at different pH conditions. The HAADOCK energies are in arbitrary units (a.u.). Binding affinities and dissociation constants are calculated from PRODIGY server

Complex (T7RNAP + Lys)	HADDOCK Energies		Number of interactions		PRODIGY Analysis	
	Internal energy complex	Binding energy	H-bond	Non bonded	Binding affinity ΔG (kcal/mol)	Dissociation constant Kd (M)
pH 5	−41913	−58488	12	149	−12.2	1.2e-09
Neutral	−42364	−58007	14	150	−10.0	5.0e-08
pH 7.9	−42435	−55193	12	179	−12.5	7.1e-10

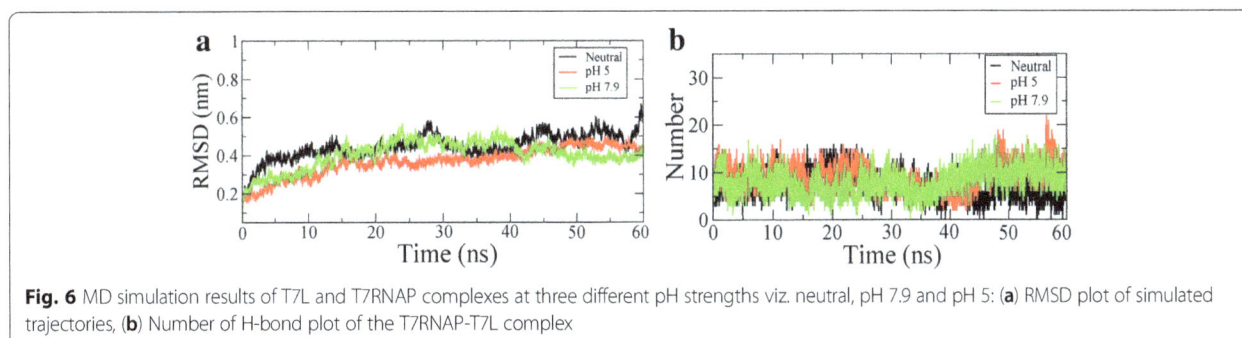

Fig. 6 MD simulation results of T7L and T7RNAP complexes at three different pH strengths viz. neutral, pH 7.9 and pH 5: (**a**) RMSD plot of simulated trajectories, (**b**) Number of H-bond plot of the T7RNAP-T7L complex

T7L at pH 5.0, the protonation and de-protonation states of Arg 2, Lys 6, Arg 8, Glu 9 and Asp 12 (2-RVQFKQRESTDA-13) and the residues Arg 60, Asp 61, Glu 62, His 68 and Lys 70 of $L_{3,2}$ region (59-GRDE-MAVGSHAKGY-72) repel the N-terminal orientation and displaced away from the $L_{3,2}$ region with the loss of hydrophobic interactions with N-terminal residues Val 2 and Phe 5. While in the case of the free energy representative structure of T7L at pH 7.9, Phe 5 maintained the interaction which prevented it from complete displacement.

Another conformational variation observed near helix-1 at pH 5 was due to the loss of helical content from residues His 36 to Gln 39. These residues were also observed with transitions and fluctuations with higher PAD degrees in a range of 42°–68° as compared to the PAD degree ranges at neutral pH (20°–65°) and pH 7.9 (20°–36°). Though the region between helix-3 and β sheet- 5 (residues Glu 115- Ala 117) showed high PAD degrees they did not contribute any conformational difference neither loss of secondary structural content at all the three pH. This observation can be explained by the maintained hydrophobic interactions by Ala 117

with Ile 14 and Tyr 114 in all three pH representatives (Additional file 5). The residues Val 125 and Ala 126 near helix-4 showed strong transitions with PAD degrees 83.7° and 77° at pH 5 as compared to other counterparts at neutral and pH 7.9 with PAD degree range 37°–45° and 48°–63° respectively with minor conformational differences. The C-terminal residues Asp 148 to Gly 150 also showed high PAD degrees at pH 5 in the range of 60° to 61°, at neutral 71° to 59° and at pH 7.9 the range was 55° to 68° with conformational differences. In this case we observed that, the hydrophobic interaction made by the C-terminal residue Leu 144 with residues of the loop connecting β4 and H3 ($L_{4,3}$) and within H3 at pH 7 (Leu 144 : Ala 93, Leu 144 : Phe 95 and Leu 144 : Met 100) and pH 7.9 (Leu 144 : Pro 97 and Leu 144 : Met 100); while these interactions were absent at pH 5 driving the C-terminal away from $L_{4,3}$ (Additional file 5).

Upon docking the individual representatives of T7L and T7RNAP at different pH strengths, it was observed that the T7L interacts with T7RNAP more strongly at both pH 5 and pH 7.9 rather than neutral pH with pH 7.9 representative having a higher K_d value. This observation was further clarified by MD simulation of all the three complexes and it was noted that the complexes at pH 5 and pH 7.9 are more stable than neutral counterpart, with pH 7.9 complex being most stable. The complexes at pH 5 and pH 7.9 also maintained the number of H-bonds, while a decrease in number was observed in the case of the neutral complex. From the MD study of T7L, we found that the T7L structure forms an open/relaxed structure at pH 5. The pH 7.9 representative also shows a comparatively relaxed conformation than neutral pH. Due to these differences, T7L is making differential interactions with T7RNAP at their individual pH. Hence, it can be proposed that the structural changes observed at both pH 5 and pH 7.9 in T7L as well as T7RNAP (Fig. 8) (Borkotoky S, Meena CK, Bhalerao GM, Murali A.: An in -silico glimpse into the pH dependent structural changes of T7 RNA polymerase: a protein with simplicity, Manuscript Submitted) are favorable for the interaction of both proteins.

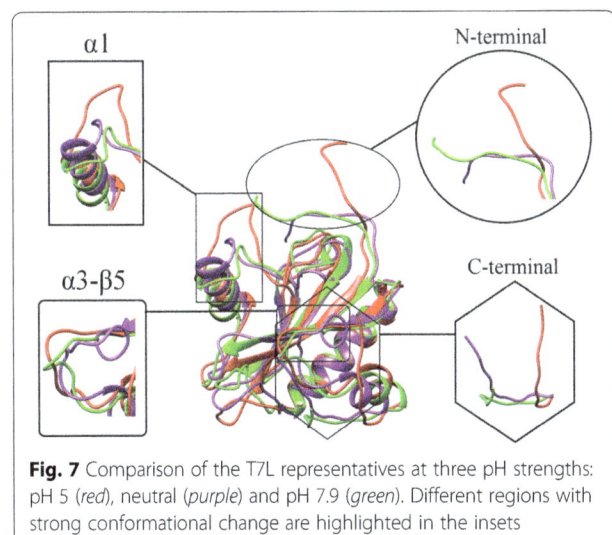

Fig. 7 Comparison of the T7L representatives at three pH strengths: pH 5 (*red*), neutral (*purple*) and pH 7.9 (*green*). Different regions with strong conformational change are highlighted in the insets

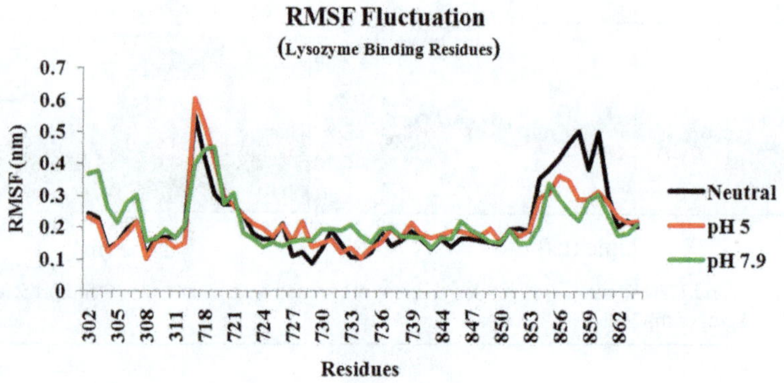

Fig. 8 Fluctuations of lysozyme binding residues of T7RNAP before binding of T7L

Conclusion

Change in physiological environments such as pH, temperature etc. are integral to structure function relationship of proteins. The present study successfully identified pH dependent structural changes in T7 lysozyme, complementing experimental studies but with a more residue level information. The MD simulation and docking studies showed that, though T7L has poor amidase activity at low pH, it does not hamper the T7RNAP interaction ability. The results obtained here can be used for mutational studies for modification of the T7L structure to control levels of inhibition of T7RNAP as well as other structural studies of related bacteriophage amidases such as T3, K11 etc.

Additional files

Additional file 1: RMSD matrices of the trajectories at a) pH 5, b) pH 7 and c) pH 7.9; Distribution of cluster ids were represented along the trajectories at d) pH 5, e) pH 7 and f) pH 7.9 at a function of time.

Additional file 2: HADDOCK docking results of T7RNAP and Lysozyme (at pH 5). A surface representation of the docked complex is shown.

Additional file 3: HADDOCK docking results of T7RNAP and Lysozyme (at neutral pH). A surface representation of the docked complex is shown.

Additional file 4: HADDOCK docking results of T7RNAP and Lysozyme (at pH 7.9). A surface representation of the docked complex is shown.

Additional file 5: Hydrophobic interactions within 5 angstroms for T7L representative structures at individual pH: a) pH 7, b) pH 7.9 and c) pH 5. The interactions were calculated by Protein Interactions Calculator (PIC) server (http://pic.mbu.iisc.ernet.in).

Acknowledgement

The authors are thankful to the University Grants Commission - New Delhi for the award of Major Research Project.

Authors' contributions

SB conceived and designed the experiments and wrote the first draft of the manuscript. SB and AM jointly made critical revisions and approved the final version.

Authors' information

Not applicable.

Competing interests

The authors declare that they have no competing interests.

References

1. Nelson DC, Schmelcher M, Rodriguez-Rubio L, Klumpp J, Pritchard DG, Dong S, Donovan DM. Endolysins as antimicrobials. Adv Virus Res. 2012;83: 299–365.
2. Schmelcher M, Loessner MJ. Bacteriophage endolysins: applications for food safety. Curr Opin Biotechnol. 2016;37:76–87.
3. de Vries J, Harms K, Broer I, Kriete G, Mahn A, During K, Wackernagel W. The bacteriolytic activity in transgenic potatoes expressing a chimeric T4 lysozyme gene and the effect of T4 lysozyme on soil- and phytopathogenic bacteria. Syst Appl Microbiol. 1999;22(2):280–6.
4. Tiwari R, Dhama K, Chakrabort S, Kapoor S. Enzybiotics: new weapon in the army of antimicrobials: a review. Asian J Anim Vet Adv. 2014;9(3):144–163.
5. Hoopes JT, Stark CJ, Kim HA, Sussman DJ, Donovan DM, Nelson DC. Use of a bacteriophage lysin, PlyC, as an enzyme disinfectant against Streptococcus equi. Appl Environ Microbiol. 2009;75(5):1388–94.
6. Cheng X, Zhang X, Pflugrath JW, Studier FW. The structure of bacteriophage T7 lysozyme, a zinc amidase and an inhibitor of T7 RNA polymerase. Proc Natl Acad Sci U S A. 1994;91(9):4034–8.
7. Sharma M, Kumar D, Poluri KM. Elucidating the pH-Dependent Structural Transition of T7 Bacteriophage Endolysin. Biochemistry-Us. 2016;55(33): 4614–25.
8. Jeruzalmi D, Steitz TA. Structure of T7 RNA polymerase complexed to the transcriptional inhibitor T7 lysozyme. EMBO J. 1998;17(14):4101–13.
9. Studier FW, Moffatt BA. Use of bacteriophage T7 RNA polymerase to direct selective high-level expression of cloned genes. J Mol Biol. 1986;189(1):113–30.
10. Kortmann M, Kuhl V, Klaffl S, Bott M. A chromosomally encoded T7 RNA polymerase-dependent gene expression system for Corynebacterium glutamicum: construction and comparative evaluation at the single-cell level. Microb Biotechnol. 2015;8(2):253–65.
11. Shis DL, Bennett MR. Synthetic biology: the many facets of T7 RNA polymerase. Mol Syst Biol. 2014;10:745.
12. Dunn JJ, Krippl B, Bernstein KE, Westphal H, Studier FW. Targeting bacteriophage T7 RNA polymerase to the mammalian cell nucleus. Gene. 1988;68(2):259–66.
13. Aoki Y, Aizaki H, Shimoike T, Tani H, Ishii K, Saito I, Matsuura Y, Miyamura T. A human liver cell line exhibits efficient translation of HCV RNAs produced by a recombinant adenovirus expressing T7 RNA polymerase. Virology. 1998;250(1):140–50.
14. Osumi-Davis PA, Sreerama N, Volkin DB, Middaugh CR, Woody RW, Woody AY. Bacteriophage T7 RNA polymerase and its active-site mutants. Kinetic, spectroscopic and calorimetric characterization. J Mol Biol. 1994;237(1):5–19.
15 Borkotoky S, Meena CK, Murali A. Interaction analysis of T7 RNA polymerase with heparin and its low molecular weight derivatives - an in silico approach. Bioinf Biol Insights. 2016;10:155–66.
16 Pronk S, Pall S, Schulz R, Larsson P, Bjelkmar P, Apostolov R, Shirts MR, Smith

JC, Kasson PM, van der Spoel D, et al. GROMACS 4.5: a high-throughput and highly parallel open source molecular simulation toolkit. Bioinformatics. 2013;29(7):845–54.

17 Oostenbrink C, Villa A, Mark AE, van Gunsteren WF. A biomolecular force field based on the free enthalpy of hydration and solvation: the GROMOS force-field parameter sets 53A5 and 53A6. J Comput Chem. 2004;25(13): 1656–76.

18 Gordon JC, Myers JB, Folta T, Shoja V, Heath LS, Onufriev A. H++: a server for estimating pKas and adding missing hydrogens to macromolecules. Nucleic Acids Res. 2005;33(Web Server issue):W368–71.

19 Amadei A, Linssen AB, Berendsen HJ. Essential dynamics of proteins. Proteins. 1993;17(4):412–25.

20 Burkoff NS, Varnai C, Wells SA, Wild DL. Exploring the energy landscapes of protein folding simulations with Bayesian computation. Biophys J. 2012; 102(4):878–86.

21 Maisuradze GG, Leitner DM. Free energy landscape of a biomolecule in dihedral principal component space: sampling convergence and correspondence between structures and minima. Proteins. 2007;67(3):569–78.

22 Daura X, Gademann K, Jaun B, Seebach D, van Gunsteren WF, Mark AE. Peptide folding: when simulation meets experiment. Angew Chem Int Edit. 1999;38(1-2):236–40.

23 Caliandro R, Rossetti G, Carloni P. Local fluctuations and conformational transitions in proteins. J Chem Theory Comput. 2012;8(11):4775–85.

24 Topno NS, Kannan M, Krishna R. Interacting mechanism of ID3 HLH domain towards E2A/E12 transcription factor–An Insight through molecular dynamics and docking approach. Biochem Biophysics Rep. 2016;5:180–90.

25 Vadlamudi Y, Muthu K, Kumar MS. Structural exploration of acid sphingomyelinase at different physiological pH through molecular dynamics and docking studies. RSC Adv. 2016;6(78):74859–73.

26 Bafunno V, Bury L, Tiscia GL, Fierro T, Favuzzi G, Caliandro R, Sessa F, Grandone E, Margaglione M, Gresele P. A novel congenital dysprothrombinemia leading to defective prothrombin maturation. Thromb Res. 2014;134(5):1135–41.

27 van Zundert GC, Rodrigues JP, Trellet M, Schmitz C, Kastritis PL, Karaca E, Melquiond AS, van Dijk M, de Vries SJ, Bonvin AM. The HADDOCK2.2 Web server: user-friendly integrative modeling of biomolecular complexes. J Mol Biol. 2016;428(4):720–5.

28 Gurung AB, Das AK, Bhattacharjee A. y Disruption of redox catalytic functions of peroxiredoxin-thioredoxin complex in Mycobacterium tuberculosis H37Rv using small interface binding molecules. Comput Biol Chem. 2017;67:69–83.

29 De Vries SJ, van Dijk M, Bonvin AMJJ. The HADDOCK web server for data-driven biomolecular docking. Nat Protoc. 2010;5(5):883–97.

30 Xue LC, Rodrigues JP, Kastritis PL, Bonvin AM, Vangone A. PRODIGY: a web server for predicting the binding affinity of protein-protein complexes. Bioinformatics. 2016;32(23):3676–8.

31 Kastritis PL, Moal IH, Hwang H, Weng Z, Bates PA, Bonvin AM, Janin J. A structure-based benchmark for protein-protein binding affinity. Protein Sci. 2011;20(3):482–91.

10

Comparison of human and mouse E-selectin binding to Sialyl-Lewis^x

Anne D. Rocheleau, Thong M. Cao, Tait Takitani and Michael R. King*

Abstract

Background: During inflammation, leukocytes are captured by the selectin family of adhesion receptors lining blood vessels to facilitate exit from the bloodstream. E-selectin is upregulated on stimulated endothelial cells and binds to several ligands on the surface of leukocytes. Selectin:ligand interactions are mediated in part by the interaction between the lectin domain and Sialyl-Lewis x (sLe^x), a tetrasaccharide common to selectin ligands. There is a high degree of homology between selectins of various species: about 72 and 60 % in the lectin and EGF domains, respectively. In this study, molecular dynamics, docking, and steered molecular dynamics simulations were used to compare the binding and dissociation mechanisms of sLe^x with mouse and human E-selectin. First, a mouse E-selectin homology model was generated using the human E-selectin crystal structure as a template.

Results: Mouse E-selectin was found to have a greater interdomain angle, which has been previously shown to correlate with stronger binding among selectins. sLe^x was docked onto human and mouse E-selectin, and the mouse complex was found to have a higher free energy of binding and a lower dissociation constant, suggesting stronger binding. The mouse complex had higher flexibility in a few key residues. Finally, steered molecular dynamics was used to dissociate the complexes at force loading rates of 2000–5000 pm/ps^2. The mouse complex took longer to dissociate at every force loading rate and the difference was statistically significant at 3000 pm/ps^2. When sLe^x-coated microspheres were perfused through microtubes coated with human or mouse E-selectin, the particles rolled more slowly on mouse E-selectin.

Conclusions: Both molecular dynamics simulations and microsphere adhesion experiments show that mouse E-selectin protein binds more strongly to sialyl Lewis x ligand than human E-selectin. This difference was explained by a greater interdomain angle for mouse E-selectin, and greater flexibility in key residues. Future work could introduce similar amino acid substitutions into the human E-selectin sequence to further modulate adhesion behavior.

Keywords: E-selectin, Receptor, Cell adhesion, Molecular dynamics, Docking, Steered molecular dynamics

Background

Selectins are a family of transmembrane adhesion molecules that mediate the inflammatory response and the cancer metastasis cascade. There are three members of the selectin family: P(latelet)-selectin, E(ndothelial)-selectin, and L(eukocyte)-selectin. All three contain an N-terminal lectin domain, epidermal-growth-factor-like (EGF) domain, a varying number of consensus repeat units, a transmembrane portion, and a cytoplasmic tail [1–3]. During inflammation, fast binding and dissociation of bonds between cells and endothelium contributes to rolling. Selectin:ligand interactions are mediated partially by the interaction between the lectin domain and Sialyl Lewis x (sLe^x), a tetra saccharide on cell surface proteins common to selectin ligands. E-selectin binds particularly well to PSGL-1, CD44, and ESL-1 [1, 4].

There is a high degree of amino acid identity between selectins of various species: about 72 and 60 % in the lectin and EGF domains, respectively [3]. Mouse E-selectin differs from human E-selectin by 29 substitutions in the lectin and EGF domains (Fig. 1). The amino acid differences between human and mouse E-selectin are fairly evenly distributed within and between the domains (Fig. 1).

* Correspondence: mike.king@cornell.edu
Meinig School of Biomedical Engineering, Cornell University, Ithaca, NY, USA

Human	1	WSYNTSTEAM	TYDEASAYCQ	QRYTHLVAIQ	NKEEIEYLNS	ILSYSPSYYW	IGIRKVNNVW
Mouse	1	WYYNASSELM	TYDEASAYCQ	RDYTHLVAIQ	NKEEINYLNS	NLKHSPSYYW	IGIRKVNNVW
Human	61	VWVGTQKPLT	EEAKNWAPGE	PNNRQKDEDC	VEIYIKRERD	VGMWNDERCS	KKKLALCY
Mouse	61	IWVGTGKPLT	EEAQNWAPGE	PNNKQRNEDC	VEIYIQRTKD	SGMWNDERCN	KKKLALCY
Human	119	TAACTNTSCS	GHGECVETIN	NYTCKCDPGF	SGLKCEQIV		
Mouse	119	TASCTNASCS	GHGECIETIN	SYTCKCHPGF	LGPNCEQAV		

Fig. 1 Sequence alignment of EGF and lectin domains of human and mouse E-selectin. The lectin domain is shown in green, and the EGF domain is shown in teal. Residue differences between species are noted in red, and the binding pocket for human E-selectin is noted in yellow and underlined

Molecule conformational changes are essential to physiological processes [5]. Selectin interdomain hinge flexibility greatly affects the on-rate of selectin:ligand binding. All the selectins have shown "open" and "closed" states that correspond to whether or not they are in complex; for instance, there is a 52° increase in the interdomain angle from unliganded P-selectin to P-selectin in complex [6]. Hydrodynamic forces in the bloodstream favor the open conformation as it can strengthen selectin:ligand bonds [7]. A flexible hinge encourages the oscillation between the two states, which facilitates greater range of motion for the lectin domain and thus provides more opportunity for binding [8, 9]. Lou et al. used molecular dynamics (MD) and site mutagenesis at the interdomain hinge of L-selectin to learn that increasing hinge flexibility via mutation caused an increase in binding on- and off-rates of selectin:ligand interactions [10]. Of particular interest are the binding site and interdomain angle, since prior dissociation studies of P-selectin:sLe^x suggest these to be important modulators of dissociation time and final conformation [11].

MD simulations are a useful tool to study the movement of a protein chain over time, given specified starting parameters [12]. The goal of this study was to determine how the structural differences between human and mouse E-selectin affect their corresponding binding and thus cell rolling behavior. MD, docking, and steered molecular dynamics (SMD) were used in conjugation with microtube rolling experiments to address this link between molecular properties and cellular scale adhesion phenomena under flow.

Methods

MD to prepare receptor (E-selectin or mutants) for docking

The lectin and EGF crystal structure of human E-selectin (1ESL) was obtained from the Protein Data Bank to provide starting atomic coordinates. The lectin and EGF domains are the effective binding unit of E-selectin.

The E-selectin:sLe^x complex crystal structure (1G1T) was not used as a starting structure as the bound complex does not allow for full flexibility of E-selectin when amino acid substitutions are made. MD, docking, and SMD simulations were performed using the YASARA (YASARA Biosciences GmbH, Vienna, Austria) package of MD programs with the YAMBER3 self-parameterizing force field. For all simulations, the temperature and pressure were held constant at 298 K and 1 atm, respectively. Other parameters used include periodic boundary conditions, the particle mesh Ewald method for electrostatic interactions, and the recommended 7.86 Å force cutoff for long-range interactions [13]. A predicted model of mouse E-selectin was created using human E-selectin as a template and substituting 29 residues.

For equilibrium simulations, human and mouse E-selectin were each solvated in a water box and neutralized by adding Na$^+$ and Cl$^-$ ions to a concentration of ~50 mM. To allow for free protein rotation, the water box was defined as a cube with sides 80 Å, at least 10 Å from the structure. The conformational stresses were removed using short steepest-descent minimizations followed by simulated annealing until sufficient convergences were reached. Free dynamics simulations were run for 10 ns. Similar equilibration simulations were run for sLe^x (taken from the 1G1T PDB structure) with a water box of size 30 × 30 × 30 Å. The average structure for each simulation run was used for further simulation steps.

Binding sLe^x to human and mouse E-selectin

Molecular docking predicts the conformation of a protein-ligand complex and enables calculation of the binding affinity [12]. sLe^x was docked to the human and mouse E-selectin structures using the AutoDock program with YAMBER3 force field. sLe^x was allowed full flexibility and E-selectin had a fixed backbone with flexible sidechains. 250 docking runs were completed, and the AutoDock scoring function sorted the runs by binding energy. Complex conformations were assumed to be different if the ligand RMSD was greater than 5 Å. Of the final conformations with positive binding energy, those for which there was no contact (5 Å or less) between the fucose residue of sLe^x and the calcium ion were eliminated as they would not be physiologically realistic. The docked complexes were solvated using the same MD steps as before with a water box of size 100 × 100 × 100 Å. The distance from the ligand to the calcium ion was analyzed over the simulation, and if it remained relatively constant, the complex was considered stable. The average free dynamics complex structures were used for the subsequent dissociation steps.

SMD to simulate dissociation under applied force
SMD was used to simulate dissociation under applied force. Constant acceleration was applied to the ligand center of mass to move it away from the receptor center of mass. The simulations were run until all the hydrogen bonds between sLe^x and E-selectin broke and the two proteins dissociated.

Microtube functionalization
Microrenathane tubes (300 μm i.d. and 50 cm long; Braintree Scientific, Braintree, MA) were sterilized with 75 % ethanol for 15 min. After three washes with PBS, the inner luminal surface was functionalized with recombinant human E-selectin (5 μg/mL) by incubating for 2 h, to allow for passive adsorption to the surface. Next, the microtubes were then incubated with dry milk powder (5 % w/v) in PBS for 1 h to prevent nonspecific adhesion. For control experiments, microtubes were prepared as indicated above except that E-selectin was replaced with BSA.

Microsphere functionalization
SuperAvidin-coated microspheres (9.94 μm diameter; CP01N, Bangs Laboratories, Fishers, ID) were washed with PBS buffer per manufacture instruction. Next, the microspheres were incubated with Sialyl-Lewis^X-biotin at specified concentrations for 1 h with gentle mixing every 15 min. Finally, the microspheres were washed twice and resuspended in flow buffer (PBS supplemented with 2 mM Ca^{2+}). The surface density of sLe^x on the microspheres was not measured in this study, however our previous work with similar sLe^x-coated microspheres and selectin surface coatings show that these materials recreate the physiological rolling behavior of leukocytes in the vasculature, with comparable rolling velocities [14].

Rolling experiment
Functionalized microspheres (2x10^6/mL) suspended in flow buffer were perfused through the microtubes using a syringe pump at 8 dyne/cm^2. Recorded videos of rolling microbeads were captured and analyzed using ImageJ similarly to prior publications [15, 16].

Results
Mouse E-selectin homology model exhibits a greater interdomain angle than human E-selectin
Human and mouse E-selectin structures were solvated and equilibrated over the course of 10-ns MD simulations. Three simulations were performed for each species; the average structures for each species over the MD simulations were examined and compared. The most prominent structural difference between the two species was the interdomain angle between the EGF and lectin geometric centers. The mean interdomain angle for human E-selectin was 93.8° and the mean for mouse E-selectin was 104.8°, a difference of 11°. Fig. 2a shows

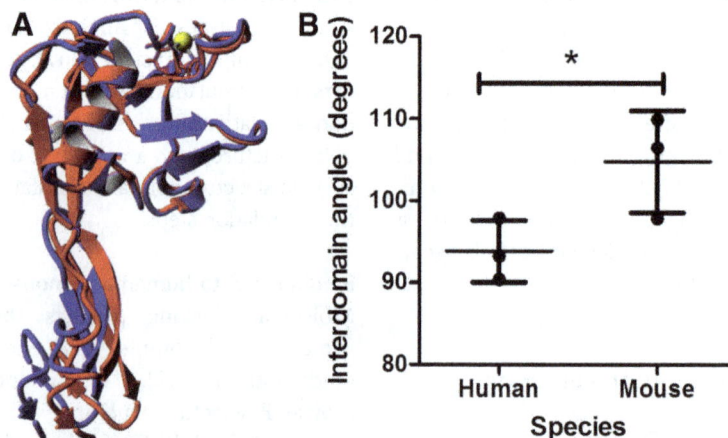

Fig. 2 Mouse E-selectin showed a greater interdomain angle than human. **a** Mouse E-selectin is shown in blue and human E-selectin is shown in red. **b** The angle is measured from geometric center of residues 1–118 to the geometric center of residues 119–157 with a hinge at the pivot. Mean and standard deviation are plotted. Calcium ion is shown in yellow. *P value < 0.05 (two-tailed t-test)

overlaid representative human and mouse structures, and Fig. 2b shows the interdomain angle quantification.

Figure 3 shows the dynamic secondary structure by residue of each simulation run. The lectin domain for each species contains two α-helices: the C-terminal end of the first α-helix is shorter by one or two residues for mouse E-selectin, and both species show some fluctuation, known as "fraying" [17], in the length of the second α-helix, particularly on the C-terminal end. The β-strands in the remainder of the lectin domain vary in length for both species. In the EGF domain, the main structural features are two antiparallel β-strands. For the human runs 1 and 2, the beta-strands show little change in their length. In the human run 3, the two β-strands became fragmented into three after 2 ns. For the mouse, the β-strands show some variation in length for runs 1 and 2 but remain mostly stable for run 3. Overall, the mouse E-selectin lectin and EGF domains contains more random coil and turns than human E-selectin.

Looking more specifically at the residue differences between species, the average backbone root mean square deviation (RMSD) by residue was compared (Fig. 4a). Mouse E-selectin exhibited a greater backbone RMSD across nearly all residues. Specifically, the regions 1–3, 6–8, 21–25, 41–42, 64–66, 79–87, 96–100, 118–121, 124–126, 139, 145–151, and 153–157 showed a difference of more than 1 Å. Each of these regions contains amino acid differences between species. Importantly, many of these regions are involved with the pivot point between the lectin and EGF domains [18]. The flexibility of each residue was compared between species by examining the root mean square fluctuation (RMSF). Fig. 4b shows the RMSF by residue for each species, averaged over the three runs. The RMSF by residue was nearly similar between human and mouse, but the mouse shows peaks at residues 21, 43, and 124 whereas the human protein does not. As expected, these are all locations where there are one or more amino acid differences between species and all are locations of increased backbone RMSD (see Fig. 4a). Residue 21 and 43 are at the C-terminal end of the first and second α-helices, respectively. As shown in Fig. 2, the length of both α-helices fluctuated over the equilibration MD simulation. Residue 124 shows the greatest increase in RMSF and is located in a section of turns and coils in the EGF domain that is roughly parallel to the main β-strands. Figure 4c shows the locations of two residues where there was the greatest difference in RSMD for the mouse E-selectin. Residue 22 is located very close to the lectin/EGF domain interface, and residue 85 is close the binding pocket in the lectin domain.

Mouse E-selectin is predicted to bind more strongly to sLe^x than human E-selectin

Equilibrated sLe^x was then docked onto the human and mouse E-selectin structures. The free energy of binding and the dissociation constant were ranked for each of the resulting complexes. Only stable complexes for which

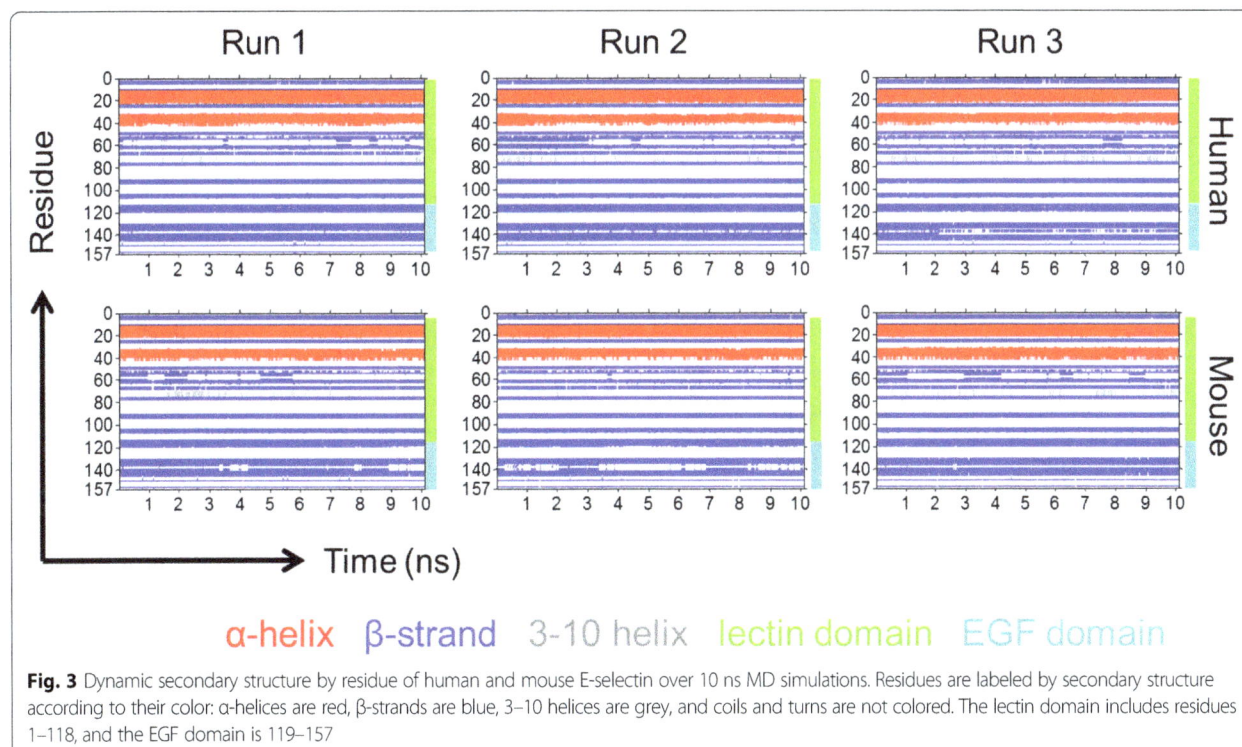

Fig. 3 Dynamic secondary structure by residue of human and mouse E-selectin over 10 ns MD simulations. Residues are labeled by secondary structure according to their color: α-helices are red, β-strands are blue, 3-10 helices are grey, and coils and turns are not colored. The lectin domain includes residues 1–118, and the EGF domain is 119–157

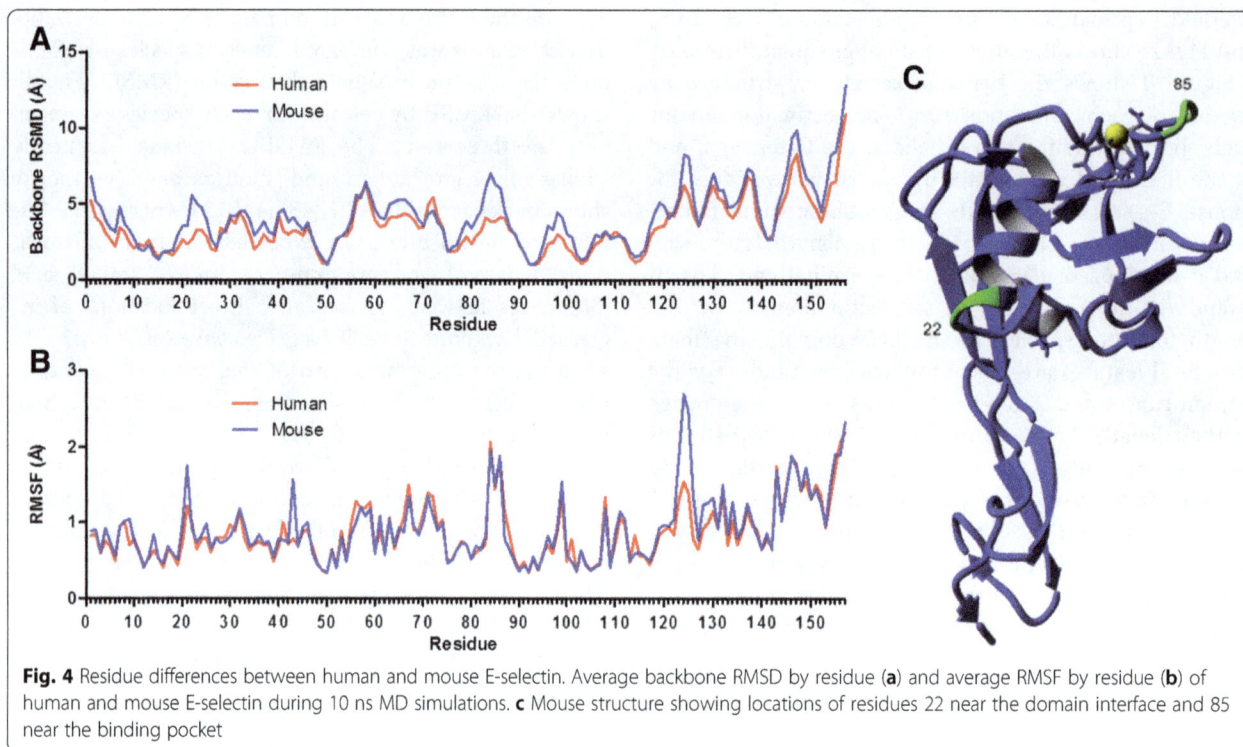

Fig. 4 Residue differences between human and mouse E-selectin. Average backbone RMSD by residue (**a**) and average RMSF by residue (**b**) of human and mouse E-selectin during 10 ns MD simulations. **c** Mouse structure showing locations of residues 22 near the domain interface and 85 near the binding pocket

there was interaction with the calcium ion were considered [19], resulting in four feasible complexes for each species, and the highest free energy complex of each species was chosen for further study [20]. The mouse E-selectin complex yielded a higher free energy of binding as well as a lower dissociation constant (Fig. 5).

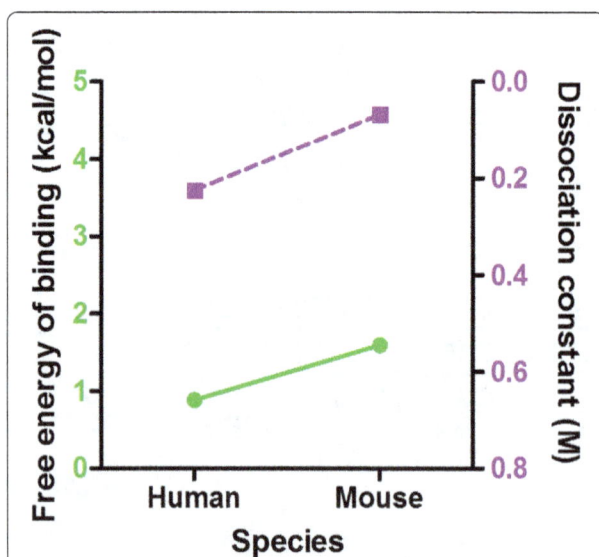

Fig. 5 Free energies of binding and dissociation constants for human and mouse E-selectin:sLe^x complexes. Free energies of binding are shown with green circles and a solid line and dissociation constants are shown with purple squares and a dashed line

Differences in dissociation time among complexes are caused more by interdomain flexibility rather than by contacts between receptor and ligand

The complexes were solvated and equilibrated for 10 ns. The average equilibrated complexes were examined prior to dissociation as per other studies of selectin binding [18, 21]. The geometric parameters analyzed included the distance and angle between the lectin and EGF domain centers of mass, the number of interdomain contacts and hydrogen bonds, the hinge distance, and the number of contacts and hydrogen bonds between the ligand and the receptor. Contacts were defined as less than 5 Å distance between two residues. As shown in Fig. 6a, the mean interdomain angle for the mouse-sLe^x complex was higher than for the human-sLe^x complex. Increased interdomain angle has been shown to increase flow-enhanced tether rate for N138G L-selectin [22], so it is predicted that mouse E-selectin will have a greater tether rate than human E-selectin. The secondary structure composition of both E-selectin species was examined (Fig. 6b). There was no significant difference in the percentage of α-helices and coil between species. However, mouse E-selectin in complex had a smaller percentage of β-strands and an increased percentage of turns compared with human E-selectin.

The secondary structure of each complex was examined over the solvated free dynamics simulation (Fig. 7). There was a notable difference in the antiparallel β-strands of the EGF domain between species. The mouse

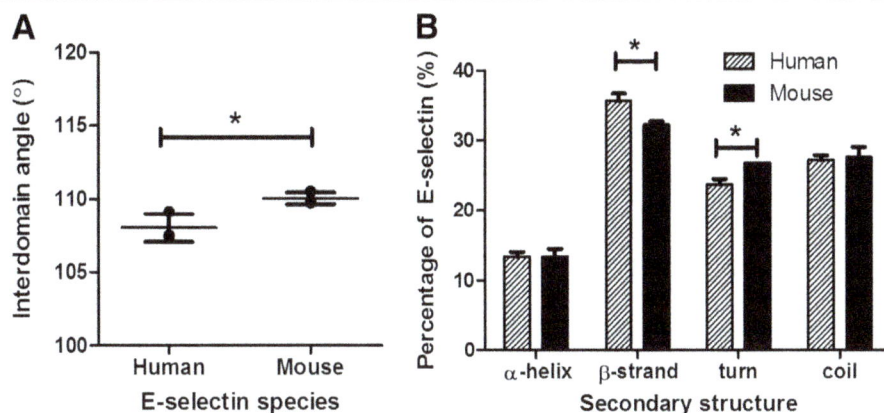

Fig. 6 Differences in domain angle and secondary structure composition between species. **a** Angle between geometric center of residues 1–118 and geometric center of residues 119–157 for human and mouse complex configurations. **b** Secondary structure composition of E-selectin by species. Mean and standard deviation are shown. *$P < 0.05$ (two-tailed t-test)

complex showed two such β-strands during each individual run and the length between the strands varied. However, all of the human complex runs oscillated between two or three short β-strands. For both species, the two α-helices in the lectin domain showed some fluctuation in the length, particularly on the C-terminal end of the second α-helix; this is similar to the trajectories of E-selectin alone (Fig. 2).

The residue flexibility of each species complex was examined by studying average RMSF values over the 10-ns free dynamics (Fig. 8a). Comparing the two species, the

mouse complex exhibited a higher RMSF at several key pivot residues, including 2, 30, and 125 (Fig. 8b). There is also an RMSF peak at residue 43, which is at the C-terminal end of the second α-helix. Adhesion is largely regulated by the interdomain hinge, so increased flexibility in this area could indicate a prolonged bond lifespan and lower off-rate [8].

The E-selectin residues in contact with sLe^x were examined for the average solvated 1G1T structure and human and mouse configuration complexes (Fig. 9). The human complex exhibited more contacts with sLe^x, defined as the

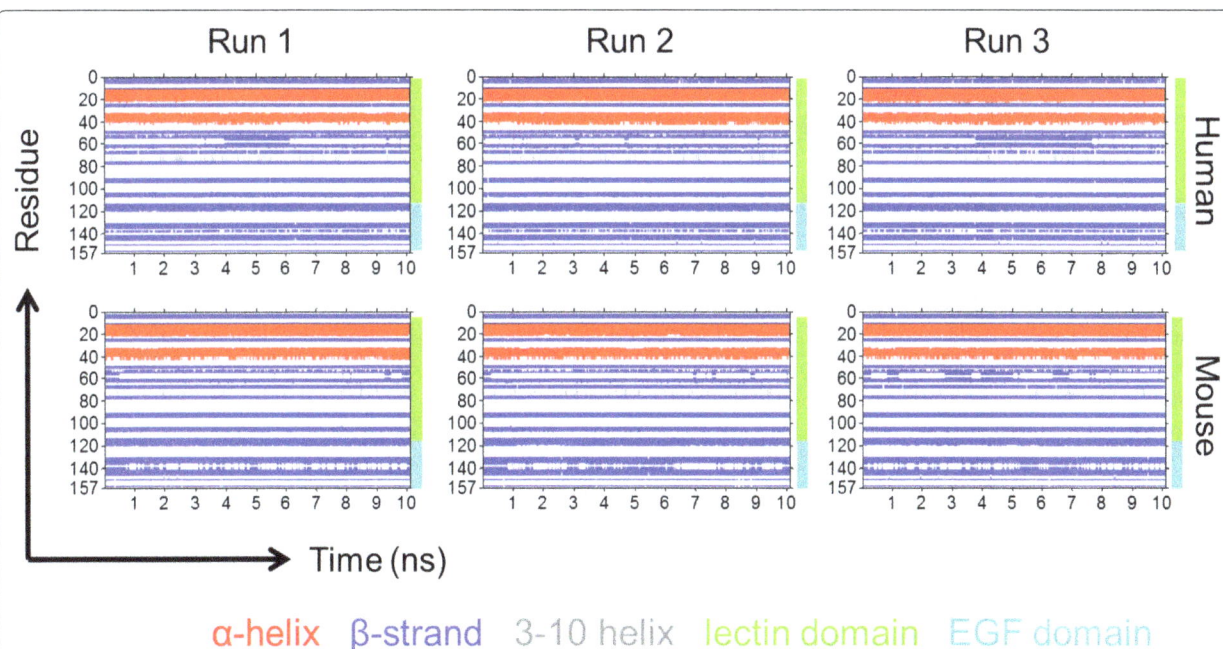

Fig. 7 Dynamic secondary structure by residue of human and mouse E-selectin complexes over 10 ns MD simulations. Residues are labeled by secondary structure according to their color: α-helices are red, β-strands are blue, 3–10 helices are grey, and coils and turns are not colored. The lectin domain includes residues 1–118, and the EGF domain is 119–157

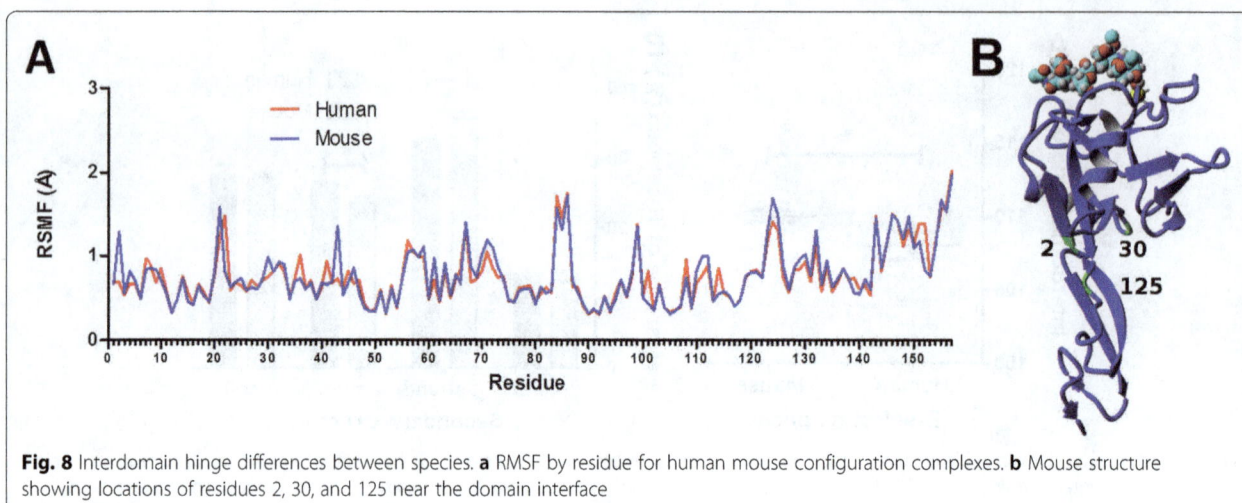

Fig. 8 Interdomain hinge differences between species. **a** RMSF by residue for human mouse configuration complexes. **b** Mouse structure showing locations of residues 2, 30, and 125 near the domain interface

number of atoms of E-selectin that were within 5 Å of any atoms of sLe^x. (Fig. 9a). The specific residues and number of contacts for each complex are shown in Fig. 9b. All of the E-selectin residues except residue 99 had RMSF values within 1 Å (Fig. 8a), indicating relatively low flexibility. This is consistent with their location within or near the binding site. Residue 82 had the most contacts, with residues 97, 105, 107, and 111 showing the next highest number of contacts. There were several contacting residues in the human complexes that had no or negligible contact for the mouse complexes, including 47, 48, 77, 78, 79, and 100. All of these residues had fewer than 50 contacts among the three runs. Conversely, two residues for which there was significantly more contact for mouse complexes than for human were 99 and 108 (Fig. 9c). Both residues 99 and 108 experienced about 100 contacts between the three mouse complexes; they are located on either end of the sLex and may serve as anchor points. Thus, despite having fewer total contacts and a similar number of residues in contact with sLe^x, the data suggest that residues 99 and 108 are of particular importance in dissociation. Residue 99 is lysine and residue 108 is arginine, both large and positively-charged amino acids. Neither of these are residues that are different between human and mouse E-selectin but both are one or two residues away from substitutions at 98, 101, and 110.

Mouse E-selectin complex takes longer to dissociate than human E-selectin

Each species complex was subjected to force loading rates between 2000 and 5000 pm/ps^2, and dissociation was determined as the point when all hydrogen bonds between the ligand and receptor were broken and did not reform. In all simulations, higher force-induced loading rates led to faster dissociation times (Fig. 10).

Under all force-induced loading rates, mouse complexes took longer on average to dissociate. However, only the rate of 3000 pm/ps^2 led to a statistically significant difference between species.

sLe^x-coated microspheres were perfused through E-selectin coated microtubes and the average rolling velocity of the microspheres on each E-selectin species were compared (Fig. 11). Microspheres were used instead of cells to eliminate effects of cell deformability or other selectin:ligand pairs not considered within the scope of this study. As expected, the microspheres rolling on mouse E-selectin showed a statistically significantly lower rolling velocity compared to microspheres perfused over human E-selectin; the average rolling velocity on human E-selectin was 11.2 μm/s and the average for mouse E-selectin was 0.63 μm/s. Rolling velocity is largely affected by off-rate [23], so the longer dissociation exhibited by simulations of the mouse E-selectin complex versus the human complex (Fig. 10) is consistent with this trend.

Discussion

Excessive leukocyte extravasation out of the bloodstream has been linked with chronic inflammation [4]. Thus, potential therapies for controlling the inflammatory response could involve inhibiting or moderating the selectin adhesion that mediates leukocyte tethering and rolling to the blood vessel walls. Homology modeling and amino acid substitutions, particularly those that affect molecular flexibility, and have been shown to be highly effective in changing adhesion and inhibitive function [24–26]. In this study, a mouse homology model comprising 29 point substitutions to the human E-selectin crystal structure greatly affected dissociation of sLe^x from the resulting complex. The adhesive

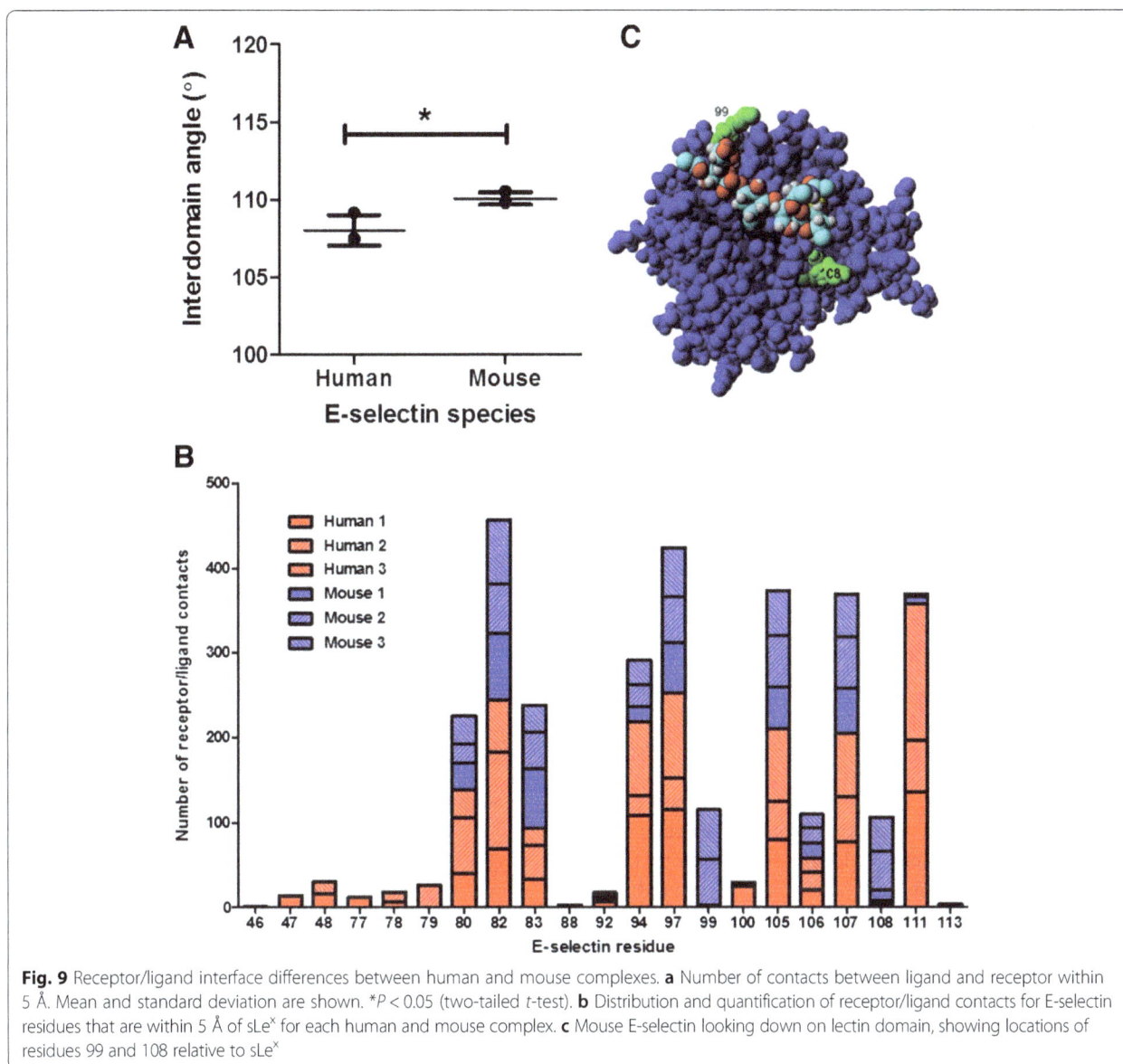

Fig. 9 Receptor/ligand interface differences between human and mouse complexes. **a** Number of contacts between ligand and receptor within 5 Å. Mean and standard deviation are shown. *$P < 0.05$ (two-tailed t-test). **b** Distribution and quantification of receptor/ligand contacts for E-selectin residues that are within 5 Å of sLe^x for each human and mouse complex. **c** Mouse E-selectin looking down on lectin domain, showing locations of residues 99 and 108 relative to sLe^x

characteristics of the mouse E-selectin homology model qualitatively match results from experiments that showed slower rolling velocity of sLe^x-coated microspheres. These results provide new insight into the connection between structure and function of species-specific E-selectin. These results suggest that differences in dissociation time result more from interdomain flexibility than by contacts between receptor and ligand.

Docking a homology model structure does accumulate more errors than using a crystal structure [27], but in this case, a crystal structure for mouse E-selectin was not available. The docking algorithm accounts for two important details: protein flexibility is a key determinant in binding, and physiologically, complexes are solvated in a salt solution [19]. The docking algorithm included flexibility in the E-selectin side chains and full flexibility in the sLe^x.

The docked structures were solvated after docking using 10-ns MD simulations to allow for more physiological conditions. Intramolecular distortion of the lectin and EGF domains was not evident for most simulations, particularly at higher force-induced loading rates. It has been shown that shear flow can have a contribution to intramolecular distortion [21], but as with most selectin:ligand dissociation simulations [13], shear flow is not directly considered in these SMD simulations.

This study demonstrates the significance of combining simulations with experimental rolling studies to gain insights into the functional differences between proteins that share sequence similarity. The differences in amino acid structure can be exploited for applications such as selectin-based leukocyte and circulation tumor cell isolation [28]. The combined methodology involving docking,

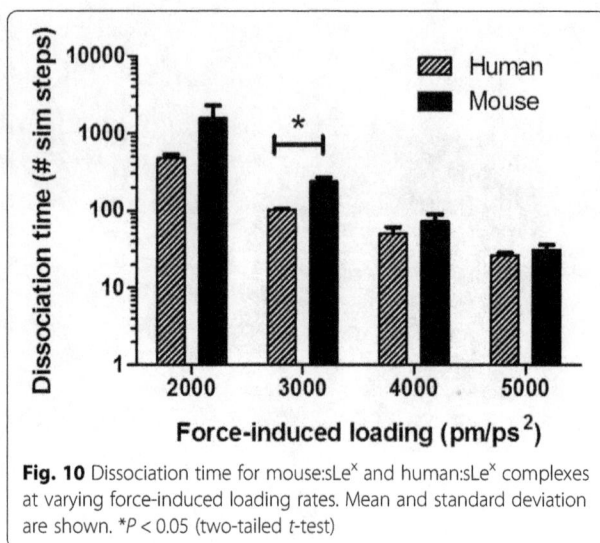

Fig. 10 Dissociation time for mouse:sLe^x and human:sLe^x complexes at varying force-induced loading rates. Mean and standard deviation are shown. *$P < 0.05$ (two-tailed t-test)

Abbreviations
BSA, bovine serum albumin; EGF, epidermal growth factor; MD, molecular dynamics; PBS, phosphate-buffered saline; PDB, protein data bank; RMSD, root mean square deviation; RMSF, root mean square fluctuation; sLe^x, sialyl Lewis x; SMD, steered molecular dynamics

Acknowledgements
None.

Funding
This work was funded by the U.S. National Institutes of Health, Grant no. HL018208 to M.R.K.

Authors' contributions
ADR performed the computational study and wrote the manuscript. TC performed the experiments. TT provided computational methods. MRK conceived of the study and edited the manuscript. All authors have read and approved the final version of the manuscript.

Competing interests
The authors declare that they have no competing interests.

SMD, and MD simulations of receptor:ligand interactions holds possibility as a means for rational drug design [29].

Conclusions
Molecular simulations were used to elucidate the binding of sLe^x to mouse and human E-selectin. Docking simulations predicted that mouse E-selectin would bind more strongly to sLe^x than human E-selectin, and SMD simulations predicted that the mouse E-selectin:sLe^x complex would exhibit a longer dissociation time. Mouse E-selectin alone and bound to sLe^x exhibited a greater interdomain angle than human E-selectin, and there were fewer receptor:ligand contacts. When tested experimentally, sLe^x-coated microspheres rolled more slowly in tubes coated with mouse E-selectin rather than human E-selectin.

References
1. McEver RP, Zhu C. Rolling cell adhesion. Annu Rev Cell Dev Biol. 2010;26:363–96.
2. Hanley WD, Wirtz D, Konstantopoulos K. Distinct kinetic and mechanical properties govern selectin-leukocyte interactions. J Cell Sci. 2004;117:2503–11.
3. Ley K. The role of selectins in inflammation and disease. Trends Mol Med. 2003;9:263–8.
4. Titz A, Marra A, Cutting B, Smieško M, Papandreou G, et al. Conformational Constraints: Nature Does It Best with Sialyl Lewis x. Eur J Org Chem. 2012; 2012:5534–9.
5. Pierse CA, Dudko OK. Kinetics and energetics of biomolecular folding and binding. Biophys J. 2013;105:L19–22.
6. Somers WS, Tang J, Shaw GD, Camphausen RT. Insights into the molecular basis of leukocyte tethering and rolling revealed by structures of P- and E-selectin bound to SLe(X) and PSGL-1. Cell. 2000;103:467–79.
7. Phan UT, Waldron TT, Springer TA. Remodeling of the lectin-EGF-like domain interface in P- and L-selectin increases adhesiveness and shear resistance under hydrodynamic force. Nat Immunol. 2006;7:883–9.
8. Lou J, Yago T, Klopocki AG, Mehta P, Chen W, et al. Flow-enhanced adhesion regulated by a selectin interdomain hinge. J Cell Biol. 2006;174:1107–17.
9. Zhu C, Yago T, Lou J, Zarnitsyna VI, Rodger P. Mechanisms for Flow-Enhanced Cell Adhesion. Ann Biomed Eng. 2008;36:604–21.
10. Lou J, Zhu C. A structure-based sliding-rebinding mechanism for catch bonds. Biophys J. 2007;92:1471–85.
11. Lü S, Long M. Forced dissociation of selectin-ligand complexes using steered molecular dynamics simulation. Mol Cell Biomech. 2005;2:161–77.
12. Hug, S., L. Monticelli, and E. Salonen. 2013. Biomolecular Simulations.
13. Cao TM, Takatani T, King MR. Effect of extracellular pH on selectin adhesion: theory and experiment. Biophys J. 2013;104:292–9.
14. King MR, Hammer DA. Multi particle adhesive dynamics. Interactions between stably rolling cells. Biophys J. 2001;81:799–813.
15. Geng Y, Yeh K, Takatani T, King MR. Three to Tango: MUC1 as a Ligand for Both E-Selectin and ICAM-1 in the Breast Cancer Metastatic Cascade. Front Oncol. 2012;2:1–8.
16. Geng Y, Chandrasekaran S, Hsu JW, Gidwani M, Hughes AD, et al. Phenotypic Switch in Blood: Effects of Pro-Inflammatory Cytokines on Breast Cancer Cell Aggregation and Adhesion. PLoS One. 2013;8:1–10.
17. Legge FS, Budi A, Treutlein H, Yarovsky I. Protein flexibility: Multiple molecular dynamics simulations of insulin chain B. Biophys Chem. 2006;119:146–57.
18. Springer TA. Structural basis for selectin mechanochemistry. Proc Natl Acad Sci U S A. 2009;106:91–6.
19. Okimoto N, Futatsugi N, Fuji H, Suenaga A, Morimoto G, et al. High-performance drug discovery: Computational screening by combining docking and molecular dynamics simulations. PLoS Comput Biol. 2009;5:e1000528.
20. Cosconati S, Forli S, Perryman AL, Harris R, David S, et al. Virtual Screening with AutoDock: Theory and Practice. Expert Opin Drug Discov. 2011;5:597–607.

Fig. 11 Rolling velocity of sLe^x-coated microspheres perfused through an E-selectin coated microtube. Mean and standard deviation shown. ***$P < 0.001$ (two-tailed t-test)

21. Kang Y, Lü S, Ren P, Huo B, Long M. Molecular dynamics simulation of shear- and stretch-induced dissociation of P-selectin/PSGL-1 complex. Biophys J. 2012;102:112–20.
22. Beste MT, Hammer DA. Selectin catch-slip kinetics encode shear threshold adhesive behavior of rolling leukocytes. Proc Natl Acad Sci U S A. 2008;105:20716–21.
23. King MR, Heinrich V, Evans E, Hammer DA. Nano-to-micro scale dynamics of P-selectin detachment from leukocyte interfaces. III. Numerical simulation of tethering under flow. Biophys J. 2005;88:1676–83.
24. Park H, Yeom MS, Lee S. Loop flexibility and solvent dynamics as determinants for the selective inhibition of cyclin-dependent kinase 4: Comparative molecular dynamics simulation studies of CDK2 and CDK4. ChemBioChem. 2004;5:1662–72.
25. Mao D, Lü S, Li N, Zhang Y, Long M. Conformational stability analyses of alpha subunit I domain of LFA-1 and Mac-1. PLoS One. 2011;6:e24188.
26. Shen J, Zhang W, Fang H, Perkins R, Tong W, et al. Homology modeling, molecular docking, and molecular dynamics simulations elucidated α-fetoprotein binding modes. BMC Bioinformatics. 2013;14 Suppl 14:S6.
27. Vakser IA. Protein-Protein Docking: From Interaction to Interactome. Biophys J. 2014;107:1785–93.
28. Hughes AD, Mattison J, Western LT, Powderly JD, Greene BT, et al. Microtube device for selectin-mediated capture of viable circulating tumor cells from blood. Clin Chem. 2012;58:846–53.
29. Alonso H, Bliznyuk AA, Gready JE. Combining docking and molecular dynamic simulations in drug design. Med Res Rev. 2006;26:531–68.

11

High-resolution structure of a type IV pilin from the metal-reducing bacterium *Shewanella oneidensis*

Manuela Gorgel[1], Jakob Jensen Ulstrup[1], Andreas Bøggild[1], Nykola C Jones[2], Søren V Hoffmann[2], Poul Nissen[1] and Thomas Boesen[1*]

Abstract

Background: Type IV pili are widely expressed among Gram-negative bacteria, where they are involved in biofilm formation, serve in the transfer of DNA, motility and in the bacterial attachment to various surfaces. Type IV pili in *Shewanella oneidensis* are also supposed to play an important role in extracellular electron transfer by the attachment to sediments containing electron acceptors and potentially forming conductive nanowires.

Results: The potential nanowire type IV pilin Pil_{Bac1} from *S. oneidensis* was characterized by a combination of complementary structural methods and the atomic structure was determined at a resolution of 1.67 Å by X-ray crystallography. Pil_{Bac1} consists of one long N-terminal α-helix packed against four antiparallel β-strands, thus revealing the core fold of type IV pilins. In the crystal, Pil_{Bac1} forms a parallel dimer with a sodium ion bound to one of the monomers. Interestingly, our Pil_{Bac1} crystal structure reveals two unusual features compared to other type IVa pilins: an unusual position of the disulfide bridge and a straight α-helical section, which usually exhibits a pronounced kink. This straight helix leads to a distinct packing in a filament model of Pil_{Bac1} based on an EM model of a *Neisseria* pilus.

Conclusions: In this study we have described the first structure of a pilin from *Shewanella oneidensis*. The structure possesses features of the common type IV pilin core, but also exhibits significant variations in the α-helical part and the D-region.

Keywords: Type IV pili, Nanowire, PilBac1, PilA, *Shewanella oneidensis*, X-Ray Crystallography, SAXS, SRCD

Background

Type IV pili are found in many Gram-negative bacteria as well as in some Gram-positive bacteria and archaea, where they function in numerous cellular processes including adhesion, DNA transfer and virulence [1-6].

Furthermore, in the metal reducing bacteria *Shewanella oneidensis* and *Geobacter sulfurreducens* type IV pili have been implicated in extracellular electron transport (EET) pathways [7-9]. Both of these organisms can respire on a variety of electron acceptors, including metals such as iron, manganese and uranium oxides, which has made these organisms attractive research targets in the fields of

environmental sciences and nanotechnology [10-15]. *S. oneidensis* and *G. sulfurreducens* can reduce extracellular electron acceptors directly through membrane bound cytochromes [16-19]; *S. oneidensis* can also produce soluble electron shuttles to transfer electrons to extracellular acceptors [20-22]. To allow for highly efficient electron transfer rates, *S. oneidensis* and *G. sulfurreducens* can form biofilms in which strong cell-cell interactions and contact between cells and insoluble electron acceptors are beneficial in certain habitats [23-25]. Such an attachment function is expected to implicate type IV pili.

Type IV pili have been associated with a more direct role in EET. Both *S. oneidensis* and *G. sulfurreducens* can form conductive filaments that transfer electrons extracellularly over multiple cell lengths from one cell to another and from a cell to an electron acceptor [7,9].

* Correspondence: thb@mbg.au.dk
[1]Department of Molecular Biology and Genetics, Aarhus University, Gustav Wieds Vej 10c, Aarhus C 8000, Denmark
Full list of author information is available at the end of the article

These filaments were collectively termed nanowires. While it was clearly shown that nanowires in *G. sulfurreducens* were made of the type IV pilin PilA, the exact subunits of nanowires in *Shewanella* have not been identified so far. Yet, there has been strong evidence that nanowires are made of proteins and studies have indicated the contribution of pili in extracellular electron transport [9,26] – whether this is due to an indirect role by attaching to electron acceptors or due to a direct role by nanowire formation, is not clear at this point. Altogether, the high overall similarity between *G. sulfurreducens* and *S. oneidensis*, including metabolic pathways, the prevalence of multiheme cytochromes and the formation of conductive filaments, makes it very likely that nanowires in both species are formed in a similar way and function based on the same principles.

Currently, two major hypotheses prevail on how nanowires transfer electrons. The metallic-like conductivity theory claims that type IV pili are the conductive units themselves [27]. The aromatic amino acids in PilA are supposedly aligned so closely that the π-electrons can be delocalized and be transferred along the pilus like in a metal lattice. According to the alternative multi-step hopping theory, type IV pili only form the backbone of nanowires to which multiheme cytochromes such as MtrC and OmcA in *S. oneidensis* and OmcS and OmcZ in *G. sulfurreducens*, respectively, attach [28,29]. The electrons can then hop from one heme of one protein to another heme of the neighboring protein. So far, the electron transfer mechanism along bacterial nanowires is not clear yet and the discussion, on which of the two mechanisms is true, is ongoing [27,28,30,31].

Type IV pilins (T4Ps) build up the polymeric pilus in a repetitive way [32-34]. Two kinds of type IV pilins have been described, type IVa and IVb pilins (T4aPs and T4bPs, respectively). These two types are primarily distinguished by the length of their leader sequences with T4aPs containing an N-terminal leader sequence of 6 to 7 residues, whereas the leader sequences in T4bPs range from 15 to 30 [3]. Generally, T4aPs are synthesized as pre-pilins in the cytoplasm and are guided to the inner membrane by their N-terminal leader sequence [35], which is then cleaved off at a conserved cleavage site by the leader peptidase PilD at the cytoplasmic face of the inner membrane [36,37]. The new N-terminus (commonly a phenylalanine) is then methylated and the processed pilins are inserted into the pilus by an inner membrane multimeric complex (including the assembly ATPase PilB) [38-40] and the assembled pilus is fully exported into the extracellular space via the outer membrane secretin PilQ [41,42]. The N-terminal transmembrane domain of type IV pilins is an approximately 20 residues long hydrophobic α-helix, which has a highly conserved sequence among different species. Downstream of this sequence motif, the sequence

variability of T4aPs is however very high and the total length of pilin proteins can vary from 90 residues (*G. sulfurreducens*) [43] to more than 150 residues (*P. aeruginosa*) [44].

T4aPs share the signature of the N-terminal leader sequence and the transmembrane α-helix with pseudopilins [45] Like a type IV pilus, a pseudopilus extends from the inner membrane into the periplasm, but it does not go beyond the outer membrane [46] (reviewed in [47,48]). Instead, it is associated with the type II secretion system and is involved in the secretion of virulence factors from the periplasm to the extracellular environment [49].

So far, more than 10 structures of different T4Ps and more than 9 structures of pseudopilins have been deposited in the protein data bank (Additional file 1: Table S1). However, only four structures of a full-length T4P are available. All other T4P structures – and all pseudopilin structures – are of N-terminally truncated constructs that do not include the N-terminal hydrophobic transmembrane α-helix. This is unfortunate, as this part is the most conserved part among T4Ps and pseudopilins. Still, all structures of T4aPs and pseudopilins exhibit a conserved central core of a long N-terminal α-helix packed against three to four antiparallel β-strands [32,50-52]. However, various structural elements can be inserted around this conserved core allowing for the diverse functions of T4Ps. In 2013 the structure of the nanowire associated pilin PilA from *G. sulfurreducens* was determined by NMR spectroscopy revealing a single, 61 residue long α-helix [43], but as yet, no structure of a T4P from *S. oneidensis* is available.

In this work, we have determined the structure of the putative nanowire associated T4P on the gene locus SO_0854 [Uniprot: q8eii5] from *S. oneidensis* by X-Ray crystallography to a resolution of 1.67 Å. This T4P from *S. oneidensis* shares the highest degree of sequence identity to PilA from *G. sulfurreducens* (48%) when comparing the first 61 residues after the cleavage site (which corresponds to the full length of PilA from *G. sulfurreducens*). (Additional file 2: Table S2, Additional file 3: Figure S1). The structure reveals the conserved fold of a type IV pilin with a parallel dimer in the asymmetric unit. We have also used Small Angle X-Ray Scattering (SAXS) and Synchrotron Radiation Circular Dichroism (SRCD) to characterize the structure and stability of this protein in solution.

Results and discussion
Sequence conservation and position in the genome
The pilin protein encoded by the gene locus SO_0854 exhibits the conserved N-terminal leader sequence (MNTLQKG) and a hydrophobic patch of 22 residues expected to form a transmembrane helix, which is the hallmark of both type IV pilins and pseudopilins [45] (Figure 1A). Additionally, it possesses two conserved

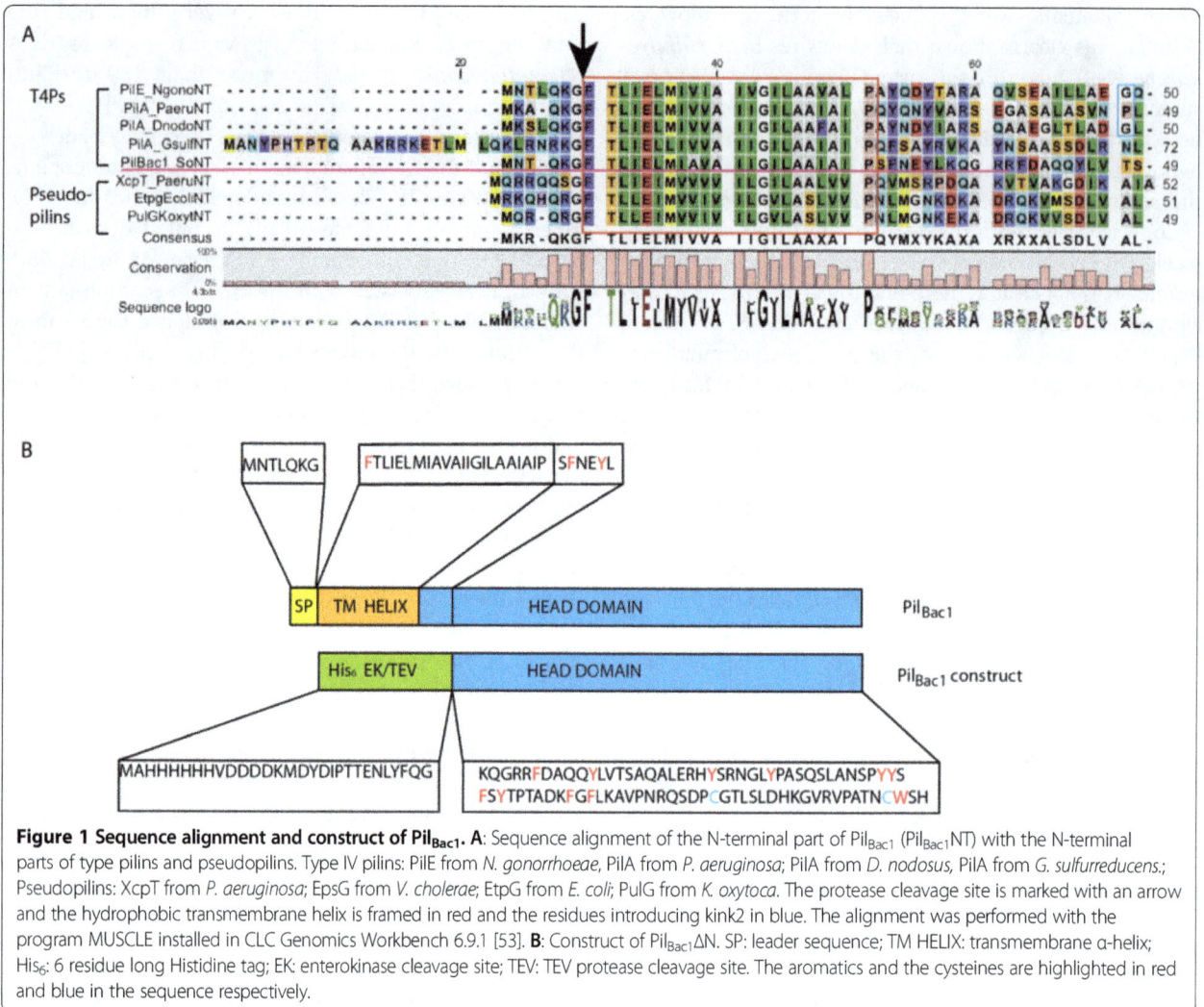

Figure 1 Sequence alignment and construct of Pil_{Bac1}. A: Sequence alignment of the N-terminal part of Pil_{Bac1} (Pil_{Bac1}NT) with the N-terminal parts of type pilins and pseudopilins. Type IV pilins: PilE from *N. gonorrhoeae*, PilA from *P. aeruginosa*; PilA from *D. nodosus*, PilA from *G. sulfurreducens*.; Pseudopilins: XcpT from *P. aeruginosa*; EpsG from *V. cholerae*; EtpG from *E. coli*; PulG from *K. oxytoca*. The protease cleavage site is marked with an arrow and the hydrophobic transmembrane helix is framed in red and the residues introducing kink2 in blue. The alignment was performed with the program MUSCLE installed in CLC Genomics Workbench 6.9.1 [53]. **B**: Construct of Pil_{Bac1}ΔN. SP: leader sequence; TM HELIX: transmembrane α-helix; His_6: 6 residue long Histidine tag; EK: enterokinase cleavage site; TEV: TEV protease cleavage site. The aromatics and the cysteines are highlighted in red and blue in the sequence respectively.

cysteines in the C-terminal part (Cys96, Cys113) which form a disulfide bridge in T4Ps. Based on these features, we classified the protein SO_0854 as a type IV pilin and added it to the subclass of T4aPs due to its short leader sequence.

So far, no consistent nomenclature for T4Ps has been established and therefore, newly described T4Ps cannot be named unambiguously. The low sequence similarity among T4Ps further complicates the naming process. For this reason, the naming of the T4P on the gene locus SO_0854 will briefly be outlined here. SO_0854 is the first open reading frame in a gene cluster consisting of three other putative type IV pilins (SO_0853, SO_0852, SO_0851) and a type IV pilin adhesin with a bactofilin motif (SO_0850). Bactofilins are fiber forming, membrane attached proteins that have been identified in many Gram-negative bacteria and they are associated with cytoskeleton related functions such as cell motility, cell morphology and cell division [54,55]. In *M. xanthus* the polymerized bactofilin BacP directly interacts with PilB and PilT which are responsible for extension

and retraction of type IV pili, respectively, and thus for the motility of the cell [56]. In *S. oneidensis* a bactofilin (SO_1662) [53] was shown to localize to the cell division ring and this bactofilin was therefore assumed to be associated with cell division [54]. Even though bactofilins constitute a recently discovered protein family and their functions have not been fully elucidated yet, the finding of this motif in the putative adhesin in this operon is intriguing. For this reason, we named the five pilin proteins on the gene loci SO_0854, SO_0853, SO_0852, SO_0851 and SO_0850 Pil_{Bac1}, Pil_{Bac2}, Pil_{Bac3}, Pil_{Bac4} and Pil_{Bac5} respectively.

Construction and purification of a soluble construct

To obtain a soluble version of Pil_{Bac1}, a construct was designed that lacks the N-terminal 35 residues including the signal peptide and the transmembrane α-helix. Instead, a His-tag and a TEV protease cleavage site were inserted to enable tag removal (leaving one N-terminal glycine) during the purification process (Figure 1B). This construct was termed $Pil_{Bac1}\Delta N$. The protein was well-expressed in *E. coli* and could be purified to homogeneity in a two-step purification procedure using two passes over a Ni-column (before and after tag removal) followed by size exclusion chromatography. Size exclusion chromatography of $Pil_{Bac1}\Delta N$ gave a monodisperse peak and, comparing the elution volume with those of globular standard proteins that were used for calibration of the size exclusion column, a molecular weight of 11 kDa was estimated, which is close to the theoretical monomeric weight of 9.9 kDa.

Thermostability of $Pil_{Bac1}\Delta N$

The stability of $Pil_{Bac1}\Delta N$ was assessed by SRCD measurements where the temperature was increased stepwise from 7 to 81°C and data were recorded from 280 nm to 190 nm (Figure 2). With increasing temperature the signal strength at 195, 210 and 222 nm fell, indicating a change or a loss of structure. Interestingly, this effect was partly reversed when re-cooling the sample back to 24°C indicating that the protein could, at least partially, refold. The change of the structure was analyzed with principle component analysis (PCA) (Figure 2B) and the contribution of the different components relative to the temperature is shown in Figure 2C. The inflection points of both curves yield an approximate melting temperature of 36 and 38°C respectively. A somewhat higher value (42°C) was obtained in a Thermofluor assay (data not shown). In the Thermofluor experiment, a steeper gradient was applied and this might have resulted in a higher melting temperature compared to the SRCD measurements. In general, a melting temperature around or below 40°C has been claimed to counteract crystallization [57].

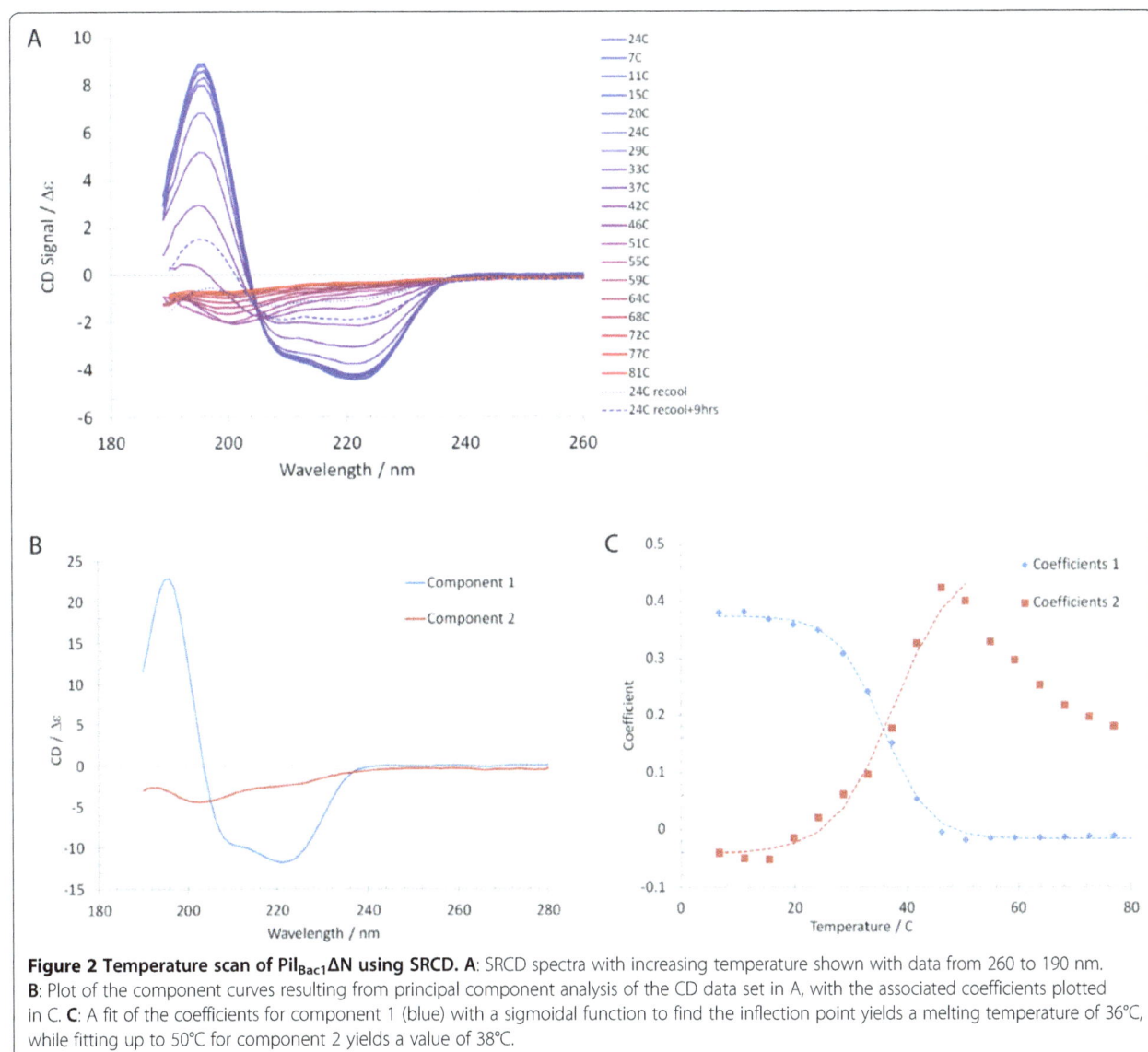

Figure 2 Temperature scan of $Pil_{Bac1}\Delta N$ using SRCD. A: SRCD spectra with increasing temperature shown with data from 260 to 190 nm. **B**: Plot of the component curves resulting from principal component analysis of the CD data set in A, with the associated coefficients plotted in C. **C**: A fit of the coefficients for component 1 (blue) with a sigmoidal function to find the inflection point yields a melting temperature of 36°C, while fitting up to 50°C for component 2 yields a value of 38°C.

X-ray crystallography

An initial hit was obtained in the Structure Screen from Molecular Dimensions in condition 42 (0.2 M $(NH_4)_2SO_4$, 30% PEG 8,000) at 15 mg/ml at 19°C. These crystals could not be reproduced when manually recreating the conditions. Introducing 100 mM CHES pH 8.6 as a buffer component into the crystallization condition yielded crystals of around 50x50x500 μm^3. Additionally, the ratio between protein and reservoir solution was increased from 1:1 to 2:1. The best looking crystals were consistently obtained at 26% PEG 8,000, 0.15 M $(NH_4)_2SO_4$ and 0.1 M CHES pH 8.6. The crystal structure of $Pil_{Bac1}\Delta N$ was initially determined by sulfur SAD at 2 Å resolution based on the anomalous signal of the two cysteines in the C-terminal domain, a bound sulfate molecule and a bound sodium ion (structure determination described in another manuscript). A high resolution data set was collected on another crystal and data were processed in the high symmetry space group I222 to a resolution of 1.67 Å using the $CC_{1/2}$ value as a cut-off [58] (Table 1). The model obtained from S-SAD was then used as a search model in molecular replacement for this data set, and all residues were modelled into clear electron density with no apparent ambiguities. The crystal form had a large solvent content (62%) yielding a high optical resolution and favorable data-to-parameter ratio at 1.67 Å resolution. Riding hydrogen atoms were included in the last model refinement and this resulted in a decrease in R_{free} of 1%.

Additional density was observed at the interface to the large solvent channels of the crystal. This density could not be attributed to solvent or an additional copy of the protein, but probably integrates features of partially associated molecules from the mother liquor or buffer solutions used for purification, such as PEG, glycerol, ions and water (Additional file 4: Figure S2).

Overall structure

$Pil_{Bac1}\Delta N$ consists of one long N-terminal α-helix packed against 4 antiparallel β-strands resembling the core fold of type IV pilins (Figure 3A). A long loop forms the αβ-loop connecting the α-helix and the first β-strand. This region is among the most variable parts in T4P structures and can contain insertions of a short α-helix or a short β-strand or simply display a random coil loop structure as it is the case for $Pil_{Bac1}\Delta N$. The two cysteines in the loops b2-b3 and the loop after b4 indeed form the conserved disulfide bridge of T4Ps that forms the disulfide-bounded loop region (D-region) and keeps strands 3 and 4 together.

$Pil_{Bac1}\Delta N$ exhibits two positively charged surface patches at the N- and at the C-terminus of the α-helix due to closely spaced arginines and lysines facing the solvent area (Figure 3B and C). A negatively charged patch is formed by residues in the C-terminus and the top of the head domain. These charged regions might act as a platform for interactions with other molecules, for example with other pilins in the pilus or receptors for attachment. For instance, in the structure of PilE from *Neisseria gonorrhoeae* docked into an EM envelope of a pilus, positively charged patches were suggested to be responsible for DNA binding [34]. For a pilin protein from *S. oneidensis* such as Pil_{Bac1}, potential binding partners could be multiheme cytochromes which were suggested to be the electron transporting components in nanowires [28,29]. In *G. sulfurreducens*, the multiheme cytochrome OmcS was shown to co-localize with nanowires suggesting a direct interaction [63] and a similar interaction can be expected from multiheme cytochromes in *S. oneidensis*.

Comparison to other type IV pilins and pseudopilins

$Pil_{Bac1}\Delta N$ is structurally similar to other T4Ps in the α-helix and the first two β-strands. However, variable regions are also characteristic of type IV pilins. These include the αβ-loop and the loops connecting the β-strands and parts of the D-region. The loops between the β-strands have been proposed to be involved in contact formation with interaction partners [34]. In $Pil_{Bac1}\Delta N$, loops b1-b2 and b2-b3 are relatively long. Compared to other type IV pilins, $Pil_{Bac1}\Delta N$ is very compact without any additional motifs or insertions, mostly due to its short sequence relative to other T4P head domains. In the reported structures of T4aPs, a disulfide bridge is usually formed between cysteines in the fourth β-strand (b4) and the last loop (Additional file 1: Table S1). In contrast, in Pil_{Bac1} the first cysteine is not situated in b4, but in the loop from b2 to b3. Another exception is the structure of PilA_4 in which the disulfide bridge is formed by two cysteines in β-strands b3 and b4 [64].

In general, structures of full-length T4Ps exhibit two kinks in the long N-terminal α-helix, one in the transmembrane part (kink 1) and one in the C-terminal part of the α-helix (kink 2) (Additional file 1: Table S1). These two kinks are due to helix breaking residues (prolines or glycines) at positions 22 and 42, which are conserved among most T4aPs. Interestingly, Pil_{Bac1} possesses a helix breaking proline at position 22, but a helix breaking residue is missing at position 42 (Figure 1). Therefore, kink 2 is missing in Pil_{Bac1} and the α-helix in the structure of $Pil_{Bac1}\Delta N$ is straight. Such a feature has commonly been observed in structures of pseudopilins and in the T4bPs PilS from *S. typhi* and TcpA from *V. cholerae* as well as in two T4a pilins, namely PilA from *G. sulfurreducens* and PilA_4 from *T. thermophilus*. The feature of a straight α-helix could suggest a different mode of packing in the pilus (see below).

Table 1 Data collection and processing statistics for the structure of Pil$_{Bac1}$ΔN

Subset	Native
Crystallization condition	26% PEG 8,000, 0.1 M CHES pH 8.6, 0.15 M (NH$_4$)$_2$SO$_4$
Beamline	BL-14.2, BESSY-II, Helmholtz Zentrum Berlin, DE
Detector	Rayonix MX225
Wavelength (Å)	0.918409
Crystal to Detector Distance (mm)	150.0
Rotation/ Frame (°)	0.5
Number of Frames	200
Data Processing statistics	
Resolution (Å)	48.23-1.67 (1.70-1.67)
Space group	I 2 2 2
Unit cell parameters (Å, °)	48.91, 96.46, 110.33; 90.0, 90.0, 90.0
No. of unique reflections	56,946 (4,148)
No. of total reflections	111,644 (8,093)
Multiplicity	1.96 (1.95)
Completeness (%)	97.0 (95.1)
R^a_{merge}	0.039 (0.641)
$R^b_{r.i.m.}$	0.054 (0.890)
Wilson B-factor (Å^2)	23.12
Mean $I/\sigma I$	13.3 (1.2)
CC$_{1/2}$	0.999 (0.578)
Refinement statistics	
R_{work}	0.1798
R_{free}	0.2075
Number of non-hydrogen atoms modelled	1631
Number of non-hydrogen protein atoms	1406
Number of ligand atoms	6
Number of solvent molecules	219
R.M.S.D. from ideal values	
Bonds (Å)	0.004
Angles (°)	1.101
Ramachandran	
Favoured (%)	99
Outliers (%)	0
Clash Score	0.73
Average B-factor (Å^2)	34.6
Average B-factor for protein (Å^2)	33.7
Average B-factor for ligands (Å^2)	61.4
Average B-factor for solvent molecules (Å^2)	39.7

Data were processed with XDS [59]. The structure was determined and built and refined with Phenix and COOT [60,61]. Values in parentheses are given for the highest-resolution shell.

^{a}: $R_{merge} = \sum_{hkl} \sum_{i} |Ii(hkl)-I(\bar{hkl})| \Big/ \sum_{hkl} \sum_{i} Ii(hkl)$ [62]; ^{b}: $R_{r.i.m.} = \sum_{hkl} [N/(N-1)]^{\frac{1}{2}} \sum_{i} |Ii(hkl)-I(\bar{hkl})| \Big/ \sum_{hkl} \sum_{i} Ii(hkl)$ [62].

To identify homologous structures to Pil$_{Bac1}$ΔN, a search with the DALI server [65] was performed (Additional file 5: Table S3). As expected, structures of type IV pilins scored the highest and among them, the highest score was seen with the T4aPs PilA_4 from *T. thermophilus* [64] (4BHR.PDB; DALI: Score: 8.7), followed by the minor pilin PilX from *N. meningitis* [66] (2OPD.PDB; DALI: Score: 8.3) and the PAK pilin from *P. aeruginosa*

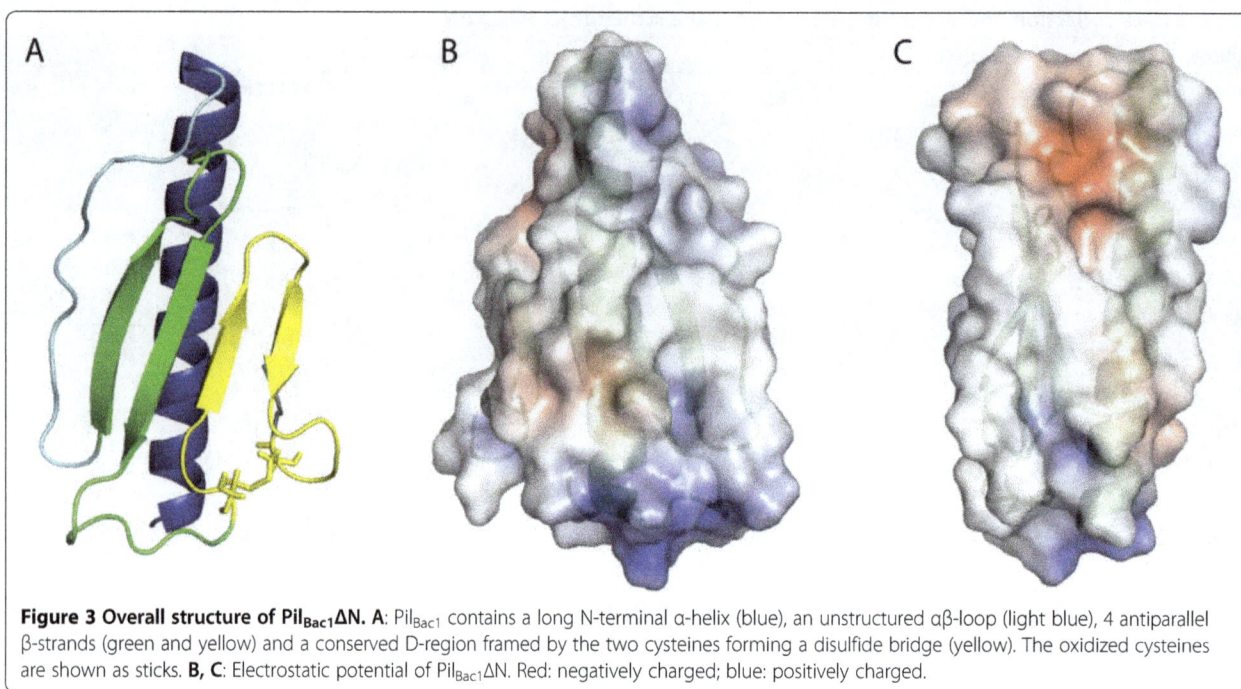

Figure 3 Overall structure of Pil$_{Bac1}$ΔN. A: Pil$_{Bac1}$ contains a long N-terminal α-helix (blue), an unstructured αβ-loop (light blue), 4 antiparallel β-strands (green and yellow) and a conserved D-region framed by the two cysteines forming a disulfide bridge (yellow). The oxidized cysteines are shown as sticks. **B, C**: Electrostatic potential of Pil$_{Bac1}$ΔN. Red: negatively charged; blue: positively charged.

[67] (1X6Z.PDB; DALI: Score: 8.1) (Figure 4A-C). PilA_4 is one of the exceptions of T4aPs with a straight, continuous α-helix and superimposes well with Pil$_{Bac1}$ΔN (Figure 4A). The biggest variation between PilA_4 and Pil$_{Bac1}$ΔN lies in the αβ-loop, which forms a short α-helix in PilA_4 and a random coil structure in Pil$_{Bac1}$ΔN.

Pil$_{Bac1}$ΔN also shares structural similarity with pseudopilins, which superimpose well in the core regions (Figure 4D). As for other type IV pilins, the biggest differences are observed in the variable regions. Additionally, pseudopilins coordinate a calcium ion in the D-region, where Pil$_{Bac1}$ΔN and other type IV pilins form a conserved disulfide bridge [68].

Pil$_{Bac1}$ΔN overlays well with the α-helix of PilA from *G. sulfurreducens* and the feature of a straight α-helix after the transmembrane part (Figure 4E); yet, this T4aP structure was not scored high by the DALI server, as PilA from *G. sulfurreducens* does not contain a head domain. Noteworthy, a gene (GSU1497) is located directly downstream of pilA that codes for a protein that was shown to be up-regulated together with PilA and a few multiheme cytochromes in microbial fuel cells [69]. Furthermore, PilA was not detected by western blotting in strains deficient for GSU1497 [70]. Therefore, this protein might constitute the missing head domain of PilA from *G. sulfurreducens* and a future structural comparison of this protein with Pil$_{Bac1}$ΔN would be very interesting.

Dimeric interface

Pil$_{Bac1}$ΔN was crystallized as a parallel dimer in the asymmetric unit in which the interface is formed by interactions between 15 and 20 residues in the α-helix from monomers A and B, respectively (forming a non-proper dimer with a screw-axis) (R.M.S.D. of 0.309 Å based on 79 out of a total of 89 C$_α$s)(Figure 5A). Previously it was noted that pilins can exist as dimers and multimers [71-73]. Many structures of type IV pilins and pseudopilins were also determined in a dimeric [32,68,74-78] or even in a trimeric state [68]. However, different to Pil$_{Bac1}$ΔN, most of them were not arranged in a physiologically relevant conformation (e.g. antiparallel, in-line). The structures of the T4P CofA from *E. coli* (3S0T. PDB) and of the pseudopilin PulG from *E. coli* (3G20. PDB) were determined as dimers, in which the monomers are further apart from each other. This dimerization was probably caused by crystal contacts. In contrast, in the structure of full-length FimA from *D. nodosus*, the two monomers are held together by extensive, intermolecular interactions. However, unlike Pil$_{Bac1}$ΔN, the dimeric interface is here formed between the transmembrane domains.

Since the N-terminal transmembrane domain in Pil$_{Bac1}$ΔN is missing in the structure, we modelled this conserved part based on the full-length structure of mature PilE from *N. gonorrhoeae*, as this region is highly conserved in T4Ps as a transmembrane α-helix (Figures 1 and 5A–D; R.M.S.D. of the α-helices and the first two β-strands: 1.7 Å). This additional domain extends the α-helices in both monomers in the crystal dimer by a further 28 residues without the introduction of any clashes (Figure 5E). Consequently, this arrangement maintains contacts between both monomers and allows for the existence of the dimer in a membrane (Figure 5E).

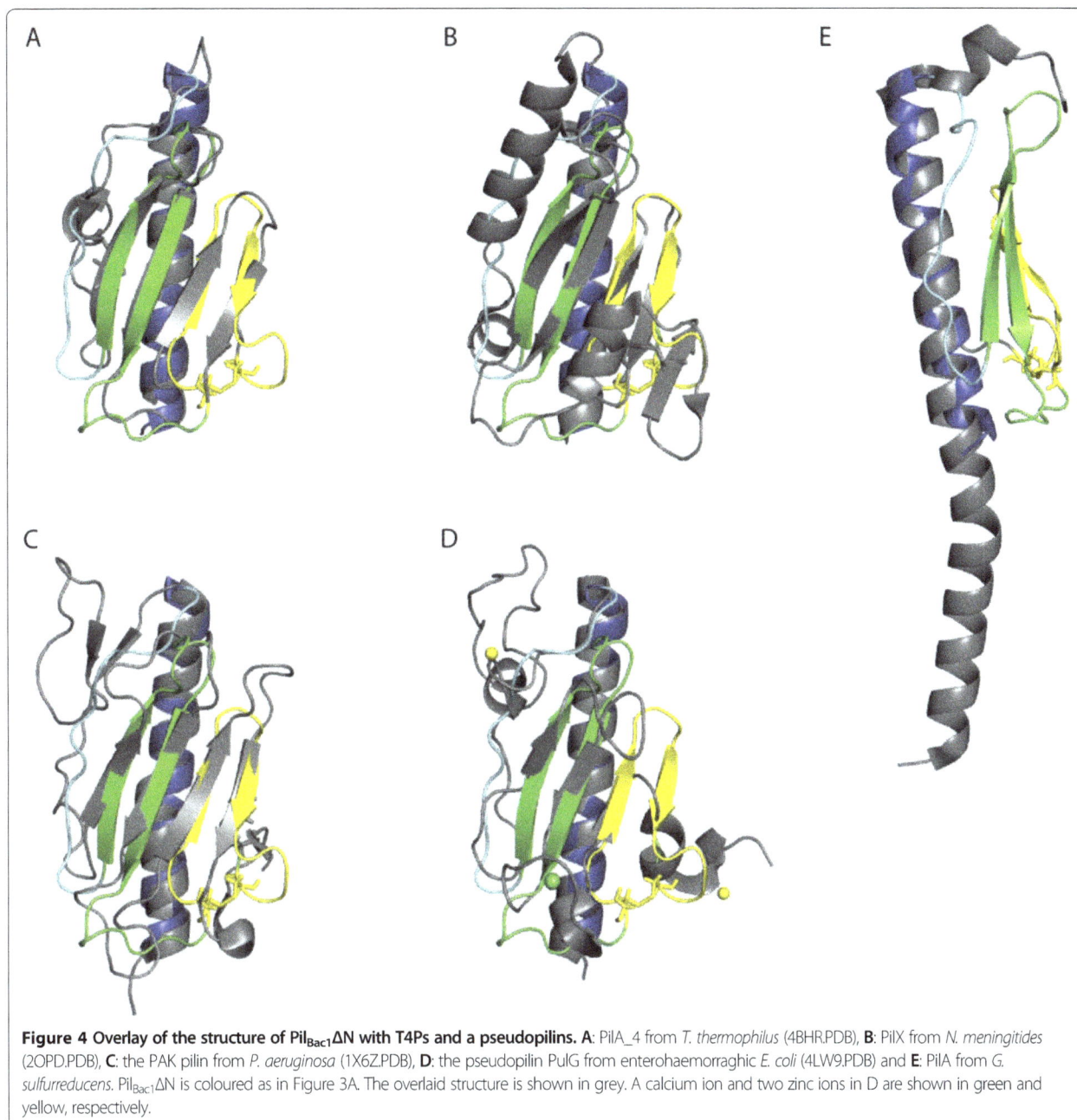

Figure 4 Overlay of the structure of Pil$_{Bac1}$ΔN with T4Ps and a pseudopilins. **A**: PilA_4 from *T. thermophilus* (4BHR.PDB), **B**: PilX from *N. meningitides* (2OPD.PDB), **C**: the PAK pilin from *P. aeruginosa* (1X6Z.PDB), **D**: the pseudopilin PulG from enterohaemorraghic *E. coli* (4LW9.PDB) and **E**: PilA from *G. sulfurreducens*. Pil$_{Bac1}$ΔN is coloured as in Figure 3A. The overlaid structure is shown in grey. A calcium ion and two zinc ions in D are shown in green and yellow, respectively.

In the modelled part of the alpha helix, a kink is introduced due to the presence of a conserved helix breaking proline in the transmembrane helix at position 22 (kink1). Such a kink has been described in all four available full-length structures of T4Ps [32,43,77,80] and separates the N- and the C-terminal parts of the helix from each other.

To assess the oligomeric state of Pil$_{Bac1}$ΔN in solution we performed SAXS studies at concentrations ranging from 1 – 16 mg/ml (Additional files 6: Table S4, Additional file 7: Figure S3). No signs of aggregation or repulsive forces were observed from the scattering data at any of these concentrations as judged by comparison of the scattering intensities at low scattering angles for all concentrations used. Guinier and Porod analysis revealed a radius of gyration of 14 and 16 Å respectively, and the molecular mass determined by Porod volume analysis indicated a molecular weight of only 6 kDa, which is below the theoretical molecular mass of 9.9 kDa. Kratky analysis indicated a well-folded molecule (Additional file 7: Figure S3). Altogether, these findings indicated that Pil$_{Bac1}$ΔN existed as a monomer in solution. A dummy atom (DA) model of Pil$_{Bac1}$ΔN was built by the program DAMMIF based on the SAXS data [81] and the crystal structure of monomeric Pil$_{Bac1}$ΔN compared to

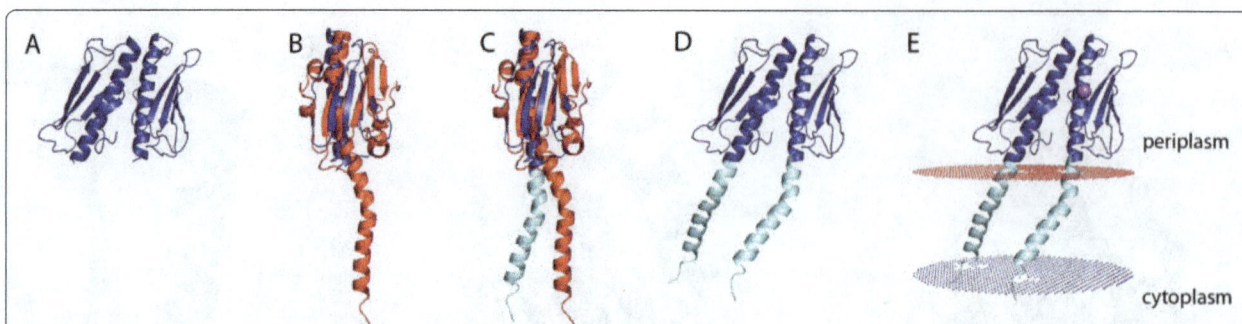

Figure 5 Dimer of Pil$_{Bac1}$ΔN. A: Dimeric interface between two Pil$_{Bac1}$ΔN molecules in the crystal. **B**: Superposition of a Pil$_{Bac1}$ΔN monomer with PilE from *N. gonorrhoeae* (2HI2.PDB). **C**: Superposition based on the head domains of a Pil$_{Bac1}$ΔN monomer with the modelled α-helix onto PilE from *N. gonorrhoeae* (2HI2.PDB). **D**: Pil$_{Bac1}$ΔN dimer with the modelled α-helices at the N-terminus based on a superposition with the α-helix from PilE from *N. gonorrhoeae* (2HI2.PDB). **E**: Potential arrangement of a Pil$_{Bac1}$ dimer in a membrane. This figure was generated with the PPM server [79]. The structure of Pil$_{Bac1}$ΔN is shown in blue, the modelled helix in cyan and PilE from *N. gonorrhoeae* is shown in red.

the DA model using the program SUPCOMB [82], yielding a good fit. In UCSF Chimera [83] a simulated map at 10 Å based on the crystal structure was fitted to a SAXS envelope based on the DA model giving a CC of 0.762 (Figure 6). A scattering curve was calculated on the basis of the crystal structure using the program CRYSOL [84] and compared to the experimentally measured scattering data. This resulted in a chi value (discrepancy between the theoretical and experimental scattering curve) of 1.82, which is indicative of good agreement.

The results from the SAXS analysis were further substantiated by size exclusion chromatography and a PISA analysis [86], which indicated that the Pil$_{Bac1}$ΔN dimer interface energy was low. However, the environment in a membrane with potentially interacting transmembrane helices is very different from that of a truncated protein in solution, the local concentration of pilins is significantly higher and the degree of translational freedom is reduced

in a membrane. Therefore we cannot exclude a possible function of dimeric Pil$_{Bac1}$ in the membrane, but further studies on the full-length Pil$_{Bac1}$ in membranes and assembled into pili will be needed to evaluate oligomeric states of Pil$_{Bac1}$.

Na$^+$-Ion binding site

In chain B clear density was observed, both in the anomalous map from the S-SAD data with a peak height of 6.5 σ (described in another manuscript) and in the 2mFo-DFc map from the high resolution data set, for an ion bound between the αβ-loop and the first β-strand. This ion was octahedrally coordinated by oxygen atoms (carbonyl oxygen of Leu36 and Phe44, delta oxygen of Asn38, 3 H_2O molecules) with average coordination distances of 2.5 Å (Figure 7A). In agreement with these coordination properties, its anomalous scattering intensity

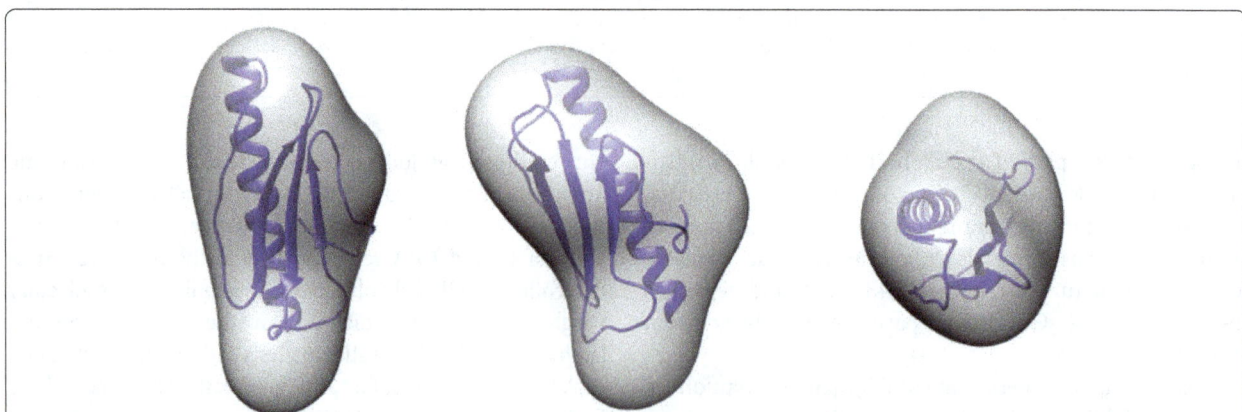

Figure 6 Crystal structure of Pil$_{Bac1}$ΔN docked into its SAXS envelope. A dummy atom model from the SAXS data was generated with DAMMIF [81] and superposed onto the crystal structure with SUPCOMB [82]. This oriented SAXS model was then converted into an envelope with the pdb2vol software from the SITUS package [81,85].

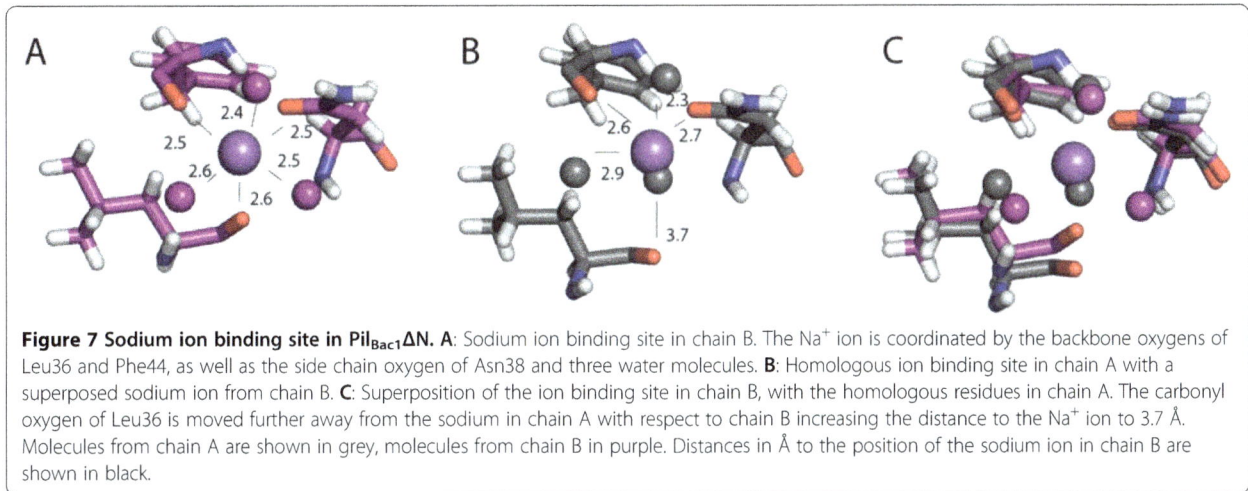

Figure 7 Sodium ion binding site in Pil$_{Bac1}$ΔN. A: Sodium ion binding site in chain B. The Na$^+$ ion is coordinated by the backbone oxygens of Leu36 and Phe44, as well as the side chain oxygen of Asn38 and three water molecules. **B**: Homologous ion binding site in chain A with a superposed sodium ion from chain B. **C**: Superposition of the ion binding site in chain B, with the homologous residues in chain A. The carbonyl oxygen of Leu36 is moved further away from the sodium in chain A with respect to chain B increasing the distance to the Na$^+$ ion to 3.7 Å. Molecules from chain A are shown in grey, molecules from chain B in purple. Distances in Å to the position of the sodium ion in chain B are shown in black.

and consistent with the presence of 100 mM NaCl in the buffer, we assigned this ion to a sodium ion (described in another manuscript).

Interestingly, this sodium ion was not observed in chain A due to a slight distortion of the ion binding site compared to chain B. The backbone of Leu36 is further away from the binding site increasing the distance from 2.6 to 3.7 Å (Figure 7B and C). Additionally, a third water molecule is missing to coordinate the ion; instead, another water molecule binds in the position where the sodium binds in chain B. Furthermore, features of positive density lie between the sodium ion binding site in chain A and a symmetry related molecule. This density is indicative of a bound molecule such as PEG which might have interfered with binding of the ion in this site.

So far, no functionally validated binding of any ions has been described for T4Ps. For pseudopilins the stabilization by calcium ions has been shown [68]. Whether the bound sodium ion in Pil$_{Bac1}$ΔN fulfills a functional role or is a crystal artefact due to the 100 mM NaCl in the buffer, is not clear yet. Hypothetically, the Na$^+$ ion could act as a regulator in filament assembly. In the extracellular space and periplasm, the sodium concentration is much higher than intracellularly and we find it possible that the pilin might bind a sodium ion in the periplasm.

Modelling of a Pil$_{Bac1}$ pilus

To investigate the putative packing in a pilus, a model composed of Pil$_{Bac1}$ subunits was generated (Figure 8). In order to do this, an atomic pilus model based on the EM structure of a PilE pilus and the crystal structures of PilE from *N. gonorrhoeae* were used as a template [34]. First, the structure of a Pil$_{Bac1}$ΔN monomer was superimposed onto monomeric PilE and the missing N-terminal 28 residues were modelled resulting in a full-length model of Pil$_{Bac1}$. This full-length chimera was then overlaid onto the

individual subunits in the PilE pilus based on the modelled, very similar transmembrane part. No significant clashes between the subunits were introduced, only minor clashes between the N-terminal part of the helix of one monomer and the C-terminal part of the helix of the neighboring monomer (Figure 8B).

As for the *Neisseria* pilus, the α-helices formed the central core with the head domains facing outwards; however, in contrast to the Neisseria pilus, the packing seemed to be less dense due to the missing kink in the C-terminal part of the helix (kink 2) and due to a smaller head domain. The positively charged patch at the N-terminus from one pilin is located closely to the negatively charged patch of the neighbouring pilin stabilizing the interaction between pilins in the pilus filament. (Figure 8C). Overall, the Pil$_{Bac1}$ pilus was a bit thinner than for *Neisseria*, because the head domain in Pil$_{Bac1}$ is less bulky with no major grooves or protrusions on the surface.

This model for a Pil$_{Bac1}$ pilus is in good agreement with the general observations that the D-region is involved in the interaction with other molecules [1,87] and should be solvent accessible. Still, it needs to be considered that this model is based on a model pilus from *N. gonorrhoeae* which in turn is based on the docking of the crystal structure of PilE into an EM envelope. Unlike PilE, Pil$_{Bac1}$ does not contain a kink in the α-helix after the transmembrane part which will orient the head domain in a slightly different angle (kink2). This will necessarily affect the packing pattern in a pilus and inevitably lead to differences to the PilE pilus and therefore, this model has to be interpreted with some caution.

Aromatic amino acids in Pil$_{Bac1}$

Malvankar and co-workers have proposed that nanowires from *G. sulfurreducens* are conductive due to the close positioning of aromatic amino acids in PilA [27].

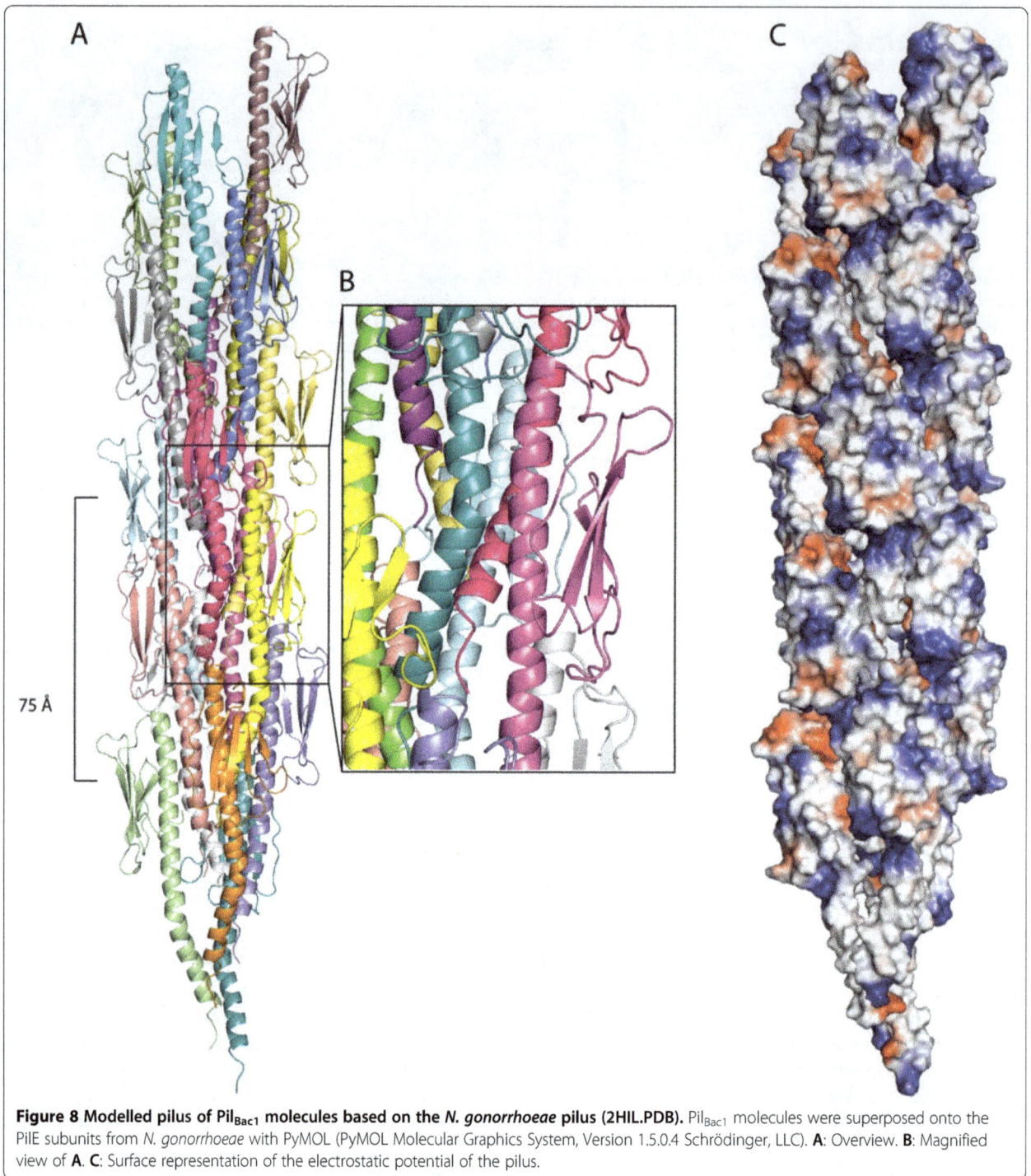

Figure 8 Modelled pilus of Pil$_{Bac1}$ molecules based on the *N. gonorrhoeae* pilus (2HIL.PDB). Pil$_{Bac1}$ molecules were superposed onto the PilE subunits from *N. gonorrhoeae* with PyMOL (PyMOL Molecular Graphics System, Version 1.5.0.4 Schrödinger, LLC). **A**: Overview. **B**: Magnified view of **A**. **C**: Surface representation of the electrostatic potential of the pilus.

The NMR structure of PilA from *G. sulfurreducens* showed that the aromatic side chains were indeed closely spaced with a maximum distance of 15 Å [43]; yet, to the best of our knowledge, the maximum distance between aromatic groups that allows for electron transfer has not been defined so far. Similar to *G. sulfurreducens*, *S. oneidensis* forms conductive nanowires and, based on the overall similarity between these two organisms including metabolic pathways and the prevalence of multiheme cytochromes, a similar electron transfer mechanism is very likely. Pil$_{Bac1}$ is the type IV pilin which is most closely related to PilA from *G. sulfurreducens* based on sequence comparisons (Additional file 2: Table S2). The full-length chimeric model of Pil$_{Bac1}$ contained 14 aromatic residues including two phenylalanines and one tyrosine in the modelled transmembrane domain. In the modelled pilus of Pil_{Bac1} subunits, the aromatic side chains were evenly spaced throughout the whole structure, with some being closer to their neighbors than others (Figure 9A). A long chain of aromatic side chains wound along the modelled filament with two clusters on each subunit in which the aromatics are closely positioned to each other with distances between 4 to 7 Å. Yet, these two clusters are separated by a gap of 11 Å which can be defined as the maximum distance between two aromatics in the pilus model. This distance compares well to PilA from Geobacter; however, the arrangement of aromatic side chains in PilE from *N. gonorrhoeae* – which has not been shown to produce conductive nanowires yet – is similar with a maximum distance between individual aromatic side chains of around 13 Å (Additional file 8: Figure S4). This may argue against the hypothesis stating that conductivity is based on a specific alignment of aromatic side chains.

Conclusions

In this study we described the high-resolution structure of the N-terminally truncated type IV pilin Pil_{Bac1} which exhibits the typical fold of type IV pilins with a long N-terminal α-helix packed against 4 antiparallel β-strands. Pil_{Bac1} was crystallized as a parallel dimer with a sodium ion bound to one of the monomers. Small-angle X-Ray scattering studies of the N-terminally truncated Pil_{Bac1} indicated that the protein exists as a monomer in solution, but further characterization of the full-length form and/or membrane bound form will be necessary to clarify the oligomeric state of Pil_{Bac1} in a cellular context. In contrast to most other T4aP head domains, Pil_{Bac1} displays a straight α-helix and a small head domain which leads to a less dense packing mode in a modelled pilus compared to other well-characterized pilins and possibly making room for interaction partners such as multiheme cytochromes.

Methods

Cloning, expression and purification

The DNA coding for residues 36 to 123 of Pil_{Bac1} from *S. oneidensis* with LIC overhangs at the 5' and 3' ends was amplified by Polymerase Chain Reaction (PCR) using genomic DNA from cell lysates as a template (forward primer: GACGACGACAAGATGGATTACGACATCCCCACTAC

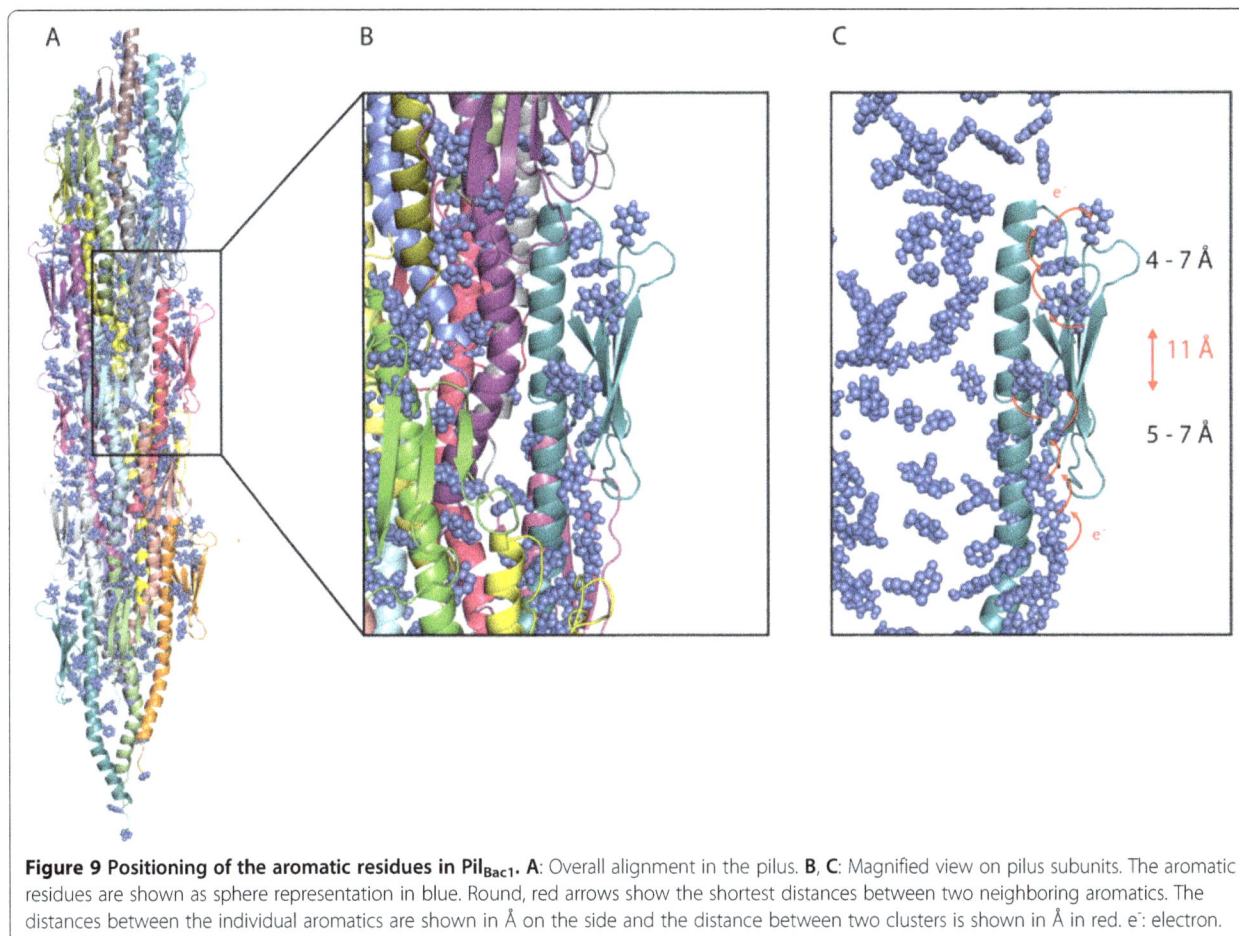

Figure 9 Positioning of the aromatic residues in Pil_{Bac1}. A: Overall alignment in the pilus. **B, C**: Magnified view on pilus subunits. The aromatic residues are shown as sphere representation in blue. Round, red arrows show the shortest distances between two neighboring aromatics. The distances between the individual aromatics are shown in Å on the side and the distance between two clusters is shown in Å in red. e⁻: electron.

T*GAGAATCTTTATTTTCAGGG*CAAGCAAGGCAGAC GCTTCGATGCGC; reverse primer: GAGGAGAAGCCC GGTTTAATGGCTCCAACAATTTGTGGCGGGG). The PCR fragment obtained was then inserted into the vector pET-46 Ek/LIC by ligation independent cloning (LIC Kit, Novagen, USA). An N-terminal His tag and an Enterokinase (EK) site are part of pET-46 Ek/LIC vector and a sequence encoding a Tobacco Etch Virus (TEV) protease cleavage site (marked in italics above) was included in the forward primer from the PCR. The correct insert of the plasmid was verified by sequencing and it was transformed into *E. coli* BL 21 DE3 Origami cells (Novagen, USA).

6–12 L of LB medium (containing 100 µg/ml ampicillin and 50 µg/ml kanamycin) were inoculated with 60–120 ml of overnight culture and grown at 37°C at 120 rpm. When an OD600 between 0.6 and 0.8 was reached, expression was induced by the addition of Isopropyl-thiogalactoside (final concentration of 1 mM) for 18 hours at 20°C. Cells were harvested by centrifugation at 8,927.1 *g* for 20 minutes. The cell pellets were resuspended in 25 ml LB/ liter culture, flash-frozen in liquid nitrogen and stored at –20°C until use.

Cell pellets were resuspended in 3 ml lysis buffer (20 mM Tris–HCl pH 7.5, 500 mM KCl, 10% Glycerol) per gram wet cell pellet and opened by sonication on ice. Unopened cells and cell debris were spun down by centrifugation at approximately 235,000 *g* at 4°C for 2 h. Imidazole was added to the supernatant to a final concentration of 10 mM and loaded onto a 5 ml nickel-chelating column (GE Healthcare, USA) that had been pre-equilibrated in lysis buffer and 10 mM imidazole. After washing with lysis buffer with 10 mM imidazole, the protein was eluted on an ÄKTA Prime with a gradient from 25 – 500 mM imidazole over 10 column volumes at 1 ml/min and fractions containing Pil_{Bac1} were pooled. The N-terminal His-tag was cleaved off the protein by the TEV protease during dialysis against 1 l lysis buffer at 4°C for 12 h. The protein was then loaded onto a 5 ml nickel-chelating column (GE Healthcare, USA) and the flow-through containing the cleaved protein was collected and concentrated on a 5 molecular weight cutoff concentrator. Pil_{Bac1} was further purified by a size exclusion step on a Superdex 75 10/300 (GE Healthcare, USA) equilibrated in 20 mM Tris–HCl pH 7.5, 100 mM NaCl. Finally, Pil_{Bac1} was concentrated to 15 mg/ml and stored at –80°C.

Synchrotron radiation circular dichroism

SRCD studies were performed at the CD1 beam line of the ASTRID synchrotron, Aarhus University, Denmark [88,89]. Light from the CD1 beam line passed through an MgF Rochon polarizer (B-Halle GmbH, Berlin) and alternating left and right circularly polarized light was produced using a photo-elastic modulator (Hinds, USA). The polarized light then passed through the sample and was detected by a photomultiplier tube (9406B, ETL, UK). SRCD spectra were taken from 280 nm to a minimum wavelength of 190 nm. To investigate the stability of the protein, spectra were recorded with increasing temperature from 7 to 81°C (5°C per step; 3 measurements at each temperature). After such a temperature scan, the sample was cooled down to 24°C and three final spectra were recorded after incubation at 24°C for 9 h.

All spectra were recorded at a concentration of 0.66 mg/ml, in 100 mM NaCl, 20 mM TRIS pH 7.5. Before and after each temperature scan, a spectrum of the buffer was recorded to check that no changes had occurred during the sample measurement (e.g. damage to the cell, changes in beam). The two buffer spectra were averaged and subtracted from the sample spectra using a spreadsheet operation.

Principal component analysis of the set of CD spectra recorded over the temperature range 7 to 81°C was performed using the Multibase 2013 (http://www.numerical-dynamics.com/) add-in for excel to yield the component curves (Figure 2B) and corresponding coefficients for each temperature (Figure 2C). The resulting coefficients were each fitted by a sigmoidal function to find the inflection point and hence the melting temperature for that component.

Thermofluor

A Thermofluor experiment was performed with $Pil_{Bac1}\Delta N$ in 100 mM NaCl, 20 mM TRIS pH 7.5 (25 µM) and with 10xSYPRO Orange (Sigma Aldrich) using a Light Cycler 480 (Roche) and the option for protein melting dynamics. The temperature was increased from 20 to 85°C with 4.4°C per s and the fluorescence was measured with 20 acquisitions per time point. Plotting the fluorescence against the temperature yielded a sigmoidal curve. The inflection point of this curve was approximated as the melting temperature of $Pil_{Bac1}\Delta N$.

X-Ray crystallography
Crystallization, data collection and processing

$Pil_{Bac1}\Delta N$ was crystallized at a concentration of 15 mg/ml in a hanging drop set-up in a 2:1 ratio with the reservoir (26% PEG 8,000, 0.15 M $(NH_4)_2SO_4$ and 0.1 M CHES pH 8.6) at 19°C. Rod shaped crystals appeared after a few days and were stored in liquid nitrogen without the addition of more cryoprotectant.

A high resolution data set was collected from crystals at beamline BL14.2, Bessy II, Berlin, [90] at 0.98 Å with 0.5 s of exposure time (Table 1). All data were processed and merged with the XDS software [59]. The Wilson B-factor was determined by the program AIMLESS [91]. The Matthews coefficient [92] and the solvent content were derived by the program XTRIAGE from the Phenix suite [60]. See also Table 1 for data collection and processing statistics.

Structure determination and analysis

The S-SAD structure of $Pil_{Bac1}\Delta N$ (PDB accession code 4US7) was used as a search model for molecular replacement using the program PHASER [93] from the Phenix suite. The model was refined with Phenix.Refine [60] with the options for xyz coordinates, TLS, individual B-factors, optimizing X-Ray stereochemical weights and ADP weights with iterative model building in Coot [61]. In the last steps, the model was refined including riding hydrogen atoms. The model was validated with the program MOLPROBITY [94] and deposited in the protein data bank with the accession code 4D40.

The electrostatics for the monomer were calculated using the PDB2PQR (version 1.8) server at http://nbcr-222.ucsd.edu/pdb2pqr_1.9.0/ with standard parameters and force field = PARSE. Files from PDB2PQR were used with the APBS [95] plugin in PyMOL 1.7.4 to generate images, coloring a range of +/- 5 kT/e by potential on the solvent accessible surface. The electrostatics for the pilus model was calculated using the built-in feature of PyMOL.

The structure of $Pil_{Bac1}\Delta N$ was superposed onto PilE from *N. gonorrhoeae* (PDB accession code 2HI2) [34] based on the C_αs of the α-helix and the first two β-strands and the first 28 residues of PilE were then added onto $Pil_{Bac1}\Delta N$ using PyMOL (PyMOL Molecular Graphics System, Version 1.5.0.4 Schrödinger, LLC). Residues 9 (V) and 23 to 28 (AYQDYT) were then mutated into the corresponding residues in Pil_{Bac1} (A; SFNFYL) and the modelled helix was subjected to energy minimization using torsion angle and Ramachandran constraints in COOT [61]. The minimized chimera structure was superposed on the first 28 residues of the subunits in the Neisseria PilE pilus model (PDB accession code 2HIL) [34] in order to generate a $Pil_{Bac1}\Delta N$ pilus model.

Small-angle X-Ray scattering
Data collection

The data collection parameters are given in Additional file 6: Table S4. Before and after a scattering profile of the protein was recorded, SAXS data of the buffer were collected. The data for the buffer were merged and subtracted from the protein (in buffer) scattering curve to yield the protein scattering curve.

Data processing and model building

The scattering data were processed with programs from the ATSAS package (PRIMUS [96], DAMMIF [81] DAMAVER [97]). The radius of gyration and the maximum diameter of the protein were calculated with the PRIMUS and GNOM programs. 12 *ab initio* models consisting of dummy atoms

were made with DAMMIF. The models from DAMMIF were evaluated with DAMAVER and for all models, the normalized spatial discrepancy (NSD) deviated no more than two standard variations from the mean value, thus no models were excluded [97]. Envelopes were made based on the DAMMIF model displaying the lowest NSD to the remaining models using the SITUS package and the pdb2vol program using default settings [85]. The X-Ray structure of $Pil_{Bac1}\Delta N$ was docked into the envelope obtained from pdb2vol using the program SUPCOMB [82] and the model was visualized in UCSF Chimera [83]. To compare the SAXS model to the crystal structure, a theoretical scattering curve was generated based on the crystal structure using CRYSOL [84].

Additional files

Additional file 1: Table S1. Structures of T4Ps and pseudopilins. "b" indicates a β-sheet, "a" an α-helix [98-103].

Additional file 2: Table S2. Sequence identities between T4Ps from *S. oneidensis* (Pil_{Bac1}So, PilESo, PilASo, MshASo, MshBSo, Pil_{Bac2}So, Pil_{Bac3}So, Pil_{Bac4}So, PilVSo, PilXSo) with PilA from *G. sulfurreducens*. The alignment was done using the program MUSCLE [53].

Additional file 3: Figure S1. Sequence alignment of T4Ps from *S. oneidensis* (Pil_{Bac1}So, PilESo, PilASo, MshASo, MshBSo, Pil_{Bac2}So, Pil_{Bac3}So, Pil_{Bac4}So, PilVSo, PilXSo) with PilA from *G. sulfurreducens*. The alignment was done using the program MUSCLE [53].

Additional file 4: Figure S2. Positive density at the interface to the solvent channels. A: Overview over the unit cell. B: Positive density at residues 3 and 5 in chain A. The 2Fo-Fc map was contoured at 1.5σ, Fo-Fc map at 3σ.

Additional file 5: Table S3. Identification of homologous proteins to $Pil_{Bac1}\Delta N$ by the DALI server [65]. The R.M.S.D. value between the structures obtained from DALI based on C_αs is given. Additionally, the number of residues in the structure is given. The three most homologous T4P structures and the most homologous pseudopilin structure are highlighted in bold.

Additional file 6: Table S4. SAXS data collection parameters and data processing statistics. The dry volume was calculated by an online server based on considerations from Harpaz et al. [104]. The molecular mass was determined by Porod volume analysis with the program AUTOPOROD [105].

Additional file 7: Figure S3. SAXS data. A: Original SAXS curves. B: SAXS curve for 16 mg/ml. C: Guinier plot for 16 mg/ml. D: Porod plot for 16 mg/ml. E: Kratky plot for 16 mg/ml. F: Pair-distance distribution plot p (r). s: momentum transfer, I: scattering intensity.

Additional file 8: Figure S4. Conservation of aromatic amino acids in Pil_{Bac1} from *S. oneidensis* (blue), PilA from *G. sulfurreducens* (yellow) and PilE from *N. gonnorhoeae* (red). A: Cartoon presentation of all three pilins. B: Cartoon presentation of PilA from *G. sulfurreducens* and the aromatics in all three pilins shown as sticks.

Abbreviations
EET: Extracellular Electron Transport; EM: Electron Microscopy; NMR: Nuclear Magnetic Resonance; ORF: Open Reading Frame; PCA: Principle Component Analysis; PISA: Proteins, Interfaces, Structures and Assemblies; SAXS: Small-Angle X-Ray Scattering; SRCD: Synchrotron Radiation Circular Dichroism; S-SAD: Sulfur-Single-wavelength Anomalous Diffraction; T4P: Type IV Pilin.

Competing interests
The authors declare that they have no competing interests.

Authors' contributions
MG and JJU performed the experiments. MG, NCJ and SVH did the SRCD analysis. MG and AB determined the structure. MG and TB wrote the manuscript and all authors commented on it. Project planning and development was carried out by MG under supervision of TB and PN. All authors read and approved the final manuscript.

Acknowledgements
We would like to thank the Graduate School of Science and Technology, Aarhus University, and the Pumpkin - Centre for membrane pumps in cells and disease for partial funding of the Ph.D. project of M. Gorgel. The work was supported by the advanced research program BIOMEMOS of the European Research Council (to PN) and by the Danish Council for Technological and Production-related Research on the project Micro-cable based bionanoelectronics (supporting TB). We are grateful to Dr. Manfred S. Weiss, Helmholtz Zentrum Berlin, for help with data collection. We thank the staff at MAX-LAB beamline I911 SAXS for help with data collection. Travel costs to Bessy and MAX-lab were financed by the Biostruct-X project 5624.20.

Author details
[1]Department of Molecular Biology and Genetics, Aarhus University, Gustav Wieds Vej 10c, Aarhus C 8000, Denmark. [2]ISA, Department of Physics and Astronomy, Aarhus University, Ny Munkegade 120, building 1525, Aarhus C 8000, Denmark.

References
1. Giltner CL, van Schaik EJ, Audette GF, Kao D, Hodges RS, Hassett DJ, et al. The Pseudomonas aeruginosa type IV pilin receptor binding domain functions as an adhesin for both biotic and abiotic surfaces (vol 59, pg 1083, 2006). Mol Microbiol. 2006;60(3):813–3.
2. Christie PJ, Atmakuri K, Krishnamoorthy V, Jakubowski S, Cascales E. Biogenesis, architecture, and function of bacterial type IV secretion systems. Annu Rev Microbiol. 2005;59:451–85.
3. Giltner CL, Nguyen Y, Burrows LL. Type IV pilin proteins: versatile molecular modules. Microbiol Mol Biol R. 2012;76(4):740–72.
4. Cehovin A, Simpson PJ, McDowell MA, Brown DR, Noschese R, Pallett M, et al. Specific DNA recognition mediated by a type IV pilin. Proc Natl Acad Sci U S A. 2013;110(8):3065–70.
5. Melican K, Michea Veloso P, Martin T, Bruneval P, Dumenil G. Adhesion of Neisseria meningitidis to dermal vessels leads to local vascular damage and purpura in a humanized mouse model. Plos Pathog. 2013;9(1):e1003139.
6. Melville S, Craig L. Type IV pili in gram-positive bacteria. Microbiol Mol Biol R. 2013;77(3):323–41.
7. Reguera G, McCarthy KD, Mehta T, Nicoll JS, Tuominen MT, Lovley DR. Extracellular electron transfer via microbial nanowires. Nature. 2005;435(7045):1098–101.
8. Reguera G, Pollina RB, Nicoll JS, Lovley DR. Possible nonconductive role of Geobacter sulfurreducens pilus nanowires in biofilm formation. J Bacteriol. 2007;189(5):2125–7.
9. Gorby YA, Yanina S, McLean JS, Rosso KM, Moyles D, Dohnalkova A, et al. Electrically conductive bacterial nanowires produced by Shewanella oneidensis strain MR-1 and other microorganisms. Proc Natl Acad Sci U S A. 2006;103(30):11358–63.
10. Myers CR, Nealson KH. Bacterial manganese reduction and growth with manganese oxide as the sole electron acceptor. Science. 1988;240(4857):1319–21.
11. Caccavo F, Lonergan DJ, Lovley DR, Davis M, Stolz JF, Mcinerney MJ. Geobacter sulfurreducens Sp-Nov, a hydrogen-oxidizing and acetate-oxidizing dissimilatory metal-reducing microorganism. Appl Environ Microb. 1994;60(10):3752–9.
12. Lonergan DJ, Jenter HL, Coates JD, Phillips EJP, Schmidt TM, Lovley DR. Phylogenetic analysis of dissimilatory Fe(III)-reducing bacteria. J Bacteriol. 1996;178(8):2402–8.
13. Bencheikh-Latmani R, Williams SM, Haucke L, Criddle CS, Wu LY, Zhou JZ, et al. Global transcriptional profiling of Shewanella oneidensis MR-1 during Cr(VI) and U(VI) reduction. Appl Environ Microb. 2005;71(11):7453–60.
14. Gregory KB, Lovley DR. Remediation and recovery of uranium from contaminated subsurface environments with electrodes. Environ Sci Technol. 2005;39(22):8943–7.
15. Nealson KH, Scott J. Ecophysiology of the Genus Shewanella. Proc Natl Acad Sci U S A. 2006;6:1133–51.
16. Shi L, Chen BW, Wang ZM, Elias DA, Mayer MU, Gorby YA, et al. Isolation of a high-affinity functional protein complex between OmcA and MtrC: Two outer membrane decaheme c-type cytochromes of Shewanella oneidensis MR-1. J Bacteriol. 2006;188(13):4705–14.
17. Ross DE, Ruebush SS, Brantley SL, Hartshorne RS, Clarke TA, Richardson DJ, et al. Characterization of protein-protein interactions involved in iron reduction by Shewanella oneidensis MR-1. Appl Environ Microb. 2007;73(18):5797–808.
18. Mehta T, Coppi MV, Childers SE, Lovley DR. Outer membrane c-type cytochromes required for Fe(III) and Mn(IV) oxide reduction in Geobacter sulfurreducens. Appl Environ Microb. 2005;71(12):8634–41.
19. Nevin KP, Kim BC, Glaven RH, Johnson JP, Woodard TL, Methe BA, et al. Anode biofilm transcriptomics reveals outer surface components essential for high density current production in Geobacter sulfurreducens fuel cells. Plos One. 2009;4(5):e5628.
20. Lies DP, Hernandez ME, Kappler A, Mielke RE, Gralnick JA, Newman DK. Shewanella oneidensis MR-1 uses overlapping pathways for iron reduction at a distance and by direct contact under conditions relevant for biofilms. Appl Environ Microb. 2005;71(8):4414–26.
21. Kotloski NJ, Gralnick JA. Flavin electron shuttles dominate extracellular electron transfer by Shewanella oneidensis. Mbio. 2013;4(1):e00553–00512.
22. Marsili E, Baron DB, Shikhare ID, Coursolle D, Gralnick JA, Bond DR. Shewanella Secretes flavins that mediate extracellular electron transfer. Proc Natl Acad Sci U S A. 2008;105(10):3968–73.
23. Reguera G, Nevin KP, Nicoll JS, Covalla SF, Woodard TL, Lovley DR. Biofilm and nanowire production leads to increased current in Geobacter sulfurreducens fuel cells. Appl Environ Microb. 2006;72(11):7345–8.
24. Lanthier M, Gregory KB, Lovley DR. Growth with high planktonic biomass in Shewanella oneidensis fuel cells. Fems Microbiol Lett. 2008;278(1):29–35.
25. McLean JS, Wanger G, Gorby YA, Wainstein M, McQuaid J, Ishii SI, et al. Quantification of electron transfer rates to a solid phase electron acceptor through the stages of biofilm formation from single cells to multicellular communities. Environ Sci Technol. 2010;44(7):2721–7.
26. Carmona-Martinez AA, Harnisch F, Fitzgerald LA, Biffinger JC, Ringeisen BR, Schroder U. Cyclic voltammetric analysis of the electron transfer of Shewanella oneidensis MR-1 and nanofilament and cytochrome knock-out mutants. Bioelectrochemistry. 2011;81(2):74–80.
27. Malvankar NS, Vargas M, Nevin KP, Franks AE, Leang C, Kim BC, et al. Tunable metallic-like conductivity in microbial nanowire networks. Nat Nanotechnol. 2011;6(9):573–9.
28. Strycharz-Glaven SM, Snider RM, Guiseppi-Elie A, Tender LM. On the electrical conductivity of microbial nanowires and biofilms. Energ Environ Sci. 2011;4(11):4366–79.
29. Bond DR, Strycharz-Glaven SM, Tender LM, Torres CI. On electron transport through geobacter biofilms. Chemsuschem. 2012;5(6):1099–105.
30. Malvankar NS, Tuominen MT, Lovley DR. Comment on "On electrical conductivity of microbial nanowires and biofilms" by S. M. Strycharz-Glaven, R. M. Snider, A. Guiseppi-Elie and L. M. Tender, Energy Environ. Sci., 2011, 4, 4366. Energ Environ Sci. 2012;5(3):6247–9.
31. Strycharz-Glaven SM, Tender LM. Reply to the 'Comment on "On electrical conductivity of microbial nanowires and biofilms"' by N. S. Malvankar, M. T. Tuominen and D. R. Lovley, Energy Environ. Sci., 2012, 5. Energ Environ Sci. 2012;5(3):6250–5. DOI: 10.1039/c2ee02613a.
32. Craig L, Taylor RK, Pique ME, Adair BD, Arvai AS, Singh M, et al. Type IV pilin structure and assembly: X-ray and EM analyses of Vibrio cholerae toxin-coregulated pilus and Pseudomonas aeruginosa PAK pilin. Mol Cell. 2003;11(5):1139–50.
33. Craig L, Pique ME, Tainer JA. Type IV pilus structure and bacterial pathogenicity. Nat Rev Microbiol. 2004;2(5):363–78.
34. Craig L, Volkmann N, Arvai AS, Pique ME, Yeager M, Egelman EH, et al. Type IV pilus structure by cryo-electron microscopy and crystallography: implications for pilus assembly and functions. Mol Cell. 2006;23(5):651–62.
35. Arts J, van Boxtel R, Filloux A, Tommassen J, Koster M. Export of the pseudopilin XcpT of the Pseudomonas aeruginosa type II secretion system via the signal recognition particle-Sec pathway. J Bacteriol. 2007;189(5):2069–76.
36. Nunn DN, Lory S. Product of the pseudomonas-aeruginosa gene pilD is a prepilin leader peptidase. Proc Natl Acad Sci U S A. 1991;88(8):3281–5.

37. Strom MS, Lory S. Amino-acid substitutions in pilin of pseudomonas-aeruginosa - effect on leader peptide cleavage, amino-terminal methylation, and pilus assembly. J Biol Chem. 1991;266(3):1656–64.

38. Nunn D, Bergman S, Lory S. Products of 3 accessory genes, pilB, pilC, and pilD, Are required for biogenesis of pseudomonas-aeruginosa pili. J Bacteriol. 1990;172(6):2911–9.

39. Chiang P, Habash M, Burrows LL. Disparate subcellular localization patterns of Pseudomonas aeruginosa type IV pilus ATPases involved in twitching motility. J Bacteriol. 2005;187(3):829–39.

40. Chiang P, Sampaleanu LM, Ayers M, Pahuta M, Howel PL, Burrows LL. Functional role of conserved residues in the characteristic secretion NTPase motifs of the Pseudomonas aeruginosa type IV pilus motor proteins PilB, PilT and PilU. Microbiol-Sgm. 2008;154:114–26.

41. Collins RF, Frye SA, Balasingham S, Ford RC, Tonjum T, Derrick JP. Interaction with type IV pili induces structural changes in the bacterial outer membrane secretin PilQ. J Biol Chem. 2005;280(19):18923–30.

42. Wolfgang M, van Putten JPM, Hayes SF, Dorward D, Koomey M. Components and dynamics of fiber formation define a ubiquitous biogenesis pathway for bacterial pili. Embo J. 2000;19(23):6408–18.

43. Reardon PN, Mueller KT. Structure of the type IVa major pilin from the electrically conductive bacterial nanowires of geobacter sulfurreducens. J Biol Chem. 2013;288(41):29260–6.

44. Nguyen Y, Jackson SG, Aidoo F, Junop M, Burrows LL. Structural characterization of novel pseudomonas aeruginosa type IV pilins. J Mol Biol. 2010;395(3):491–503.

45. Nunn D. Bacterial type II protein export and pilus biogenesis: more than just homologies? Trends Cell Biol. 1999;9(10):402–8.

46. Durand E, Bernadac A, Ball G, Lazdunski A, Sturgis JN, Filloux A. Type II protein secretion in Pseudomonas aeruginosa: the pseudopilus is a multifibrillar and adhesive structure. J Bacteriol. 2003;185(9):2749–58.

47. Sandkvist M. Biology of type II secretion. Mol Microbiol. 2001;40(2):271–83.

48. Campos M, Cisneros DA, Nivaskumar M, Francetic O. The type II secretion system - a dynamic fiber assembly nanomachine. Res Microbiol. 2013;164(6):545–55.

49. Sandkvist M. Type II secretion and pathogenesis. Infect Immun. 2001;69(6):3523–35.

50. Parge HE, Forest KT, Hickey MJ, Christensen DA, Getzoff ED, Tainer JA. Structure of the fiber-forming protein pilin at 2.6-angstrom resolution. Nature. 1995;378(6552):32–8.

51. Kohler R, Schafer K, Muller S, Vignon G, Diederichs K, Philippsen A, et al. Structure and assembly of the pseudopilin PulG. Mol Microbiol. 2004;54(3):647–64.

52. Yanez ME, Korotkov KV, Abendroth J, Hol WGJ. The crystal structure of a binary complex of two pseudopilins: EpsI and EpsJ from the type 2 secretion system of vibrio vulnificus. J Mol Biol. 2008;375(2):471–86.

53. Edgar RC. MUSCLE: multiple sequence alignment with high accuracy and high throughput. Nucleic Acids Res. 2004;32(5):1792–7.

54. Kuhn J, Briegel A, Morschel E, Kahnt J, Leser K, Wick S, et al. Bactofilins, a ubiquitous class of cytoskeletal proteins mediating polar localization of a cell wall synthase in Caulobacter crescentus. Embo J. 2010;29(2):327–39.

55. Koch MK, McHugh CA, Hoiczyk E. BacM, an N-terminally processed bactofilin of Myxococcus xanthus, is crucial for proper cell shape. Mol Microbiol. 2011;80(4):1031–51.

56. Bulyha I, Lindow S, Lin L, Bolte K, Wuichet K, Kahnt J, et al. Two small GTPases Act in concert with the bactofilin cytoskeleton to regulate dynamic bacterial cell polarity. Dev Cell. 2013;25(2):119–31.

57. Dupeux F, Rower M, Seroul G, Blot D, Marquez JA. A thermal stability assay can help to estimate the crystallization likelihood of biological samples. Acta Crystallogr D. 2011;67:915–9.

58. Karplus PA, Diederichs K. Linking crystallographic model and data quality. Science. 2012;336(6084):1030–3.

59. Kabsch W. Xds. Acta Crystallogr D. 2010;66:125–32.

60. Afonine PV, Grosse-Kunstleve RW, Echols N, Headd JJ, Moriarty NW, Mustyakimov M, et al. Towards automated crystallographic structure refinement with phenix.refine. Acta Crystallogr D. 2012;68:352–67.

61. Emsley P, Cowtan K. Coot: model-building tools for molecular graphics. Acta Crystallogr D. 2004;60:2126–32.

62. Weiss MS. Global indicators of X-ray data quality. J Appl Crystallogr. 2001;34:130–5.

63. Leang C, Qian XL, Mester T, Lovley DR. Alignment of the c-type cytochrome OmcS along pili of geobacter sulfurreducens. Appl Environ Microb. 2010;76(12):4080–4.

64. Karuppiah V, Collins RF, Thistlethwaite A, Gao Y, Derrick JP. Structure and assembly of an inner membrane platform for initiation of type IV pilus biogenesis. Proc Natl Acad Sci U S A. 2013;110(48):E4638–47.

65. Holm L, Rosenstrom P. Dali server: conservation mapping in 3D. Nucleic Acids Res. 2010;38:W545–9.

66. Helaine S, Dyer DH, Nassif X, Pelicic V, Forest KT. 3D structure/function analysis of PilX reveals how minor pilins can modulate the virulence properties of type IV pili. Proc Natl Acad Sci U S A. 2007;104(40):15888–93.

67. Dunlop KV, Irvin RT, Hazes B. Pros and cons of cryocrystallography: should we also collect a room-temperature data set? Acta Crystallogr D. 2005;61:80–7.

68. Korotkov KV, Gray MD, Kreger A, Turley S, Sandkvist M, Hol WGJ. Calcium is essential for the major pseudopilin in the type 2 secretion system. J Biol Chem. 2009;284(38):25466–70.

69. Holmes DE, Chaudhuri SK, Nevin KP, Mehta T, Methe BA, Liu A, et al. Microarray and genetic analysis of electron transfer to electrodes in Geobacter sulfurreducens. Environ Microbiol. 2006;8(10):1805–15.

70. Richter LV, Sandler SJ, Weis RM. Two isoforms of Geobacter sulfurreducens PilA have distinct roles in pilus biogenesis, cytochrome localization, extracellular electron transfer, and biofilm formation. J Bacteriol. 2012;194(10):2551–63.

71. Watts TH, Worobec EA, Paranchych W. Identification of pilin pools in the membranes of pseudomonas-aeruginosa. J Bacteriol. 1982;152(2):687–91.

72. Pugsley AP. Multimers of the precursor of a type IV pilin-like component of the general secretory pathway are unrelated to pili. Mol Microbiol. 1996;20(6):1235–45.

73. Petrov A, Lombardo S, Audette GF. Fibril-mediated oligomerization of pilin-derived protein nanotubes. J Nanobiotechnology. 2013;11:24.

74. Audette GF, Irvin RT, Hazes B. Crystallographic analysis of the Pseudomonas aeruginosa strain K122-4 monomeric pilin reveals a conserved receptor-binding architecture. Biochemistry-Us. 2004;43(36):11427–35.

75. Yanez ME, Korotkov KV, Abendroth J, Hol WGJ. Structure of the minor pseudopilin EpsH from the type 2 secretion system of Vibrio cholerae. J Mol Biol. 2008;377(1):91–103.

76. Lim MS, Ng D, Zong ZS, Arvai AS, Taylor RK, Tainer JA, et al. Vibrio cholerae El Tor TcpA crystal structure and mechanism for pilus-mediated microcolony formation. Mol Microbiol. 2010;77(3):755–70.

77. Hartung S, Arvai AS, Wood T, Kolappan S, Shin DS, Craig L, et al. Ultrahigh Resolution and Full-length Pilin Structures with Insights for Filament Assembly, Pathogenic Functions, and Vaccine Potential. J Biol Chem. 2011;286(51):44254–65.

78. Kolappan S, Roos J, Yuen ASW, Pierce OM, Craig L. Structural characterization of CFA/III and Longus type IVb Pili from Enterotoxigenic Escherichia Coli. J Bacteriol. 2012;194(10):2725–35.

79. Lomize MA, Pogozheva ID, Joo H, Mosberg HI, Lomize AL. OPM database and PPM web server: resources for positioning of proteins in membranes. Nucleic Acids Res. 2012;40(Database issue):D370–6.

80. Forest KT, Dunham SA, Koomey M, Tainer JA. Crystallographic structure reveals phosphorylated pilin from Neisseria: phosphoserine sites modify type IV pilus surface chemistry and fibre morphology. Mol Microbiol. 1999;31(3):743–52.

81. Franke D, Svergun DI. DAMMIF, a program for rapid ab-initio shape determination in small-angle scattering. J Appl Crystallogr. 2009;42:342–6.

82. Kozin MB, Svergun DI. Automated matching of high- and low-resolution structural models. J Appl Crystallogr. 2001;34:33–41.

83. Pettersen EF, Goddard TD, Huang CC, Couch GS, Greenblatt DM, Meng EC, et al. UCSF Chimera–a visualization system for exploratory research and analysis. J Comput Chem. 2004;25(13):1605–12.

84. Svergun D, Barberato C, Koch MHJ. CRYSOL - a program to evaluate x-ray solution scattering of biological macromolecules from atomic coordinates. J Appl Crystallogr. 1995;28:768–73.

85. Wriggers W. Using Situs for the integration of multi-resolution structures. Biophys Rev. 2010;2(1):21–7.

86. Krissinel E, Henrick K. Inference of macromolecular assemblies from crystalline state. J Mol Biol. 2007;372(3):774–97.

87. Harvey H, Habash M, Aidoo F, Burrows LL. Single-residue changes in the C-terminal disulfide-bonded loop of the pseudomonas aeruginosa type IV Pilin influence pilus assembly and twitching motility. J Bacteriol. 2009;191(21):6513–24.

88. Miles AJ, Hoffmann SV, Tao Y, Janes RW, Wallace BA. Synchrotron radiation circular dichroism (SRCD) spectroscopy: New beamlines and new applications in biology. Spectrosc-Int J. 2007;21(5–6):245–55.

89. Miles AJ, Janes RW, Brown A, Clarke DT, Sutherland JC, Tao Y, et al. Light flux density threshold at which protein denaturation is induced by

synchrotron radiation circular dichroism beamlines. J Synchrotron Radiat. 2008;15:420–2.

90. Mueller U, Darowski N, Fuchs MR, Forster R, Hellmig M, Paithankar KS, et al. Facilities for macromolecular crystallography at the Helmholtz-Zentrum Berlin. J Synchrotron Radiat. 2012;19:442–9.

91. Evans PR, Murshudov GN. How good are my data and what is the resolution? Acta Crystallogr D. 2013;69:1204–14.

92. Matthews BW. Solvent content of protein crystals. J Mol Biol. 1968;33(2):491.

93. Mccoy AJ, Grosse-Kunstleve RW, Adams PD, Winn MD, Storoni LC, Read RJ. Phaser crystallographic software. J Appl Crystallogr. 2007;40:658–74.

94. Chen VB, Arendall WB, Headd JJ, Keedy DA, Immormino RM, Kapral GJ, et al. MolProbity: all-atom structure validation for macromolecular crystallography. Acta Crystallogr D. 2010;66:12–21.

95. Baker NA, Sept D, Joseph S, Holst MJ, McCammon JA. Electrostatics of nanosystems: application to microtubules and the ribosome. Proc Natl Acad Sci U S A. 2001;98(18):10037–41.

96. Konarev PV, Volkov VV, Sokolova AV, Koch MHJ, Svergun DI. PRIMUS: a Windows PC-based system for small-angle scattering data analysis. J Appl Crystallogr. 2003;36:1277–82.

97. Volkov VV, Svergun DI. Uniqueness of ab initio shape determination in small-angle scattering. J Appl Crystallogr. 2003;36:860–4.

98. Hazes B, Sastry PA, Hayakawa K, Read RJ, Irvin RT. Crystal structure of Pseudomonas aeruginosa PAK pilin suggests a main-chain-dominated mode of receptor binding. J Mol Biol. 2000;299(4):1005–17.

99. Balakrishna AM, Saxena AM, Mok HY, Swaminathan K. Structural basis of typhoid: Salmonella typhi type IVb pilin (PilS) and cystic fibrosis transmembrane conductance regulator interaction. Proteins. 2009;77(2):253–61.

100. Helaine S, Carbonnelle E, Prouvensier L, Beretti JL, Nassif X, Pelicic V. PilX, a pilus-associated protein essential for bacterial aggregation, is a key to pilus-facilitated attachment of Neisseria meningitidis to human cells. Mol Microbiol. 2005;55(1):65–77.

101. Alphonse S, Durand E, Douzi B, Waegele B, Darbon H, Filloux A, et al. Structure of the Pseudomonas aeruginosa XcpT pseudopilin, a major component of the type II secretion system. J Struct Biol. 2010;169(1):75–80.

102. Franz LP, Douzi B, Durand E, Dyer DH, Voulhoux R, Forest KT. Structure of the minor pseudopilin XcpW from the Pseudomonas aeruginosa type II secretion system. Acta Crystallogr D. 2011;67:124–30.

103. Korotkov KV, Hol WG. Structure of the GspK-GspI-GspJ complex from the enterotoxigenic Escherichia coli type 2 secretion system. Nat Struct Mol Biol. 2008;15(5):462–8.

104. Harpaz Y, Gerstein M, Chothia C. Volume changes on protein folding. Structure. 1994;2(7):641–9.

105. Petoukhov MV, Konarev PV, Kikhney AG, Svergun DI. ATSAS 2.1 - towards automated and web-supported small-angle scattering data analysis. J Appl Crystallogr. 2007;40:S223–8.

12

Rosetta Broker for membrane protein structure prediction: concentrative nucleoside transporter 3 and corticotropin-releasing factor receptor 1 test cases

Dorota Latek🆔

Abstract

Background: Membrane proteins are difficult targets for structure prediction due to the limited structural data deposited in Protein Data Bank. Most computational methods for membrane protein structure prediction are based on the comparative modeling. There are only few de novo methods targeting that distinct protein family. In this work an example of such de novo method was used to structurally and functionally characterize two representatives of distinct membrane proteins families of solute carrier transporters and G protein-coupled receptors. The well-known Rosetta program and one of its protocols named Broker was used in two test cases. The first case was de novo structure prediction of three N-terminal transmembrane helices of the human concentrative nucleoside transporter 3 (hCNT3) homotrimer belonging to the solute carrier 28 family of transporters (SLC28). The second case concerned the large scale refinement of transmembrane helices of a homology model of the corticotropin-releasing factor receptor 1 (CRFR1) belonging to the G protein-coupled receptors family.

Results: The inward-facing model of the hCNT3 homotrimer was used to propose the functional impact of its single nucleotide polymorphisms. Additionally, the 100 ns molecular dynamics simulation of the unliganded hCNT3 model confirmed its validity and revealed mobility of the selected binding site and homotrimer interface residues. The large scale refinement of transmembrane helices of the CRFR1 homology model resulted in the significant improvement of its accuracy with respect to the crystal structure of CRFR1, especially in the binding site area. Consequently, the antagonist CP-376395 could be docked with Autodock VINA to the CRFR1 model without any steric clashes.

Conclusions: The presented work demonstrated that Rosetta Broker can be a versatile tool for solving various issues referring to protein biology. Two distinct examples of de novo membrane protein structure prediction presented here provided important insights into three major areas of protein biology. Namely, the dynamics of the inward-facing hCNT3 homotrimer system, the structural changes of the CRFR1 receptor upon the antagonist binding and finally, the role of single nucleotide polymorphisms in both, hCNT3 and CRFR1 proteins, were investigated.

Keywords: Membrane proteins, Solute carrier transporters, SLC28, Concentrative nucleoside transporter 3, Single nucleotide polymorphisms, G protein-coupled receptors, Corticotropin-releasing factor receptor 1, Rosetta Broker, GPCRM, NAMD

Correspondence: dlatek@chem.uw.edu.pl
Faculty of Chemistry, University of Warsaw, Pasteur St. 1, 02-093 Warsaw, Poland

Background

Structure prediction of small (up to 150 amino acids) globular proteins has improved so much that it has become nearly as accurate as low resolution experimental methods [1]. However, there is still a serious bottleneck in membrane protein structure prediction. The number of membrane protein structures deposited in Protein Data Bank (PDB) is much smaller than that for globular proteins. As a consequence, PDB provides relatively weak statistics for membrane proteins. There are two families of membrane proteins which still lack adequate characterization though they represent important drug targets. The first family is a well-known family of G protein-coupled receptors (GPCRs) which share a common structural motif of seven transmembrane helices. The second one is less known and more structurally diverse family of solute carrier transporters (SLCs).

For many years G protein-coupled receptors have been drug targets for many diseases including neurological, cardiovascular, endocrinological disorders. Structure prediction of GPCRs using template structures from the same GPCR subfamily, e.g., rhodopsin-like, frizzled or secretin GPCRs, proved to be accurate enough for drug design in many cases (see, e.g., results of GPCR Dock competitions [1–3]). Furthermore, there are many tools and web services for automatic structure prediction of GPCRs, e.g., GPCRM [4, 5], GPCR-Tasser [6], GPCRMod-sim [7], GOMODO [8], etc. In general, the GPCR homology modeling includes the following steps: selection of a template structure providing a proper deformation of transmembrane helices (kinks, bulges, etc.), alignment generation and finally loop refinement which greatly affects ligand binding [5]. Despite the recent progress in the GPCR modeling, reliable structure prediction of GPCRs based on distant homology (the SMO receptor case in GPCR Dock 2013 [1]) or prediction of the opposite activation state (the 5-HT$_{2B}$ receptor case in GPCR Dock 2013), is still out of reach for the majority of researchers.

Structures of 52 human families of SLCs consisting of 386 proteins are less known than GPCRs. SLCs are integral transmembrane proteins through which endogenous (i.a. ions, nucleotides, peptides) and exogenous substances (i.a. xenobiotics, drugs and their metabolites) are transported down or against the electrochemical gradient by coupling the transport with the flow of Na + or H+ [9]. SLCs are involved in drug absorption, distribution, metabolism and excretion (ADME). What is more, they can be drug targets themselves, e.g., in cancer and antibacterial pharmacotherapies [9]. SLCs influence the effectiveness of pharmacotherapy and the occurrence of drug side effects and drug-drug interactions [9, 10]. Recently, it was also proved that SLCs are important for pharmacogenomics studies [9–11]. For

example, single nucleotide polymorphisms (SNPs) observed in the SLC47 family of transporters affect the pharmacotherapy of diabetes type II in the ethnical group of Latin Americans [12]. Depending on the localization of their expression SLCs can play different roles. For example, SLCs expressed in intestinal cells, hepatocytes and cells of the brain-blood barrier play an important role in nutrition absorption and protection against xenobiotics [9, 10]. On the other hand, SLCs localized in kidneys and liver cells are involved in excretion of drugs and their metabolites [9, 10].

All SLCs are alpha-helical membrane proteins but structure and sequence similarity among members of different SLC families is limited [9]. For example, multidrug and toxin extrusion protein 1 (MATE1) from SLC47 share only 19.0%, sodium-coupled neutral amino acid transporter 1 (SLC38) 15.9% and riboflavin transporter (SLC52) 13.6% sequence identity with hCNT3 from SLC28 (data obtained with Clustal-Omega [13]. On the other hand, TM-score [14] between the corresponding crystal structures of close homologs of vcCNT (PDB id: 3TIJ) and MATE1 (PDB id: 4HUK) is only 0.288 with only 213 aligned residues (out of total 459 residues of 4HUK) with heavy atom RMSD equal to 6.85 Å. What is more, SLCs are too large to be studied only by de novo methods [15, 16]. For that reason, recent theoretical studies [16–18] on SLCs involve only homology modeling with template structures from bacterial organisms sharing at least 20% sequence identity with targets or additional data from experiments [19, 20]. Another problem in the homology modeling of SLCs is a significant structural difference between their inward-facing occluded conformation and the outward-facing conformation. The change between inward and outward-facing conformation during the transport process through the cell membrane requires not only transmembrane helices deformation like in the case of the GPCR activation but also a large change in the transporter topology (see so-called inverted topology [21]). Forrest et al. proposed the first solution to this problem called the repeat swap technique [22], first applied to the LeuT transporter from the SLC6 family [23]. Recently, her group also calibrated that technique using known crystal structures of inward- and outward-facing conformations of the GltPh transporter and successfully used it for studying the elevator-like model of the vcCNT transport mechanism [22].

Difficulties in structure prediction of GPCRs and SLCs experienced when there was no close homologous structure in PDB prompted the scientific community to search for new computational methods. A versatile example of such novel method is Broker (also known as Topology Broker) implemented as an extension of the well-known Rosetta program [24, 25]. In this manuscript, Broker together with GPCRM [4, 5] and MODELLER [26] (see Fig. 1) was used to model de novo transmembrane helices of the SLC transporter and to impose native-like deformations of transmembrane helices of the G protein-coupled receptor.

Two test cases were selected: the human CNT3 transporter in the inward-facing conformational state and the CRFR1 receptor in its inactive state. CNT3 together with CNT1 and CNT2 form the SLC28 family of concentrative nucleoside transporters. CNTs actively transport nucleosides or nucleoside-derived drugs, e.g., anticancer gemcitabine or antiviral ribavirin, by coupling their transport to the movement of Na + ions towards inside of the cell [9, 27]. The validity of the hCNT3 homotrimer model constructed in this work was assessed with the 100 ns molecular dynamics (MD) simulation with explicit membrane. The second case tested here was the corticotropin-releasing factor receptor type 1 (CRFR1). CRFR1 is a G protein-coupled receptor from the secretin-like GPCR family (the former class B of GPCR receptors) [28]. It mediates the stress response and is known as a molecular target in the treatment of depression and anxiety. The model of CRFR1 was compared to the crystal structure of CRFR1 (PDB id: 4K5Y) and used in a small molecule docking experiment. Finally, genetic variations associated with the presented protein structures were discussed. Namely, in both, hCNT3 and CRFR1 models several functionally important residues associated with single nucleotide polymorphisms (SNPs) were localized. A potential impact of SNPs on the functioning of hCNT3 and CRFR1 proteins was hypothesized.

Methods
CNT3 model building
To build the hCNT3 model the standard automodel routine of MODELLER-9v11 [26] and the vcCNT

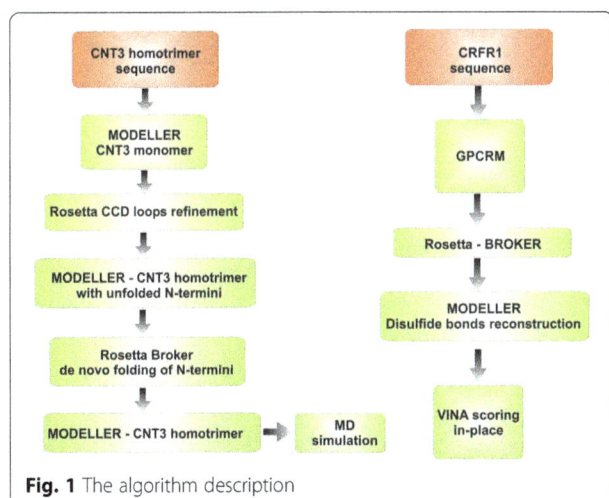

Fig. 1 The algorithm description

template structure (PDB id: 3TIJ) were used. A small molecule ligand uridine, a sodium ion and two water molecules which were present in the binding site of the crystal structure of vcCNT (see Fig. 2d) were also added. Thus, the proper orientation of side chains inside the hCNT3 binding site was preserved during the model building procedure. To build the model of the hCNT3 monomer only the fragment of the full 691-residue long sequence of hCNT3 (Uniprot id: Q9HAS3) was used. Namely, the N and C-terminus which were predicted to be outside the membrane (see Uniprot) were cut out leaving the 522-residue hCNT3 sequence (see Fig. 2e) corresponding to the residue range 91 – 612 from the Q9HAS3 entry. The lowest energy model, according to the DOPE energy function, of the hCNT3 monomer out of 100 generated was selected and used in the subsequent loop refinement. The refinement of the hCNT3 monomer loops was performed in Rosetta3 using the cyclic coordinate descent algorithm (CCD) [29]. To preserve efficiency of sampling of conformational space loop refinement simulations were divided in three separate categories. The first one was dedicated to the loop refinement of the 185 - 194 sequence region, the second one to the 128 - 136, 234 - 237, 258 - 266 and 317 - 341 sequence regions and the third one to the 486 - 493 sequence region. In each category 1000 loop models were generated. All 1000 models generated in each loop category were subjected to the clustering analysis with the Rosetta cluster application. From each category 20 cluster representatives, each of which had the lowest total Rosetta score within its cluster, were selected. All the cluster representatives were combined with each other to generate 8000 (20 × 20 × 20) possible loops combinations. Each loop combination was used to build one model of the hCNT3 homotrimer using the vcCNT template structure (PDB id: 3TIJ) and the MODELLER procedure described above. Here, the 3-fold symmetry of the hCNT3 homotrimer was kept. The DOPE potential was used to select the best model of hCNT3 out of all 8000 generated. That 1566-residue long hCNT3 model (all three subunits: 3 × 522 residues) was cut to the 1350-residue long model by removing N-termini of the subunits B and C. That 1350-residue long model of hCNT3 was subjected to de novo folding of N-terminus of the subunit A with Rosetta Broker [25]. For the Broker simulation all standard settings for Rosetta3 were used (see Additional file 1: Table S1–S2). Namely, implicit membrane energy terms described in details in [30] and the fragment library (3- and 9-residue long fragments) obtained with Robetta (http://robetta.baker lab.org/fragmentsubmit.jsp) were used. The consensus membrane topology predictor TOPCONS [31] and the hCNT3 Uniprot entry (id: Q9HAS3) were used to

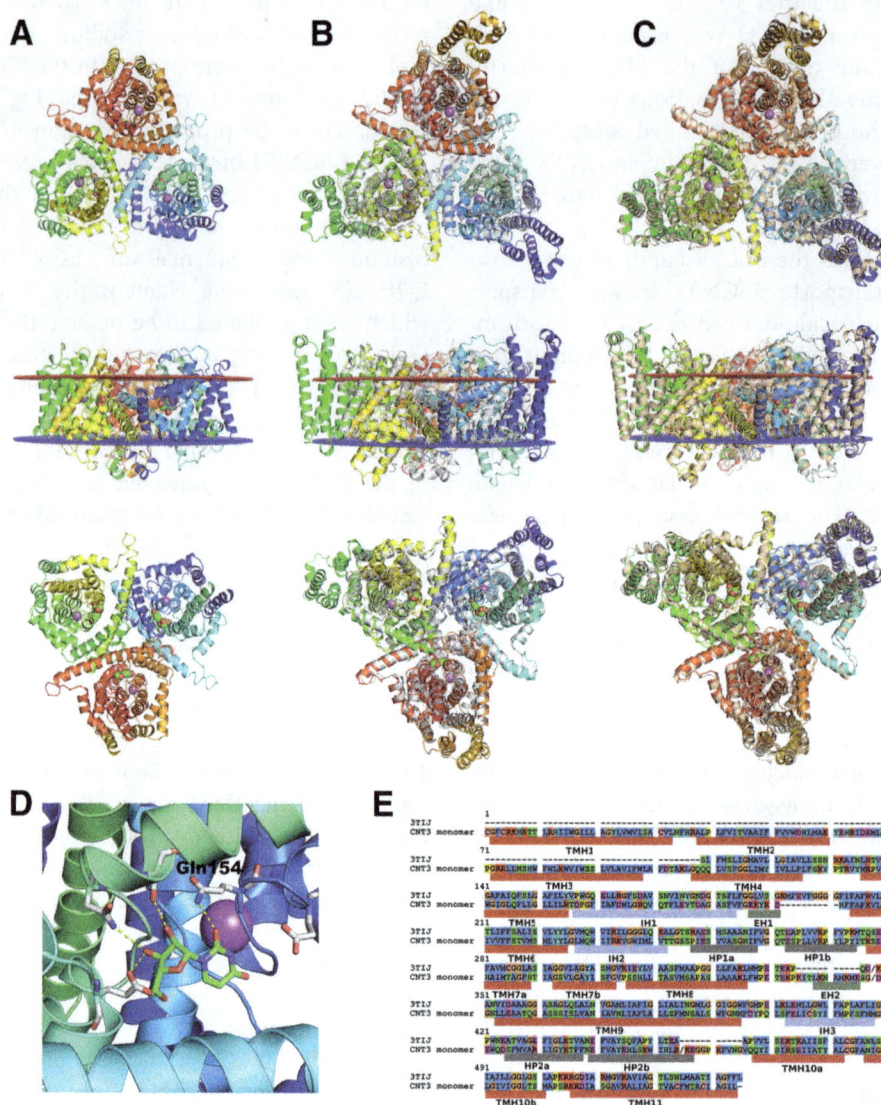

Fig. 2 a The crystal structure of the vcCNT homotrimer (PDB id: 3TIJ) shown in the extracellular, membrane and intracellular view, respectively. **b** A homology model of the hCNT3 homotrimer superposed on the crystal structure of vcCNT (grey) shown in the extracellular, membrane and intracellular view, respectively. **c** A homology model of the hCNT3 homotrimer superposed on the low-energy structure obtained from the 1956 frame out of all 5000 frames of the 100 ns MD simulation, shown in the extracellular, membrane and intracellular view, respectively. **d** The binding site of the uridine molecule (shown in green) and the sodium ion (shown as a violet sphere) located inside the crystal structure of vcCNT. The polar contacts between uridine and the transporter were depicted with yellow dashed lines. The indicated Gln154 in vcCNT corresponds to Gln251 in the model of hCNT3. **e** The sequence alignment of the template sequence (vcCNT) and the target sequence (hCNT3). Transmembrane helices (TMHs) are shown in red, extracellular and short helices (EH) in green, amphipathic helices (IH) are shown in blue and finally helices outside the lipid bilayer (HP) are shown in grey

detect positions of three N-terminal transmembrane helices (TMHs) (see Fig. 2e). Additionally, the sequence profile-based lipophilicity prediction was performed and used in the Broker simulation. During the Broker simulation only the N-terminal 108-residue long fragment in the first subunit A with the predicted three TMHs was kept flexible. The rest of the homotrimer was kept as a rigid body. Nevertheless, various approaches were tested (data not shown) before the final modeling protocol was decided. Namely, longer N-terminal fragments, 198- and 247-residue long, including the 90- and 139-residue long membrane regions of hCNT3 were folded de novo without the rest of the hCNT3 homotrimer. Also, the short, 108-residue long N-termini only in the presence of the subunit A structure was folded. Yet, it turned out that the best option for the Broker simulation was folding of the short, 108-residue long N-termini of the subunit A with the presence of other subunits B and C forming the whole 1350-residue long hCNT3 homotrimer. 10,000 models were generated and clustered using the Rosetta3 cluster application. Top ten low-energy models from the most populated cluster of the hCNT3 models according to the Rosetta total score were selected and visually inspected. One selected model was used as a template to build the final hCNT3 homotrimer model with the described above MODELLER procedure. The N-terminal region with three TMHs predicted de novo was repeated in all three subunits to ensure the 3-fold symmetry of the homotrimer. A total number of 20 hCNT3 homotrimer models were generated and the lowest energy model according to DOPE was subjected to the further analysis and the MD simulation.

Molecular dynamics simulation

The MD simulation was performed using the GPU-accelerated NAMD [32] software with the CHARMM27 [33] all-atom force field and periodic boundary conditions. Electrostatic interactions were computed using the particle-mesh Ewald method (PME) with a real space cutoff of 1.0 nm. The Lennard-Jones interactions were also cut off at 1.0 nm. The hCNT3 homotrimer model was inserted in a pre-equilibrated palmitoyloleoylphosphatidylcholine (POPC) membrane with VMD [34, 35]. The final lipid membrane was composed of 349 lipids. The system was solvated using the TIP3P water model (41,236 water molecules) and neutralized by adding 35 chloride counterions. Aspartic acid, arginine, glutamic acid, and lysine residues were used in their physiological protonation states. Neither uridine nor sodium ion molecules which were present in the vcCNT template structure were added to the system. The final system contained a total number of 195,438 atoms. The equilibration phase started with the 1 ns long melting of lipid tails while the rest of the system remained fixed. Then, after the steepest descent system minimization only protein coordinates were harmonically restrained and the 2 ns equilibration of the whole system was performed. Finally, the harmonic constraints were released and the further equilibration of the whole system lasted for 2 ns. The size of the final periodic box after the equilibration phase was 14.8 nm × 14.5 nm × 105 nm. The 100 ns production run was executed using a 2 fs time step with a snapshot of the system conformation and its energy saved every 20 ps and 10 ps, respectively. The pressure control was provided by using a modified Nosé-Hoover method in which Langevin dynamics is used to control fluctuations in the barostat. The thermostat was provided by Langevin dynamics with damping coefficient of 1/ps. The simulation was conducted at the conditions of 300 K and 1 atm. RMSD plots (see Figs. 3, 4, 5 and 6) describing the hCNT3 behavior during the MD simulation were prepared with VMD.

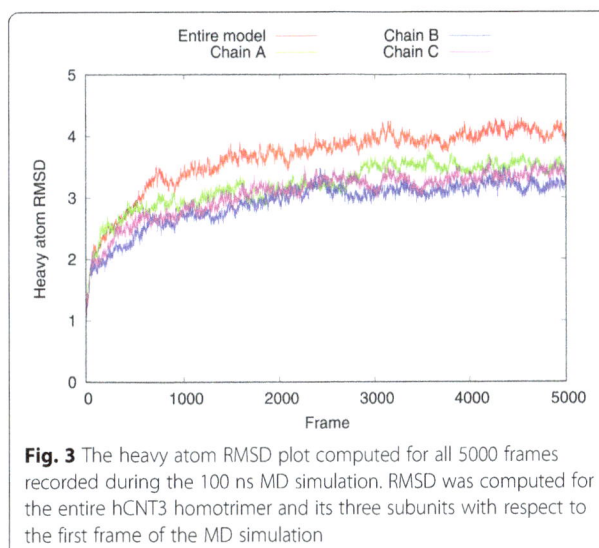

Fig. 3 The heavy atom RMSD plot computed for all 5000 frames recorded during the 100 ns MD simulation. RMSD was computed for the entire hCNT3 homotrimer and its three subunits with respect to the first frame of the MD simulation

CRFR1 model building

To build the CRFR1 model a standalone version of GPCRM described previously [5] was used. The human glucagon receptor (GCGR) structure (PDB id: 4L6R) [36] from the secretin-like branch of the GPCR family was selected as a template. To generate the CRFR1 model a PDB sequence was used (PDB id: 4K5Y, Uniprot entry: P34998, isoform 2 – CRF-R2). The isoform 2 differs from the canonical CRF-R1 sequence only in such way that a part of the sequence is missing. GPCRM generated 3000 models. Only one out of the ten best models proposed by GPCRM was selected for the next stage based on the RMSD criterion referring to the crystal CRFR1 structure (PDB id: 4K5Y). The membrane topology prediction for the Rosetta Broker input was

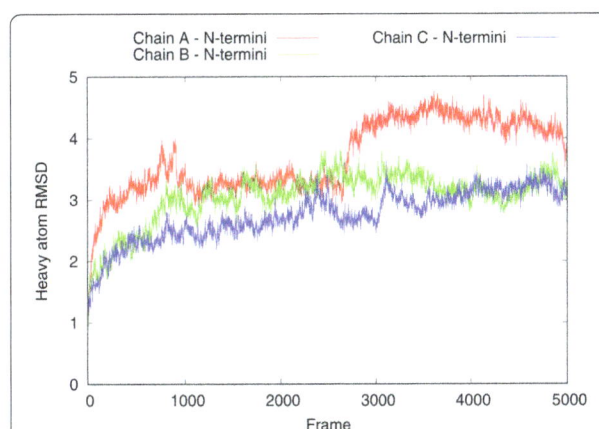

Fig. 4 The heavy atom RMSD plot computed for all 5000 frames recorded during the 100 ns MD simulation. RMSD was computed for N-terminal regions including three transmembrane helices located in all three hCNT3 subunits with respect to the first frame of the MD simulation. At the end of the simulation all N-terminal regions are of 3.5 Å RMSD

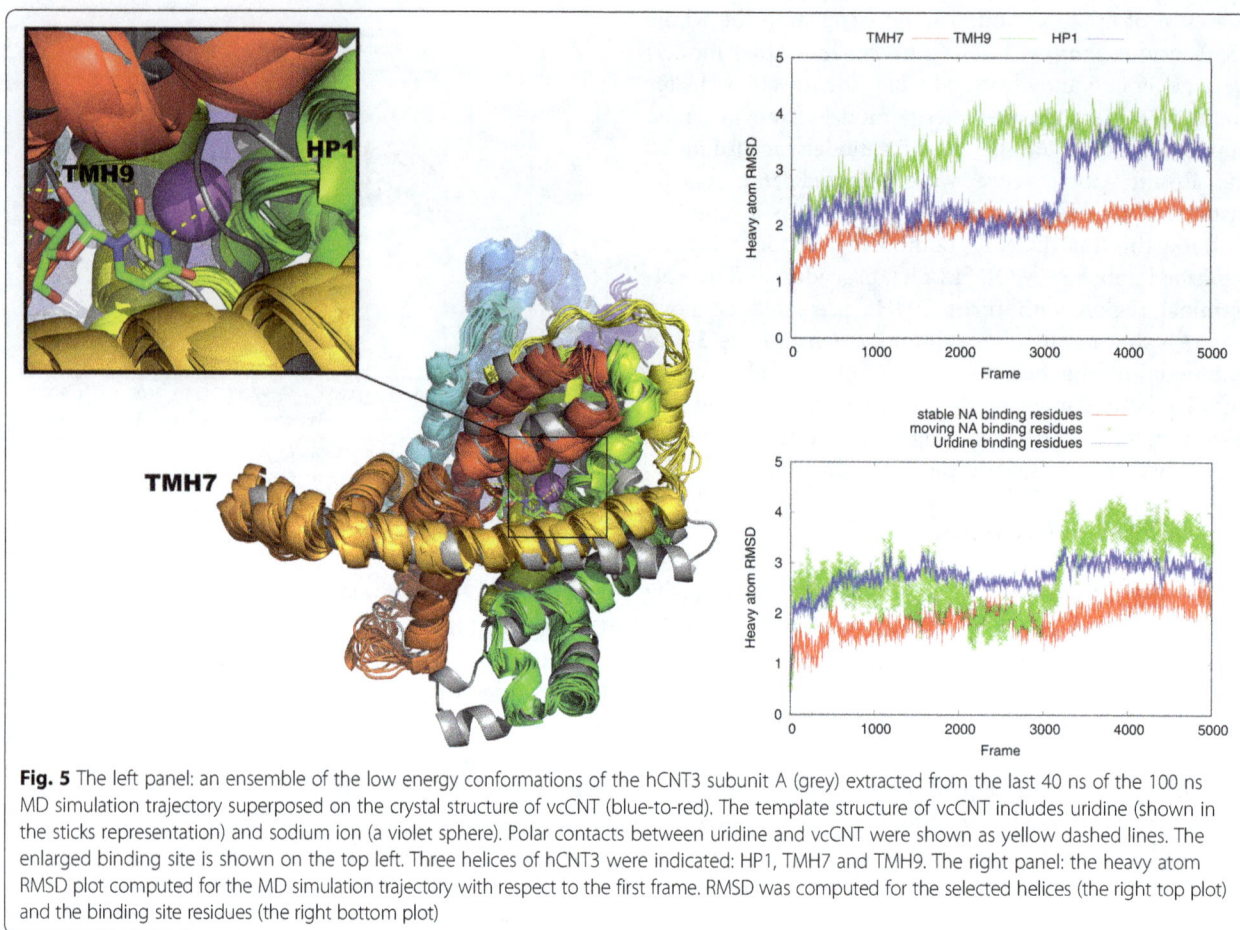

Fig. 5 The left panel: an ensemble of the low energy conformations of the hCNT3 subunit A (grey) extracted from the last 40 ns of the 100 ns MD simulation trajectory superposed on the crystal structure of vcCNT (blue-to-red). The template structure of vcCNT includes uridine (shown in the sticks representation) and sodium ion (a violet sphere). Polar contacts between uridine and vcCNT were shown as yellow dashed lines. The enlarged binding site is shown on the top left. Three helices of hCNT3 were indicated: HP1, TMH7 and TMH9. The right panel: the heavy atom RMSD plot computed for the MD simulation trajectory with respect to the first frame. RMSD was computed for the selected helices (the right top plot) and the binding site residues (the right bottom plot)

extracted directly from the CRFR1 model. The Broker simulations were divided into 3 stages. In the first stage, only the N-terminal fragment of the transmembrane helix 1 (TMH1) was reconstructed (2000 models) and the lowest RMSD model was selected. In the next step, TMH2, TMH3, TMH4, TMH5, TMH7 were rebuilt (30,000 models) and again the lowest RMSD model with respect to the crystal structure of CRFR1 was selected. In the final step of the Broker simulation TMH6 was reconstructed to fit the native structure [28] of CRFR1 (20,000 models). As it was tested before [5] the best way to impose disulfide bonds in a GPCR model is to use MODELLER. For that reason, the last modeling stage was devoted to the MODELLER reconstruction of disulfide bonds which were slightly deformed during the Broker simulation (100 models). The lowest MODELLER objective function model was selected for the antagonist docking in Autodock VINA [37].

As it was mentioned above, the main selection criterion in all the CRFR1 modeling stages was RMSD with respect to the CRFR1 crystal structure (PDB id: 4K5Y). The reason for that was the main purpose of the current work. Namely, the current work was not focused on the assessment of the Rosetta Broker force field accuracy. The

accuracy of knowledge-based force fields in the membrane protein structure prediction is an important topic [38] but outside the scope of this study. Here, only the best possible results which could be obtained with the current force field and the current sampling algorithm implemented in Broker were examined. That is why only the RMSD criterion was used and not the energy criterion for the CRFR1 models selection.

Small molecule docking

The binding mode of the CRFR1 antagonist CP-376395 is well described in [28] and the current study was not focused on the antagonist docking itself. Instead, this work was focused on the assessment of the quality of the CRFR1 homology model in the binding site area and detection of possible atom clashes. For that reason, the CP-376395 molecule was placed exactly in the same position inside the CRFR1 homology model as in the crystal CRFR1 structure. What is more, only the local refinement of the binding site was performed with Autodock VINA [37] before computing the value of the empirical docking scoring function which estimated the free energy of the ligand binding. The free energy of the antagonist binding which reflected steric clashes between atoms [37] was

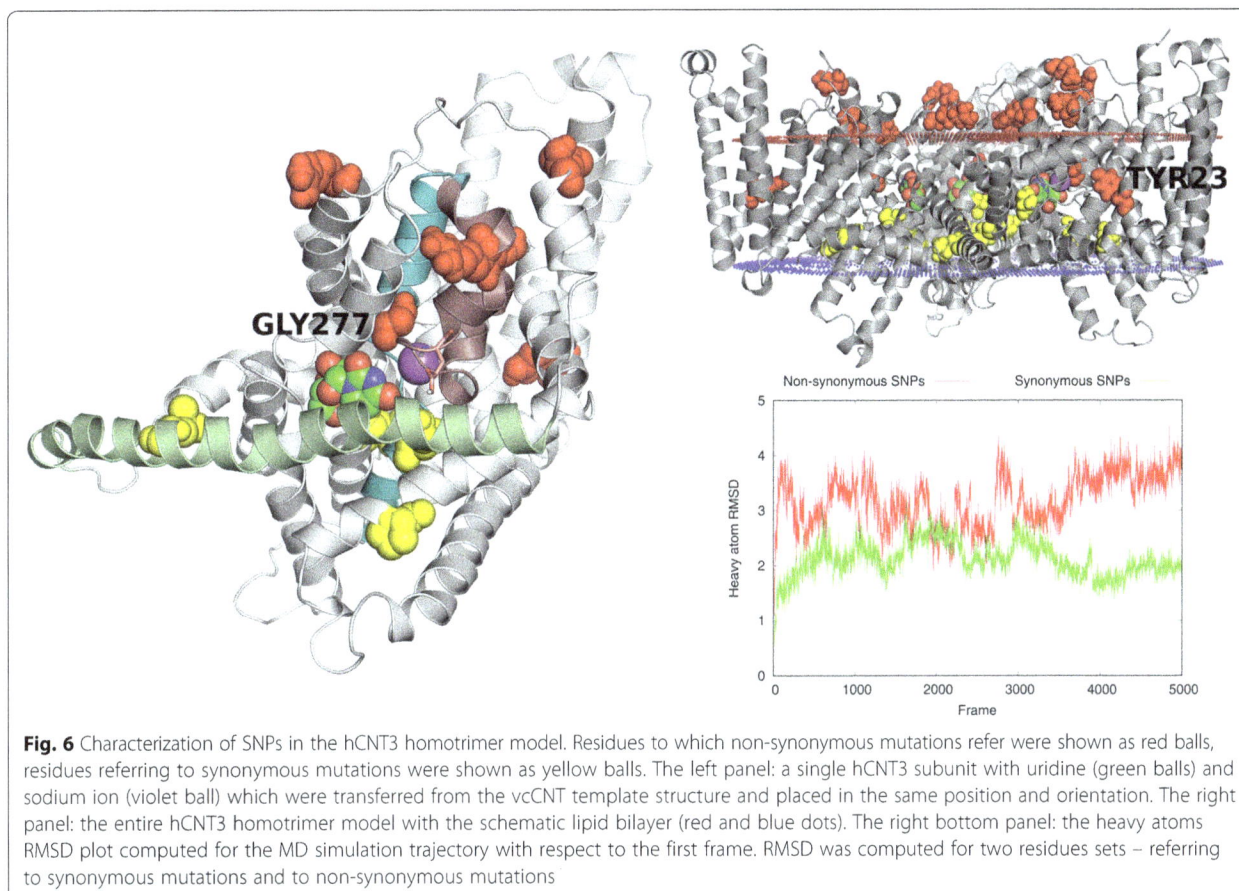

Fig. 6 Characterization of SNPs in the hCNT3 homotrimer model. Residues to which non-synonymous mutations refer were shown as red balls, residues referring to synonymous mutations were shown as yellow balls. The left panel: a single hCNT3 subunit with uridine (green balls) and sodium ion (violet ball) which were transferred from the vcCNT template structure and placed in the same position and orientation. The right panel: the entire hCNT3 homotrimer model with the schematic lipid bilayer (red and blue dots). The right bottom panel: the heavy atoms RMSD plot computed for the MD simulation trajectory with respect to the first frame. RMSD was computed for two residues sets – referring to synonymous mutations and to non-synonymous mutations

provided for three cases. The first case was the crystal structure of the CRFR1 complex with the CP-376395 antagonist (PDB id: 4K5Y). The second case was the template-based CRFR1 model built by GPCRM with CP-376395 transferred from the crystal CRFR1 structure and placed exactly in the same position and orientation. The third case was the CRFR1 model built by GPCRM but refined with the Broker algorithm with CP-376395 transferred from the crystal CRFR1 structure (PDB id: 4K5Y). In the all three cases the standard Autodock VINA settings were used together with the local_only option and the 20Åx20Åx20Å searching space size.

Single nucleotide polymorphisms

Single nucleotide polymorphisms (SNPs) for hCNT3 were downloaded from the UCSF Pharmacogenetics of Membrane Transporters (PMT) database (http://pharmacogenetics.ucsf.edu) (HGNC id: 16,484, HGNC symbol: SLC28A3) [39]. SNPs for the CRFR1 receptor were obtained from the National Institute of Health Short Genetic Variations database (dbSNP) [40] (id: 1394) and refer to the isoform 1 (CRF-R1). Nevertheless, sequence numbering for SNPs was adjusted to fit the isoform 2 sequence (CRF-R2) which was used to build the CRFR1 model and was included in the PDB entry for that receptor (PDB id: 4K5Y).

Results and discussion
CNT3 transporter

In 2012 the first crystal structure of the transporter from the SLC28 family was solved [9]. It was the structure of the bacterial vcCNT transporter isolated from Vibrio cholerae (PDB id: 3TIJ). In 2014 another crystal structure of vcCNT was released in PDB [27]. However, the current study had been started before releasing of the 2014 structure so the 2012 structure was used as a template. The vcCNT transporter is crucial for toxin excretion and plays an important role in antibiotic resistance. Most probably, CNTs change their conformations during the transport according to the elevator-like mechanism [41]. The PDB entry 3TIJ represents an inward-facing occluded conformation of the vcCNT transporter and thus represents a suitable template to build a model of the inward-facing conformation of the human CNT3 transporter (hCNT3). Human hCNT3 and bacterial vcCNT sequences share 39.46% sequence identity (according to the Clustal Omega web service [42]). Therefore, the homology modeling of hCNT3 using vcCNT as the template structure can be described as relatively easy [43, 44]. An important difference in the binding site area between hCNT3 and vcCNT is the presence of two cysteine residues Cys471 and Cys512 in hCNT3 which are so close to each other that they could form a disulfide bond. However, Slugoski et al. [45] showed that the cysteinless hCNT3 transporter still retained wild-type functional activity yet with the increased K50 dissociation constant for the sodium ion binding. What is more, CNT1 and CNT2 human homologs of CNT3 lack cysteines in that sequence region (data not shown). So, most probably that disulfide bridge is not present in the hCNT3 binding site as it was not preserved evolutionary. Therefore, the model of the hCNT3 homotrimer presented in this work was built with two non-bonded cysteines inside the binding site.

The vcCNT template structure did not correspond to the full human CNT3 sequence. There were three transmembrane helices in the N-terminal part of the hCNT3 sequence (see Fig. 2e) which were not present in vcCNT. In this work, those three TM helices were reconstructed de novo with Rosetta Broker [24, 25]. In principle, such de novo algorithm as Rosetta Broker is able to generate protein structures from families which are absent or poorly populated in PDB. That is especially important for the case of membrane protein family. The protein representation used in Broker and in other Rosetta protocols relies on internal coordinates (bond lengths, angles and torsions) which makes Rosetta a highly scalable algorithm [24]. Broker is also very efficient regarding the computational time and the conformational space sampling. That is due to a specific design of its programming architecture managing the interplay between variety of sampling strategies (so-called Movers) and the central broking mechanism [24]. The specific broking mechanism enables to simulate large protein systems consisting of many domains or monomers of various symmetry types (see the "fold-and-dock" Rosetta protocol [25, 46]). The complex architecture of Rosetta Broker enables also to efficiently combine de novo folding simulations with homology modeling, protein-protein or peptide docking and experimental data [24, 25]. Importantly, there is massive data concerning the Rosetta usage, settings and common problems encountered by users deposited in help files, FAQ or the Rosetta Commons Forum. Among very few other de novo methods for structure prediction of membrane proteins it is worth mentioning methods which are based on the identification of residue-residue contacts from multiple sequence alignments [47–49]. There are also fold-recognition methods for membrane proteins such as I-TASSER [50] or FILM3 [51] and a large number of homology-based methods often dedicated to only one protein family, e.g., G protein-coupled receptors (GPCR-I-TASSER [6], GPCRM [5], GOMoDo [8], GPCR-ModSim [7]) [52].

The accuracy of the generated hCNT3 model in its N-terminal part could be assessed, e.g., by scoring it against empirical data such as cross-linking data [53]. Unfortunately, no such data was available for the current study.

Therefore, the validity of the N-terminal part and of the whole hCNT3 homotrimer model was assessed by molecular dynamics (MD). A 100 ns MD simulation was performed and heavy atoms RMSD with respect to the first frame was computed (see Figs. 3, 4, 5 and 6). The entire hCNT3 homotrimer structure and each monomer subunit (A, B and C) separately were stable during the whole 100 ns MD simulation with RMSD below 4 Å (see Fig. 3). Also, the three transmembrane helices which were predicted de novo by Broker were maintained in all three subunits during the MD simulation (see Fig. 4) though with slight deformations resulting in RMSD of the last simulation frame below 4 Å. The above results support the validity of de novo prediction of the N-terminal fragment of hCNT3.

Additionally, the stability of the selected helices during the MD simulation was examined in details (see Fig. 5). Two helices located in the binding site area: TMH7 and HP1 and one helix TMH9 located at the homotrimer interface were selected for the detailed analysis. The binding site residues were grouped in two sets after superposing our model on the vcCNT crystal structure which contained uridine and sodium ion inside the binding site. The first set, including Ile281, Thr280, Val249 and Asn246, Gly277, Gln251, Gly250, surrounded the sodium ion. The second set of residues consisted of Glu253, Thr252, Gln251, Glu429, Asn475, Ser478, Phe430, Phe473, Gly250, Phe278 and on TMH9: Leu356, Ile360, Asn359. That second set of residues surrounded the uridine ligand. Three residues: Thr280, Ile281 and Gly277 from the sodium ion binding residues set could be distinguished as quite stable during the MD simulation (RMSD values fluctuating around 2 Å). Residues from the uridine binding cluster were more flexible and were subjected to larger conformational changes (higher RMSD values around 3 Å were observed) (see Fig. 5, the right bottom panel). TMH9 located at the homotrimer interface was also flexible with RMSD reaching 4 Å (see the Fig. 5, the right top panel), as it was observed earlier [54]. Unfolding of HP1 in the absence of the sodium ion [54] was also observed (see Fig. 5, the left top panel), especially in the region between Val242 and Ser254. That corresponded to a significant change of RMSD from 2 to 3 Å around the frame no. 3200 (64 ns) for that region of HP1 (see Fig. 5, the right top panel). TMH7 was stable during the whole MD simulation with RMSD values fluctuating around 2 Å (see Fig. 5, the right top panel) as it was observed earlier [54].

In Fig. 6 SNPs for hCNT3 were displayed graphically. Most residues associated with non-synonymous amino acid changes (shown in red) were located in the extracellular region of the hCNT3 sequence which was quite flexible (RMSD values fluctuating around 4 Å – see the right bottom panel). Residues associated with synonymous changes (shown in yellow) were located in the middle part of the protein between lipid bilayers, yet a bit closer to the cytoplasmic side. Positions of those residues were stable during the simulation (RMSD values fluctuating around 2 Å – see Fig. 6, the right bottom panel)). In Fig. 6 one example SNP was shown in details (Gly277). Gly277 could be replaced with Arg. In such a case, a hydrogen bond could be formed with adjacent Gln251. That most probably could hamper binding of the ligand because Gln251 forms a hydrogen bond with uridine (see Fig. 2d). Biological studies confirmed the important phenotypic effect of that SNP. Namely, it was observed that mutation G↔A causing the mentioned above amino acid replacement Gly↔Arg resulted in the reduced uptake of inosine and thymidine in oocytes [39]. Another interesting example of non-synonymous SNP referred to Tyr23 located in the N-terminal part of TMH1 for which de novo structure prediction with Broker was performed. That SNP refers to the mutation A↔G which leads to the amino acid change Tyr↔Cys on the protein level. On the phenotype level that SNP has been related to ribavirin induced anemia, so most probably it causes the minor transporter activity, as it was suggested by Allegra et al. [55]. Referring back to the hCNT3 model presented in the current work, there was a close cysteine residue (Cys31) located in the bottom of TMH1 which could form a disulfide bridge with Cys23 replacing Tyr23. That could cause unfolding of that bottom part of TMH1. Another hypothesis explaining the structural effect of that SNP involved other residues: Phe248 (HP1), Phe163 (IH1) and Phe147 (TMH5)

which were quite close to Tyr23. Interaction Cys-Phe is one of the strongest interactions in membrane proteins [56] so it is plausible that the amino acid replacement Tyr↔Cys could enable the new interaction Cys-Phe. What is more, Phe248 is located near the sodium ion, close to the middle part of HP1. As it was mentioned above in the description of the MD simulation, a noticeable movement of the middle region of HP1 away from the binding site towards TMH1 was observed in several low-energy frames recorded at the end of the simulation (see Fig. 5, the left top panel). It is plausible, that the amino acid change Tyr↔Cys enabling the mentioned above interaction Cys23-Phe248 could stabilize such a movement of HP1. That movement of HP1 away from the binding site might worsen the ligand - HP1 interactions and thus decrease the transporter activity. Nevertheless, more detailed studies regarding the impact of SNPs on the functioning of the hCNT3 transporter are certainly needed to confirm the above findings.

CRFR1 receptor

Currently, the closest template for the CRFR1 receptor is the glucagon receptor GCGR (PDB id: 4L6R [36]). Although sequence identity is quite high (34%), the binding site of CRFR1 is located much deeper inside the receptor, than in the case of the GCGR binding site (see Fig. 7). Consequently, the binding site of CRFR1 with TMH6 moved away from the center of receptor is much more spacious than the GCGR binding site. For that reason, the typical homology modeling procedure would fail if the

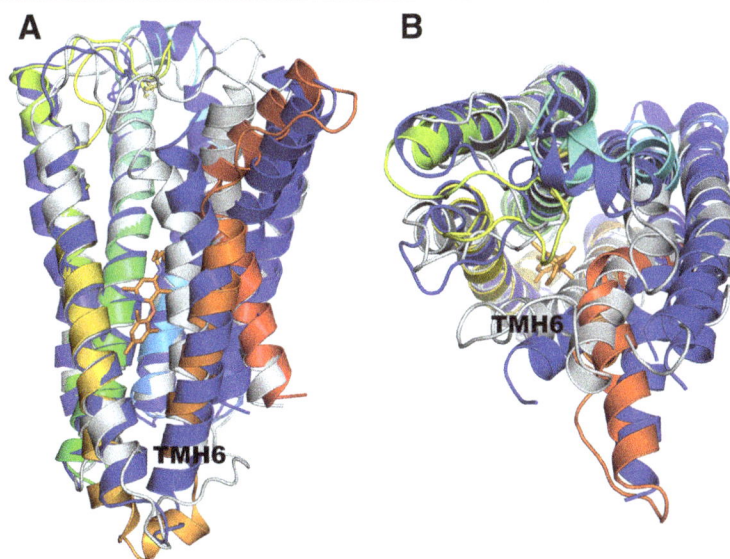

Fig. 7 Comparison of three structures of CRFR1 receptor: a crystal structure (PDB id: 4K5Y) shown in dark blue with the antagonist CP-376395 shown in orange, a homology model of CRFR1 generated with MODELLER using the glucagon GPCR receptor template structure (PDB id: 4L6R) shown in grey, a Broker-refined model of CRFR1 shown in a blue-to-red color scheme. Here, the side view (**a**) and the top, extracellular view (**b**) of the receptor was shown

CRFR1 model was built using GCGR as a template. Indeed, the CRFR1 model generated with the typical homology modeling procedure (GPCRM) and the GCGR template structure provided the inaccurate CRFR1 model. The heavy atom RMSD with respect to the CRFR1 crystal structure (PDB id: 4K5Y) of the 8-residue fragment of TMH6 located inside the binding site was equal to 7.99 Å (see Table 1). The Broker large scale refinement of transmembrane helices in the CRFR1 model provided a more accurate conformation of TMH6 with RMSD equal to 3.41 Å (see Table 1). That large scale reconstruction of transmembrane helices in the CRFR1 homology model removed several atom clashes which were detected while docking the antagonist CP-376395 to that model (see Fig. 8 and Table 1). Those steric clashes were the main reason for the high repulsive energy of the ligand binding estimated with Autodock VINA (see Table 1). It is worth noting, that the Broker refinement of TMHs also made forming of the crucial polar contact between CP-376395 and Asn283 [28] possible. Here, the numbering of residues fits the PDB entry of CRFR1 (PDB id: 4K5Y). Additionally to the Broker large scale refinement of TMHs also several amino acid side chains surrounding the ligand (Leu287, Phe284, Leu280, Tyr327, Phe203, Asn202, Leu320, Glu209, Met206, Gln355, Leu323) were refined during the ligand docking. That approach further improved the quality of the CRFR1 binding site. The residues which were kept flexible during the ligand docking were depicted in the wire (before docking) and sticks (after docking) representation in Fig. 8.

In Fig. 9 several residues in TMH6 were marked with the ball representation. They refer to SNPs reported in the literature for CRF-R1 (isoform 1 of the receptor) but were adjusted to the sequence of isoform 2 which was included in the PDB entry for CRFR1 (PDB id: 4K5Y). Here, other SNPs located in other TMHs were not shown for the sake of clarity. Residues associated with non-synonymous amino acid mutations (red) were located in the extra and intracellular part but also in the middle of TMH6. Residues associated with synonymous mutations were located only in the extra and intracellular part of TMH6. Two residues: Thr316 and Val318 associated with

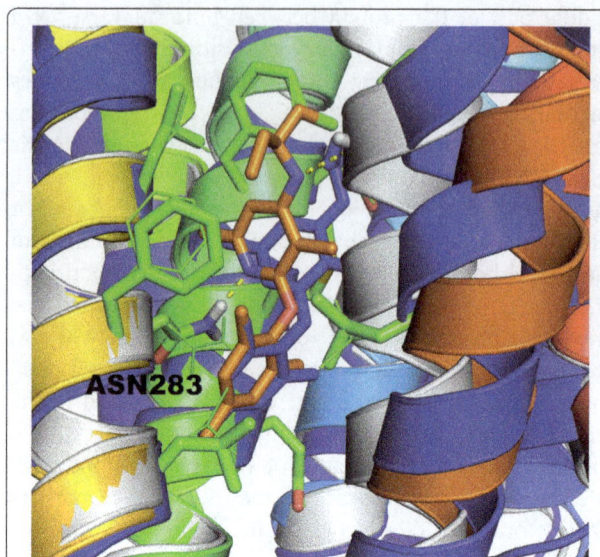

Fig. 8 Comparison of the CP-376395 antagonist binding site of the CRFR1 crystal structure (dark blue), the homology model generated with MODELLER (grey), the homology model refined with Broker (blue-to-red color scheme with the ligand shown in orange). A few selected residues of the homology model refined with Broker were shown as green sticks. Those residues were selected to be flexible during the Autodock VINA docking. The starting side chain conformations of those residues (before docking) were shown with the wire representation while the resulting conformation (after docking) was shown with sticks. The polar contacts between the antagonist and the receptor were depicted with yellow dashed lines

a mutation causing a frame shift (blue) were located in the middle of TMH6. The side chain of Thr316 was facing towards the CRFR1 binding site and was close to the ligand molecule. It was reported in the literature that it could be mutated to non-polar Ile316. Thus, the polar contact with the antagonist would be lost. Another residue, Val318 was facing the membrane, so most probably, if it was mutated to polar Cys a slight deformation of TMH6 in that region could be observed as a structural effect of that SNP. Polar Lys314 associated with non-synonymous amino acid mutation was located in the middle of TMH6 and was facing the membrane. It could be mutated to Arg or Asn. Most probably, such mutations could have an impact only on the TMH6 deformation induced by the change in the amino acid charge. To sum up, it could be hypothesized that mutations of Val318 and Lys314 could alter the bending of TMH6 and thus could change the size of the space accessible for the ligand binding. Thus, the strength of the ligand-receptor interactions could be changed. That could be relevant, for example, for the individual response to pharmacotherapy. Indeed, the described above SNPs are believed to be associated with the varied individual response to the treatment of asthma with inhaled corticosteroids (see the 225,965 entry in the ClinVar NCBI's database).

Table 1 Autodock VINA affinity scores for the antagonist CP-376395 – CRFR1 complex

CRFR1 receptor structure	Heavy-atom RMSD of 8-residue fragment of 6TMH [Å]	Autodock VINA affinity score [kcal/mol]
Crystal	0.00	−9.46
Template-based	7.99	18.16
Broker-refined	3.41	−6.79

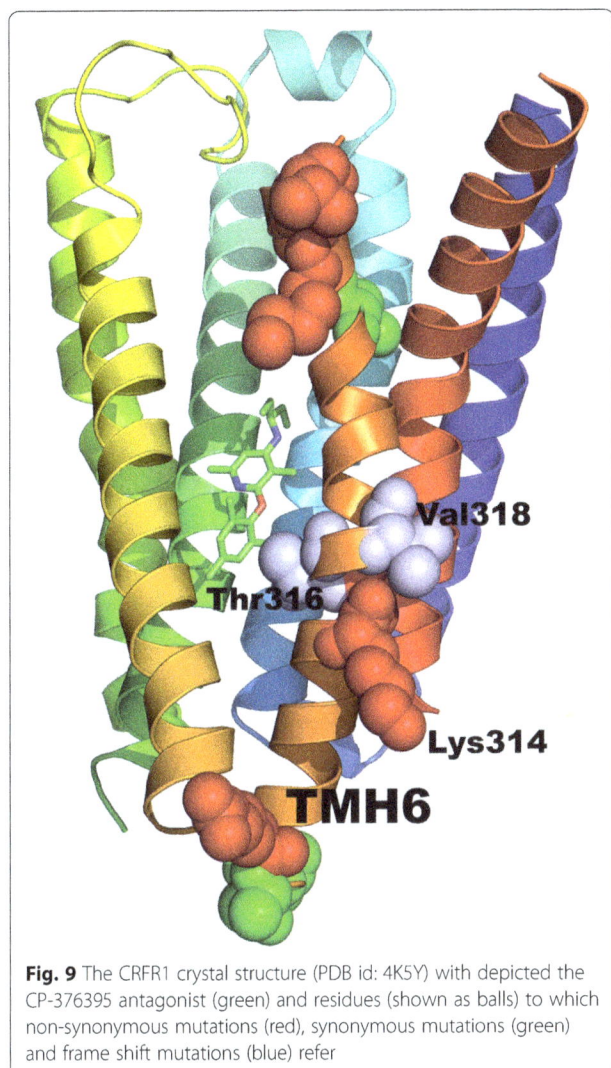

Fig. 9 The CRFR1 crystal structure (PDB id: 4K5Y) with depicted the CP-376395 antagonist (green) and residues (shown as balls) to which non-synonymous mutations (red), synonymous mutations (green) and frame shift mutations (blue) refer

Building homology models for many of the receptors from the GPCR family is impeded by the lack of close homologs which could serve as templates. Here, one of the ways to work around that problem was shown. Namely, the large scale refinement of transmembrane helices in the CRFR1 homology model was shown to improve its overall accuracy and also its usefulness in ligand docking.

Theoretical models of hCNT3 and CRFR1 membrane proteins generated in this study were used for the analysis of SNPs. The role of SNPs in changing the protein structure or protein-ligand interactions was discussed and hypothetical structural changes caused by amino acid mutations were proposed. Nevertheless, biological studies should be performed to confirm those findings and to derive conclusions important for pharmacogenomics.

Abbreviations
CNT: Concentrative nucleoside transporter; CRFR1: Corticotropin-releasing factor receptor 1; GCGR: Glucagon receptor; GPCR: G protein-coupled receptor; MD: Molecular dynamics; SLCs: Solute carriers; SNP: Single nucleotide polymorphism; TMH: Transmembrane helix

Acknowledgements
I thank Szymon Niewieczerzal for his help with the MD simulation setup. I acknowledge Prof. Andrej Sali and Adrian Stecula from the University of California, San Francisco (UCSF) for their help, valuable discussions referring to the work on the hCNT3 transporter and access to the computational resources during the EMBO fellowship.

Funding
The study was financed by National Science Centre in Poland, the SONATA grant no. DEC-2012/07/D/NZ1/04244. I also would like to acknowledge the EMBO short term fellowship ref. ASTF 329 - 2013.

Authors' contributions
DL contributed to the manuscript idea and the first and final manuscript draft. DL performed computations and data analysis. DL read and approved the final manuscript.

Competing interests
The author declares that she has no competing interests.

Conclusions

The current study supports that Rosetta Broker framework can be a versatile tool for de novo building and large scale reconstruction of transmembrane helices. It was able not only to reconstruct TMHs in such a way that kink angles were changed but also move away the whole TM helix away from the CRFR1 receptor center. Although no validation so far has been provided for the current de novo prediction of N-terminal TMHs in hCNT3, the MD simulation showed that such prediction was plausible.

It is certain that another detailed study regarding the hCNT3 transporter is needed to broaden the knowledge about its mechanism of action, the role of its single nucleotide polymorphisms and small molecules interactions. The current study is only one of the first steps in understanding the role of solute carriers transporters from the SLC28 family. The hCNT3 model built during this study could be used in future to manage sparse experimental data such as cross-links [53].

References
1. Kufareva I, Katritch V, Stevens RC, Abagyan R. Advances in GPCR modeling evaluated by the GPCR dock 2013 assessment: meeting new challenges. Structure. 2014;22(8):1120–39.
2. Michino M, Abola E, Brooks CL 3rd, Dixon JS, Moult J, Stevens RC. Community-wide assessment of GPCR structure modelling and ligand docking: GPCR dock 2008. Nat Rev Drug Discov. 2009;8(6):455–63.
3. Kufareva I, Rueda M, Katritch V, Stevens RC, Abagyan R. Status of GPCR modeling and docking as reflected by community-wide GPCR dock 2010 assessment. Structure. 2011;19(8):1108–26.
4. Latek D, Bajda M, Filipek S. A hybrid approach to structure and function modeling of G protein-coupled receptors. J Chem Inf Model. 2016;56(4):630–41.
5. Latek D, Pasznik P, Carlomagno T, Filipek S. Towards improved quality of GPCR models by usage of multiple templates and profile-profile comparison. PLoS One. 2013;8(2):e56742.
6. Zhang J, Yang J, Jang R, Zhang Y. GPCR-I-TASSER: a hybrid approach to G protein-coupled receptor structure modeling and the application to the human genome. Structure. 2015;23(8):1538–49.
7. Esguerra M, Siretskiy A, Bello X, Sallander J, Gutierrez-de-Teran H. GPCR-ModSim: a comprehensive web based solution for modeling G-protein coupled receptors. Nucleic Acids Res. 2016;44(W1):W455–62.

8. Sandal M, Duy TP, Cona M, Zung H, Carloni P, Musiani F, Giorgetti A. GOMoDo: a GPCRs online modeling and docking webserver. PLoS One. 2013;8(9):e74092.

9. Johnson ZL, Cheong CG, Lee SY. Crystal structure of a concentrative nucleoside transporter from Vibrio cholerae at 2.4 a. Nature. 2012;483(7390):489–93.

10. Geier EG, Schlessinger A, Fan H, Gable JE, Irwin JJ, Sali A, Giacomini KM. Structure-based ligand discovery for the large-neutral amino acid transporter 1, LAT-1. Proc Natl Acad Sci U S A. 2013;110(14):5480–5.

11. Schlessinger A, Geier E, Fan H, Irwin JJ, Shoichet BK, Giacomini KM, Sali A. Structure-based discovery of prescription drugs that interact with the norepinephrine transporter, NET. Proc Natl Acad Sci U S A. 2011;108(38):15810–5.

12. Chen L, Pawlikowski B, Schlessinger A, More SS, Stryke D, Johns SJ, Portman MA, Chen E, Ferrin TE, Sali A, et al. Role of organic cation transporter 3 (SLC22A3) and its missense variants in the pharmacologic action of metformin. Pharmacogenet Genomics. 2010;20(11):687–99.

13. Sievers F, Wilm A, Dineen D, Gibson TJ, Karplus K, Li W, Lopez R, McWilliam H, Remmert M, Soding J, et al. Fast, scalable generation of high-quality protein multiple sequence alignments using Clustal omega. Mol Syst Biol. 2011;7:539.

14. Zhang Y, Skolnick J. TM-align: a protein structure alignment algorithm based on the TM-score. Nucleic Acids Res. 2005;33(7):2302–9.

15. Latek D, Kolinski A. CABS-NMR–de novo tool for rapid global fold determination from chemical shifts, residual dipolar couplings and sparse methyl-methyl NOEs. J Comput Chem. 2011;32(3):536–44.

16. Matsson P, Bergstrom CA. Computational modeling to predict the functions and impact of drug transporters. In Silico Pharmacol. 2015;3(1):8.

17. Colas C, Grewer C, Otte NJ, Gameiro A, Albers T, Singh K, Shere H, Bonomi M, Holst J, Schlessinger A. Ligand discovery for the Alanine-serine-Cysteine transporter (ASCT2, SLC1A5) from homology modeling and virtual screening. PLoS Comput Biol. 2015;11(10):e1004477.

18. Ung PM, Song W, Cheng L, Zhao X, Hu H, Chen L, Schlessinger A. Inhibitor discovery for the human GLUT1 from homology modeling and virtual screening. ACS Chem Biol. 2016;11(7):1908–16.

19. Colas C, Smith DE, Schlessinger A. Computing substrate selectivity in a peptide transporter. Cell Chem Biol. 2016;23(2):211–3.

20. Schlessinger A, Yee SW, Sali A, Giacomini KM. SLC classification: an update. Clin Pharmacol Ther. 2013;94(1):19–23.

21. Duran AM, Meiler J. Inverted topologies in membrane proteins: a mini-review. Comput Struct Biotechnol J. 2013;8:e201308004.

22. Vergara-Jaque A, Fenollar-Ferrer C, Kaufmann D, Forrest LR. Repeat-swap homology modeling of secondary active transporters: updated protocol and prediction of elevator-type mechanisms. Front Pharmacol. 2015;6:183.

23. Forrest LR, Zhang YW, Jacobs MT, Gesmonde J, Xie L, Honig BH, Rudnick G. Mechanism for alternating access in neurotransmitter transporters. Proc Natl Acad Sci U S A. 2008;105(30):10338–43.

24. Porter JR, Weitzner BD, Lange OF. A framework to simplify combined sampling strategies in Rosetta. PLoS One. 2015;10(9):e0138220.

25. DiMaio F, Leaver-Fay A, Bradley P, Baker D, Andre I. Modeling symmetric macromolecular structures in Rosetta3. PLoS One. 2011;6(6):e20450.

26. Sali A, Blundell TL. Comparative protein modelling by satisfaction of spatial restraints. J Mol Biol. 1993;234(3):779–815.

27. Johnson ZL, Lee JH, Lee K, Lee M, Kwon DY, Hong J, Lee SY. Structural basis of nucleoside and nucleoside drug selectivity by concentrative nucleoside transporters. elife. 2014;3:e03604.

28. Hollenstein K, Kean J, Bortolato A, Cheng RK, Dore AS, Jazayeri A, Cooke RM, Weir M, Marshall FH. Structure of class B GPCR corticotropin-releasing factor receptor 1. Nature. 2013;499(7459):438–43.

29. Canutescu AA, Dunbrack RL Jr. Cyclic coordinate descent: a robotics algorithm for protein loop closure. Protein Sci. 2003;12(5):963–72.

30. Yarov-Yarovoy V, Schonbrun J, Baker D. Multipass membrane protein structure prediction using Rosetta. Proteins. 2006;62(4):1010–25.

31. Tsirigos KD, Peters C, Shu N, Kall L, Elofsson A. The TOPCONS web server for consensus prediction of membrane protein topology and signal peptides. Nucleic Acids Res. 2015;43(W1):W401–7.

32. Phillips JC, Braun R, Wang W, Gumbart J, Tajkhorshid E, Villa E, Chipot C, Skeel RD, Kale L, Schulten K. Scalable molecular dynamics with NAMD. J Comput Chem. 2005;26(16):1781–802.

33. MacKerell AD Jr, Banavali N, Foloppe N. Development and current status of the CHARMM force field for nucleic acids. Biopolymers. 2000;56(4):257–65.

34. Humphrey W, Dalke A, Schulten K. VMD: visual molecular dynamics. J Mol Graph. 1996;14(1):33–8. 27-38.

35. Kandemir-Cavas C, Cavas L, Alyuruk H. The Topology Prediction of Membrane Proteins: A Web-Based Tutorial. Interdiscip Sci. 2016. doi:10.1007/s12539-016-0190-7.

36. Siu FY, He M, de Graaf C, Han GW, Yang D, Zhang Z, Zhou C, Xu Q, Wacker D, Joseph JS, et al. Structure of the human glucagon class B G-protein-coupled receptor. Nature. 2013;499(7459):444–9.

37. Trott O, Olson AJ. AutoDock Vina: improving the speed and accuracy of docking with a new scoring function, efficient optimization, and multithreading. J Comput Chem. 2010;31(2):455–61.

38. Koehler Leman J, Ulmschneider MB, Gray JJ. Computational modeling of membrane proteins. Proteins. 2015;83(1):1–24.

39. Badagnani I, Chan W, Castro RA, Brett CM, Huang CC, Stryke D, Kawamoto M, Johns SJ, Ferrin TE, Carlson EJ, et al. Functional analysis of genetic variants in the human concentrative nucleoside transporter 3 (CNT3; SLC28A3). Pharmacogenomics J. 2005;5(3):157–65.

40. Sherry ST, Ward MH, Kholodov M, Baker J, Phan L, Smigielski EM, Sirotkin K. dbSNP: the NCBI database of genetic variation. Nucleic Acids Res. 2001;29(1):308–11.

41. Colas C, Ung PM, Schlessinger A. SLC transporters: structure, function, and drug discovery. Medchemcomm. 2016;7(6):1069–81.

42. Li W, Cowley A, Uludag M, Gur T, McWilliam H, Squizzato S, Park YM, Buso N, Lopez R. The EMBL-EBI bioinformatics web and programmatic tools framework. Nucleic Acids Res. 2015;43(W1):W580–4.

43. Krieger E, Nabuurs SB, Vriend G. Homology modeling. Methods Biochem Anal. 2003;44:509–23.

44. Krieger E, Joo K, Lee J, Raman S, Thompson J, Tyka M, Baker D, Karplus K. Improving physical realism, stereochemistry, and side-chain accuracy in homology modeling: four approaches that performed well in CASP8. Proteins. 2009;77(Suppl 9):114–22.

45. Slugoski MD, Smith KM, Mulinta R, Ng AM, Yao SY, Morrison EL, Lee QO, Zhang J, Karpinski E, Cass CE, et al. A conformationally mobile cysteine residue (Cys-561) modulates Na+ and H+ activation of human CNT3. J Biol Chem. 2008;283(36):24922–34.

46. Das R, Andre I, Shen Y, Wu Y, Lemak A, Bansal S, Arrowsmith CH, Szyperski T, Baker D. Simultaneous prediction of protein folding and docking at high resolution. Proc Natl Acad Sci U S A. 2009;106(45):18978–83.

47. Wang Y, Barth P. Evolutionary-guided de novo structure prediction of self-associated transmembrane helical proteins with near-atomic accuracy. Nat Commun. 2015;6:7196.

48. Nugent T. De novo membrane protein structure prediction. Methods Mol Biol. 2015;1215:331–50.

49. Hopf TA, Colwell LJ, Sheridan R, Rost B, Sander C, Marks DS. Three-dimensional structures of membrane proteins from genomic sequencing. Cell. 2012; 149(7):1607–21.

50. Roy A, Kucukural A, Zhang Y. I-TASSER: a unified platform for automated protein structure and function prediction. Nat Protoc. 2010;5(4):725–38.

51. Nugent T, Jones DT. Accurate de novo structure prediction of large transmembrane protein domains using fragment-assembly and correlated mutation analysis. Proc Natl Acad Sci U S A. 2012;109(24):E1540–7.

52. Busato M, Giorgetti A. Structural modeling of G-protein coupled receptors: an overview on automatic web-servers. Int J Biochem Cell Biol. 2016;77(Pt B):264–74.

53. Webb B, Lasker K, Velazquez-Muriel J, Schneidman-Duhovny D, Pellarin R, Bonomi M, Greenberg C, Raveh B, Tjioe E, Russel D, et al. Modeling of proteins and their assemblies with the integrative modeling platform. Methods Mol Biol. 2014;1091:277–95.

54. Feng Z, Hou T, Li Y. Transport of nucleosides in the vcCNT facilitated by sodium gradients from molecular dynamics simulations. Mol BioSyst. 2013;9(8):2142–9.

55. Allegra S, Cusato J, De Nicolo A, Boglione L, Gatto A, Cariti G, Di Perri G, D'Avolio A. Role of pharmacogenetic in ribavirin outcome prediction and pharmacokinetics in an Italian cohort of HCV-1 and 4 patients. Biomed Pharmacother. 2015;69:47–55.

56. Gomez-Tamayo JC, Cordomi A, Olivella M, Mayol E, Fourmy D, Pardo L. Analysis of the interactions of sulfur-containing amino acids in membrane proteins. Protein Sci. 2016;25(8):1517–24.

13

Investigation of allosteric coupling in human β₂-adrenergic receptor in the presence of intracellular loop 3

Canan Ozgur[1], Pemra Doruker[2] and E. Demet Akten[3*]

Abstract

Background: This study investigates the allosteric coupling that exists between the intra- and extracellular parts of human β₂-adrenergic receptor (β₂-AR), in the presence of the intracellular loop 3 (ICL3), which is missing in all crystallographic experiments and most of the simulation studies reported so far. Our recent 1 μs long MD run has revealed a transition to the so-called *very inactive* state of the receptor, in which ICL3 packed under the G protein's binding cavity and completely blocked its accessibility to G protein. Simultaneously, an outward tilt of transmembrane helix 5 (TM5) caused an expansion of the extracellular ligand-binding site. In the current study, we performed independent runs with a total duration of 4 μs to further investigate the *very inactive* state with packed ICL3 and the allosteric coupling event (three unrestrained runs and five runs with bond restraints at the ligand-binding site).

Results: In all three independent unrestrained runs (each 500 ns long), ICL3 preserved its initially packed/closed conformation within the studied time frame, suggesting an inhibition of the receptor's activity. Specific bond restraints were later imposed between some key residues at the ligand-binding site, which have been experimentally determined to interact with the ligand. Restraining the binding site region to an open state facilitated ICL3 closure, whereas a relatively constrained/closed binding site hindered ICL3 packing. However, the reverse operation, i.e. opening of the packed ICL3, could not be realized by restraining the binding site region to a closed state. Thus, any attempt failed to free the ICL3 from its locked state due to the presence of persistent hydrogen bonds.

Conclusions: Overall, our simulations indicated that starting with *very inactive* states, the receptor stayed almost irreversibly inhibited, which in turn decreased the overall mobility of the receptor. Bond restraints which represented the geometric restrictions caused by ligands of various sizes when bound at the ligand-binding site, induced the expected conformational changes in TM5, TM6 and consequently, ICL3. Still, once ICL3 was packed, the allosteric coupling became ineffective due to strong hydrogen bonds connecting ICL3 to the core of the receptor.

Keywords: β₂-adrenergic receptor, Intracellular loop 3 (ICL3), G protein-coupled receptor, Allosteric coupling, Transmembrane helix

Background

Human β₂-adrenergic receptor (β₂-AR) is a member of the G-protein coupled receptor (GPCR) superfamily that is responsible in the eukaryotic signal transduction, responding to hormones adrenaline and noradrenaline to mainly induce the smooth muscle relaxation in the lung tissue. As all members of GPCRs, β₂-AR shares the 7TM structural motif, which consists of seven transmembrane-spanning alpha helices connected by loop regions at the intra- and extracellular sides of the membrane.

As the first hormone-activated GPCR structure to be reported by X-ray crystallography [1, 2], the high resolution structural information was obtained through the elimination of the third intracellular loop (ICL3) replaced with the protein T4 lysozyme (T4L) and also the C-terminal tail in order to increase both the proteolytic stability and crystallizability. ICL3 links the cytoplasmic ends of transmembrane helices V and VI (TM5 and TM6) and has a functional role for both the recognition

* Correspondence: demet.akten@khas.edu.tr
[3]Department of Bioinformatics and Genetics, Faculty of Natural Sciences and Engineering, Kadir Has University, Cibali, 34083 Istanbul, Turkey
Full list of author information is available at the end of the article

and the activation of G proteins [3, 4]. Unlike other intracellular loop regions, ICL3 has a highly variable length and sequence among the members of the GPCR superfamily, even the closely related subtypes. It is believed that a GPCR's selectivity to different G proteins originated in the structural uniqueness of ICL3.

One experimental study conducted by West et al. [5] investigated the ligand-dependent perturbation of the conformational ensemble for β_2-AR, which incorporated the ICL3 region through hydrogen/deuterium exchange (HDX) coupled with mass spectroscopy. HDX data suggested ICL3's role as a molecular switch, where antagonist and inverse agonist binding shifted the equilibrium toward inactive states, which is characterized by protection to exchange (i.e., stabilization) in the ICL3 loop and flanking regions of TM helices V and VI. In contrast, binding of a full agonist shifted equilibrium toward higher accessibility and/or destabilization of ICL3 region. The lack of mobility or the stabilization of ICL3 was also observed in our previous work by Ozcan et al.[6], where the presence of ICL3 had a significant impact on the overall dynamics of the receptor, especially the arrangements of some key residues at both the ligand-binding site and the G-protein binding site. In that study, in addition to the *loop* model where the missing ICL3 region was generated in fully atomistic detail, a second model called *clipped* model was created with the two open ends of TM5 and TM6 covalently bonded to each other. During the 1 μs long MD run, the *loop* model found a very inactive conformation towards 600 ns, when TM6 moved towards the receptor's core region with ICL3 packing underneath the membrane and blocking the G-protein's binding site. ICL3 preserved its closed conformation and consequently, the receptor's overall dynamics has decreased significantly with only minor fluctuations for the remaining 400 ns, and became similar to the *clipped* model's dynamics, which showed no major variations.

Early experimental peptide studies showed that the peptides synthesized with the primary sequence 259–273 corresponding to ICL3 region of G_s-coupled β_2-AR, selectively bind with G proteins, stimulate their functional activity, trigger signaling cascade in the absence of hormonal stimulus, and decrease the regulatory effects of β_2-AR agonists [7–9]. Another peptide study revealed that the ICL3-derived peptides can form helical structures and contain clusters of positively charged residues exposed to one helix side which is crucial for interacting with the negatively charged receptor binding site of the $G\alpha_s$ subunit (Shinagawa et al. [10], Okuda et al, [11]). The last result is in good agreement with the simulation study by Ozcan et al. [6] during which the unstructured loop region generated through modeling techniques adopted a few turns of helices before blocking the G-protein binding cavity.

Yet, despite its functional significance, many experimental and simulation studies conducted so far have neglected its presence [12–15]. The present work focuses on the effect of the intracellular loop region ICL3 on the intrinsic dynamics of the receptor. The intrinsic dynamics is the key determinant of the receptor's function, whereas the tertiary structure encodes the dynamic behavior. Thus, it is important to have a complete 3D structure of the receptor in order to understand the system's function to a greater degree.

The study presented here adopted the same atomistic model of the receptor that incorporates the ICL3 region generated in our previous study [6]. First, we start by investigating the stability of the alternative inactive state of the receptor, which was observed during the last half of the 1 μs long MD simulation. The most distinguishing features of the alternative state was the closure of ICL3 that completely blocked the G protein binding site at the intracellular region, and the simultaneous enlargement of the ligand-binding site at the extracellular part. Both regions of the receptor changed their conformation almost simultaneously which suggested a strong allosteric coupling. In two independent continuation MD runs (500 ns long each), the receptor preserved its stable ICL3 conformation as well as the extracellular part of the receptor.

In addition to the presence of ICL3, the allosteric effect was further investigated through specific bond restraints between key residues at the ligand-binding site, which led to an alternative closed state. Also, the probability of open-to-closed or closed-to-open transition in ICL3 conformation was revealed for the first time. Addition of restraints mimics the presence/interaction of the ligand and possibly accelarates the closure event. So, the conformational shift that would take place in the presence of the ligand would require much longer simulations, whereas the restraints have provided an enhanced event sampling by providing an exaggerated driving force or perturbation at the binding site.

Even though the allosteric coupling that exists between the intracellular and the extracellular regions of the receptor is a well-known feature of β_2-AR and many other GPCRs, the presence of ICL3, which directly influences the overall dynamics of the receptor, was not taken into consideration when the coupling behavior was investigated. Therefore, the present study will be the first in providing this correlated motion that exists between the extracellular ligand-binding site and the intracellular G-protein binding site that incorporates the ICL3 region. Through imposing distance restraints between some key residues at the ligand-binding site, we were able to trigger a series of conformational changes in the transmembrane helices that led to the close packing of ICL3.

Methods
Protein-membrane system preparation
The system under study was adopted from Ozcan's work [6] where the initial conformation was the x-ray crystallographic structure of human β_2AR in complex with T4 lysozyme and the partial inverse agonist carazolol (PDB id: 2RH1) at 2.4 Å resolution [2]. The anchor protein T4L was covalently attached to two ends of helices 5 and 6 in order to facilitate crystallization. After removal of T4L, the missing intracellular loop region ICL3 was added between residues 230 and 262 after being modeled as an unstructured loop of 32 residues long via homology modeling tool, MODWEB [16]. Also, the bound ligand at the ligand-binding site was removed and the apo form of the receptor was used as an initial conformation.

The receptor was then embedded in the double-layered 1-palmitoyl-2-oleoyl-phosphatidylcholine (POPC) phospholipid cell membrane generated with VMD's Plug-in tool [17]. The receptor was positioned with an oblique angle of 8° between its main principal component along the membrane and the z-axis [18]. After solvating the protein-lipid system with VMD's solvate module, Na$^+$ and Cl$^-$ ions were added to neutralize the total charge of the system, which is necessary for Particle-Mesh Ewald summation method used in electrostatic energy calculations [19]. The resulting periodic box dimensions were (86x86x100) in Angstrom. The total number of atoms in the system was 68.001, of which 5.055 belong to protein, 20.770 to lipids, 42.135 to water molecules, and 41 to ions.

The system's equilibration consisted of several stages such as melting of lipid tails, minimization and equilibration with protein restrained, equilibration with protein released and lastly the production runs [6]. Nanoscale Molecular Dynamics (NAMD) v2.8 software tool was used for all our MD runs [20]. The force fields used were CHARMM27 [21, 22] for lipids, CHARMM22 [23, 24] for proteins and TIP3P model for water in the system [25]. In this work, the isothermal-isobaric (NPT) ensemble was employed using Langevin dynamics in order to keep the temperature constant with a Langevin damping coefficient (gamma) of 5/ps for all non-hydrogen atoms. The pressure was kept constant at 1 atm using a Nose-Hoover Langevin piston with 100 fs period and 50 ps damping timescale [26]. Long-range electrostatic interactions were treated by particle mesh Ewald (PME) method, with a grid point density of over 1 Å. A cutoff of 12 Å was used for van der Waals and short-range electrostatics interactions with a switching function. Time step was set to 2 fs by using SHAKE algorithm for bonds involving hydrogens [27] and the data was recorded at every 200 ps. The number of time steps between each full electrostatics evaluation was set to 2. Short-range non-bonded interactions were calculated at every time step.

Independent MD runs with and without restraints
All MD runs are listed in Table 1. The first one is actually from our previous study and will be used as reference here. Two independent 500 ns runs named as "MD1µs_cont1" and "MD1µs_cont2" were continuations of the first run. They were based on the final snapshot of the original run "MD1µs", which corresponds to a *very inactive* state of the receptor with a closed ICL3 that packed underneath the receptor [6]. No restraints were applied to the binding site in these continuation runs. The goal here was to observe how long ICL3 would preserve its closed state, which also indicate its conformational stability.

In order to study the effect of restraints on the overall dynamics of β_2AR, especially of ICL3, additional bond energy terms were applied to seven pairs of key residues located at the ligand-binding cavity that are known to be critical in binding signaling molecules (See Table 2). All extra bonded terms were harmonic potentials of the form $U(x) = k(x - x_{ref})^2$ where k is the spring constant and x_{ref} is the restraint distance value. A total of five independent MD runs named as *rstr1 to rstr5*, under different restraints and with different starting conformations were performed as listed in Table 1. One additional 500 ns long MD run was performed on the last snapshot of *rsrt4* (Run #8) with all distance restraints removed in order to investigate the stability of the packed conformation of ICL3 in a restraint-free environment.

The restraint distances that we selected for restraint MD runs are provided in Table 2. Seven of these distances are between Asp113 on TM3 and residues on TM5 (Ser203, Ser204, Ser207), TM6 (Phe289, Asn293) and TM7 (Asn312). One other is between residues on TM6 and TM7. Mainly, the size of the ligand-binding site is determined by the position of TM5, TM6 and TM7 with respect to the more stationary TM3 as depicted in Fig. 1. The distance ranges previously observed for the active

Table 1 Details of several MD runs with and without restraints

Run#	Run Name	Total time (ns)	Restrained (Yes/No)	Initial structure
1	MD1µs[a]	1000	No	inactive crystal structure, 2RH1
2	MD1µs_cont1	500	No	last frame of *Run #1*
3	MD1µs_cont2	500	No	last frame of *Run #1*
4	rstr1	500	Yes	initial frame of *Run #1*
5	rstr2	500	Yes	initial frame of *Run #1*
6	rstr3	500	Yes	frame @ 470th ns of *Run #1*
7	rstr4	500	Yes	frame @ 470th ns of *Run #1*
8	MD500ns	500	No	last frame of *Run #7 (rstr4)*
9	rstr5	500	Yes	last frame of *Run #1*

[a]This run was performed prior to this work in Ozcan's study [6]. The remaining runs #2 through #10 were based on the same system created for run #1

Table 2 Restraint distances in all seven MD runs and their corresponding values in experimentally reported active and inactive states

Residue pair	Exper.[b] (Å)	Distances in crystallographic structures (Å)		Bond Restraints (Å)				
		Inactive (PDB id: 2RH1)	Active (PDB id: 3SN6)	rstr1	rstr2	rstr3	rstr4	rstr5
Ser203O$^{\gamma a}$-Asp113C$^{\gamma a}$	8.0–10.0	11.2	10.3	17	8	17	17	8
Ser204O$^{\gamma}$-Asp113C$^{\gamma}$	8.0–10.0	14.2	12.4	14	10	14	14	10
Ser207O$^{\gamma}$-Asp113C$^{\gamma}$	8.0–10.0	11.5	10.4	11.7	8	11.7	11.7	8
Ser207C$^{\alpha}$-Asp113C$^{\alpha}$	N/A	12.2	12.0	-	-	-	17	-
Asn293C$^{\beta1}$-Asp113C$^{\beta}$	8.0–10.0	13.6	14.0	14	15	14	14	8
Phe289C$^{\beta}$-Asp113C$^{\beta}$	8.0–8.4	11.7	12.3	13	12	13	13	8
Asn312 C$^{\beta}$-Asp113C$^{\beta}$	8.0–8.4	9.1	8.6	10	9	10	10	8
Phe289C$^{\beta}$-Asn312C$^{\beta}$	8.0–8.4	5.5	5.5	5.5	5.5	5.5	5.5	8

[a]γ Oxygen, and β and γ Carbon atoms of the side chains were taken into consideration
[b]These are the distance ranges observed previously in various experimental studies [28–33]

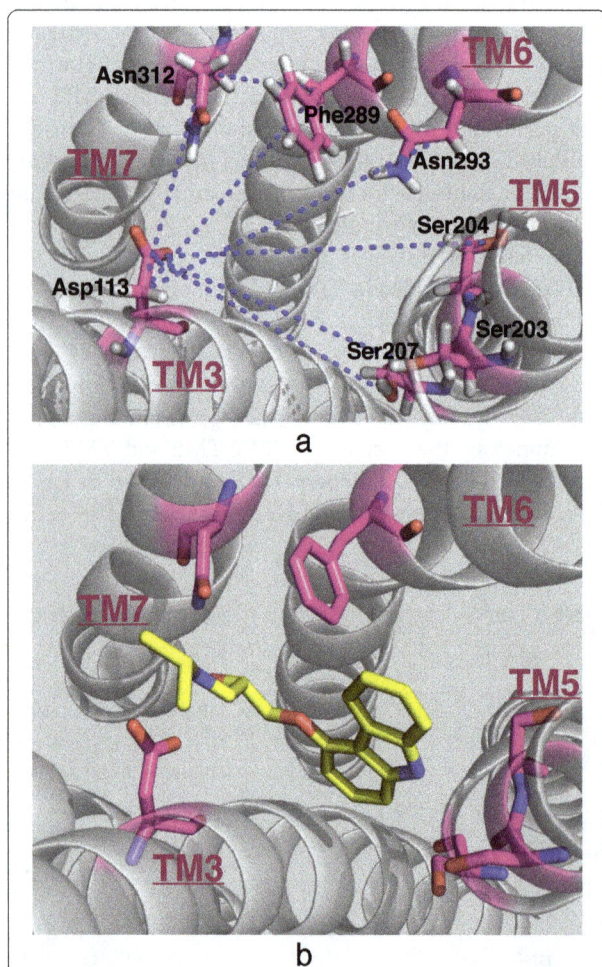

Fig. 1 Extracellular view of the ligand-binding site. **a** Only key residues and the seven restrained distances are highlighted (**b**) From the same angle as in (**a**), the ligand carazolol as bound in the crystal structure of the inactive state (PDB: 2RH1)

state of the receptor in several different experimental studies [28–33] are also listed in the second column.

Table 2 also lists the same distances in both active and inactive x-ray structures (PDB ids: 3SN6 and 2RH1). For the inactive state, all distances are higher than the experimentally observed ranges, as expected. Surprisingly, for the active state, almost all distance values are slightly out of range. Yet, most of them are smaller than those in the inactive state, especially those between three serine residues on TM5 and Asp113 on TM3. Therefore, we will focus on the first three distances (rows in Table 2), which clearly distinguish the active states from the inactive ones.

In the first restrained MD run (*rstr1*), Ser203-Asp113 distance was set to 16 Å for the first 300 ns and then increased to 17 Å for the remaining 200 ns. Other distances were set to values closer to those observed in the inactive crystal structure. Here, the value of 17 Å was explicitly selected as it was recorded in the last frame of the original MD run (*MD1μs*), when ICL3 closed upon G-protein binding site.

In the second restrained MD run (*rstr2*), the same three critical distances between serine residues and Asp113 were restrained to 8 Å, 10 Å and 8 Å, which all fall within the experimental range of the active state. Similarly, the other distances were set closer to those in the crystal structures. The third run (*rstr3*) almost share the same set of restraints as in *rstr1* and *rstr2*, except that it starts with a different initial conformation of the receptor (See Table 1). In the fourth run (*rstr4*), the same initial conformation was used as in *rstr3*, with an additional restraint distance set between two backbone alpha-carbon atoms in Ser207 and Asp113. Finally, in the last run (*rstr5*), the final frame of the original MD run (*MD1μs*) was used as a starting conformation and the ligand-binding site was severely constricted.

Results and discussion
Analysis of two continuation runs indicates stability of packed ICL3

Two 500 ns long MD runs were performed as an extension of the original *MD1μs* simulation (Ozcan et al., [6]), as previously described in Methods section. Initially in *MD1μs* run, ICL3 was in an extended conformation, and highly mobile as illustrated in the upper portion of Fig. 2c (red ribbon). At around 600 ns of *MD1μs*, it started to pack under β₂AR and kept its stationary state until the end of the simulation. The aim here was to investigate how long this stationary, and relatively restricted state would carry on. Both extended simulations, the so-called *MD1μs_ctd1* and *MD1μs_ctd2*, selected the final snapshot of the original run as their initial conformation. *MD1μs_ctd1* started with the same velocities as in *MD1μs*'s final state, whereas *MD1μs_ctd2* was carried out with a different velocity distribution, in order to enhance the sampling.

During the first continuation run, ICL3 stayed in its close form with only minor fluctuations (~2 Å) as shown in the

RMSD profiles illustrated in Fig. 2a. All the RMSD calculations were carried out after the alignment of each MD snapshot in the trajectory to the initial snapshot based on the transmembrane region, as it is the least mobile part of the receptor. The RMSD profiles labeled as *All*, *Core*, and *Tmemb* represent the RMSD of the whole receptor, the core which is the receptor without ICL3 and the transmembrane region which consists of seven alpha helices located inside the membrane, respectively (See Additional file 1: Figure S1). This complete blockage of the G protein's binding site suggests an inhibition of the receptor's activity.

In the second continuation run using a different velocity distribution, a temporary increase up to 5 Å in the RMSD value was observed halfway through the trajectory, which led to a slight opening of ICL3 as reflected by the white colored snapshot in Fig. 2b. However, this opening was only temporary and lasted for about 20 ns. This was followed by a sharp decrease in RMSD to 4 Å caused by the closure of ICL3 to a position slightly

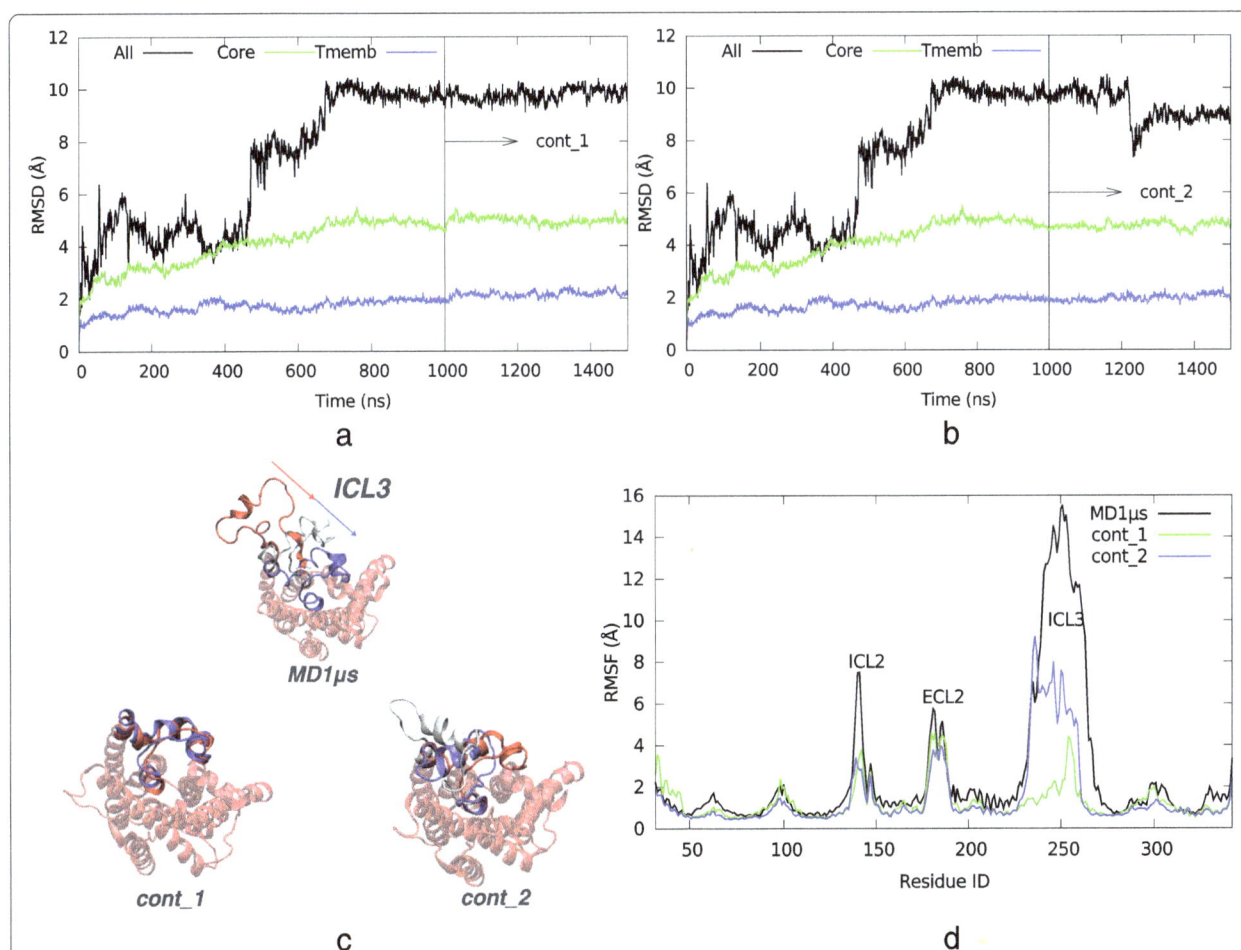

Fig. 2 Results of original and continuation runs without restraints. RMSD profiles of (**a**) the first *MD1μs_cont1* and (**b**) the second *MD1μs_cont2* continuation runs (500 ns each) shown together with the *original* 1 μs MD run. **c** Intracellular view of the initial (red), intermediate (white) and final (blue) snapshots superimposed for each run. **d** RMSF profiles for the *original, MD1μs_cont1* and *MD1μs_ctd2* runs

different than the initial state. ICL3 stayed there for the rest of the simulation.

In both continuation runs, the closed state of ICL3 was preserved, representing an extreme inactive state, where the G-protein binding site was completely blocked. Experimental studies revealed both active and inactive states of the receptor, but none of these structures incorporated the ICL3 region. Here, it is the presence of ICL3 that caused the receptor to adopt such a novel inactive state, which was found to be noticeably stable.

The fluctuation of each residue averaged over the whole trajectory was determined for each simulation and illustrated in Fig. 2d. As expected, in both continuation runs, the mobility of ICL3 stayed at a much lower level than in the original MD simulation. Moreover, a slight decrease of mobility was observed in every part of the receptor, especially on two important loop regions, ICL2 (intracellular) and ECL2 (extracellular), in conjunction with the decrease in ICL3's mobility.

The stability of ICL3 was further investigated by a detailed analysis of the hydrogen bond network in the loop conformation. Figure 3 illustrates the profile of the residues involved in hydrogen bonding along the trajectory, which was focused on ICL3 and its neighborhood region. By the time ICL3 closure is completed at around 600–700 ns, a total of 8 stable hydrogen bonds has been observed between ICL3 and the rest of the receptor (core region), which was maintained throughout the simulation. It is noteworthy that this stable network of hydrogen bonds was located mostly at the two junctions of ICL3-TM5 and ICL3-TM6. In the first continuation run, nearly all 8 hydrogen bonds were preserved, whereas in the second continuation run, half of them was lost when a slight opening was observed, but still, an alternative close state of ICL3 was adopted later towards the end of the simulation with most of the hydrogen bonds recovered.

In our simulation studies, the closure of ICL3 was strongly coupled with the lower part of TM6, which exhibited an inward movement of 7.5 Å, in the opposite direction of the outward movement of 14 Å observed during activation (experimentally measured at the Cα carbon of Glu 268 [1]). The RMSD profiles of the intracellular part of TM6 with respect to the active state (PDB id:3SN6) illustrated at the top section of Additional file 2: Figure S2 show that as ICL3 started to change its conformation to a closed state, TM6's intracellular part shifted to the opposite direction of activation and stayed there for both continuation runs. On the other hand, the intracellular part of TM5 attached to ICL3 at the other end, seemed unaffected by these conformational variations. As illustrated at the lower section of Additional file 1: Figure S1, TM5 stabilized at around 2 Å during the original MD as well as both continuation runs.

Two of the key residues at the binding site are Asp113 on TM3 and Ser207 on TM5, which are known to interact both with agonists and antagonists, via hydrogen bonds or close contacts. They are situated at the two distant corners of the binding site and when the ligand is favorably bound, Ser207 is near the ligand's aromatic moiety, while Asp113 usually makes multiple hydrogen bonds with the ligand's polar end group (See Fig. 1b). Therefore, the distance between these two residues directly controls the binding capability of the ligand. Experimental measurements already determined an approximate distance range of [8 Å -10 Å] between the two side chain atoms, O^{γ} of Ser207 and C^{γ} of Asp113, when the receptor was found in its active state [31, 32].

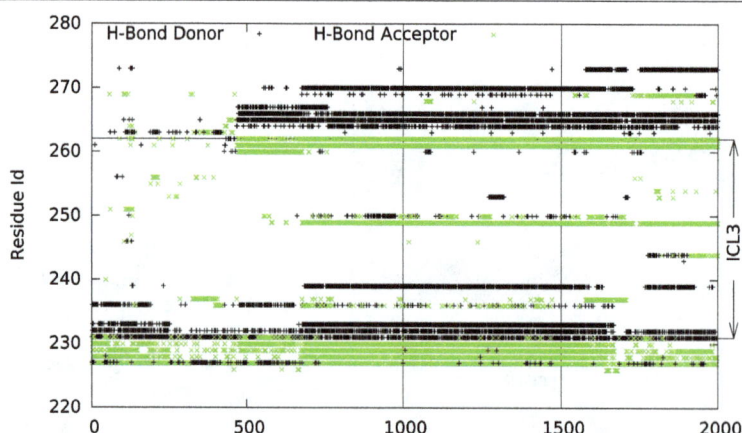

Fig. 3 Hydrogen bond profiles. The first and the second continuation runs (500 ns each), MD1ms_cont1 and MD1ms_cont2, covering a time range of [1000:1500] and [1500-2000] ns, respectively. The first 1000 ns corresponds to the original run. Donor and acceptor groups are illustrated by red and green dots, respectively

As the receptor passes from an active state to an inactive one, the same distance also increases and stabilizes roughly at around 11 Å -12 Å. Thus, as an indicator of activation/inactivation, the same distance was monitored for both continuation runs.

In our *original* MD run, a close correspondence between this value and the conformational state of the lower part of TM6 was established as shown in Fig. 4; as ICL3 exhibited its major conformational shift from an open to a closely packed state, the lower part of TM6 shifted towards the core of the receptor (See Fig. 4d) and at the same time, the Ser207-O^γ and Asp113-C^γ distance started to increase up to 17 Å - 18 Å, which is majorly caused by the outward shift of TM5 (See Fig. 4c). In both continuation runs, the same distance fluctuated within a range of 13 Å - 18 Å, which is still above 11 Å -12 Å of the crystal structure of the inactive state [1, 2] (See Fig. 4a and b). Moreover, no significant conformational change in the lower part of TM6 was observed, which is mainly caused by the stationary ICL3.

Rapid closure of ICL3 is observed as restraints expand the ligand-binding site

The goal here was to investigate the allosteric coupling between the intra- and extracellular parts of the receptor, by applying some distance restraints to several key residues at the extracellular ligand-binding site region (See Fig. 1). As listed in Table 1 (See Methods section), there exist seven distances between side chain atoms that were experimentally observed within a certain range when the receptor adopted an active state [28–33]. In our first constrained simulation (*rstr1*), one of those distances which exists between $S203O^\gamma$ and $D113C^\gamma$, was restrained to 16 Å for 300 ns and later increased to 17 Å for another 200 ns, while the remaining six were restrained to those observed in the inactive crystal structure (PDB id: 2RH1) for the whole 500 ns long simulation (See Table 2).

The high value of 17 Å was especially selected for $S203O^\gamma$-$D113C^\gamma$ distance in order to reveal the same allosteric response of the intracellular part of TM6 and ICL3 observed previously in the original MD run. As expected,

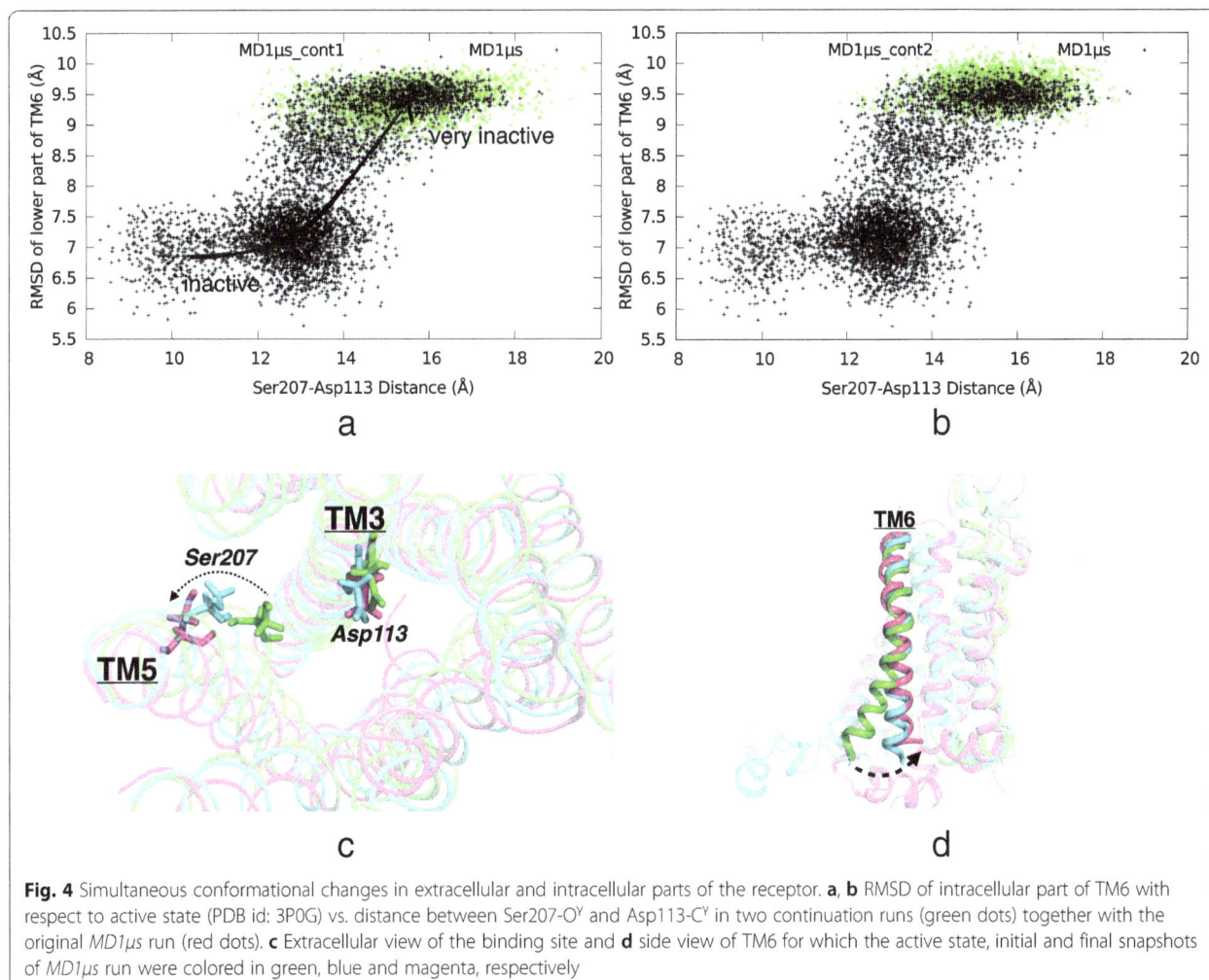

Fig. 4 Simultaneous conformational changes in extracellular and intracellular parts of the receptor. **a, b** RMSD of intracellular part of TM6 with respect to active state (PDB id: 3P0G) vs. distance between Ser207-O^γ and Asp113-C^γ in two continuation runs (green dots) together with the original *MD1µs* run (red dots). **c** Extracellular view of the binding site and **d** side view of TM6 for which the active state, initial and final snapshots of *MD1µs* run were colored in green, blue and magenta, respectively

a close correspondence was observed between the extra-cellular and intracellular parts of the receptor, as ICL3 packed towards the core of the receptor by the end of 200 ns, which is about 400 ns earlier than in the original *MD1µs*.

The closure of ICL3 was monitored through the x and y coordinates of its center of mass, as illustrated with colored points corresponding to different time regimes in Fig. 5a. The interacting alpha helical part of G protein was shown as a straight line connecting all its x-y coordinates, simply to give an idea about its position with respect to ICL3. Also, in Fig. 5b and c, a total of 20 snapshots gradually changing color from red to white and finally to blue well demonstrate the closure of ICL3 towards the core of the receptor during simulation in different angles.

ICL3 preserves its open conformation as restraints narrow the ligand-binding site

A second restrained MD run (*rstr2*) was performed with the same initial frame as used in the first run. This time, the ligand-binding site region was narrowed down via bond restraints between three serines (Ser203-O^Y, Ser204-O^Y, Ser207-O^Y) and Asp113-C^Y to 8 Å, 10 Å and 8 Å, respectively. The simulation was performed for a total of 500 ns. ICL3 preserved its initially open conformation throughout the simulation, in agreement with the allosteric coupling behavior between the intra- and extracellular parts. Similar to the first restrained run, the position of ICL3's center of mass was monitored and all 20 snapshots were illustrated from the side and the intracellular views as in Fig. 6.

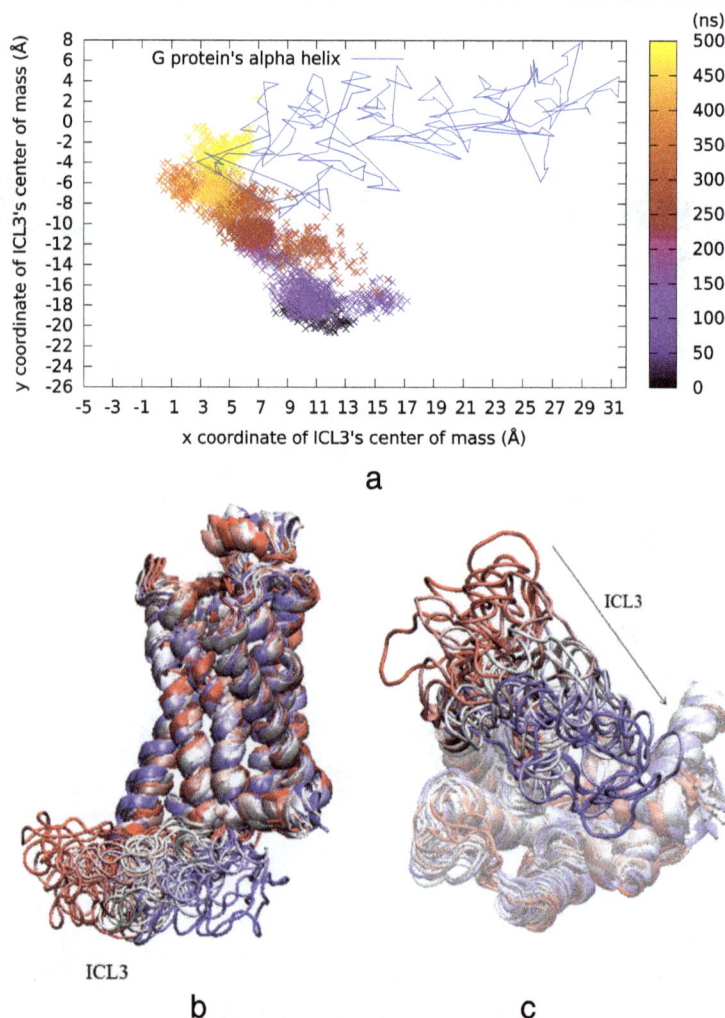

Fig. 5 Results of 500 ns long *rstr1* run. **a** ICL3's center of mass (x and y only) color-coded by time step. Lines represent the G protein's α helix x and y coordinates extracted from the active state's crystal structure (PDB id: 3SN6). **b** Side and **c** intracellular views of 20 snapshots colored from red (initial), to white (intermediate), to blue (final) during simulation

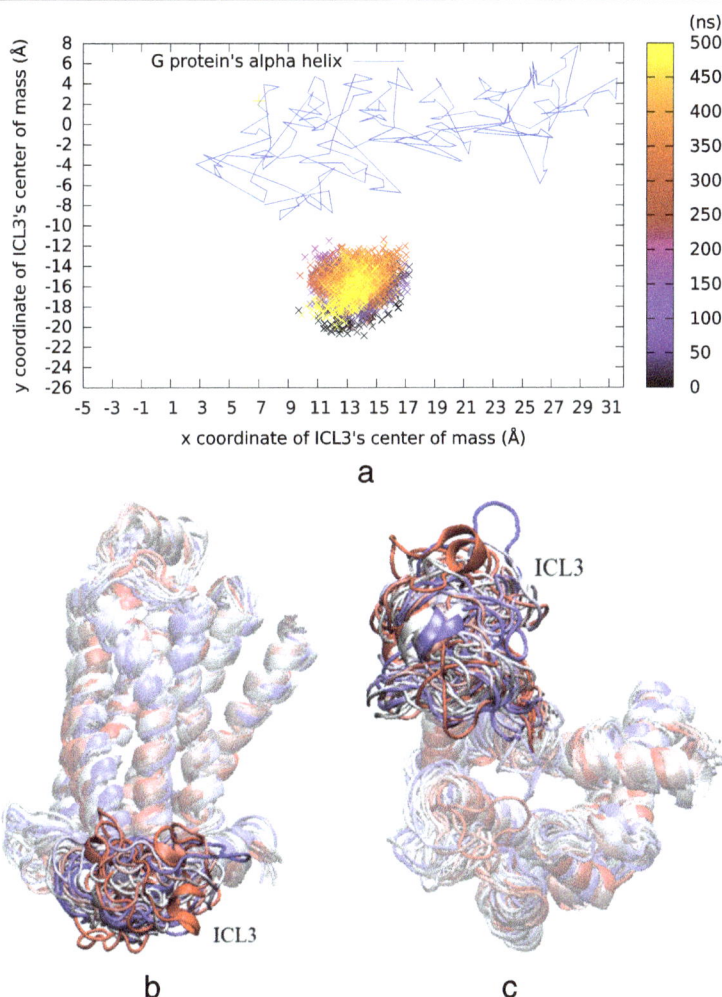

Fig. 6 Results of 500 ns long *rstr2* run. ICL3 preserves its initial open state as bond restraints narrows the ligand-binding site, in agreement with the allosteric coupling behavior between the intra- and extracellular parts of the receptor. **a** ICL3's center of mass (x and y only) color-coded by time step. Lines represent the G protein's a helix x and y coordinates extracted from the active state's crystal structure (PDB id: 3SN6). **b** Side and (**c**) intracellular views of 20 snapshots colored from red (initial), to white (intermediate), to blue (final) during simulation

ICL3 closure necessitates the outward tilt of TM5

One important finding about the closure of ICL3 in the first restrained run and also the original run was the simultaneous outward tilt of TM5 towards the lipid bilayer, which is crucial in initiating the conformational changes along TM5 and TM6 and consequently on ICL3 (See Fig. 4c). In both runs, the distance restraints applied to residues on TM3 and TM5 shifted TM5 but not the more stationary TM3. Consequently, this desired outward tilt in the extracellular part of TM5 was followed by the inward tilt of the intracellular part of TM5 and also of TM6, which induced the expected ICL3 closure (See Fig. 4d).

The necessity of TM5's outward tilt was demonstrated in another 500 ns long restrained run (*rstr3* in Table 2) that used similar restraints as in the first restrained run, but an alternative initial conformation, in which ICL3 was in an extended form but slightly packed and oriented towards the core of the receptor (See Additional file 3: Figure S3). The applied restraints simply necessitated an *expanded* ligand-binding site, which was expected to induce the closure of ICL3. However, no closure was observed in ICL3, which covered a wide range of alternative states nearby G-protein binding site and towards the end of 500 ns, ended up close to its initial position (See Additional file 4: Figure S4). When the conformational change in TM5 was observed, it was clear that as a result of the distance restraint, the outward tilt in TM5 was not notable since both TM3 and TM5 moved apart at the extracellular side (See Additional file 5: Figure S5). Furthermore, no major conformational change in the intracellular part of TM5 was observed. Consequently, ICL3's motion stayed random between open and close states, and no closure was observed. This result shows that the inward tilt of TM6 at the intracellular side was not enough to induce the closure of ICL3, which necessitates the inward tilt in both TM5 and TM6.

Our next attempt in *rstr4* was to impose an additional bond restraint that will bring out the desired outward tilt in TM5, which was not obvious in our previous run (*See* Table 2). Since the backbone atoms' fluctuations are usually minor compared to those of side chain atoms, the bond restraint of 17 Å was imposed between two backbone atoms, C^{α} atom of Ser207 and C^{α} atom Asp113. This time, the new additional restraint was expected to cause the important outward tilt in the extracellular part of TM5. Indeed, both the expected ICL3 closure and the desired outward tilt in TM5 were observed. In addition, ICL3 closure was accomplished under 100 ns, which was two times faster than the first restrained run (See Fig. 7). Another difference was the final position of ICL3, which was shifted about 5 Å in the *x*-axis

with respect to the previously observed positions and located towards the middle of the G-protein binding cavity. In order to further investigate the stability of ICL3 in this alternative closed state, another 500 ns long MD run (*MD500ns*) was performed with all restraints removed (*run #8* in Table). ICL3 preserved its closed state as illustrated with the center of mass profile in Additional file 6: Figure S6.

Packed ICL3 could not be opened by constricting the ligand-binding site

The final restrained run (*rstr5*) was set up to observe the allosteric effect caused by narrowing the ligand-binding site region. The final snapshot of the original MD run (*MD1μs*) was taken as the initial conformation. Here, the ICL3 was fully packed, blocking the G-protein binding site. The ligand-binding site was severely constricted

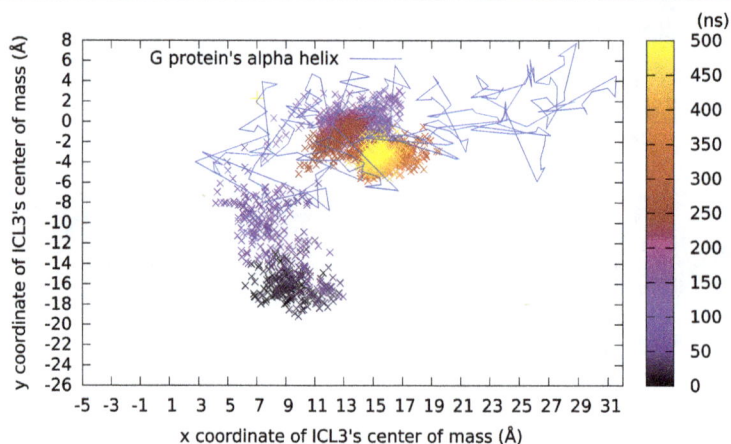

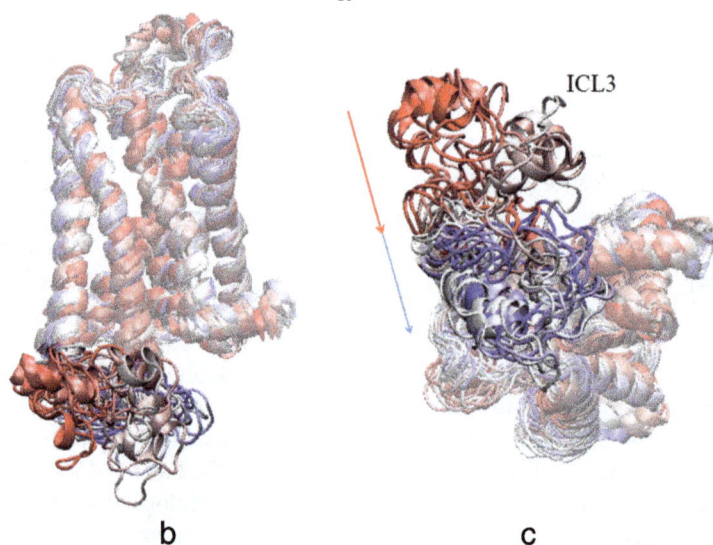

Fig. 7 Results of 500 ns long *rstr4* run. The expected ICL3 closure was observed under 100 ns, when an additional bond restraint was imposed between the backbone atoms in the ligand-binding site. **a** ICL3's center of mass (x and y only) color-coded by time step. Lines represent the G protein's a helix x and y coordinates extracted from the active state's crystal structure (PDB id: 3SN6). **b** Side and (**c**) intracellular views of 20 snapshots colored from red (initial), to white (intermediate), to blue (final) during simulation

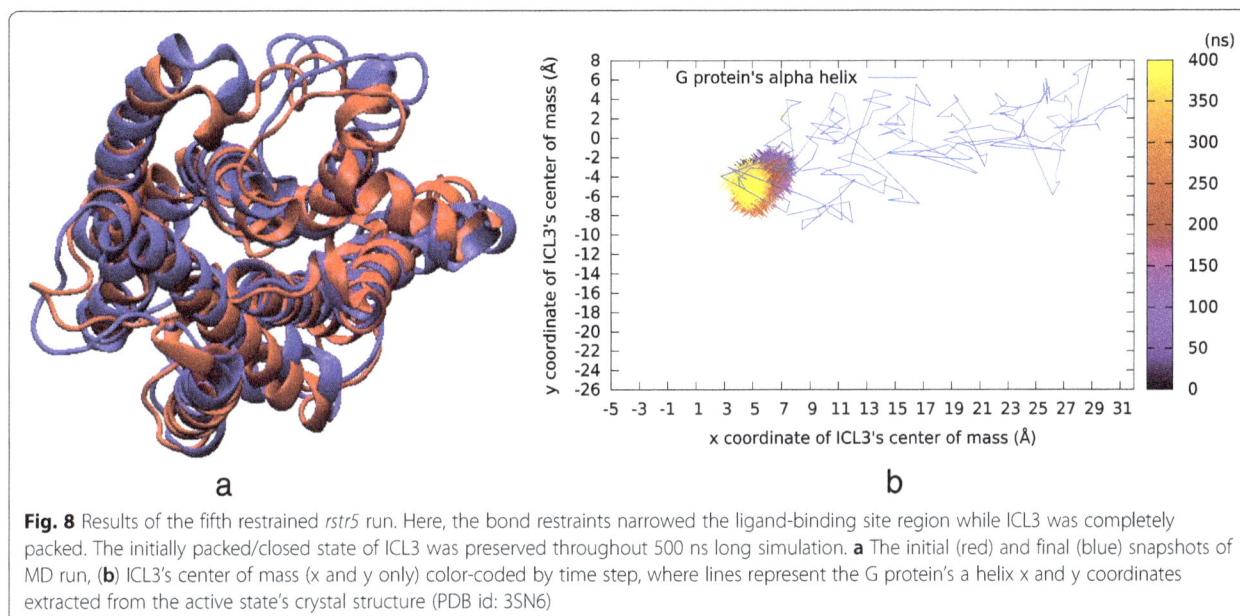

Fig. 8 Results of the fifth restrained *rstr5* run. Here, the bond restraints narrowed the ligand-binding site region while ICL3 was completely packed. The initially packed/closed state of ICL3 was preserved throughout 500 ns long simulation. **a** The initial (red) and final (blue) snapshots of MD run, (**b**) ICL3's center of mass (x and y only) color-coded by time step, where lines represent the G protein's a helix x and y coordinates extracted from the active state's crystal structure (PDB id: 3SN6)

with distance values of 8 Å between almost all pairs of atoms (See Table 2), yet any attempt failed to free the ICL3, which only covered a very confined space during 500 ns long MD run (See Fig. 8). This last result simply point to an important aspect of the receptor's dynamics. It is rather easy to induce the packing of a loose ICL3 by expanding the extracellular binding site region. Yet, it is almost impossible to unpack an already packed or a half packed ICL3 by simply narrowing the extracellular binding site region. Clearly, the energetic barrier to unpack the ICL3 and consequently open the G-protein binding site is too high to be overcome by a few restraints applied at a far region of the receptor. This energetic barrier is most likely due the existence of several hydrogen bonds that exist between ICL3 and the adjacent ends of TM5 and TM6. Thus, the outward tilt of TM6 including the ICL3, which exposes the G-protein binding site, needs to be induced by some exterior forces acting directly on that specific region only.

One activation mechanism proposed by Dror et al. [34] also supports this finding. They have shown that in its basal form, the receptor's intracellular part of TM6 fluctuated between open or half open (intermediate) states, and adopted a fully open active state only when a G protein was bound from the intracellular region and pushed the binding site to an open form. If an agonist was bound at the extracellular binding site, then this active state was stabilized. On the other hand, when G protein was released from this agonist-bound state, it was observed that the receptor's intracellular part of TM6 quickly returned to its inactive state obstructing the G-protein binding site. This finding indicated that the active state cannot be induced by some agonists alone and can only be reached by some exterior forces.

Conclusions

The very inactive state of the receptor [6], was further investigated by two 500 ns long MD runs, which presented a highly stable packed state of ICL3. Although a slight tendency for its opening was observed in one of these simulations (Fig. 2), the closed state was adopted shortly afterwards. The hydrogen bond network analysis revealed several hydrogen bonds connecting ICL3 with adjacent TM5 and TM6 regions, thus stabilizing this novel packed state.

Several distance restraints were applied to key residues at the extracellular ligand-binding site in order to investigate their effect on the intracellular G-protein binding site including ICL3. Bond restraints caused either an expansion or constriction of the ligand-binding site. Key distances that majorly control the size of the cavity were between Asp113 on TM3 and three serine residues (S203, S204 and S207) on TM5. When the ligand-binding site was forced to an expanded/open state via these restraints (*rstr1*), ICL3 closure took place following a straight pathway towards the G-protein binding site, as in the original *MD1μs* run. On the other hand, when the same ligand-binding site was forced to a constricted state (*rstr2*), no change at the intracellular part was observed as ICL3 preserved its initial open conformation. These two observations were both in agreement with the 'pincer-like' behavior of the receptor, where the intracellular part becomes wider as the extracellular part becomes narrower, and vice versa [35].

In both runs *MD1μs* and *rstr1*, closure of ICL3 was observed to closely couple with the inward tilt of both TM5's and TM6's intracellular parts and also the outward tilt in TM5's extracellular part, where the signal propagation starts. In our third restrained run (*rstr3*) which started with an alternative state of the receptor, directed closure of ICL3 was not observed. When initial and final snapshots were aligned, it was clear that the distance restraints forced TM3 rather than TM5 to be displaced and consequently, only a minor outward tilt in TM5's upper half was observed with no significant displacement at its intracellular region. Addition of an extra restraint between backbone atoms of the same residue pair (S203-D113) produced the desired outward tilt of TM5 at the upper half in *rstr4* run. Consequently both TM5 and TM6 were displaced at the intracellular part and ICL3 closed instantaneously on G-protein binding site within 100 ns, which was the most rapid closure to be observed so far. Here, ICL3 adopted an alternative packed state, which was further observed to be stable for another 500 ns when the restraints were removed.

Another set of restraints that constrict the ligand-binding site was applied on the closed ICL3 state (*rstr5*) in order to free the loop from its interactions with the receptor. However, our attempt to open up ICL3 failed during the time scale of our runs. To summarize, our current study revealed alternative packed states of ICL3, which are stabilized by several hydrogen bonds between ICL3 and the rest of the receptor. Furthermore, in contrast to the persistence of ICL3 in its locked position, it was almost always straightforward to bring ICL3 from a loose, free state to a locked one by simply applying a few distance restraints that expand the extracellular ligand-binding site.

Starting with such *very inactive* states, the receptor stayed almost irreversibly inhibited during our runs, which in turn decreased the overall mobility of the receptor. Experimental support is currently lacking for the locked, inactive state of β_2AR, due to the fact that ICL3 has been missing in most studies including the crystal structures.

The bond restraints imposed in our study simply represent the restrictions caused by ligands of various sizes bound at the ligand-binding site. Small agonist molecules tend to fit to a narrow region, whereas the antagonists and inverse agonists of larger size induce a more expanded binding site. As a result of allosteric coupling between intra- and extracellular regions, which is mediated through the transmembrane helices, particularly TM5 and TM6, the population of conformational states of ICL3 between unpacked and packed positions and thereby the binding of G-protein were modulated.

Additional files

Additional file 1: Figure S1. Representations of (a) Core and (b) Tmemb regions highlighted in blue color in the receptor.

Additional file 2: Figure S2. Root mean square deviations of the intracellular parts of TM5 (at the bottom) and TM6 (at the top) with respect to the active state (PDB id: 3P0G) for the original *1 μs* long MD run and two 500 ns long continuation runs (*MD1μs_ctd1* and *MD1μs_ctd2*).

Additional file 3: Figure S3. Initial (in magenta) and intermediate (in cyan) snapshots of the original *1 μs* MD run from (a) side and (b) intracellular views. Here, the intermediate state was taken as the starting conformation for the third restrained run (*rstr3* in Table 2).

Additional file 4: Figure S4. ICL3's center of mass (*x* and *y* coordinates only) during the third 500 ns long restrained run (*rstr3* in Table 2). Stationary G protein's α helix was also represented with lines connecting its *x* and *y* coordinates extracted from the active state's crystal structure

Additional file 5: Figure S5. Initial and final snapshots of the third 500 ns long restrained run (*rstr3* in Table 2). Top and bottom views are the extracellular and the intracellular sides, respectively.

Additional file 6: Figure S6. ICL3's center of mass (*x* and *y* coordinates only) during the 8th run (*MD500ns*), where all the restraints were released. Stationary G protein's α helix was also represented with lines connecting its *x* and *y* coordinates extracted from the active state's crystal structure

Abbreviations
ANM, anisotropic network model; ECL, extracellular loop; GPCRs, G protein coupled receptors; ICL, intracellular loop; MD, molecular dynamics; PCA, principal component analysis; POPC, palmitoyloleoyl-phosphatidylcholine; RMSD, root mean square deviation; RMSF, root mean square fluctuation; T4L, T4-lysozyme; TM, transmembrane helix; TMEMB, transmembrane region; β_2AR, β_2-adrenergic receptor.

Acknowledgements

This work has been fully supported by The Scientific and Technological Research Council of Turkey (TÜBİTAK, Project #213 M544).

Authors' contributions
CO has carried out the molecular dynamics simulations. All authors have participated in the analysis and interpretation of MD trajectory, and writing of the manuscript. All authors read and approved the final manuscript.

Competing interests
The authors declare that they have no competing interests.

Author details
[1]Computational Science and Engineering Program and Polymer Research Center, Bogazici University, Istanbul, Turkey. [2]Department of Chemical Engineering and Polymer Research Center, Bogazici University, Istanbul, Turkey. [3]Department of Bioinformatics and Genetics, Faculty of Natural Sciences and Engineering, Kadir Has University, Cibali, 34083 Istanbul, Turkey.

References
1. Rasmussen SG, Choi HJ, Rosenbaum DM, Kobilka TS, Thian FS, Edwards PC, Burghammer M, Ratnala VR, Sanishvili R, Fischetti RF, Schertler GF, Weis WI, Kobilka BK. Crystal structure of the human beta2 adrenergic G-protein-coupled receptor. Nature. 2007;450:383–7.
2. Cherezov V, Rosenbaum DM, Hanson MA, Rasmussen SG, Thian FS, Kobilka TS, Choi HJ, Kuhn P, Weis WI, Kobilka BK, Steven RC. High-resolution crystal structure of an engineered human beta2-adrenergic G protein-coupled receptor. Science. 2007;318:1258–65.
3. Kobilka BK, Deupi X. Conformational complexity of G-protein-coupled receptors. Trends Pharmacol Sci. 2007;28:397–406.
4. Rasmussen SG, DeVree BT, Zou Y, Kruse AC, Chung KY, Kobilka TS, Thian FS, Chae PS, Pardon E, Calinski D, Mathiesen JM, Shah ST, Lyons JA, Caffrey M, Gellman SH, Steyaert J, Skiniotis G, Weis WI, Sunahara RK, Kobilka BK. Crystal structure of the beta2 adrenergic receptor-Gs protein complex. Nature. 2011;477:549–55.

5.   West GM, Chien EY, Katritch V, Gatchalian J, Chalmers MJ, Stevens RC, Griffin PR. Ligand-dependent perturbation of the confor- mational ensemble for the GPCR β2 adrenergic receptor revealed by HDX. Structure. 2011;19(10):1424–32.

6.   Ozcan O, Uyar A, Doruker P, Akten ED. Effect of intracellular loop 3 on intrinsic dynamics of human β₂-adrenergic receptor. BMC Struct Biol. 2013;13:29.

7.   Palm D, Munch G, Dees C, Hekman M. Mapping of β-adrenoceptor coupling domains to Gs-protein by site- specific synthetic peptides. FEBS Lett. 1989;254:89–93.

8.   Munch G, Dees C, Hekman M, Palm D. Multisite contacts involved in coupling of the β-adrenergic receptor with the stimulatory guanine-nucleotide-binding regulatory protein. Structural and functional studies by β-receptor-site- specific synthetic peptides. Eur J Biochem. 1991;198(2):357–64.

9.   Okamoto T, Murayama Y, Hayashi Y, Inagaki M, Ogata E, Nishimoto I. Identification of a Gs activator region of the β2-adrenergic receptor that is autoregulated via protein kinase A-dependent phosphorylation. Cell. 1991; 67(4):723–30.

10. Shinagawa K, Ohya M, Higashijima T, Wakamatsu K. Circular dichroism studies of the interaction between synthetic peptides corresponding to intracellular loops of ß-adrenergic receptors and phospholipid vesicles. J Biochem. 1994;115:463–68.

11. Okuda A, Matsumoto O, Akaji M, Taga T, Ohkudo T, Kobayashi Y. Solution structure of intracellular signal-transducing peptide derived from human ß -adrenergic receptor. Biochem Biophys Res Commun. 2002;291:1297–1301.

12. Kim TH, Chung KY, Manglik A, Hansen AL, Dror RO, Mildorf TJ, Shaw DE, Kobilka BK, Prosser RS. The Role of Ligands on the Equilibria Between Functional States of a G Protein-Coupled Receptor. J Am Chem Soc. 2013;135:9465–74.

13. Bhattacharya S, Vaidehi N. Differences in Allosteric Communication Pipelines in the Inactive and Active States of a GPCR. Biophys J. 2014;107:422–34.

14. Niesen MJM, Bhattacharya S, Vaidehi N. The Role of Conformational Ensembles in Ligand Recognition in G-Protein Coupled Receptors. J Am Chem Soc. 2011;133:13197–204.

15. Liu JJ, Horst R, Katritch V, Stevens RC, Wüthrich K. Biased Signaling Pathways in β₂-adrenergic Receptor Characterized by ¹⁹F-NMR. Science. 2012;335:1106–10.

16. Pieper U, Eswar N, Webb BM, Eramian D, Kelly L, Barkan DT, Carter H, Mankoo P, Karchin R, Marti-Renom MA, Davis FP, Sali A. MODBASE, a Database of Annotated Comparative Protein Structure Models and Associated Resources. Nucleic Acids Res. 2009;37(Database issue):D347–54.

17. Humphrey W, Dalke A, Schulten K. VMD - Visual Molecular Dynamics. J Mol Graph. 1996;14:33–8.

18. Lomize MA, Lomize AL, Pogozheva ID, Mosberg HI. OPM: orientations of proteins in membranes database. Bioinformatics. 2006;22:623–5.

19. Essmann U, Perera L, Berkowitz ML, Darden T, Lee H, Pedersen LG. A Smooth Particle Mesh Ewald Method. J Chem Phys. 1995;103(19):8576–93.

20. Phillips JC, Braun R, Wang W, Gumbart J, Tajkhorshid E, Villa E, Chipot C, Skeel RD, Kale L, Schulten K. Scalable Molecular Dynamics with NAMD. J Comput Chem. 2005;26(16):1781–802.

21. Schlenkrich M, Brickmann J, MacKerell AD Jr, Karplus M: An empirical potential energy function for phospholipids: criteria for parameter optimization and applications. In Biological Membranes: A Molecular Perspective from Computation and Experiment. 1st edition. Edited by Merz KM Jr, Roux B. Birkhauser, Boston; 1996.

22. Feller SE, Yin D, Pastor RW, MacKerell Jr AD. Molecular dynamics simulation of unsaturated lipids at Low hydration: parametrization and comparison with diffraction studies. Biophys J. 1997;73:2269–79.

23. Mackerell AD, Bashford D, Bellott M, Dunbrack RL, Evanseck JD, Field MJ, Fischer S, Gao J, Guo H, Ha S, Joseph D, Kuchnir L, Kuczera K, Lau FTK, Mattos C, Michnick S, Ngo T, Nguyen DT, Prodhom B, Roux B, Schlenkrich M, Smith J, Stote R, Straub J, Watanabe M, Wiorkiewicz-Kuczera J, Yin D, Karplus M. Self-consistent parameterization of biomolecules for molecular modeling and condensed phase simulations. FASEB J. 1992;6:A143–A143.

24. Mackerell AD, Bashford D, Bellott M, Dunbrack RL, Evanseck JD, Field MJ, Fischer S, Gao J, Guo H, Ha S, Joseph D, Kuchnir L, Kuczera K, Lau FTK, Mattos C, Michnick S, Ngo T, Nguyen DT, Prodhom B, Roux B, Schlenkrich M, Smith J, Stote R, Straub J, Watanabe M, Wiorkiewicz-Kuczera J, Yin D,

Karplus M. All-atom empirical potential for molecular modeling and dynamics studies of proteins. J Phys Chem B. 1998;102:3586–616.

25. Jorgensen WL, Chandrasekhar J, Madura JD, Impey RW, Klein ML. Comparison of simple potential functions for simulating liquid water. J Chem Phys. 1983;79:926–35.

26. Feller SE, Zhang Y, Pastor RW. Computer-Simulation of Liquid/Liquid Interfaces II. Surface-Tension Area Dependence of a Bilayer and Monolayer. J Chem Phys. 1995;103:10267–76.

27. Ryckaert JP, Ciccotti G, Berendsen HJC. Numerical integration of Cartesian equations of motion of a system with constraints – molecular dynamics of N-alkanes. J Comput Phys. 1977;23:327–41.

28. Elling CE, Frimurer TM, Gerlach LO, Jorgensen R, Holst B, Schwartz TW. Metal ion site engineering indicates a global toggle switch model for seven-transmembrane receptor activation. J Biol Chem. 2006;281:17337–46.

29. Elling CE, Thirstrup K, Holst B, Schwartz TW. Conversion of agonist site to metal-ion chelator site in the beta(2)-adrenergic receptor. Proc Natl Acad Sci U S A. 1999;96:12322–7.

30. Liapakis G, Ballesteros JA, Papachristou S, Chan WC, Chen X, Jav- itch JA. The forgotten serine. A critical role for Ser-2035.42 in ligand binding to and activation of the beta 2-adrenergic receptor. J Biol Chem. 2000;275:37779–88.

31. Sato T, Kobayashi H, Nagao T, Kurose H. Ser203 as well as Ser204 and Ser207 in fifth transmembrane domain of the human beta2-adrenoceptor contributes to agonist binding and receptor activation. Br J Pharmacol. 1999;128:272–4.

32. Gouldson PR, Winn PJ, Reynolds CA. A molecular dynamics approach to receptor mapping: application to the 5HT3 and beta 2-adrenergic receptors. J Med Chem. 1995;38:4080–6.

33. Wieland K, Zuurmond HM, Krasel C, IJzerman AP, Lohse MJ. Involvement of Asn-293 in stereospecific agonist recognition and in activation of the beta 2-adrenergic receptor. Proc Natl Acad Sci U S A. 1996;93:9276–81.

34. Dror RO, Arlow DH, Maragakis P, Mildorf TJ, Pan AC, Xu H, Borhani DW, Shaw DE. Activation mechanism of the β2-adrenergic receptor. PNAS. 2011;108:18684–9.

35. Bokoch MP, Zou Y, Rasmussen SG, Liu CW, Nygaard R, Rosenbaum DM, Fung JJ, Choi H, Thian FS, Kobilka TS, Puglisi JD, Weis WI, Pardo L, Prosser RS, Mueller L, Kobilka BK. Ligand-specific regulation of the extracellular surface of a G-protein-coupled receptor. Nature. 2010;463:108–12.

14

Modeling of the OX$_1$R–orexin-A complex suggests two alternative binding modes

Lasse Karhu, Ainoleena Turku and Henri Xhaard[*]

Abstract

Background: Interactions between the orexin peptides and their cognate OX$_1$ and OX$_2$ receptors remain poorly characterized. Site-directed mutagenesis studies on orexin peptides and receptors have indicated amino acids important for ligand binding and receptor activation. However, a better understanding of specific pairwise interactions would benefit small molecule discovery.

Results: We constructed a set of three-dimensional models of the orexin 1 receptor based on the 3D-structures of the orexin 2 receptor (released while this manuscript was under review), neurotensin receptor 1 and chemokine receptor CXCR4, conducted an exhaustive docking of orexin-A$_{16-33}$ peptide fragment with ZDOCK and RDOCK, and analyzed a total of 4301 complexes through multidimensional scaling and clustering. The best docking poses reveal two alternative binding modes, where the C-terminus of the peptide lies deep in the binding pocket, on average about 5–6 Å above Tyr$^{6.48}$ and close to Gln$^{3.32}$. The binding modes differ in the about 100° rotation of the peptide; the peptide His26 faces either the receptor's fifth transmembrane helix or the seventh helix. Both binding modes are well in line with previous mutation studies and partake in hydrogen bonding similar to suvorexant.

Conclusions: We present two binding modes for orexin-A into orexin 1 receptor, which help rationalize previous results from site-directed mutagenesis studies. The binding modes should serve small molecule discovery, and offer insights into the mechanism of receptor activation.

Keywords: Orexin-A, OX$_1$ receptor, Peptide docking, G protein-coupled receptor, Pose selection, Multidimensional scaling, GPCR

Background

The orexinergic system is composed of two receptor subtypes, named orexin 1 and 2 receptors (OX$_1$R and OX$_2$R respectively), and of two agonistic peptide ligands, orexin-A and orexin-B [1]. Orexin receptors are mainly found in the central nervous system, but also in the periphery (gastrointestinal track, pancreas, adrenal gland and adipose tissue) [2]. Certain cancer cell lines also express OX$_1$ receptors, whose activation induces apoptosis [3]. The endogenous orexin peptides induce feeding and wakefulness, and malfunctions of the orexin system are one of the reasons behind narcolepsy in mice, dogs and humans [2]. Small molecules (i.e. not peptides) have been developed to act as orexin receptor antagonists [4]. As expected, antagonists have opposing effects to orexin peptides; reduced feeding [5] and induction of sleep [4].

The first drug targeting the orexin receptors, the antagonist suvorexant (Belsomra®), has recently reached the market in the United States of America and in Japan.

Orexin peptides and receptors were discovered independently in 1998 by two research groups. Sakurai and co-workers discovered two peptides that produced robust Ca^{2+} elevation through activation of two receptors which they expressed in CHO cells [1]. The two peptides were named orexin-A and -B, and the receptor subtypes were designated as OX$_1$ and OX$_2$ receptors according to the Greek word for appetite, *oreksis* (ὄρεξις), since the peptides induced feeding in mice. De Lecea and co-workers discovered about simultaneously a mRNA sequence expressed in hypothalamus that encodes the precursor of the two peptides [6]. They named the peptides hypocretins 1 and 2.

The orexin peptides are produced as a 131-amino acid (in human) precursor, prepro-orexin, which is enzymatically cleaved to produce one unit of each peptide [1].

* Correspondence: henri.xhaard@helsinki.fi
Division of Pharmaceutical Chemistry and Technology, Faculty of Pharmacy, University of Helsinki, P.O. Box 56, 00014, Finland

Human orexin-A is a 33-amino acid peptide containing two intramolecular disulfide bridges (Cys6–Cys12, Cys7–Cys14), an N-terminal pyroglutamoyl residue, and an amidated C-terminus [1]. Human orexin-B is composed of 28 residues and is amidated on its C-terminus like orexin-A, but lacks the disulfide bridges [1]. While the N-termini of the peptides are different, the C-termini are near-identical (11 out of 15 amino acids are identical). The receptor-bound conformations are not known, but NMR-structures for both peptides in buffered water solution have been solved [7,8]. Orexin-B comprises two helical parts (helix I: Leu7–Gly19 and helix II: Ala23–Met28) joined with a short linker or hinge (Asn20–Ala22) [7], whereas orexin-A has three helical sections (helix I: Leu16–Ala23, helix II: Asn25–Thr32 and helix III: Cys6–Gln9) [8]. The peptides were observed in multiple conformations: orexin-A is either in bent or straight conformation across the set of 30 NMR models [8], while the single model derived for orexin-B shows the hinge bent opposite to the conformation of orexin-A [7] (Additional file 1).

Mutations on the orexin peptides have shown that the C-terminal residues and the amidation of the C-terminus are the most important factors for receptor activation [9-12]. The N-terminus is not as important, as both peptides retained activity when truncated down to a C-terminal fragment of 19 residues [9-11]. Further truncation lowered the maximal response, but fragments as short as 12 amino acids still retained some activity [9-11]. No key residues have been found in the N-terminal part of the peptide.

The orexin receptors OX_1R and OX_2R are G protein-coupled receptors (GPCRs) that in human are composed of 425 amino acids (OX_1R) and 444 amino acids (OX_2R) [1]. As GPCRs, the overall structure of orexin receptors consists of seven helical transmembrane segments (TM1–7) connected by three intra- and three extracellular loops (ICL1–3 and ECL1–3 respectively), an extracellular N-terminus and an intracellular C-terminus, confirmed by the recent crystal structure of OX_2R [PDB:4S0V] [13] that was solved while this manuscript was under review. The OX_2 receptor has the conserved disulfide bridge connecting TM3 and ECL2, as was expected based on the receptor sequences. Most likely the OX_1R will also have this bridge formed by $Cys119^{3.25}$ and $Cys202^{xl2.50}$.[a] Both receptors have also suitable cysteines for C-terminal palmitoylation (Cys375 and Cys376 in OX_1R), which is observed in most crystallized GPCRs. The human OX_1R and OX_2R share a full-length pairwise sequence identity of 64%, and without terminals and ICL3, the sequence identity of the TM bundle is up to 80%. Orexin-A is equipotent towards both receptor subtypes, whereas orexin-B is equipotent with orexin-A towards OX_2R but 10-fold less potent in activating OX_1R [10,11].

The receptors have been mutated [14,15] and chimeras of OX_1R and OX_2R have been constructed [15,16] to study the contributions of different amino acids to interactions with ligands (Figure 1). Alanine mutations of OX_1R residues $Gln126^{3.32}$, $Val130^{3.36}$, $Asp203^{xl2.51}$, $Trp206^{xl2.54}$, $Tyr215^{5.38}$, $Phe219^{5.42}$, $Tyr224^{5.47}$, $Tyr311^{6.48}$, $His344^{7.39}$, and $Tyr348^{7.43}$ decreased the potency and/or maximum response of orexin-A [14]. A similar study conducted on OX_2R discovered that mutations of $Thr231^{5.46}$ and $Asn324^{6.55}$ (corresponding to Thr223 and Asn318 in OX_1R) to alanine led to a 10-fold decrease in orexin-A potency [15]. This indicated that the orexin receptor ligand binding pocket is lined by residues from TMs 3, 5, 6 and 7 as well as ECL2, which was confirmed by the crystal structure of OX_2R bound to suvorexant [13].

Computational modeling is a powerful tool to gain insight in the binding of the orexin peptides and the interactions leading to receptor activation. The prospective GPCR Dock studies [17-19] have shown that the transmembrane region of GPCRs can be reliably modeled and that computational tools are getting better at recreating receptor–ligand complexes. However, peptide docking without a known bioactive conformation remains challenging in part due to the inherent flexibility of peptides. In GPCR Dock 2010, the task of modeling chemokine receptor CXCR4 in complex with a synthetic 16-residue cyclic peptide proved difficult, since available templates had only distant homology to CXCR4 and the binding interactions were poorly characterized [18]. Recently, peptide docking software such as HADDOCK [20] (originally designed for protein–protein docking), Rosetta FlexPepDock [21], and DynaDock [22] have been developed. These software were tested with peptides ranging from 2 to 16 residues, often binding into a shallow groove on the protein surface [20-22]. Buried binding sites and helical peptides have been problematic [20,21]. Concerning GPCRs, peptides are docked with multiple methods; a rigid docking can be followed by a short molecular dynamics simulation [23-26], or semi-flexible methods can be used, such as Glide with induced fit [27] or GOLD, which allows rotamer-library-based side-chain rotation for selected residues [28]. Genetic algorithms can be used to produce changes to peptide backbone conformation [29]. In this study, we have used ZDOCK in combination with RDOCK to perform an exhaustive mapping of the OX_1R binding site while allowing limited peptide and receptor flexibility. ZDOCK and RDOCK were originally developed for protein–protein docking and refinement [30,31], but they are also usable in peptide docking, which became evident in the GPCR Dock 2010 assessment, where one of the best peptide-docking results came from a group utilizing ZDOCK [18].

Previously, Heifetz and co-workers have aimed to establish a binding mode for orexin peptides to orexin

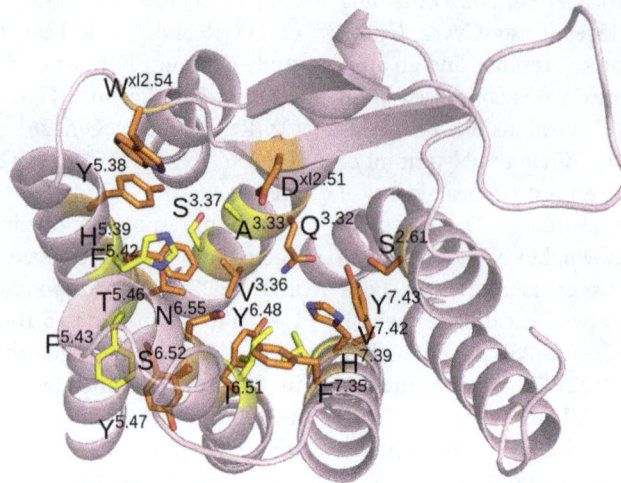

Figure 1 Point-mutated residues on the orexin receptors. Orange: mutation impaired the orexin-induced receptor activation in one or both subtypes; yellow: mutation did not alter the receptor function significantly [14,15].

receptors [28]. In their study, the dopamine D3 receptor served as a template for orexin receptor modeling. To account for protein flexibility, receptor conformations for docking were harvested from a short molecular dynamics simulation, and certain side chains in both receptor and ligand were allowed to adopt different rotamers. However, recent crystal structures for peptide-binding GPCRs have shown features such as the β-hairpin in the ECL2 that their models lack, and thus their results need to be updated.

Here, combining the data from the mutational studies conducted on orexinergic system and the crystal structures of peptide-binding GPCRs neurotensin receptor 1 (NTSR1), chemokine receptor CXCR4, and the recent crystal structure of the OX$_2$R, we have constructed 3D-models of the OX$_1$R. An exhaustive docking algorithm allowed mapping of the available space for orexin-A within the receptor cavity. Based on the molecular interactions observed in the docking results, we propose two alternative binding modes for orexin-A into OX$_1$R. Studying these binding interactions will increase the understanding on the mechanisms by which the orexin peptides activate their cognate receptors, and provide a general framework to understand peptide-binding GPCRs.

Methods
Structural alignment of GPCRs
In order to identify structurally conserved regions, we superposed a total of 19 GPCR crystal structures available on RCSB Protein Data Bank (PDB) with Discovery Studio 3.5 [32]. Lysozyme chains were removed. A sequence alignment was derived from the superposition (Additional files 2 and 3). OX$_1$R sequence was added manually to the alignment based on conserved motifs on each transmembrane helix [33]. We initially based the orexin receptor sequence

alignment in the ECL2-region on the observation that all available crystallized peptide binding GPCRs — chemokine receptor CXCR4, neurotensin receptor 1 (NTSR1) and the four opioid receptors (mu, kappa, delta, and nociceptin) [34-39] — incorporate a similar β-hairpin fold of ECL2, composed of two five-residue β strands (in OX$_1$R residues 184–188 and 200–204, see arrows in Figure 2 and in Additional file 2) and a turn of variable length (4–10 residues) between them. For OX$_2$R, this hairpin structure was confirmed by the crystal structure, although not all amino acids in the turn were resolved [13]. In the crystal structures, the first β strand follows directly the TM4 and the second ranges from xl2.48 to xl2.52. The conserved disulfide bridge between TM3 and ECL2 constrains the second β strand, and therefore the β hairpin, above TM3.

Template selection for homology modeling
Based on the structural alignment, the phylogenetic analysis of GPCRs [40], and the shapes of the observed binding pockets, we initially selected the crystal structure of the neurotensin receptor 1 [PDB:4GRV] [35] as a primary template for homology modeling. At the time, NTSR1 was the only crystallized receptor from the β branch of rhodopsin-like GPCRs where orexin receptors are found [40]. Like orexin receptors, the NTSR1 is also naturally activated by a peptide ligand, neurotensin. Neurotensin$_{8-13}$ fragment has been co-crystallized with the receptor, but there is no G protein (or an antibody mimicking it), and thus the receptor conformation is not fully that of an activated GPCR [35]. While this article was under review, the crystal structure of OX$_2$R in complex with the antagonist suvorexant was released [13]. To incorporate these recent data into our study, we utilized also the OX$_2$R crystal structure as a template for homology modeling.

Figure 2 Sequence alignment used in homology modeling. Target sequence: OX₁R [UniProt:O43613]. Template sequences: OX₂R [PDB:4S0V], NTSR1 [PDB:4GRV] and CXCR4 [PDB:3ODU]. Orange: orexin receptor residues found to be important by site-directed mutagenesis. Cyan: NTSR1 residues that interact with neurotensin₈₋₁₃. Boxed: OX₂R residues within 4 Å of suvorexant. Italics: helix 8 from dopamine D3 receptor [PDB:3PBL]. Cylinders and arrows: TM helices and β strands seen in template structures. Numbering refers to OX₁R. Triangle: x.50 residue. *: TM3–ECL2 disulfide bridge. Gray: identical residues between OX₁R and templates. Illustrated with Alscript [57].

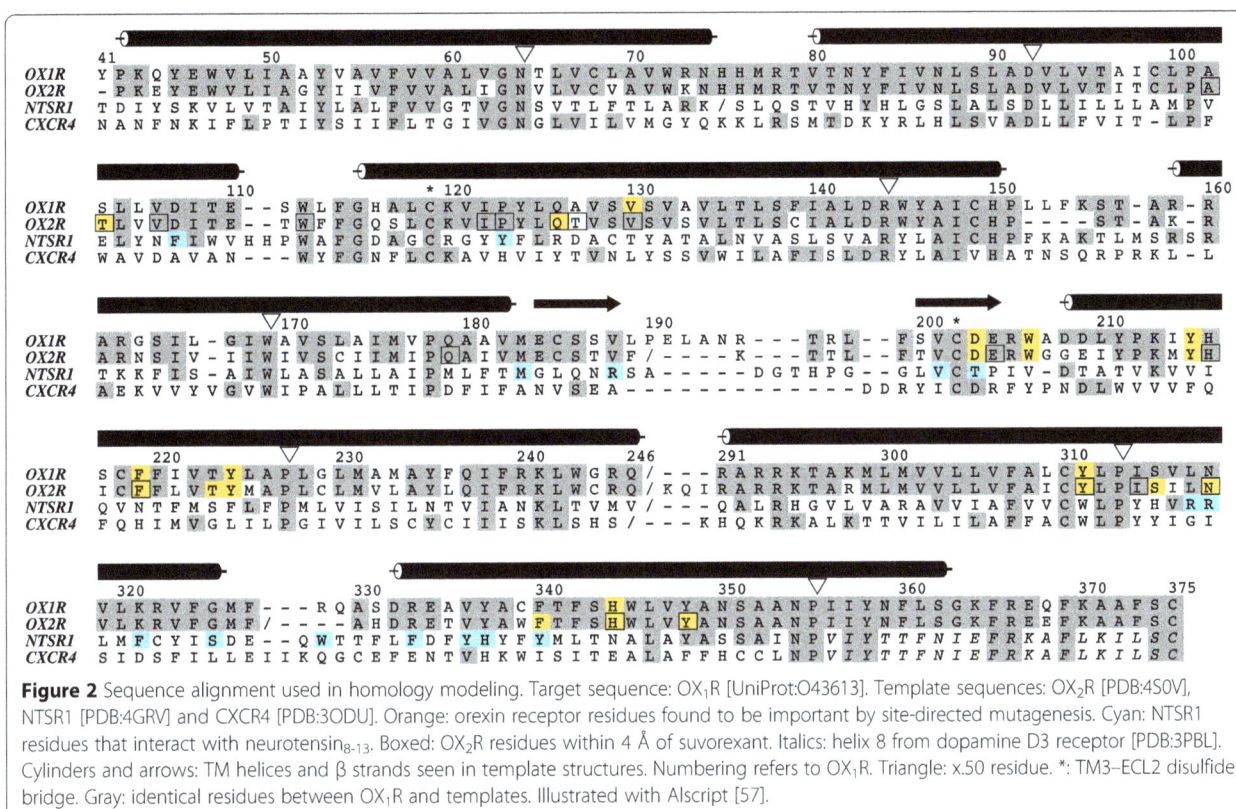

The NTSR1 crystal structure entails a binding cavity constricted by the TM6; the extracellular end of the helix is tilted towards the binding cavity, narrowing the cavity and limiting the exposure of the TM5 residues to the binding cavity. Therefore, we built two secondary homology models with more open binding cavities. One secondary model was based on the chemokine receptor CXCR4 [PDB:3ODU] [34], which naturally binds a small protein, although the receptor was crystallized with synthetic ligands. The CXCR4 crystal structure shows a more open binding cavity than the NTSR1. For the other secondary model, we constructed a modified NTSR1 structure template (NTSR1_TM6) by rotating the TM6 in the NTSR1 to occupy the same space as TM6 in CXCR4; this was done with Maestro 9.3.5 [41].

Neither selected crystal structure shows the 8th helix parallel to membrane plane observed in many other GPCR crystals. We selected dopamine D3 receptor [PDB:3PBL] [42] as a template for the 8th helix. Residues after Pro$^{7.50}$ in NTSR1, CXCR4 and NTSR1_TM6 were replaced by those of dopamine D3 after careful superposition of TMs 1 and 7 of crystal structures. This was a cosmetic step that most likely does not affect the docking results. In retrospect, a more recent X-ray crystal structure of the NTSR1 [PDB:4BUO] [43] shows an intracellular assembly with the canonical TM8, as also does the recent crystal structure of the OX₂R [13].

Model building

Models of OX₁R consisting of the residues Tyr41$^{1.27}$–Gln246$^{5.69}$ and Arg291$^{6.28}$–Cys375 were built using the four templates mentioned above. The N- and C-termini, and ICL3 were omitted as there were no suitable templates. Homology modeling was carried out with MODELLER 9v8 [44], a comparative protein modeling program, using default settings. Pairwise alignment of OX₁R with the templates was fine-tuned in tandem with model building (Figure 2). Ten models were derived from each template structure.

Model evaluation

The 40 models were evaluated visually to eliminate unreasonable constructs and to select four models for docking, each displaying an open binding cavity and resulting from a different template. Modeller DOPE scores did not differ significantly between models of same origin. We selected the models based on the conformations of ECL2 and ECL3. The ECL2, especially the turn between the strands of the β hairpin, was required to show a secondary structure similar, and occupying roughly the same space, as those observed in the crystal structures of peptide binding receptors. The ECL3 was required not to constrict the entrance of the binding cavity.

Orexin peptide conformation for docking

We used the straight α-helical conformation of orexin-A (the second NMR model in [PDB:1WSO] [8]) in this study, as the bent conformation did not fit the predicted binding site in a preliminary docking. Instead of the full orexin-A peptide, a fragment comprising of residues 16–33 was used. This fragment retains biological activity [10], and using it avoids the problem of the N-terminus of the intact orexin-A colliding with the extracellular loops of receptor models and limiting the conformational space.

Docking of orexin peptides with ZDOCK and RDOCK

Prior to docking, the receptor models and the peptide fragment were converted to CHARMm atom types as required by the docking program. ZDOCK [30], an exhaustive initial-stage docking algorithm for protein–protein complexes, was used with default settings. We filtered the docking poses by accepting only the poses where the ligand C-terminal residues (shown to be crucial for activity, see Background) were part of the receptor–ligand interface and the ligand did not traverse between the TM helices into the space occupied by the cell membrane. The poses were refined with RDOCK [31] using default settings. RDOCK is a CHARMm force field based refinement algorithm that performs limited molecular dynamics to fine-tune receptor–ligand complexes from ZDOCK. RDOCK uses a two-stage scoring function; van der Waals energy is first calculated to discard docking poses with clashes and then the poses are scored based on desolvation and electrostatic energies. Accelrys Discovery Studio 3.5 [32] was used as an interface to ZDOCK and RDOCK and to visualize the results.

Data analysis on the docking poses

We clustered the refined docking poses modelwise using an algorithm devised by Daura and co-workers [45], implemented in MATLAB [46]. In short, a matrix of all pairwise root mean square deviations (RMSD) of the peptide α carbons is calculated. The pose with most neighbors (here RMSD < 3 Å) is flagged as the cluster seed, and the neighbors are included in the cluster and removed from the pool of poses. The process is repeated until no two poses are closer than the cutoff. For cluster scoring, we used the median RDOCK score of the poses in each cluster.

For multidimensional scaling, we pooled all docking poses across models and calculated all pairwise RMSD values. MATLAB was used to reduce dimensions to two (*mdscale* function) and to visualize the outcome. Solvent accessible surface area for the peptide ligand was calculated with Naccess [47] (default settings). For measurements of ligand depth, the z-coordinate (z-axis normal to the membrane plane) of the Leu33 α carbon (Cα) was used. The zero-plane was set to the Cα's of Thr223$^{5.46}$, Tyr311$^{6.48}$ and Tyr348$^{7.43}$.

We assessed the rotation of the peptide ligand around its helical axis by drawing a vector towards the side chain of His26 (from Ala28 Cα to His26 Cα) in xy-plane (the plane parallel to the membrane). By using a common initial point for the vectors, preferences in ligand orientation could be seen.

The contact mapping was carried out with MATLAB by calculating the pairwise distances between ligand atoms and atoms in the receptor residues in the binding cavity. Any pairwise distance between atoms closer than 4 Å was considered a contact. No differentiation between side-chain and main-chain atoms was done at this point.

Results and discussion

Homology modeling

The crystal structures of class A (rhodopsin-like) GPCRs show clear conservation within TM segments and short loops, as illustrated by the structural alignment. The sequence alignment within the TM region is unambiguous (Additional file 2). The ECL2 and ECL3 vary both in length and in conformation between receptors, but closely related receptors often show similarities; for example all peptide binding receptors (NTSR1, CXCR4, and the four opioid receptors mu, kappa, delta, and nociceptin, and also the recent OX$_2$R structure [13,34-39]) show similar β-hairpin structures in the ECL2, although the segment between the β strands varies in length: three amino acids in CXCR4, nociceptin, delta, and mu opioid receptors, five in kappa opioid receptor, and nine in NTSR1 (see Additional file 2). The orexin receptors have a segment of 11 amino acids between the β strands, but five of them were not solved in the recent OX$_2$R crystal structure. Also ECL3 differs in length among the OX$_1$R and the crystallized receptors.

Orexin receptors have most class A GPCR-specific motifs; instead of the conserved Tyr3.51 (in the "DRY" motif) and Trp6.48 (at the bottom of the binding cavity, the "CWxP" motif), orexin receptors have Trp145$^{3.51}$ and Tyr311$^{6.48}$. As both substitutes are aromatic residues, the structural functions are likely to be conserved. In the extracellular half of the orexin receptor TM3, Pro123$^{3.29}$ is present. This feature is common in the β branch of rhodopsin-like GPCRs, and a comparison retrospective to this work between the OX$_2$R structure and the other crystallized class A GPCRs shows that the conformation of the TM3 remains unaltered by the proline.

Templates originally used in this study (NTSR1 and CXCR4) both have sequence identities of 23.6% (70 identical residues out of alignment length of 296) to human OX$_1$R transmembrane bundle (Tyr41$^{1.27}$–Gln246$^{5.69}$ and Arg291$^{6.28}$–Cys375). This level of sequence identity is usually considered poor for homology modeling, but the overall fold shared by the crystallized class A GPCRs was likely to be conserved also in the orexin receptors, which was confirmed by the OX$_2$R crystal structure [13]. Within the transmembrane bundle, NTSR1 and CXCR4 both have six alignment gaps in comparison to OX$_1$R. For NTSR1, all gaps fall into loops (Figure 2). In contrast, CXCR4, together with opioid receptors, shows a gap at 2.57 (2x551 according to the structure-based residue numbering proposed by the GPCRDB [48]), which results in the absence of a bulge shown by other crystallized class A GPCRs. CXCR4 has also a bulge-inducing insertion at 4.47 (4x471) (Additional file 2), while other alignment gaps occur in the loops. These are present in our CXCR4–OX$_1$R sequence alignment used for homology modeling (Figure 2). As the TMs 2 and 4 are only marginally exposed to the interhelical cavity, the effect of possible misalignment on the binding site of the CXCR4-based model is negligible.

Considering the conserved TM3–ECL2 disulfide bridge, human orexin receptors have two cysteine residues in the ECL2: Cys185/193 and Cys202/210 (OX$_1$R/OX$_2$R). Based on the sequence alignment, and the fact that rat OX$_2$R has arginine instead of Cys193 [UniProt:P56719], we assumed that the Cys202/210 would be involved in the disulfide bridge with Cys119/127$^{3.25}$. The crystal structure of OX$_2$R indeed shows the disulfide bridge between Cys127 and Cys210 (corresponding to Cys119 and Cys202 in OX$_1$R).

Homology models

Originally three models were selected, one from each template, among the 30 generated models. Later a fourth model, based on the recent OX$_2$R structure, was included from a set of ten constructed models (the models are available as Additional file 4). Overall, the main chains superimpose well among the models, and in retrospect also to OX$_2$R crystal structure, but some differences arise especially in the loops and in the TM6 (Figure 3, Additional file 5). The side chain conformations show more variance, but the difference in the backbone conformation is more significant to the docking, as the applied docking protocol is capable of adjusting the side chains but not the protein backbone.

Our original primary model, based on NTSR1, has a narrow cavity due to the inward tilt of the TM6 (volume of ca. 1400 Å^3, calculated with 3V-web server [49]). In retrospect, the overall shape and size of the cavity in the NTSR1-based model closely resembles that of the OX$_2$R-based model, which in turn is near-identical in

conformation to the OX$_2$R crystal structure (pairwise heavy atom RMSD 1.07 Å). The ECL2 of the NTSR1-based model adopts a β-hairpin structure similar to the OX$_2$R-based model, but the turn between the strands varies in conformation due to Modeller loop modeling (Figure 3A). The transmembrane bundle of the NTSR1-based model superimposes well to the OX$_2$R-based model, although the side-chain rotamers vary (Figure 3B,C). The heavy atom RMSD between NTSR1- and OX$_2$R-based models for binding-site-facing residues is 3.4 Å.

Location of the TM6 is a major difference between the OX$_2$R-based model and the NTSR1_TM6- and CXCR4-based models (Additional file 5). Due to the outward-leaning TM6, the binding cavities of these two models are more open and spacious (ca. 2000 Å^3). Also these models show the β-hairpin structure in the ECL2 (Additional file 5), and the TMs 2–5 superimpose well to the OX$_2$R-based model, again with varying side-chain rotamers. The different conformation of the TM6 in the NTSR1_TM6- and CXCR4-based models leads to poor superimposition of binding site residues over the OX$_2$R-based model. The TM7 is similarly located in all models, but in the CXCR4-based model the TM7 shows a counterclockwise rotation of ~50° around the helical axis in comparison to the other models, which slightly alters the set of residues that face the binding cavity. The heavy atom RMSD of binding-site-facing residues of the NTSR1_TM6- and CXCR4-based models in comparison to OX$_2$R-based model is 4.0 Å and 4.7 Å respectively.

Docking results

Docking into the OX$_2$R-based model produced 1099 docking poses, and to the NTSR1-based model 1164 docking poses. Secondary models based on CXCR4 and NTSR1_TM6 produced 1180 and 858 poses respectively. The poses were clustered into 53, 50, 68 and 48 clusters based on pairwise RMSD.

In the OX$_2$R-based model, the docking poses form a tight "bouquet" (Figure 4A, Additional file 6), with some poses leaning over to the TM5-side of the cavity. Top-scoring clusters occupy a tight space vertically in the middle of the binding cavity, again with some clusters leaning over to TM5 (Figure 4B). In the NTSR1-based model, the available space for the peptide ligand is fan-shaped (Figure 5A, Additional file 6), which is a result of the narrow interhelical cavity. The top-scoring clusters tend to show a vertical ligand orientation with C-terminus deep in the cavity (Figure 5B). Few clusters show poses higher and slanted towards TM1.

The more spacious binding cavities of the secondary models result in wider distributions of docked poses. The CXCR4-based model has a more open binding site, which leads to a wide bouquet-like distribution of

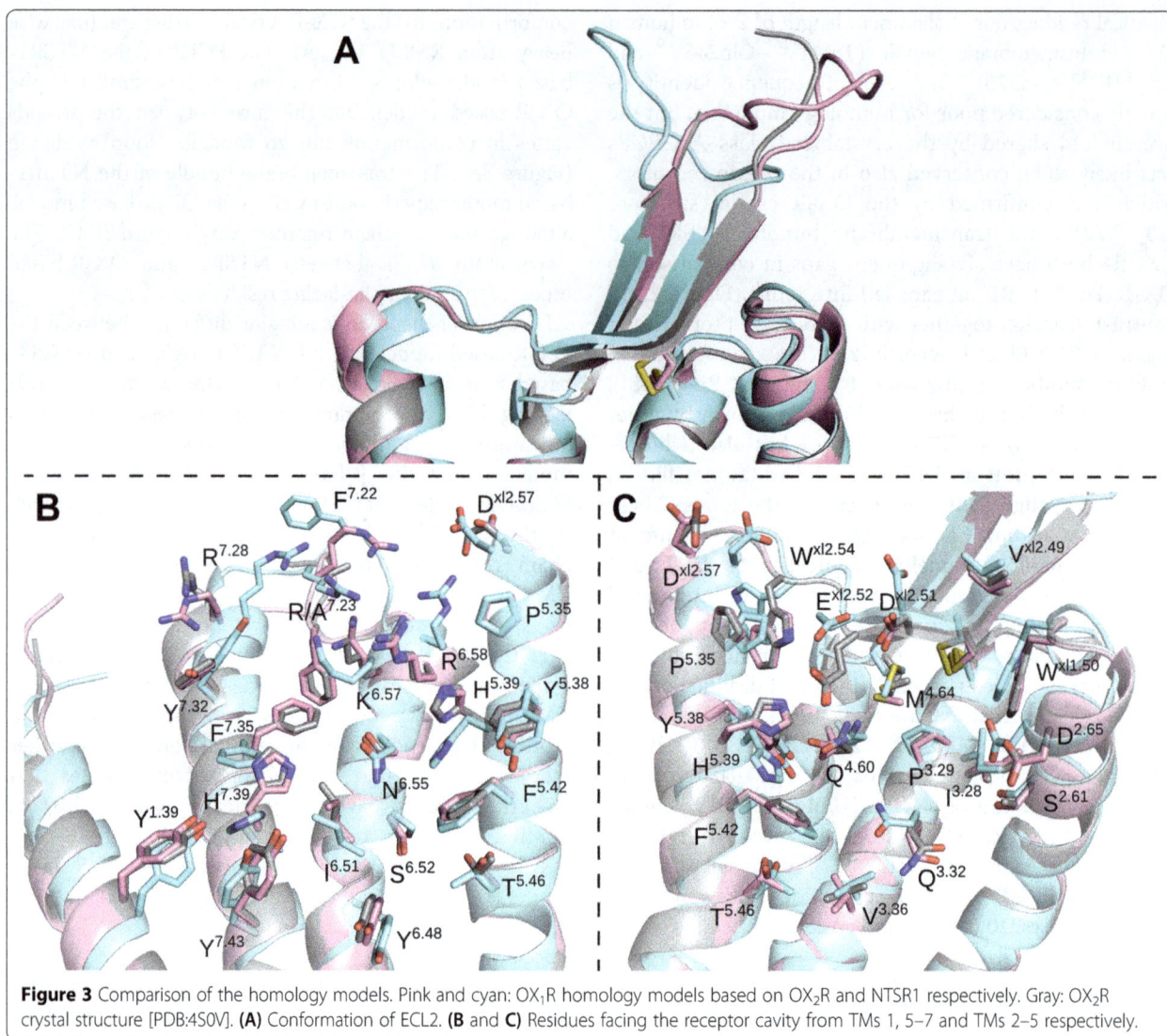

Figure 3 Comparison of the homology models. Pink and cyan: OX$_1$R homology models based on OX$_2$R and NTSR1 respectively. Gray: OX$_2$R crystal structure [PDB:4S0V]. **(A)** Conformation of ECL2. **(B** and **C)** Residues facing the receptor cavity from TMs 1, 5–7 and TMs 2–5 respectively.

docking poses (Additional file 6). The same distribution is seen with the ten largest and the top-ten-scoring clusters. However, top-ranking clusters reveal no preferences in ligand position. For the NTSR1_TM6-based model, the docking poses fall into two groups; the poses residing at the TM5-side of the cavity, and the poses leaning towards the TM1-side of the cavity (Additional file 6). The ten largest clusters show similar distribution, but the TM5-side and an upright orientation is favored by the top-ten-scoring clusters.

Scores for the individual top-scoring poses varied modelwise. The NTSR1-based model produced the highest scores (the best score -17.86), and 35 docking poses had RDOCK score of -10 or less. The OX$_2$R-based model produced 11 poses with RDOCK score below -10 (best score -14.47), whereas the NTSR1_TM6 and CXCR4-based secondary models have 8 and 6 poses with

scores < -10, best scores being -12.15 and -11.83 respectively. The narrow binding cavities of the NTSR1- and OX$_2$R-based models may enable the formation of more favorable interactions than the secondary models with more open binding cavities. The average docking pose scores show only minor differences (-1.50, -1.13 -0.68 and -1.05 for NTSR1-, OX$_2$R-, NTSR1_TM6- and CXCR4-based models respectively).

RDOCK has originally been designed to refine and score protein–protein complexes, not docked peptide ligands. The scoring, however, relies on calculated desolvation and electrostatic energies, so it should also be applicable to peptide docking. In our study, the connection between the 3D-location of the docking pose and the score can be seen in the score differences among clusters, and in the distribution of top-ranked 5 and 10% of the docking poses into clusters (Additional file 7).

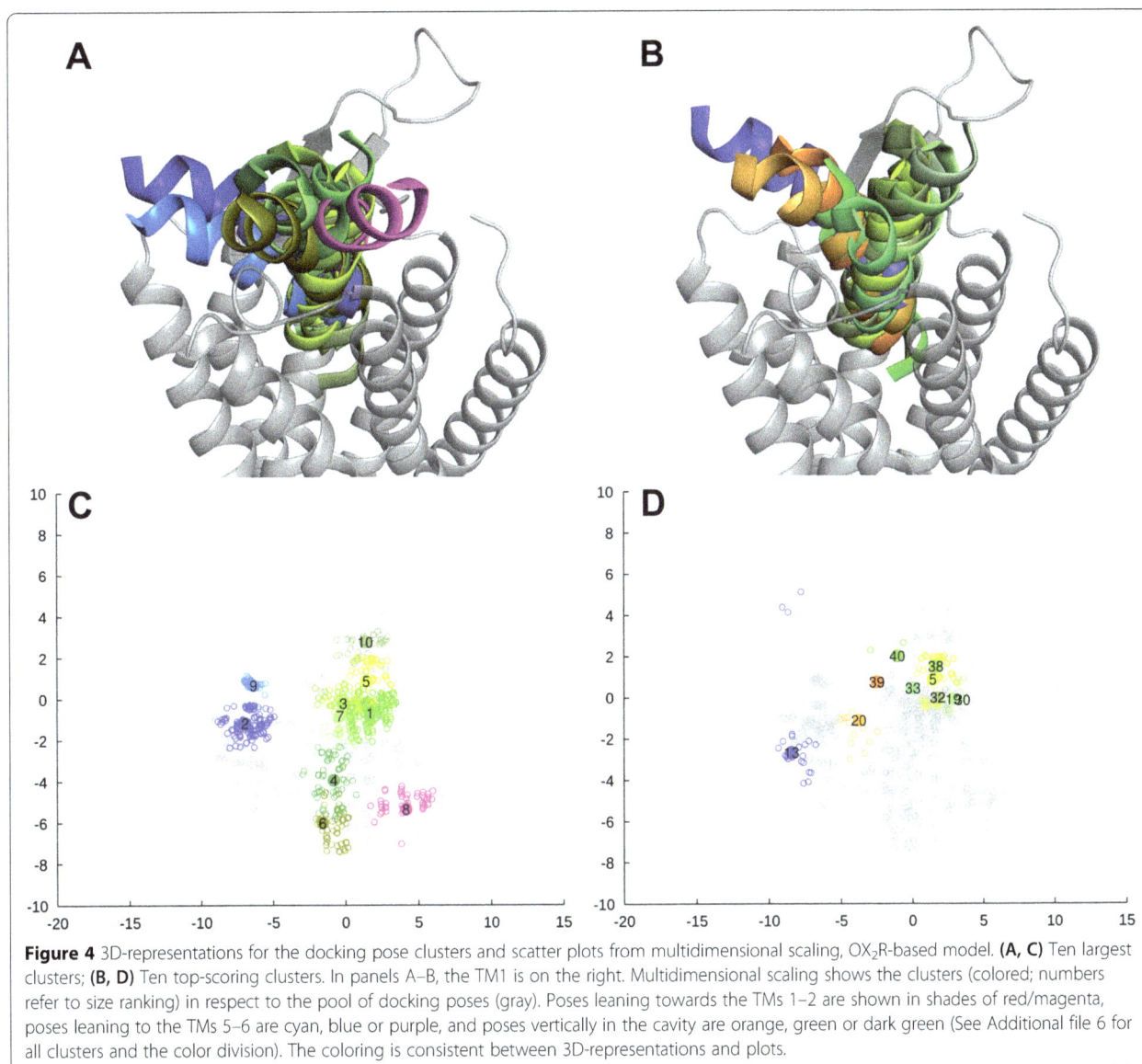

Figure 4 3D-representations for the docking pose clusters and scatter plots from multidimensional scaling, OX₂R-based model. **(A, C)** Ten largest clusters; **(B, D)** Ten top-scoring clusters. In panels A–B, the TM1 is on the right. Multidimensional scaling shows the clusters (colored; numbers refer to size ranking) in respect to the pool of docking poses (gray). Poses leaning towards the TMs 1–2 are shown in shades of red/magenta, poses leaning to the TMs 5–6 are cyan, blue or purple, and poses vertically in the cavity are orange, green or dark green (See Additional file 6 for all clusters and the color division). The coloring is consistent between 3D-representations and plots.

One-way variance analysis (ANOVA) shows that the differences between the cluster scores are statistically significant (data not shown). The scoring shows no bias towards deep ligand binding, as it appears to be uncorrelated with both the solvent accessibility of the peptide ligand and the depth of binding (Additional file 7). Therefore it appears reasonable to focus further analysis on the top-ranking individual poses.

Top-ranking poses

For each model, top-ranking poses were selected for closer examination. For the NTSR1-based model, the 35 docking poses that had RDOCK score of -10 or lower were used. As the top-scores for the other models were in general worse, a filter of RDOCK score < -8 was applied to

yield 29, 38 and 29 docking poses from the OX₂R-, CXCR4- and NTSR1_TM6-based models respectively. In all four models, the majority of top-ranking poses show the peptide ligand about vertically fairly deep in the binding cavity (Figure 6). The NTSR1-based model shows ligand depth of 3.7–9.9 Å (median 6.1 Å, zero-level at Tyr311$^{6.48}$ Cα, see Methods), whereas the OX₂R-based model favors deeper binding (median 5.0 Å, 2.8–14.7 Å). Regarding the secondary models, the best poses from the NTSR1_TM6-based model are more diverse, and depths range from 2.9 Å to 16.6 Å (median 5.5 Å). The best docking poses from the CXCR4-based model are a bit higher, 5.5–11.4 Å (median 8.8 Å).

The rotational orientation was assessed as the direction of bulky residues close to C-terminus (His26,

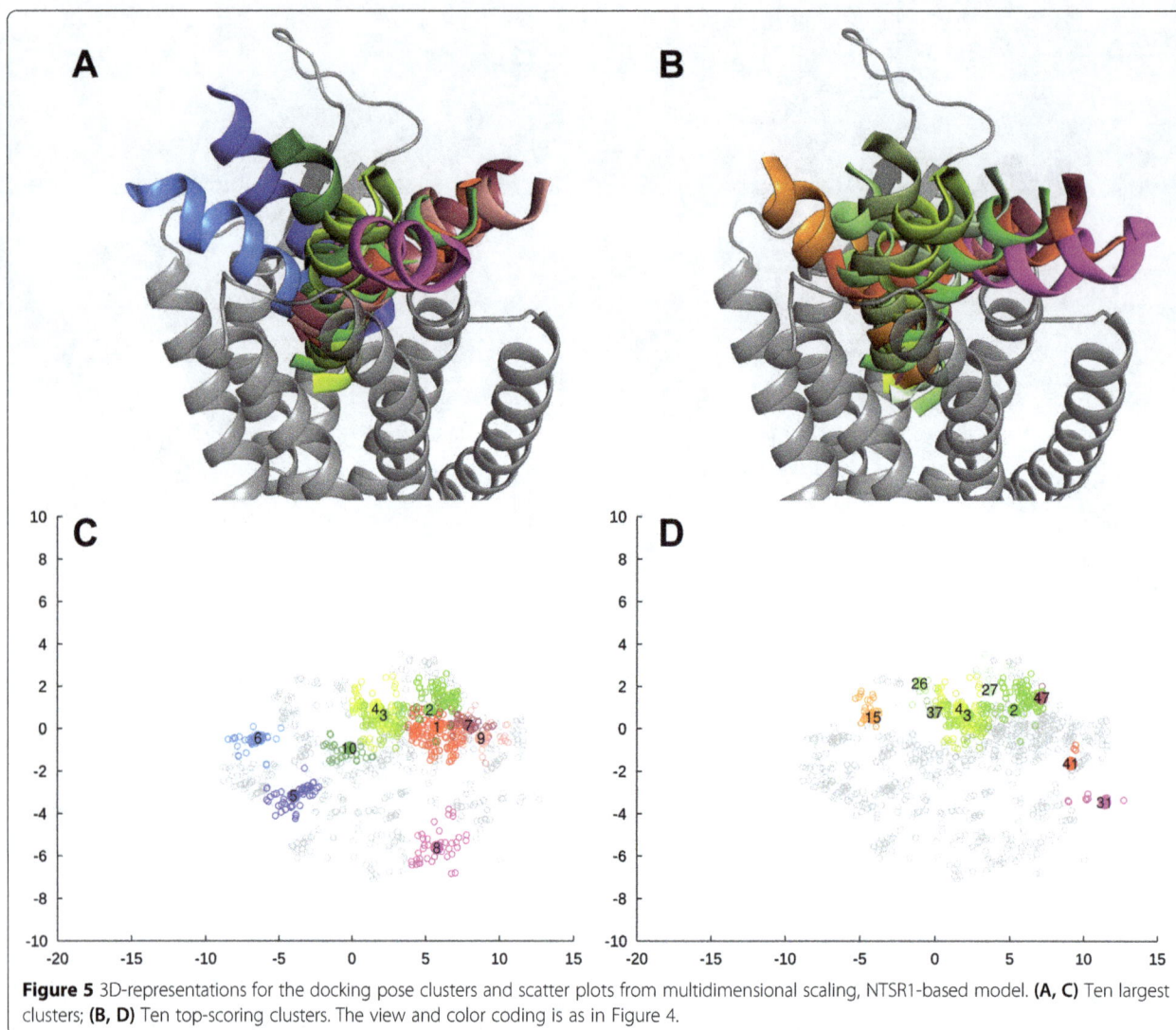

Figure 5 3D-representations for the docking pose clusters and scatter plots from multidimensional scaling, NTSR1-based model. **(A, C)** Ten largest clusters; **(B, D)** Ten top-scoring clusters. The view and color coding is as in Figure 4.

Leu30, and Leu33) in respect to the receptor (Figure 6). In each model, definite preferences are seen, although these preferences are not the same for all models. In the top-ranking poses from the NTSR1-based model and the NTSR1_TM6-based secondary model, the bulky residues face the TM5-side of the cavity in 70% of the poses. For the remaining top-scoring poses, the bulky residues face TM7. The preferences in the NTSR1_TM6-based model are not as strict as in the NTSR1-based model. The OX_2R-based model also shows these two groups of docking poses, but the preference is reversed; the majority of the top-ranking poses (69%) shows the bulky ligand residues facing the TM7, whereas the TM5-facing poses are a minority (24%). The docking poses in the CXCR4-based secondary model have a different preference, where the majority of the poses has the bulky residues on the TM1-side of the cavity, or facing towards TM2–3-side of the receptor. This difference in the preferred orientation is not surprising, given that the docking

poses in CXCR4-based model are in average ~2 Å closer to extracellular surface and thus have access to different areas of the binding cavity. This is likely caused by the bulky residues of TM7, especially His344[7.39], which in the CXCR4-based model face the cavity more prominently due to the 50° counterclockwise rotation of the TM7. These modelwise preferences in orientation are clearly mirrored in the mapping of contact frequencies between ligand and receptor residues (Additional file 8).

Two alternative binding modes

The top-ranking docking poses from the OX_2R- and NTSR1-based models were divided into two categories based on the peptide rotational state. The binding mode with TM5-facing bulky residues ("TM5-mode") was adopted by 31 poses (7 + 24 poses from the OX_2R- and NTSR1-based models respectively), whereas 30 poses (20 + 10) show the bulky residues towards the TM7 ("TM7-mode"). The OX_2R-based model shows two

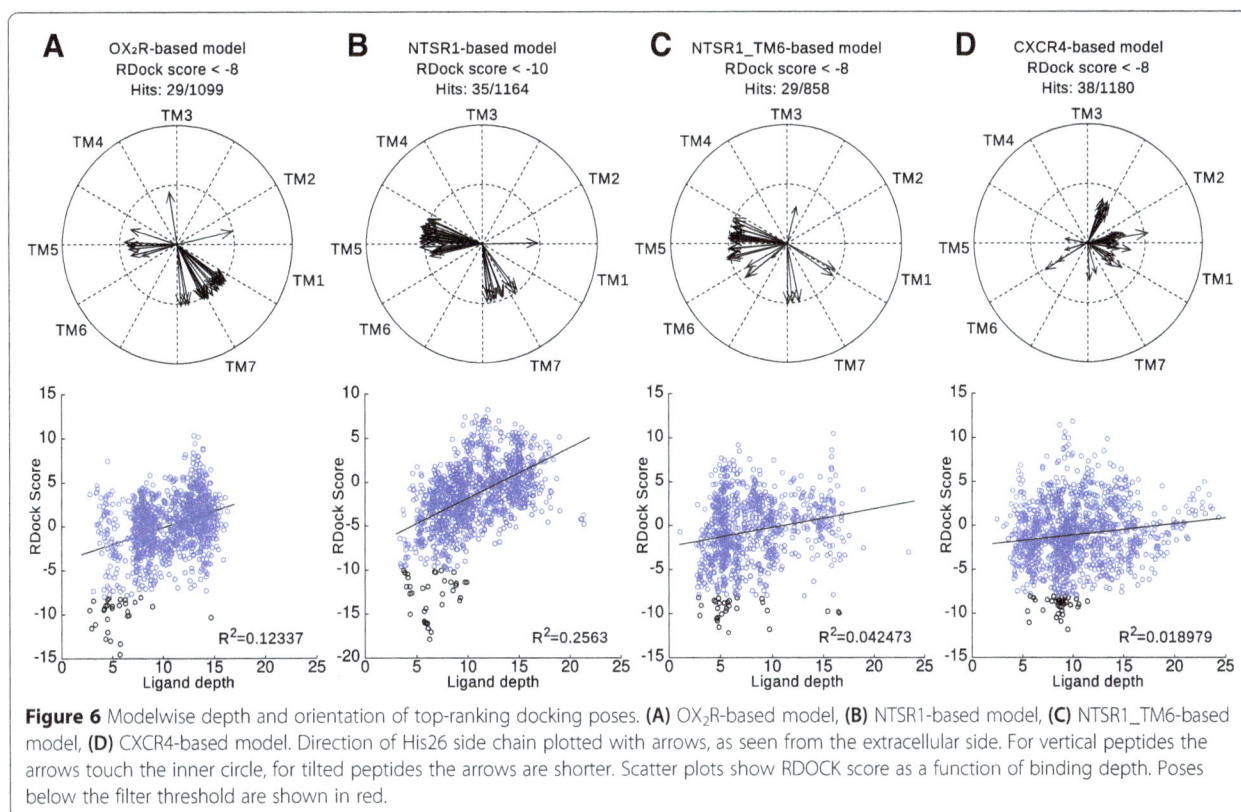

Figure 6 Modelwise depth and orientation of top-ranking docking poses. **(A)** OX$_2$R-based model, **(B)** NTSR1-based model, **(C)** NTSR1_TM6-based model, **(D)** CXCR4-based model. Direction of His26 side chain plotted with arrows, as seen from the extracellular side. For vertical peptides the arrows touch the inner circle, for tilted peptides the arrows are shorter. Scatter plots show RDOCK score as a function of binding depth. Poses below the filter threshold are shown in red.

outliers that do not fall into these two categories, while the NTSR1-based model shows one (Figure 6).

In both binding modes, the peptide C-terminus lies deep in the interhelical pocket, and forms reasonable interactions that take advantage of important amino acids (discussed in detail below). In the peptide N-terminus, the TM5-mode shows apparent better complementarity of hydrophobic and hydrophilic residues between the peptide and the receptor than the TM7-mode (Figure 7). Especially the hydrophobic amino acids in the peptide N-terminus (L16, L19, and L20) make a drastic difference between the binding modes. The TM5-mode shows these amino acids close to the ECL2 hairpin, partially shielded from the solvent (Figure 7B), whereas in the TM7-mode these amino acids are exposed to the solvent (Figure 7D). This exposure would remain the same with full orexin-A peptide as the disulfide-bridge-stabilized N-terminus lies on the opposite side of the peptide than the hydrophobic group of L16, L19 and L20 (Figure 7A). However, our models lack the receptor N-terminus, and both the conformation of the turn structure in the receptor ECL2 hairpin and the ligand N-terminus could be different. These factors could have extensive effect on the solvent exposure of hydrophobic ligand residues.

Both binding modes appear to be compatible with full-length orexin-A. The disulfide-bridge-stabilized N-terminus of orexin-A in the straight conformation would

be close above ECL3 in the TM5-mode, whereas in the TM7-mode it would be near the hairpin-turn of the ECL2. In contrast, the bent conformation, which is more frequently seen in the solution NMR-studies, would not fit these binding modes, as the peptide N-terminus would clash into TM7 or receptor N-terminus in the case of the TM5-mode and into the ECL2 for the TM7-mode.

Binding interactions

For both binding modes, a representative pose was selected to illustrate binding interaction at the atomic level (Figures 8 and 9). The interactions are summarized in Table 1. In general, orexin-A presents two large hydrophobic surfaces, one close to each terminus. The polar side chains, the peptide backbone at the flexible hinge region and, at the last helical turn, the exposed carbonyls and the amidated C-terminus offer sites for hydrogen bonding and electrostatic interactions.

We compared orexin-A C-terminal interactions to suvorexant binding in the OX$_2$R crystal structure [13]. Suvorexant binds deep in the cavity with multiple hydrophobic interactions, while the triazole ring is sandwiched within hydrogen-bonding distance between Gln$^{3.32}$ and Asn$^{6.55}$ and the amide carbonyl could hydrogen bond to Asn$^{6.55}$ and His$^{7.39}$ (water-mediated) (Figure 10B). This binding mode does not disturb intramolecular receptor

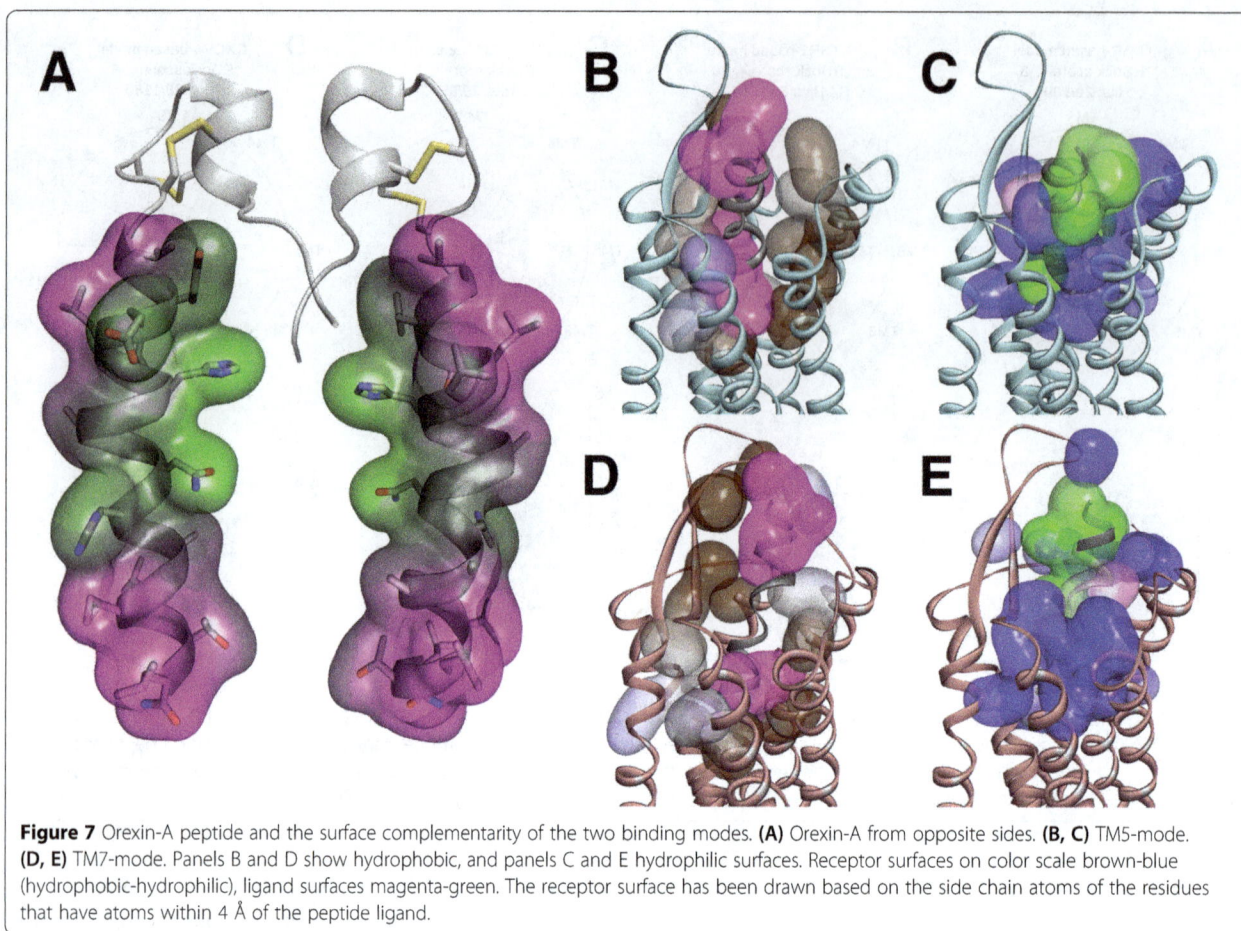

Figure 7 Orexin-A peptide and the surface complementarity of the two binding modes. **(A)** Orexin-A from opposite sides. **(B, C)** TM5-mode. **(D, E)** TM7-mode. Panels B and D show hydrophobic, and panels C and E hydrophilic surfaces. Receptor surfaces on color scale brown-blue (hydrophobic-hydrophilic), ligand surfaces magenta-green. The receptor surface has been drawn based on the side chain atoms of the residues that have atoms within 4 Å of the peptide ligand.

interactions lining the binding pocket, namely $Asp^{xl2.51}$–$Arg^{6.59}$, $Glu^{xl2.52}$–$His^{5.39}$ and $Asp^{2.65}$–$His^{7.39}$.

In our TM7-binding mode, the ligand C-terminus closely follows the hydrogen bonding of suvorexant to $Asn^{6.55}$ and $Gln^{3.32}$ (Figure 10A). The TM7-mode also features His26 close to receptor $His344^{7.39}$ and $Asp107^{2.65}$, which is especially interesting in the light of recent results suggesting that orexin-A binding to OX_1R is calcium-dependent [50], as histidine/aspartic acid clusters are known to participate in the hexadentate co-ordination of metal ions. The ligand Leu31 is close enough to break the $His216^{5.39}$–$Glu204^{xl2.52}$ salt bridge. Flexibility and small side chains at the peptide hinge region would permit hydrogen bonds from $Asp203^{xl2.51}$ and $Arg322^{6.59}$ to the peptide backbone. The phenolic oxygen in the $Tyr311^{6.48}$ lies 4.8 Å away from the C-terminal carbonyl, but could reach hydrogen bonding distance with a different rotamer. In other GPCRs, the corresponding $Trp^{6.48}$ is often thought as a key residue for receptor activation.

The TM5-mode, on the other hand, does not replicate the suvorexant hydrogen-bonding pattern (Figure 10C). However, it displays hydrogen bonding to receptor $Gln126^{3.32}$ (ligand T32 carbonyl or C-terminus) and the

peptide C-terminus often comes close enough to hydrogen bond to $Tyr311^{6.48}$ (roughly half of the top-ranking poses that adopt the TM5-mode show $Tyr311^{6.48}$ within 4 Å, but the representative pose in Figure 9 does not). The ligand His26 is close to $His216^{5.39}$ and $Glu204^{xl2.52}$, which could serve as a metal binding site as well, and again the hinge region offers hydrogen-bonding sites for $Asp203^{xl2.51}$ and $Arg322^{6.59}$.

These interactions are reminiscent of the activation determinants of other GPCRs. Adrenoceptors, for example, show an active state where the binding site contraction is stabilized by ligand binding between the transmembrane helices, namely hydrogen bonding to $Asp^{3.32}$, $Ser^{5.42}$, $Ser^{5.46}$ and $Asn^{7.39}$ [51-53]. However, even though our binding modes show the orexin peptide deeper than neurotensin in NTSR1, the peptide does not fully reach the depth of the adrenoceptor agonists. Contacts to $Phe219^{5.42}$ are formed by Leu31 in the TM7-mode, and Leu33 in the TM5-mode, but only in few poses within the more open binding cavity of the NTSR1_TM6-based model, the orexin peptide penetrates deep enough to bind to $Thr223^{5.46}$.

Closer to the extracellular surface, the orexin peptide forms interactions which are more like those seen

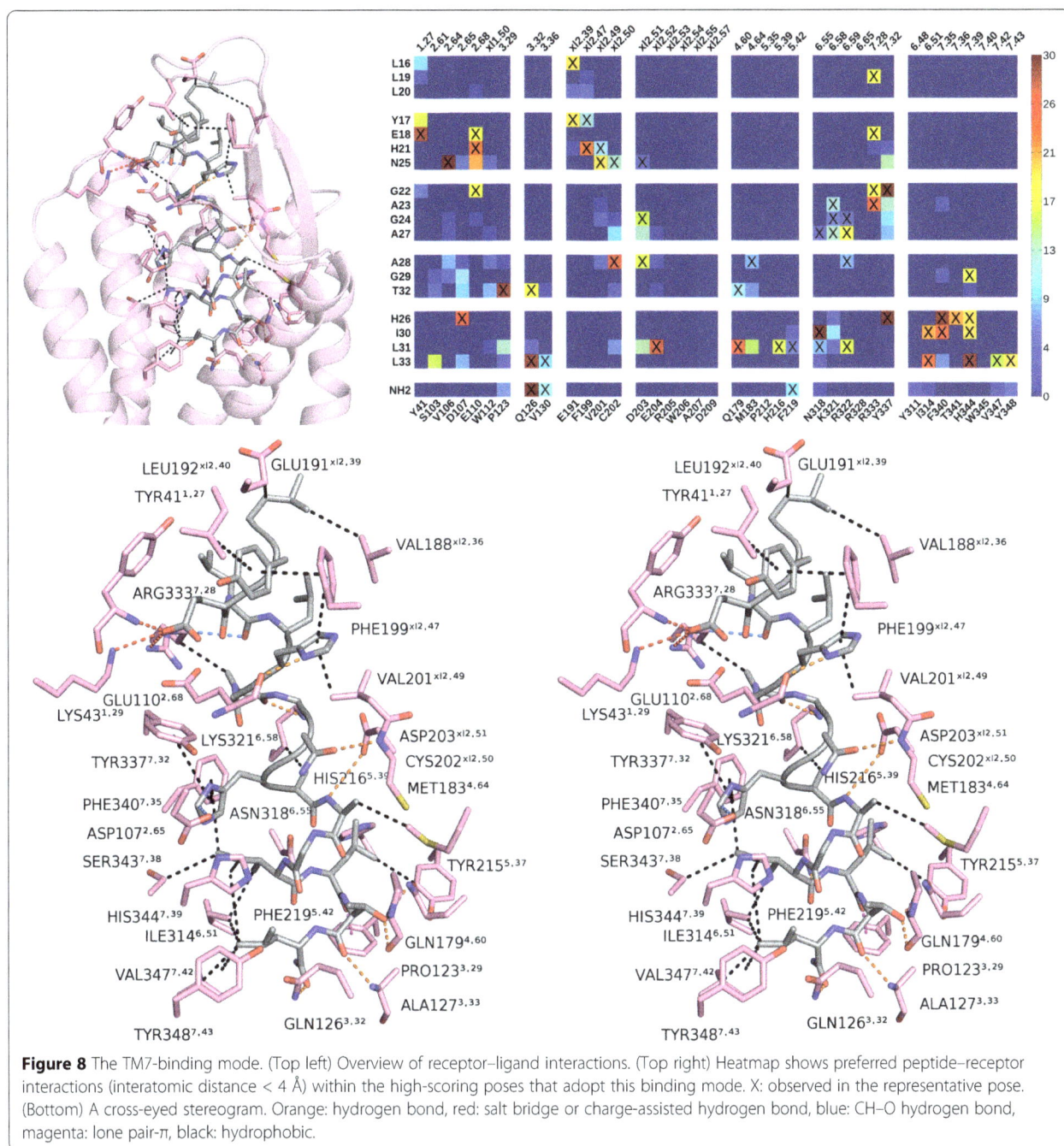

Figure 8 The TM7-binding mode. (Top left) Overview of receptor–ligand interactions. (Top right) Heatmap shows preferred peptide–receptor interactions (interatomic distance < 4 Å) within the high-scoring poses that adopt this binding mode. X: observed in the representative pose. (Bottom) A cross-eyed stereogram. Orange: hydrogen bond, red: salt bridge or charge-assisted hydrogen bond, blue: CH–O hydrogen bond, magenta: lone pair-π, black: hydrophobic.

between neurotensin$_{8-13}$ and NTSR1 [35]. Polar ligand residues in the TM7-mode interact with the receptor N-terminus and extracellular loops 1 and 3 (Glu18 with Lys43$^{1.29}$ and Arg333$^{7.28}$, His21 with Glu110$^{2.68}$, Asn25 with Cys202$^{xl2.50}$, and His26 with Asp107$^{2.65}$), whereas in the TM5-mode interaction to the ECL2 and ECL3 take place (His26 to Glu204$^{xl2.52}$, Asn25 to Lys321$^{6.58}$ and Tyr337$^{7.32}$, and His21 to Lys321$^{6.58}$ and to Arg333$^{7.28}$ or Arg328$^{6.65}$). These interactions put together could change the binding site conformation and result in the activation of the receptor.

Comparison to the receptor point mutations and neurotensin binding

The point-mutation studies on orexin receptors have indicated residues Gln126$^{3.32}$, Val130$^{3.36}$, Asp203$^{xl2.51}$, Trp206$^{xl2.54}$, Tyr215$^{5.38}$, Phe219$^{5.42}$, Thr223$^{5.46}$, Tyr224$^{5.47}$, Tyr311$^{6.48}$, Asn318$^{6.55}$, His344$^{7.39}$ and Tyr348$^{7.43}$ to be relevant for the orexin-peptide-triggered receptor activation (see Background). Of these residues, we already discussed Gln126$^{3.32}$, Asp203$^{xl2.51}$, Phe219$^{5.42}$, Thr223$^{5.46}$, Tyr311$^{6.48}$, Asn318$^{6.55}$ and His344$^{7.39}$ above. Concerning the remaining amino acids, the TM7 binding mode was

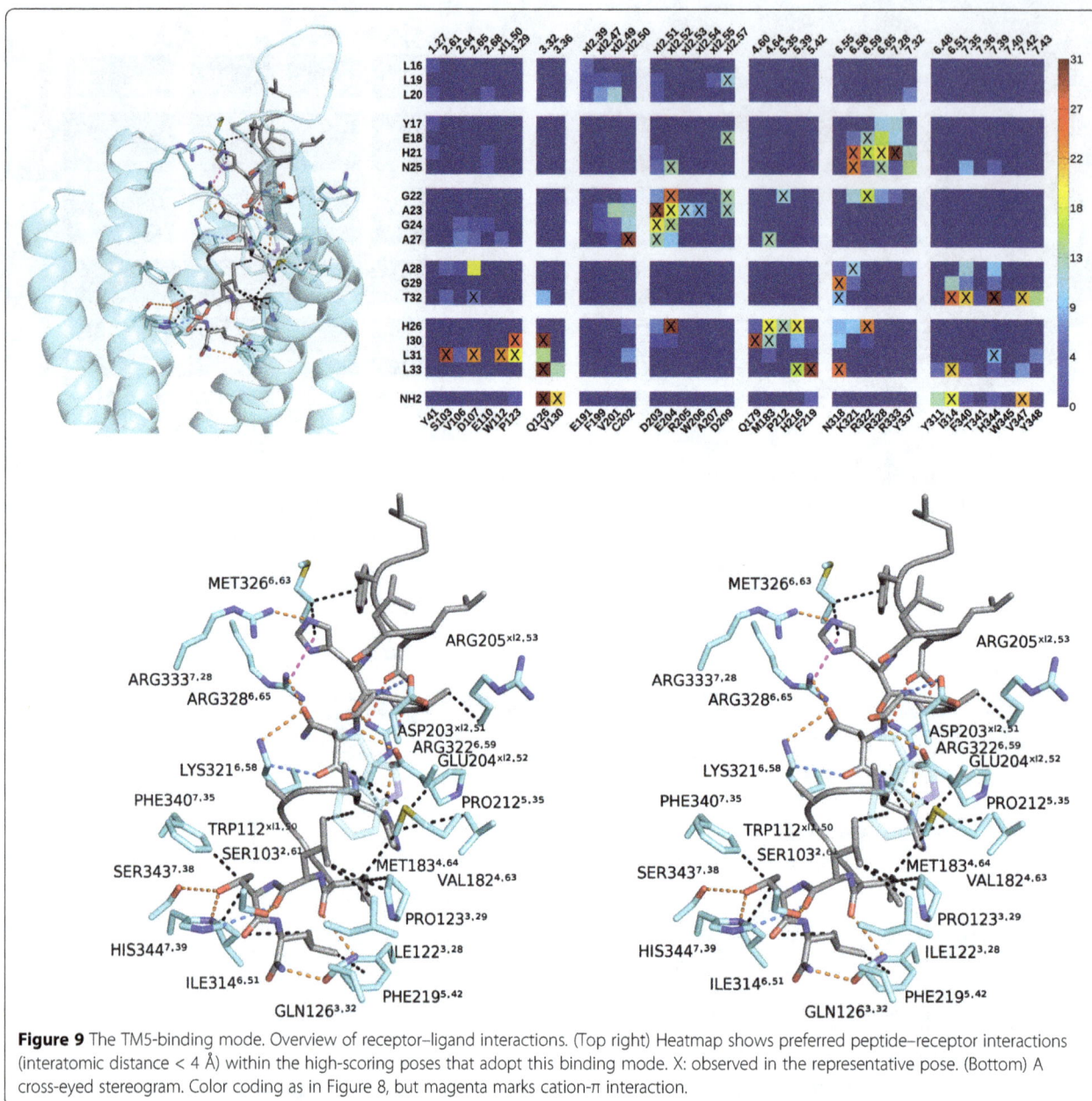

Figure 9 The TM5-binding mode. Overview of receptor–ligand interactions. (Top right) Heatmap shows preferred peptide–receptor interactions (interatomic distance < 4 Å) within the high-scoring poses that adopt this binding mode. X: observed in the representative pose. (Bottom) A cross-eyed stereogram. Color coding as in Figure 8, but magenta marks cation-π interaction.

found to interact also with Tyr215$^{5.38}$ and Tyr348$^{7.43}$ (Figure 8, Table 1). The side chain of Trp206$^{x12.54}$ lies between TMs 4 and 5, lining the binding cavity, but a different rotamer could bring the side chain closer to the ligand. The side chain of Tyr224$^{5.47}$ lies between TMs 5 and 6 in all models, where it is not exposed to the ligand. Val130$^{3.36}$ at the bottom of the binding cavity is often within 4 Å of the peptide ligand C-terminus. In total, the TM7-mode shows interactions to eight of these residues, whereas the TM5-mode interacts with five.

Residues at the bottom of the binding cavity (Val130$^{3.36}$, Thr223$^{5.46}$, Tyr311$^{6.48}$ and Tyr348$^{7.43}$) are difficult for the ligand to reach in our models. Water molecules are often seen to take part in ligand–receptor

interactions, but the applied docking protocol handles water implicitly, so water molecule mediated interactions cannot be addressed. It is also noteworthy that site-directed mutagenesis is an indirect method, and that the indicated residues might not take part directly in the ligand binding, but are part of the receptor activation cascade, or otherwise crucial for the receptor function.

The data from mutation studies was used to direct our docking efforts towards the binding site formed both by the cavity between TM helices and by the loops. This approach has proven effective in the case of neurotensin receptor 1, where extensive mutation and modeling studies had predicted that neurotensin would interact mainly with the ECL3 and upper parts of TMs 6 and 7

Table 1 Binding interactions of the two presented binding modes

Ligand residue	Interactions with receptor residues	
	TM7-mode	TM5-mode
Leu16	Alkyl - Val188$^{xl2.36}$, Glu191$^{xl2.39}$	-
Tyr17	Aromatic - Phe199$^{xl2.47}$	Alkyl-π - Met326$^{6.63}$
Glu18	Salt bridge - Arg333$^{7.28}$, Lys43$^{1.29}$	Salt bridge - Arg322$^{6.59}$
	H-bond - Tyr41$^{1.27}$ (backbone N)	
Leu19	CH–O to backbone - Arg333$^{7.28}$	-
Leu20	-	-
His21	Aromatic - Phe199xl$^{2.47}$	H-bond - Arg333$^{7.28}$
	Alkyl-π - Val201$^{xl2.49}$	Cation-π - Arg328$^{6.65}$
	H-bond - Glu110$^{2.68}$ (backbone carbonyl)	Alkyl-π - Met326$^{6.63}$
		H-bond to backbone - Arg322$^{6.59}$
Gly22	-	-
Ala23	Alkyl - Arg333$^{7.28}$	Alkyl - Arg205$^{xl2.53}$
	H-bond to backbone - Lys321$^{6.58}$	
Gly24	-	CH–O hydrogen bond - **Asp203$^{xl2.51}$**
Asn25	H-bond - Cys202$^{xl2.50}$ (backbone nitrogen)	H-bond - Arg328$^{6.65}$, Lys321$^{6.58}$ (putative)
		H-bond to backbone - Glu204$^{xl2.52}$, Lys321$^{6.58}$ (conventional or CH–O)
His26	Aromatic - Tyr337$^{7.32}$, Phe340$^{7.35}$	Alkyl-π - Arg322$^{6.59}$, Val182$^{4.63}$, Pro212$^{5.35}$
	CH–O hydrogen bond - Asp107$^{2.65}$	H-bond - Glu204$^{xl2.52}$
		CH–O hydrogen bond - Pro212$^{5.35}$ (carbonyl)
Ala27	Alkyl - Lys321$^{6.58}$	Alkyl - Met183$^{4.64}$
Ala28	Alkyl - Met183$^{4.64}$	-
	H-bond to backbone - **Asp203$^{xl2.51}$**	
Gly29	-	H-bond to backbone - **Asn318$^{6.55}$** (requires rotamer change)
Ile30	Alkyl-π - Phe340$^{7.35}$, **His344$^{7.39}$**	Alkyl - Pro123$^{3.29}$
	Alkyl - Ile314$^{6.51}$, Ser323$^{7.38}$	H-bond to backbone - **Gln126$^{3.32}$**
	H-bond to backbone - Asn318$^{6.55}$	
Leu31	Alkyl-π - **Tyr215$^{5.37}$**, His216$^{5.39}$	Alkyl-π - Trp112$^{xl1.50}$
	Lone pair-π from backbone - **Phe219$^{5.42}$**	Alky - Ile122$^{3.28}$, Pro123$^{3.29}$
		H-bond to backbone - Ser103$^{2.61}$
		CH–O hydrogen bond to backbone - **His344$^{7.39}$**
Thr32	H-bond - Gln179$^{4.60}$, Pro123$^{3.29}$ (backbone carbonyl)	Alkyl-π - Phe340$^{7.35}$, **His344$^{7.39}$**
		Alkyl - Ile314$^{6.51}$
		H-bond - Ser343$^{7.38}$, **His344$^{7.39}$** (either to threonine hydroxyl or backbone carbonyl)
Leu33	Alkyl-π - **His344$^{7.39}$**, Tyr348$^{7.43}$	Alkyl-π - **Phe219$^{5.42}$**
	Alkyl - Ile314$^{6.51}$, Val347$^{7.42}$	Alkyl - Ile314$^{6.51}$
	H-bond to backbone - **Gln126$^{3.32}$**	
NH2	H-bond - **Tyr311$^{6.48}$** (requires rotamer change)	Close to **Gln126$^{3.32}$** (unfavorable geometry for H-bond)

Interactions divided by type. Unless otherwise noted, the interacting atoms are side-chain atoms. "Requires rotamer change" denotes putative interactions which would take place if a receptor residue adopted a slightly different rotamer. Receptor residues whose mutation has been shown to be detrimental to orexin peptide binding are in bold.

[54,55]. The crystal structure was found to be well in line with this prediction [35]. It shows the neurotensin$_{8-13}$ binding in a way consistent with the mutation experiments, fairly high in the cavity (neurotensin$_{8-13}$ Leu13 Cα is ~12 Å above Tyr359$^{7.43}$ Cα). The orexin peptides, however, are considerably larger than

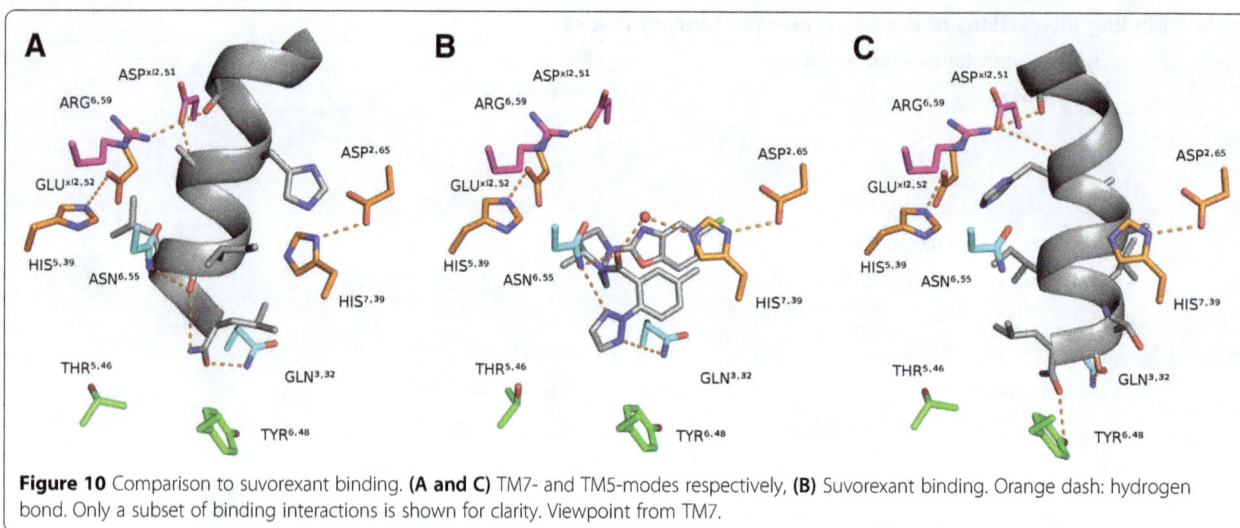

Figure 10 Comparison to suvorexant binding. **(A and C)** TM7- and TM5-modes respectively, **(B)** Suvorexant binding. Orange dash: hydrogen bond. Only a subset of binding interactions is shown for clarity. Viewpoint from TM7.

neurotensin, and mutational data suggests deeper binding and different interacting residues on the receptor (Figure 3). It is noteworthy that homologous binding-site-facing residues are often smaller in orexin receptors than in NTSR1, permitting the entry of a larger ligand. Changes such as $Tyr^{3.29}$ to proline, $Arg^{3.32}$ to glutamine, $Tyr^{6.51}$ to isoleucine, and $Arg^{6.55}$ to asparagine create a more spacious binding cavity in OX_1R than observed in NTSR1.

Unfortunately, the mutation studies on the orexin receptors are not as extensive as on the neurotensin receptor 1. Point mutations on orexin receptors have been focused on the residues deep in the cavity, with the exception of few ECL2 residues, whereas the ECL3 has so far been neglected. In addition, in the mutation studies only few residues are reported not to be important, making mutation based comparison and validation of binding modes more difficult.

Conclusions

In this work, we present two alternative binding modes for orexin-A to OX_1R, each with their own merits. The receptor models, based on the framework of the neurotensin receptor 1 and the orexin 2 receptor, which was published while this work was under consideration, provide accurate representations of the transmembrane bundle, and the conformation of the extracellular domain. Our docking protocol allows for side chain movements, which should smooth out small-scale inaccuracies in the conformation. The binding modes are consistent with what is known of GPCR activation in general, and fit well to the mutational data. The available mutation data only partially covers the predicted binding site, but we hope our work will direct further mutation studies, especially towards the ECL3. Due to the high sequence identity

between the orexin receptor subtypes and similarity of the peptide C-termini, these results should also be transferable to OX_2R and orexin-B. These alternative binding modes for the orexin-A into OX_1R, produced by computational modeling and docking, should benefit further characterization of orexin receptor interactions and therapeutic small molecule discovery.

Endnote

[a]Residues in the transmembrane helices are numbered according to Ballesteros and Weinstein [33]. The most conserved residue of each transmembrane helix is defined as N.50 where N is the ordinal number of the helix counting from the N-terminus. Residues in the ECL2 are numbered similarly so that the bridge-forming cysteine is designated as xl2.50 [56]. In addition, the structure-based residue numbering proposed by the GPCRDB is used when there are differences in the bulges or constrictions within the helices [48].

Additional files

Additional file 1: The conformations of orexin peptides in aqueous solution. Orexin-A has been reported with multiple conformations that fall into two categories; a bent conformation and a straight conformation. For orexin-B, one conformation has been reported.

Additional file 2: Structural alignment of crystallized class A GPCRs, and the sequence alignment of human OX_1R.

Additional file 3: The structural alignment from Additional file 2, in FASTA format. Chain breaks have not been annotated.

Additional file 4: PDB-file of the OX_1R homology models together with the representative docking poses for the models. 1: OX_2R-based model, 2: NTSR1-based model, 3: CXCR4-based model, 4: NTSR1_TM6-based model.

Additional file 5: Comparison of the secondary models to the OX_2R crystal structure. Corresponds to Figure 2, but shows the CXCR4- and NTSR1-based models.

Additional file 6: Additional figures on the docking pose clusters.
Additional figure 6.1: 3D-representations for all docking pose clusters modelwise. AF6.2, AF6.3: Docking pose clusters in the CXCR4- and NTSR1_TM6-based secondary models, corresponding to Figures 4 and 5. AF6.4: Superclustering across models.

Additional file 7: Statistics for the docking poses, modelwise. Five plots are shown for each model. 1) A box plot of RDOCK score vs. clusters; 2) Scatter plot of RDOCK score vs. ligand depth; 3) Scatter plot of RDOCK score vs. ligand solvent accessible surface area; 4) Scatter plot of ligand solvent accessible surface area vs. ligand depth; and 5) Distribution of high-scoring poses into clusters.

Additional file 8: Representative docking poses and interaction heatmaps, modelwise. Additional figures corresponding to Figures 8 and 9, but for all top-scoring docking poses for each model. Addition figure 8.1 shows the representative high-scoring pose for the OX₂R-based model, AF8.2 for the NTSR1-based model, AF8.3 for the CXCR4-based model, and AF8.4 for the NTSR1_TM6-based model.

Abbreviations
TM: Transmembrane segment; GPCR: G protein-coupled receptor; OX$_1$R: Orexin receptor 1; OX$_2$R: Orexin receptor 2; NTSR1: Neurotensin receptor 1; CXCR4: Chemokine receptor CXCR4; Cα: α carbon; ICL: Intracellular loop; ECL: Extracellular loop; PDB: (RCSB) Protein Data Bank; RMSD: Root-mean-square deviation.

Competing interests
The authors declare that they have no competing interests.

Authors' contributions
LK designed the study, carried out the modeling and docking, analyzed the data, prepared figures, and drafted the manuscript. AT assisted in the evaluation of models and in the analysis of data, and participated in the manuscript editing. HX participated in the design of the study, guided all the work, and assisted in finalizing the manuscript. All authors read and approved the final manuscript.

Acknowledgements
We acknowledge CSC – IT Center for Science Ltd. for providing computational resources, the Drug Discovery and Chemical Biology Consortium (DDCB), especially Dr. Leo Ghemtio, for maintaining the computational infrastructure in the laboratory, and the European COST Action CM1207 (GLISTEN) for organizing a European network for GPCR-researchers. We thank the National Doctoral Programme in Informational and Structural Biology (ISB) and the FinPharma Doctoral Program (FPDP) for organizing graduate studies. This work was supported by the Research Foundation of the University of Helsinki, the Finnish Cultural Foundation, the FinPharma Doctoral Program and the Orion Research Foundation.

References
1. Sakurai T, Amemiya A, Ishii M, Matsuzaki I, Chemelli RM, Tanaka H, et al. Orexins and orexin receptors: a family of hypothalamic neuropeptides and G protein-coupled receptors that regulate feeding behavior. Cell. 1998;92:573–85.
2. Sakurai T, Mieda M, Tsujino N. The orexin system: roles in sleep/wake regulation. Ann N Y Acad Sci. 2010;1200:149–61.
3. Laburthe M, Voisin T. The orexin receptor OX1R in colon cancer: a promising therapeutic target and a new paradigm in G protein-coupled receptor signalling through ITIMs. Br J Pharmacol. 2012;165:1678–87.
4. Scammell TE, Winrow CJ. Orexin receptors: pharmacology and therapeutic opportunities. Annu Rev Pharmacol Toxicol. 2011;51:243–66.
5. Haynes AC, Jackson B, Chapman H, Tadayyon M, Johns A, Porter RA, et al. A selective orexin-1 receptor antagonist reduces food consumption in male and female rats. Regul Pept. 2000;96:45–51.
6. De Lecea L, Kilduff TS, Peyron C, Gao X-B, Foye PE, Danielson PE, et al. The hypocretins: hypothalamus-specific peptides with neuroexcitatory activity. Proc Natl Acad Sci U S A. 1998;95:322–7.
7. Lee J-H, Bang E, Chae K-J, Kim J-Y, Lee DW, Lee W. Solution structure of a new hypothalamic neuropeptide, human hypocretin-2/orexin-B. Eur J Biochem. 1999;266:831–9.
8. Takai T, Takaya T, Nakano M, Akutsu H, Nakagawa A, Aimoto S, et al. Orexin-A is composed of a highly conserved C-terminal and a specific, hydrophilic N-terminal region, revealing the structural basis of specific recognition by the orexin-1 receptor. J Pept Sci. 2006;12:443–54.
9. Darker JG, Porter RA, Eggleston DS, Smart D, Brough SJ, Sabido-David C, et al. Structure-activity analysis of truncated orexin-A analogues at the orexin-1 receptor. Bioorg Med Chem Lett. 2001;11:737–40.
10. Ammoun S, Holmqvist T, Shariatmadari R, Oonk HB, Detheux M, Parmentier M, et al. Distinct recognition of OX1 and OX2 receptors by orexin peptides. J Pharmacol Exp Ther. 2003;305:507–14.
11. Lang M, Söll RM, Dürrenberger F, Dautzenberg FM, Beck-Sickinger AG. Structure-activity studies of orexin A and orexin B at the human orexin 1 and orexin 2 receptors led to orexin 2 receptor selective and orexin 1 receptor preferring. J Med Chem. 2004;47:1153–60.
12. Asahi S, Egashira S-I, Matsuda M, Iwaasa H, Kanatani A, Ohkubo M, et al. Development of an orexin-2 receptor selective agonist, [Ala (11), D-Leu (15)] orexin-B. Bioorg Med Chem Lett. 2003;13:111–3.
13. Yin J, Mobarec JC, Kolb P, Rosenbaum DM. Crystal structure of the human OX2 orexin receptor bound to the insomnia drug suvorexant. Nature. 2015;519:247–50.
14. Malherbe P, Roche O, Marcuz A, Kratzeisen C, Wettstein JG, Bissantz C. Mapping the binding pocket of dual antagonist almorexant to human orexin 1 and Orexin 2 receptors: comparison with the selective OX1 antagonist SB-674042 and the selective OX2 antagonist N-Ethyl-2-[(6-methoxy-pyridin-3-yl)-(toluene-2-sulfonyl)-amino]-N-py. Mol Pharmacol. 2010;78:81–93.
15. Tran D-T, Bonaventure P, Hack M, Mirzadegan T, Dvorak C, Letavic M, et al. Chimeric, mutant orexin receptors show key interactions between orexin receptors, peptides and antagonists. Eur J Pharmacol. 2011;667:120–8.
16. Putula J, Kukkonen JP. Mapping of the binding sites for the OX1 orexin receptor antagonist, SB-334867, using orexin/hypocretin receptor chimaeras. Neurosci Lett. 2012;506:111–5.
17. Michino M, Abola E, GPCR Dock 2008 participants, Brooks CL, Dixon JS, Moult J, et al. Community-wide assessment of GPCR structure modelling and ligand docking: GPCR Dock 2008. Nat Rev Drug Discov. 2009;8:455–63.
18. Kufareva I, Rueda M, Katritch V, GPCR Dock 2010 participants, Stevens RC, Abagyan R. Status of GPCR modeling and docking as reflected by community-wide GPCR Dock 2010 assessment. Structure. 2011;19:1108–26.
19. Kufareva I, Katritch V, Participants of GPCR Dock 2013, Stevens RC, Abagyan R. Advances in GPCR modeling evaluated by the GPCR dock 2013 assessment: meeting new challenges. Structure. 2014;22:1120–39.
20. Trellet M, Melquiond ASJ, Bonvin AMJJ. A unified conformational selection and induced fit approach to protein-peptide docking. PLoS One. 2013;8:e58769.
21. Raveh B, London N, Schueler-Furman O. Sub-angstrom modeling of complexes between flexible peptides and globular proteins. Proteins. 2010;78:2029–40.
22. Antes I. DynaDock: a new molecular dynamics-based algorithm for protein-peptide docking including receptor flexibility. Proteins. 2010;78:1084–104.
23. Prusis P, Schiöth HB, Muceniece R, Herzyk P, Afshar M, Hubbard RE, et al. Modeling of the three-dimensional structure of the human melanocortin 1 receptor, using an automated method and docking of a rigid cyclic melanocyte-stimulating hormone core peptide. J Mol Graph Model. 1997;15:307–17.
24. De Wachter R, De Graaf C, Keresztes A, Vandormael B, Ballet S, Rognan D, et al. Synthesis, biological evaluation, and automated docking of constrained analogues of the opioid peptide H-Dmt- D -Ala-Phe-Gly- 5-tetrahydro-2-benzazepin-3-one Scaffold. J Med Chem. 2011;54:6538–47.
25. Chandrashekaran IR, Rao GS, Cowsik SM. Molecular modeling of the peptide agonist-binding site in a neurokinin-2 receptor. J Chem Inf Model. 2009;49:1734–40.
26. Ganjiwale AD, Rao GS, Cowsik SM. Molecular modeling of neurokinin B and tachykinin NK$_3$ receptor complex. J Chem Inf Model. 2011;51:2932–8.
27. Matsoukas M-T, Potamitis C, Plotas P, Androutsou ME, Agelis G, Matsoukas J, et al. Insights into AT1 receptor activation through AngII binding studies. J Chem Inf Model. 2013;53:2798–811.
28. Heifetz A, Barker O, Morris GB, Law RJ, Slack M, Biggin PC. Toward an understanding of agonist binding to human Orexin-1 and Orexin-2 receptors with G-protein-coupled receptor modeling and site-directed mutagenesis. Biochem. 2013;52:8246–60.

29. Rodrigo J, Pena A, Murat B, Trueba M, Durroux T, Guillon G. Mapping the binding site of Arginine Vasopressin to V 1a and V 1b Vasopressin receptors. Mol Endocrinol. 2007;21:512–23.

30. Chen R, Li L, Weng Z. ZDOCK: an initial-stage protein-docking algorithm. Proteins. 2003;52:80–7.

31. Li L, Chen R, Weng Z. RDOCK: refinement of rigid-body protein docking predictions. Proteins. 2003;53:693–707.

32. Discovery Studio. Version 3.5. San Diego, CA: Accelrys Software Inc; 2012.

33. Ballesteros JA, Weinstein H. Integrated methods for the construction of three-dimensional models and computational probing of structure-function relations in G protein-coupled receptors. Methods Neurosci. 1995;25:366–428.

34. Wu B, Chien EYT, Mol CD, Fenalti G, Liu W, Katritch V, et al. Structures of the CXCR4 chemokine GPCR with small-molecule and cyclic peptide antagonists. Science. 2010;330:1066–71.

35. White JF, Noinaj N, Shibata Y, Love J, Kloss B, Xu F, et al. Structure of the agonist-bound neurotensin receptor. Nature. 2012;490:508–13.

36. Manglik A, Kruse AC, Kobilka TS, Thian FS, Mathiesen JM, Sunahara RK, et al. Crystal structure of the µ-opioid receptor bound to a morphinan antagonist. Nature. 2012;485:321–6.

37. Wu H, Wacker D, Mileni M, Katritch V, Han GW, Vardy E, et al. Structure of the human κ-opioid receptor in complex with JDTic. Nature. 2012;485:327–32.

38. Granier S, Manglik A, Kruse AC, Kobilka TS, Thian FS, Weis WI, et al. Structure of the δ-opioid receptor bound to naltrindole. Nature. 2012;485:400–4.

39. Thompson AA, Liu W, Chun E, Katritch V, Wu H, Vardy E, et al. Structure of the nociceptin/orphanin FQ receptor in complex with a peptide mimetic. Nature. 2012;485:395–9.

40. Fredriksson R, Lagerström MC, Lundin L-G, Schiöth HB. The G-protein-coupled receptors in the human genome form five main families: phylogenetic analysis, paralogon groups, and fingerprints. Mol Pharmacol. 2003;63:1256–72.

41. Maestro. Version 9.4. New York, NY: Shrödinger, LLC; 2013.

42. Chien EYT, Liu W, Zhao Q, Katritch V, Han GW, Hanson MA, et al. Structure of the human dopamine D3 receptor in complex with a D2/D3 selective antagonist. Science. 2010;330:1091–905.

43. Egloff P, Hillenbrand M, Klenk C, Batyuk A, Heine P, Balada S, et al. Structure of signaling-competent neurotensin receptor 1 obtained by directed evolution in Escherichia coli. Proc Natl Acad Sci U S A. 2014;111:E655–62.

44. Săli A, Blundell TL. Comparative protein modelling by satisfaction of spatial restraints. J Mol Biol. 1993;234:779–815.

45. Daura X, Gademann K, Jaun B, Seebach D, Van Gunsteren WF, Mark AE. Peptide folding: when simulation meets experiment. Angew Chemie Int Ed. 1999;38:236–40.

46. Matlab. Version R2013a. Natick, MA: MathWorks Inc; 2013.

47. Hubbard SJ, Thornton JM. Naccess. London, England: Department of Biochemistry and Molecular Biology, University College London; 1993.

48. Isberg V, de Graaf C, Bortolato A, Cherezov V, Katritch V, Marshall FH, et al. Generic GPCR residue numbers – aligning topology maps while minding the gaps. Trends Pharmacol Sci. 2015;36:22–31.

49. Voss NR, Gerstein M. 3V: cavity, channel and cleft volume calculator and extractor. Nucleic Acids Res. 2010;38(Web Server issue):W555–62.

50. Putula J, Pihlajamaa T, Kukkonen JP. Calcium affects OX1 orexin (hypocretin) receptor responses by modifying both orexin binding and the signal transduction machinery. Br J Pharmacol. 2014;171:5816–28.

51. Warne T, Moukhametzianov R, Baker JG, Nehme R, Edwards PC, Leslie AGW, et al. The structural basis for agonist and partial agonist action on a β1-adrenergic receptor. Nature. 2011;469:241–4.

52. Rasmussen SGF, Choi H-J, Fung JJ, Pardon E, Casarosa P, Chae PS, et al. Structure of a nanobody-stabilized active state of the β(2) adrenoceptor. Nature. 2011;469:175–80.

53. Xhaard H, Rantanen V-V, Nyronen T, Johnson MS. Molecular evolution of adrenoceptors and dopamine receptors: implications for the binding of catecholamines. J Med Chem. 2006;49:1706–19.

54. Kitabgi P. Functional domains of the subtype 1 neurotensin receptor (NTS1). Peptides. 2006;27:2461–8.

55. Härterich S, Koschatzky S, Einsiedel J, Gmeiner P. Novel insights into GPCR-peptide interactions: mutations in extracellular loop 1, ligand backbone methylations and molecular modeling of neurotensin receptor 1. Bioorg Med Chem. 2008;16:9359–68.

56. Xhaard H, Nyrönen T, Rantanen V-V, Ruuskanen JO, Laurila J, Salminen T, et al. Model structures of α2-adrenoceptors in complex with automatically docked antagonist ligands raise the possibility of interactions dissimilar from agonist ligands. J Struct Biol. 2005;150:126–43.

57. Barton GJ. ALSCRIPT: a tool to format multiple sequence alignments. Protein Eng. 1993;6:37–40.

15

DynaDom: structure-based prediction of T cell receptor inter-domain and T cell receptor-peptide-MHC (class I) association angles

Thomas Hoffmann, Antoine Marion and Iris Antes[*]

Abstract

Background: T cell receptor (TCR) molecules are involved in the adaptive immune response as they distinguish between self- and foreign-peptides, presented in major histocompatibility complex molecules (pMHC). Former studies showed that the association angles of the TCR variable domains (Vα/Vβ) can differ significantly and change upon binding to the pMHC complex. These changes can be described as a rotation of the domains around a general Center of Rotation, characterized by the interaction of two highly conserved glutamine residues.

Methods: We developed a computational method, DynaDom, for the prediction of TCR Vα/Vβ inter-domain and TCR/pMHC orientations in TCRpMHC complexes, which allows predicting the orientation of multiple protein-domains. In addition, we implemented a new approach to predict the correct orientation of the carboxamide endgroups in glutamine and asparagine residues, which can also be used as an external, independent tool.

Results: The approach was evaluated for the remodeling of 75 and 53 experimental structures of TCR and TCRpMHC (class I) complexes, respectively. We show that the DynaDom method predicts the correct orientation of the TCR Vα/Vβ angles in 96 and 89% of the cases, for the poses with the best RMSD and best interaction energy, respectively. For the concurrent prediction of the TCR Vα/Vβ and pMHC orientations, the respective rates reached 74 and 72%. Through an exhaustive analysis, we could show that the pMHC placement can be further improved by a straightforward, yet very time intensive extension of the current approach.

Conclusions: The results obtained in the present remodeling study prove the suitability of our approach for interdomain-angle optimization. In addition, the high prediction rate obtained specifically for the energetically highest ranked poses further demonstrates that our method is a powerful candidate for blind prediction. Therefore it should be well suited as part of any accurate atomistic modeling pipeline for TCRpMHC complexes and potentially other large molecular assemblies.

Keywords: T-cell recognition, TCR structural modeling, Epitope prediction, Glutamine side chain prediction, Protein domain association angles, Immunoinformatics, Adoptive T-cell therapy, Vaccine design

* Correspondence: antes@tum.de
Department of Biosciences and Center for Integrated Protein Science
Munich, Technische Universität München, Emil-Erlenmeyer-Forum 8, 85354
Freising, Germany

Background

An early event in the T cell mediated immune response is the recognition of pathogenic peptides contained in major histocompatibility complex (MHC) molecules. The capability of the vertebrate immune system to distinguish between a vast variety of pathogenic- and self-peptides is achieved by a tremendous population of different T cell variants (i.e., in a magnitude estimated from 10^6 to 10^7), which differ from each other in the T cell receptor (TCR) [1–3]. Such a diversity results from the combination of two membrane anchored TCR chains (α and β), which are encoded by gene segments joined in a process known as v(d)j recombination [4]. As depicted in Fig. 1, each chain consists of two immunoglobulin-like domains, the variable domain (further referred to as

Vα and Vβ) and the constant domain. The v(d)j combination process occurs during the T cell maturation in the thymus, where variable (v) and joining (j) gene segments are combined while nucleotides are randomly introduced within the variable domains (V). In the case of the Vβ domain, an additional short segment is inserted in between the v and j segments, further increasing the TCR diversity (d). The binding interface of the TCR to the peptide-MHC molecule complex (pMHC) is formed by loops named as complementary determining regions (CDR), and each chain of TCR contains three CDRs. While the primary structure of CDR1 and CDR2 loops evolved together with the MHC molecules [5], the sequence of CDR3 loops is determined by the v(d)j recombination and thus exhibits a higher diversity [6].

The number of resolved bound and unbound TCR structures has drastically increased to 200 in the Protein Data Bank [7] during the past few years. Nevertheless, considering the vast variety of TCRs and the high polymorphism of the MHC molecules, the development of reliable structural methods is of crucial importance in order to complement time consuming experimental structural techniques [8]. Such modeling approaches can help in the field of rational TCR design/optimization (e.g., adoptive T cell cancer therapy) [9, 10], in the context of vaccine design [11, 12], and in the development of a consistent theory for T cell signal transduction, which is still not fully understood [13].

Over the past two decades, many theoretical methodologies have been developed and applied to model and predict TCRpMHC interactions.

The main focus in the area has been on the prediction of the MHC/peptide interaction without explicit consideration of the T-cell receptor as the experimental study of MHC-peptide binding has been a very active field since the mid-90s whereas the systematic investigation of the T cell response started about a decade later. In addition, MHC-peptide binding is a necessary prerequisite for the T cell response and thus has by itself already a highly predictive value. Therefore various sequence and structure based prediction tools have been developed of MHC-peptide binding in the past decades [14, 15]. Next to MHC-peptide specific structure-based prediction methods such as EpiDock, PREDEP, pDOCK, DynaPred, or DockTope [16–20], also general molecular docking approaches were applied [21, 22].

The first atomistic model of a TCRpMHC complex was built in 1995 by Almagro et al. using homology modeling and molecular dynamics techniques [23], before the first X-ray structures of a TCR (1tcr [24]) and of a TCRpMHC complex (1ao7 [25]) were solved in 1996. Later, Michielin et al. realized a homology model of the T1 TCR structure bound to the photoreactive PbSC peptide and to the murin K^d MHC class I molecule, using

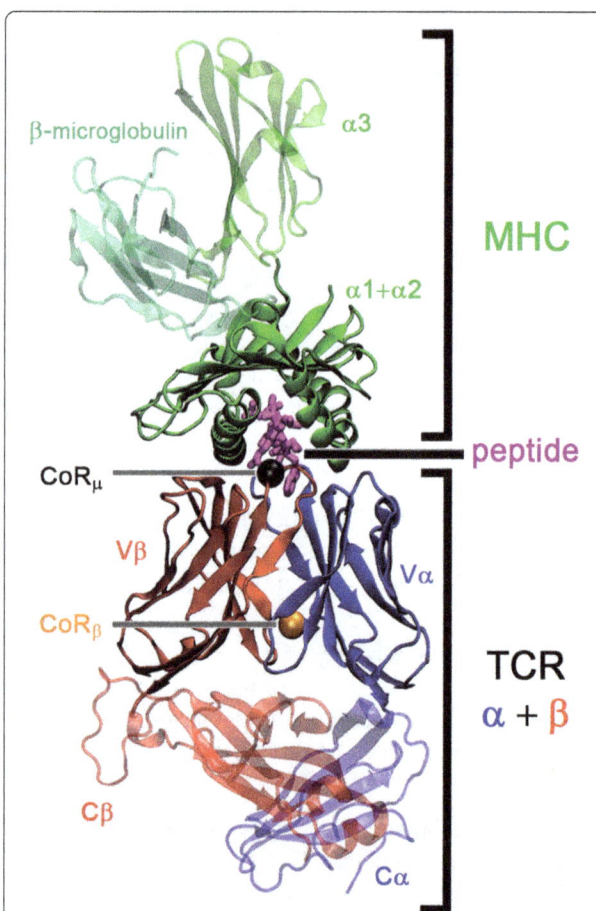

Fig. 1 Representation of the TCRpMHC complex (PDB-ID 2bnq). The MHC class I molecule is depicted in *green* (i.e., α1, α2, and α3 chains). The β-microglobulin is colored in *cyan* and the peptide bound to MHC in *magenta*. The two chains of TCR, α and β, are represented in *blue* and *red* colors, respectively. In the present application, the domains shown as transparent are removed from the structure, and only the two variable domains of TCR (i.e., Vα and Vβ), the α1 and α2 chains of MHC as well as the peptide are modeled. In addition, the two centers of rotations $CoR_β$ and $CoR_μ$ are respectively represented by an *orange* and a *black* colored ball

the 1ao7 crystal structure of the TCRpMHC complex as a template [8]. The authors applied a methodology combining the MODELLER program with simulated annealing techniques [26], and suggested a rational homology model, which was refined based on previous mutation studies [27]. Further developments of the approach led to the TCRep 3D method [28], which was recently applied in the context of rational TCR design [10]. In addition, Haidar et al. enhanced the affinity of the A6 TCR to TAX:HLA-A2 for about 100-fold using a structure-based model [29]. More recently, Pierce et al. [30] developed an approach based on their scoring function ZAFFI and on the Rosetta interface mutagenesis tool [31] to identify relevant point mutations that could increase the affinity of a TCR to a pMHC complex in the field of therapeutic immunology. The method allowed to optimize the DMF5 TCR to bind the ELAGI-GILTV:HLA-A2 complex with a remarkable ~400-fold higher affinity. The same group also developed TCRFlex-Dock, a method to model a pMHC ligand onto a TCR that takes advantage of the Monte Carlo-based Rosetta-Dock protocol [32, 33]. For a benchmark test set of twenty structures [33], the prediction of near native models was reached in 80% of the cases. The TCRFlex-Dock method was recently applied to predict models of TCRs bound to MHC like ligands such as CD1 and MR1 [34]. In that work, the authors showed that the use of multiple docking starting positions significantly improves the performance of the prediction.

In order to achieve an accurate molecular model of TCRpMHC complexes, it is necessary to consider several topological aspects of this sophisticated system. First, a precise description on an atomistic level is required, since small alterations in the TCR's or in the ligand's sequence can drastically affect the transduced signal [35]. As it was shown in other studies, mutations in the receptor or in the ligand can modify the binding affinity and thus the relative placement of the two units of the complex [36–39].

A second aspect to consider is the variation of the Vα/Vβ inter-domain angle within the TCR, as this is a system specific feature, and as it can adapt upon binding of the pMHC. The analysis of the inter-domain angle between the Vα and Vβ TCR domains as well as its influence on the binding of pMHC was analyzed in several computational studies, which compared broad sets of TCR structures. Notably, by applying the pseudo-dyad method, McBeth et al. suggested that the resulting observed differences between the free and the MHC bound forms of TCRs constitute a feature of the receptor to adapt to different ligands, thus allowing cross reactivity [40]. Dunbar et al. analyzed a non-redundant set of TCRs with the ABangle methodology [41], which describes both the Vα- and the Vβ-orientation in an

absolute manner, by considering a torsion angle, four bend angles and one distance as descriptors [42]. In the context of rational TCR-like antibody design, the authors found that antibodies adopt angles comprised in a different range than the one observed for TCRs. In our previous work [43], we analyzed the relative Vα- and Vβ-orientation by reducing the variable domains to cuboids, which served as basis for a distance based clustering. We observed that TCRs belonging to the same clonotype associate in the same angular cluster. Furthermore, we identified a Center of Rotation (further referred to as CoR_β and depicted in Fig. 1) and determined its location in the middle of a conserved interaction between two glutamine residues, one in the Vα and one in the Vβ domain. The various inter-domain angles in the evaluation set could be obtained through a rotation around this center. Recent studies, including ours, further emphasized the large range of values that the TCR Vα/Vβ inter-domain angle can adopt [40, 43, 44] and thus its influence on the positioning of the ligand binding CDR loops. These results suggest that next to the orientation of the pMHC ligand with respect to the TCR [24, 25, 36, 45, 46], also the Vα/Vβ inter-domain angle should be explicitly taken in account to assess an accurate homology modeling of TCRpMHC complexes. This last comment is in agreement with recent observations about the dynamics of the TCRpMHC system and the influence of the TCR on the pMHC structure [44, 47]. In addition, it was shown for antibodies that the consideration of the V_H/V_L angles for homology modeling can increase the accuracy considerably [41, 48, 49]. In this context, Dunbar et al. identified key structural parameters, which provide a comprehensive description of the movement of the V_H and V_L domains with respect to each other [41]. Based on these features and on their respective values in the available antibody structures, Bujotzek et al. trained a predictor for the association of the two domains [48]. The authors further concluded that the consideration of the association angles is crucial for the prediction of highly accurate homology models of antibodies [49].

Along the course of the present study, we pointed out a third topological aspect that can have an impact on the success of TCRpMHC complexes modeling. The Vα/Vβ orientation directly depends on the proper interaction of two specific glutamine residues. During protein structure elucidation by X-ray crystallography, the ambiguous electron densities of nitrogen and oxygen atoms can hamper the correct assignment of these two elements. In the case of asparagine and glutamine residues, this often leads to misassigned atoms in the carboxamide group of the side chain. The detailed investigation of high-resolution structures shows that approximately 20% of these residues are assigned in a wrong flip state, leading to a non-optimal hydrogen bond network [50–

53]. The respective orientation of asparagine and glutamine residues has a dramatic impact on most of molecular modeling techniques [53], and should be corrected by considering their direct environment. Due to this significance, several approaches have been developed in order to address this issue. Among those, the most popular ones are HBPLUS (X-PLOR package) [52], NETWORK (WHAT IF package) [53], Reduce (Mol-Probity package) [50, 54, 55], NQ Flipper [51, 56, 57], the Independent Cluster Decomposition Algorithm (ICDA) [58], Protonate 3D [59], Protoss [60, 61], and the Computational Titration method [62].

Despite the great improvements in TCRpMHC complex modeling achieved during the past decades, some of the critical aspects described above are still not taken into account. To the best of our knowledge, none of the currently available methodologies explicitly include the adaption of the Vα/Vβ inter-domain angles, although these have a direct impact on the disposition of the CDR loops, and as a consequence, on the contact between the TCR and the pMHC ligand.

In what follows, we present a new method, DynaDom, for the prediction of TCR Vα/Vβ inter-domain and TCR/pMHC association angles. We implemented our approach into the DynaCell suite [63], a general force-field-based molecular modeling program developed in our group. Our new method uses an extendable multidimensional rigid body optimization approach based on the work by Mirzaei et al. [64]. The implementation is specifically designed in a way that allows for an arbitrary definition of rigid bodies and for the inclusion of local flexibility on different levels (e.g., from the domain to the residue level) into the modeling pipeline (Fig. 2). As a first application, we evaluate here the DynaDom method for the remodeling of a large set of TCR and TCRpMHC complexes. This evaluation intends to determine the general capability of a rotation-based algorithm and the relevance of our CoR-concept [43] for the successful prediction of association angles. Notably, we demonstrate here that it is possible to distinguish between correct and wrong models by solely using the force-field-based interaction energy computed between the different units of the complex. This is indeed very promising for future blind homology modeling of TCRpMHC complexes and others, especially if a sufficient amount of experimental data is not available for the training of an application-specific, knowledge–based scoring function.

Methods

The new DynaDom prediction method presented here is based on the concepts developed for our previous analysis of the structural features of TCRpMHC complexes [43] and uses the same theoretical framework as defined therein. In ref. [43], we performed a comprehensive and systematic analysis of the Vα/Vβ inter-domain angles in a set of 85 structures, by representing each domain as a unified cuboid (for a brief summary of the methodology, see Additional file 1: Text S1A). The main results of that former work can be summarized as follows: i) we showed that the TCR complexes of the analysis set can be grouped into six structural clusters, by solely using the Vα/Vβ inter-domain angle as a descriptor; ii) we identified a conserved center of rotation that determines the orientation of the Vβ domain with respect to Vα (further referred to as CoR$_\beta$); iii) we pointed out that this center of rotation (CoR$_\beta$) is characterized by the interaction of two highly conserved glutamine residues (Q; one per variable domain), forming a stable hydrogen bond linkage between Vα and Vβ.

In the present work, we intend to translate this structural knowledge gained in ref. [43] into a computational pipeline to model TCR and TCRpMHC complexes. We hereafter detail our strategy by first describing the general concepts of DynaDom and the extension of the center of rotation concept to the case of TCRpMHC complexes. Then, we describe the theoretical framework of our rigid body optimization algorithm and give a detailed description of the overall prediction pipeline. Finally, we define the particular data set used in the present test application of DynaDom for the remodeling of TCR and TCRpMHC complexes.

General concepts of DynaDom

The DynaDom modeling approach is based on the unified cuboid description of a given molecular assembly by assigning one cuboid to each structural domain, as applied to the TCR Vα and Vβ domains in ref. [43]. In the case of TCRpMHC complexes the Vα, Vβ, and pMHC units are represented by three independent cuboids. To further reduce the dimensionality of the problem, we use the Vα domain as internal, fixed coordinate frame (see ref. [43], Fig. 2, and Text S1.A of Additional file 1 for details). In this frame, the placement of the Vβ and pMHC units can be simply described by a series of translation and rotation operations around a given point of the cuboid. For the placement of Vβ with respect to Vα, we here chose this point as the previously identified center of rotation, CoR$_\beta$ (Fig. 1). In a similar manner, we define here a center of rotation for the placement of pMHC units, which we shall refer to as CoR$_\mu$.

The binding of the pMHC ligands onto TCRs has been suggested to occur in a generally diagonal mode [46], based on the analysis of the early structures of this complex [24, 25, 36]. More recently, Rudolph et al. introduced a general unified method to measure the binding angle of TCRs with respect to their ligands and determined the angular range of 24 complexes [45]. The latter method,

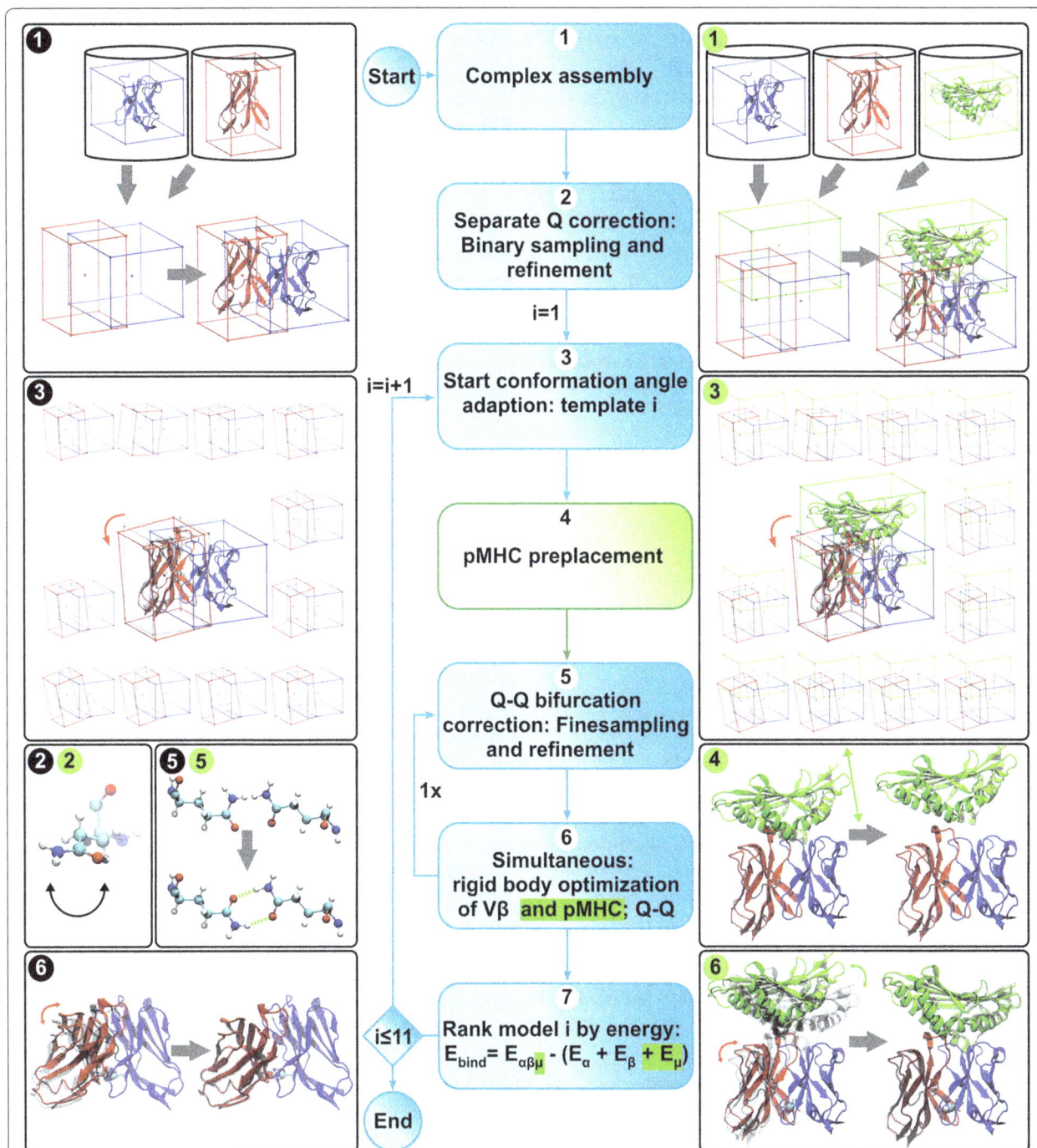

Fig. 2 TCR and TCRpMHC complexes modeling pipeline. Center column: standard pipeline (see Methods) for the remodeling of the TCR Vα/Vβ association angles and for the pMHC positioning with respect to the TCR. *Blue* highlighted steps are performed in both modeling pipelines: only Vβ and combined Vβ/pMHC placement. *Green* highlighted steps are performed only if the pMHC is included in the remodeling process. The left and the right columns illustrate the individual steps of the pipeline. Steps with numbers circled in *black*: TCR Vα/Vβ association angle modeling pipeline, steps with numbers circled in *green*: combined Vβ/pMHC modeling pipeline. Steps 3 to 7 are performed for each of the 11 starting conformations. The protein domains represented in *blue*, *red*, and *green* color correspond to the Vα, Vβ, and pMHC units, respectively. In step 1 (both for TCR and TCRpMHC modeling), the different protein domains are described by unified cuboids and assembled. The illustration of steps 2 and 5 show the Q-flip correction/optimization. At step 2, each glutamine residue is optimized with respect to its direct environment only (only the corresponding variable domain is accounted for). Whereas in step 5, the two glutamine residues are optimized simultaneously, thus accounting for the whole TCR environment. In step 4 (only for TCRpMHC modeling), the pMHC unit is pre-placed, translated away from the TCR and optimized with respect to the fixed TCR variable domains. At step 6 (both for TCR and TCRpMHC modeling), the position of all cuboids as well as the orientation of the glutamine residues are optimized concurrently. The latter illustrations show the structure before and after optimization, with the target crystal structure depicted in *gray*

based on a general rotational axis, is however too general to describe all the transformations of a pMHC complex in a three-dimensional space. We thus adapted our previously introduced cuboid method [43] and measured the three Euler angle components of the TCR/pMHC orientation. Equivalently to the determination of CoR_β (see Additional file 1: Text S1A for details), we define here a center of rotation for the orientation of the pMHC cuboid relative to the Vα domain (CoR_μ; Fig. 1). Unlike CoR_β, CoR_μ does not correspond any conserved residue and lies at the center of the peptide binding groove of the pMHC complex. As CoR_β and CoR_μ are solely defined by the Vβ and pMHC coordinates, we further use their location as rotational centers for the rigid body optimization. These locations will be named as CoR_β- and CoR_μ-based rotational centers, respectively.

Our strategy for the modeling of TCRpMHC complexes intuitively resembles the plausible biological process: i.e., first the association of the two TCR variable domains, then the approach of the pMHC complex towards the TCR CDR loops (Fig. 2). We thus assume here that the general orientation of the TCR variable domains is determined prior to the binding of the pMHC ligand, and then further adjusts upon binding. As a consequence, comprehensive sampling of possible Vα/Vβ orientations is crucial and can determine the success of the modeling attempt. Based on our former analysis [43], we define here 11 starting orientations for the association of the two TCR variable domains. Such number of different initial conditions is intended to cover the large range of Vα/Vβ inter-domain angles and to avoid artificial local minima. It also increases the probability that at least one of the obtained Vα:Vβ complexes is close enough to the final bound conformation such that it can effectively bind the pMHC ligand (for details about the choice of the 11 starting orientations see Additional file 1: Text S1B).

To determine the starting orientation of the pMHC cuboid, we analyzed the Vα/pMHC angles associated with all structures considered in our set (see subsection Structural data sets). The crystal structure 3e3q [65] showed the lowest angular deviation with respect to the others and was chosen as reference structure. Based on the normal vector to the plane defined by the MHC β-sheet backbone atoms of 3e3q we derived a translation axis for the pMHC.

The 11 starting structures for the TCRpMHC modeling pipeline are obtained such that for each of the 11 starting Vα/Vβ orientations, the pMHC ligand is placed in a general position (based on the pMHC orientation in 3e3q) and afterwards translated away from the TCR along the 3e3q-based translation axis. For each of these structures one cuboid is defined around each domain, i.e. Vα, Vβ, and pMHC.

The relative position of these cuboids is then optimized iteratively along a succession of operations

defined by our pipeline algorithm, for each of the 11 starting conformations of a given complex (Fig. 2). Our pipeline is built in a modular manner, such that beside the interaction between the different subunits, it is possible to explicitly model the flexibility of some relevant parts of the molecular system. As we detailed in the Introduction, the center of rotation between Vα and Vβ (CoR_β) is characterized by the specific interaction of two glutamine residues (Q). However, because of the ambiguous character of the Q side chain, the assignment of the atoms (Q-flip state) in crystal structures often happens to be wrong. Furthermore, the central location of these two residues in the complex makes them particularly sensitive to variations in the Vα/Vβ inter-domain angles. For these reasons, our pipeline also includes a Q-flip correction/optimization module, which is applied alongside with the general optimization of the whole complex.

For the modeling of one TCRpMHC complex, our pipeline algorithm thus results in a total of 11 structures, each of them originating from the corresponding starting orientation of the TCR variable domains. These structures are optimized and finally ranked according to their interaction energy.

Rigid body optimization

We implemented our method within the DynaCell suite of programs [63] using a rigid body energy minimization approach based on the work by Mirzaei et al. [64] together with the Broyden-Fletcher-Goldfarb-Shanno (BFGS) algorithm as implemented in the GNU scientific library (libGSL; version 1.15 double) [66]. Details about the applied settings of the algorithm and a discussion about its convergence are presented in Additional file 1: Text S1C. Mirzaei et al. introduced the original algorithm focusing on the RBEM problem for molecular docking [64]. The approach is specifically designed for an efficient rotation of the rigid bodies around a center of rotation and is particularly well suited for our application. However, the original method only allows for a simultaneous optimization of the relative position of only two rigid bodies. We therefore extended it such that the simultaneous optimization of the orientation of an arbitrary number of rigid bodies is possible.

We implemented the method in a generalized, modular way, allowing for the individual design of application specific optimization pipelines, based on a given combination of the different functions during runtime. Each pipeline step consists in the assembly of sub-process operators (SOs), which evaluate an objective function and the corresponding gradient to further perform the resulting coordinates transformations. Each SO manipulates the coordinates of a subset of atoms and calculates the value of the objective function within a given environment (i.e., including the whole system or only part of it). So far, we implemented three different families of SOs. We shall briefly describe them below, while a more detailed presentation can be found as Supporting Information (Additional file 2: Text S2).

The first family of SOs consists of the basic operators for the rigid body rotation and translation. The objective function of these operators is computed from the non-bonded interactions between the rigid bodies of interest (not the intra-cuboid interactions). The operators modify the three parameters for the rotation and one for the translation of the rigid body, either simultaneously or independently.

Next to these general operators, we implemented a specific carboxamide group rotation operator. This operator is valid for both asparagine and glutamine residues. It only modifies one parameter, namely the dihedral angle that defines the orientation of the side chain's carboxamide group. The rotation axis is set along the carbon-carbon bond next to this functional group (i.e., $C\beta$-$C\gamma$ and $C\gamma$-$C\delta$ for an asparagine and a glutamine residue, respectively). The objective function accounts for all bonded energy terms within the corresponding side chain and non-bonded energies within the system (i.e., including the intra-cuboid interactions). This sub-process operator can be used within our prediction pipeline algorithm or independently, and we shall refer to as Q-flip correction tool in the following.

Finally, we implemented a rigid body position restraint operator to prevent unrealistically large translational motions and hence to avoid irrelevant conformations. The objective function in this case consists of a harmonic potential applied on the distance between a given reference and a mobile point. The harmonic penalty is applied if the distance is greater than the defined threshold. For the present modeling of TCRpMHC complexes, CoR_β and CoR_μ are used to restrain the positions of the $V\beta$ and pMHC cuboids, respectively. The threshold values are set to 7.5 Å for $V\beta$ and to 13.0 Å for pMHC.

TCRpMHC prediction pipeline

The standard modeling pipeline for the prediction of the $V\beta$ orientation in $V\alpha/V\beta$ complexes of TCRs as well as the orientation of both $V\beta$ and pMHC in TCRpMHC complexes is summarized in the central panel of Fig. 2. The illustration of the steps (left and right panels of the figure) applied during the modeling of the TCR alone are shown in black circles, while the steps used for the modeling of the TCRpMHC complexes are circled in green color. In addition, an animation of the modeling process is available as Supporting Information (Additional file 3: Movie S1). We hereafter provide a detailed description of this pipeline for the modeling of one given TCRpMHC complex, based on the steps depicted in Fig. 2.

A modeling attempt starts with the representation of each subunit ($V\alpha$, $V\beta$, and pMHC) as a cuboid and their placement in the reference coordinate frame (Fig. 2, step 1: complex assembly).

After this initial assembly, the Q-flip state of the central glutamine residues located at the CoR_β is corrected independently in each of the two TCR variable domains, by accounting only for the interactions within the corresponding domain, $V\alpha$ or $V\beta$ (Fig. 2, step 2: separate Q correction). At that stage, we only intend to correct the possibly wrong assignment of the Q-flip state in the crystal structure. The interaction with the Q in the opposite domain is not yet considered, as the correct Q-Q assembly is crucially dependent on the final $V\alpha/V\beta$ association angle, which is still unknown at this stage of modeling. As a consequence, only two orientations are considered, the original one and a rotation of 180°. The orientation presenting the lowest energy is selected and further refined by performing 30 steps of BFGS energy minimization. As here only one parameter needs to be optimized, the optimization process is straightforward and 30 steps are sufficient to reach convergence (for a more detailed discussion of the chosen BFGS settings see Additional file 1: Text S1C).

Based on these two preparation steps, 11 starting orientations of the TCR variable domains are constructed (Fig. 2, step 3: conformational angle adaption) and the pMHC ligand is placed in a general position and translated away from the TCR (Fig. 2, step 4: pMHC pre-placement). Details about this step are discussed in the "General concepts of DynaDom" sub-section and in the Additional file 1: Text S1A/B. The position of the pMHC is pre-optimized with respect to the fixed $V\alpha$ and $V\beta$ domains, for a maximum of 150 BFGS steps. This step intends to mimic an approach of the pMHC from far towards an already formed TCR assembly. As the position of only one rigid body is optimized here, most of the optimizations converge in less than 150 steps. For the few optimizations that do not meet the BFGS convergence criteria, the energy still drops in few steps and reaches a stable plateau (for a more detailed discussion of the chosen BFGS settings see Additional file 1: Text S1C). As this step constitutes a preparation for the main optimization (step 6), these structures are considered as converged and the modeling pipeline proceeds.

Next, the orientation of the central Q residues is explicitly sampled in the context of the whole TCR assembly (i.e., accounting for the intra- and inter-subsystems interactions) with a fine step of 18°, leading to 400 different orientations (Fig. 2, step 5: Q-Q bifurcation correction). The orientation with the lowest energy is then selected and further minimized for a maximum of 30 steps and used in the next step. This ensures a proper orientation of the Q residues with respect to each other for the current $V\alpha/V\beta$ orientation. Here again, considering the straightforward parameter space to be optimized and the explicit sampling performed ahead with a fine angle increment, 30 steps are sufficient to reach convergence.

The core step of the DynaDom algorithm takes place after these preparation steps. At this stage, the position of Vβ and pMHC as well as the orientation of the two glutamine residues are concurrently optimized (Fig. 2., step 6: Simultaneous rigid body optimization). The minimization is first conduced for a preliminary 50 steps. Then step 5 is repeated to ensure optimal Q-Q placement and interaction at the center of rotation CoR$_β$ with respect to the adjusted global conformation of the complex. Step 6 is finally repeated for a maximum of 2950 steps. Here again, most of the minimizations converge in a few hundred steps. The rare cases in which the BFGS algorithm did not converge were systematically analyzed, showing that in each case, the energy strongly decreases in a few steps and oscillates around a minimum value (see Additional file 1: Text S1C for a detailed discussion). Therefore, these rare cases were also considered as converged in the present work.

Finally, the quality of the current model (i.e., originating from the ith starting conformation out of 11 in the present application) is evaluated by computing the complex binding energy ($E_{i,bind}$) as:

$$E_{i,bind} = E_{i, complex} - (E_α + E_β + E_μ),$$

where $E_{i, complex}$ is the total energy of the complex and $E_α$, $E_β$, and $E_μ$ are the energy terms of the individual complex components Vα, Vβ, and pMHC (notice that these last quantities are constant for each prediction run and are thus computed only once). The energy is evaluated using the OPLS-AA force field [67, 68]. As the current application is a remodeling attempt, we additionally computed the all-atom positional root mean square deviation (RMSD) with respect to the crystal structure for each of the 11 final models.

The ranking of the 11 final models is performed according to their binding energy. For the current remodeling application, we define an energy criterion to assess the success of the remodeling attempt as C_E. The energy criterion is fulfilled if the model having the best binding energy also bears an RMSD lower than 2 Å with respect to the original crystal structure. To gain more insight into the performance of our DynaDom method, we define a second success criterion based on the RMSD, C_R, which allows us to evaluate the performance of the structural modeling by assessing the structural deviation of the model from the corresponding experimental structure. This structural criterion is fulfilled if at least one of the 11 final models has an RMSD lower than 2 Å with respect to the original crystal structure.

Structural data sets
We selected 75 biological units (BUs) originating from 48 different crystal structures contained in the set that we previously analyzed in ref. [43].

In that study, we observed that the different BUs within a given crystal structure can slightly differ from each other (RMSD < 1 Å), especially in the exact location of side chain atoms. This is presumably due to the relatively high intrinsic flexibility of the complexes or the limited resolution in some of the structures (differs from 1.5 to 3.5 Å). To evaluate the robustness of our method and its capability to tackle such inaccuracies, we included all BUs in our two main datasets. The inclusion of all BUs also results in a larger data set and, as no training of a scoring function is performed (DynaDom is a force-field based approach as described in the previous subsections), introduces no bias to the method itself. In addition, the current datasets only contain structures in which all atoms that are involved in the modeling process were experimentally resolved. Although these atoms or residues could be easily modeled, this would potentially introduce a bias in the set, which we prefer to avoid here. As summarized in Table 1, the TCRpMHC crystal structures selected for this work belong to two different species (i.e., 17 murine and 31 human) and 22 different TCR clono-types (mutations not accounted). The coordinates of each structure were aligned with respect to the conserved residues of the Vα domain, as described in ref. [43] and in the previous subsections. The TCR constant domains, the MHC α$_3$ domain, and the β-microglobulin were systematically removed from the structures as well as all non-protein atoms (the discarded domains are represented with transparent colors in Fig. 1). Hydrogen atoms were added and topologies were created for the OPLS-AA force field [67, 68] using the pdb2gmx tool (Version 4.5.6) [69].

We further derived three different data sets. First, to evaluate the performance of our method for the remodeling of the association angle of the TCR Vα and Vβ domains, we removed the pMHC ligand in each structure. This resulted in a set of 75 Vα/Vβ complexes, which we shall refer to as DS$_T$ in the following. In addition, we created a second data set, in which only the first BU in the PDB file of the corresponding structure was included (48 structures, further referred to as DS$_T^*$). Then, to perform the remodeling of TCRpMHC complexes, we selected among the 75 BUs only the structures containing an MHC class I molecule. The resulting third data set, named as DS$_C$, contains a total of 53 TCRpMHC complexes. We disregarded MHC class II molecules in the DS$_C$ set to ensure a proper comparison between the samples. A third set could have been dedicated to MHC class II molecules. However, we sustained from remodeling also that set as the results obtained for the MHC class I complexes already showed that further optimization of the pipeline, beyond this publication, is necessary for accurate pMHC placement. In addition, the size of the MHC class II set (22 structures) would

Table 1 Description of the structural dataset DS$_T$ and the subset DS$_C$

PDB	DS[a]	TCR-Name	S[b]	L[c]	R[d]
1ao7	T/C	A6	H	I	[25]
1fo0	T/C	BM3.3	M	I	[75]
1fyt	T	HA1.7	H	II	[76]
1j8h	T	HA1.7	H	II	[77]
1kj2	T/C	KB5-C20	M	I	[78]
1mi5	T	LC13	H	I	[79]
1mwa	T/C	2C	M	I	[80]
1nam	T/C	BM3.3	M	I	[81]
1oga	T/C	JM22	H	I	[82]
1qse	T	A6	H	I	[35]
1u3h	T	TCR172.10	M	II	[83]
2bnq	T/C	1G4	H	I	[37]
2bnr	T/C	1G4	H	I	[37]
2e7l	T/C	2C m6 [T7]	M	I	[84]
2esv	T/C	KK50.4	H	I	[85]
2f53	T/C	1G4 c49c50	H	I	[38]
2f54	T/C	1G4 AV-wt	H	I	[38]
2gj6	T/C	A6	H	I	[73]
2iam	T	E8	H	II	[86]
2ian	T	E8	H	II	[86]
2nx5	T/C	ELS4	H	I	[87]
2oi9	T/C	2C [T7-wt]	M	I	[84]
2ol3	T/C	BM3.3	M	I	[88]
2p5e	T/C	1G4 c58c61	H	I	[39]
2p5w	T/C	1G4 c58c62	H	I	[39]
2pxy	T	1934.4	M	II	[89]
2pye	T/C	1G4 c5c1	H	I	[39]
2vlk	T/C	JM22	H	I	[90]
2vlr	T/C	JM22	H	I	[90]
3c5z	T	B3K506	M	II	[91]
3c60	T	YAe62	M	II	[91]
3c6l	T	2 W20	M	II	[91]
3d39	T/C	A6	H	I	[92]
3d3v	T/C	A6	H	I	[92]
3dxa	T/C	DM1	H	I	[93]
3e2h	T/C	2C m67 [T7]	M	I	[65]
3e3q	T/C	2C m13 [T7]	M	I	[65]
3ffc	T/C	cf34	H	I	[94]
3gsn	T/C	RA14	H	I	[95]
3h9s	T/C	A6	H	I	[96]
3kpr	T/C	LC13	H	I	[97]
3kps	T/C	LC13	H	I	[97]
3kxf	T/C	SB27	H	I	[98]

Table 1 Description of the structural dataset DS$_T$ and the subset DS$_C$ *(Continued)*

3mbe	T	TCR 21.30	M	II	[99]
3mv8	T/C	TK3 Q55H	H	I	[100]
3pwp	T/C	A6	H	I	[101]
3qiu	T	226 TCR	M	II	[102]
3qiw	T	226 TCR	M	II	[102]

a) T: Structure only in dataset DS$_T$. T/C: Structure in both datasets, DS$_T$ and DS$_C$
b) Species (S): *H* human, *M* mouse
c) Ligand type (L): MHC class I or II. See Additional file 4: Table S1 of the Supporting Information for details about the MHC alleles and the different peptides
d) References

have been very small for a robust analysis. Further details about each data set are listed in Table S1 of the Supporting Information (Additional file 4: Table S1).

Results and discussion

The structural prediction of immunologically relevant molecular assemblies has focused the interest of a wide range of methodological developments over the past decades, especially in the field of antibody-antigen interactions [41, 48, 49, 70]. Compared to the effort made so far in antibody modeling, the number of predicted TCRpMHC structures is still relatively small, as we discussed in the Background section. In the case of antibodies, it was recently shown by some of us, that statistical learning techniques can efficiently predict the V_H/V_L association angles [49]. Such very appealing approaches are based on experimentally observed structural features and require a large amount of existing data. In the particular case of antibodies, over 2000 crystal structures are already available, thus allowing the application of such knowledge-based methodologies. Considering the relatively small amount of TCR structures referenced in the Protein Data Bank (i.e., about 200), such a road can unfortunately not be envisaged for the prediction of association angles in TCR complexes. As a consequence, we developed here a solely force-field based optimization strategy for TCR and TCRpMHC complexes modeling. Such a force-field based approach can potentially be applied to other similar systems, even if a sufficient amount of experimental data is not available for the training of a specific scoring function.

As we extensively described in the Background and in the Methods sections, this new algorithm, named as DynaDom, is derived from our previous comprehensive analysis of the Vα/Vβ TCR variable domain association angles [43]. The main conclusions that arose from this former work can be summarized as follows: i) TCR complexes can be classified into structural clusters, differing significantly in their Vα/Vβ inter-domain angles, ii) the angular differences between the structural clusters can be described by a simple rotation around a center of rotation (CoR_β, see the Methods section and Fig. 1 for details), and iii) the CoR_β is characterized by two highly conserved glutamine residues, which contribute to the interaction between the TCR $V\alpha$ and $V\beta$ domains via a stabilizing hydrogen bond network.

For the remodeling of TCRpMHC complexes, the DynaDom method uses a unified cuboid description of the three different units of this complex (i.e., $V\alpha$, $V\beta$, pMHC). The optimization of the total system is performed by a rotation-based algorithm, which is based on our Center of Rotation concept (i.e., CoR_β and CoR_μ, as described in the Methods section). In our previous analysis study, we observed that the $V\alpha/V\beta$ association angle spectrum is much larger in unbound TCRs than in structures bound to the pMHC [43]. Pierce et al. further emphasized that for the prediction of TCRpMHC complexes from unbound units [33] the side chains of the CDR loops must adapt to their environment in order to allow for a proper interaction between the different units of the complex. Therefore, the inclusion of local side chain flexibility at the domain interface would most likely be a necessary extension for the prediction of TCRpMHC structures from unbound or homologous TCR and pMHC structures by homology modeling techniques. Our generalized, modular implementation ensures that the additional inclusion of local flexibility is straightforward. However, the adaptation of the algorithm would require additional extensive evaluation efforts, which would go beyond the scope of the present work and will be part of future investigations. Nevertheless, in the present work we already tested such a feature by the inclusion of local side chain flexibility for the two Q-Q residues at the CoR_β, which we found to be crucial for the prediction success as we shall discuss in the following subsections.

The current version of our pipeline algorithm, results from an extensive series of evaluations intended to assess the effect of the different parameters. Hereafter, we present and discuss our main findings together with the actual evaluation of the method. We first discuss the optimization of the orientation of the glutamine residues (Q-flip correction) located at the interface between the two TCR variable domains (i.e., at the Center of Rotation CoR_β of $V\beta$ with respect to $V\alpha$), based on the original experimental structures. Then, we analyze the effect of such a Q-flip correction together with the use of restraints on the remodeling of $V\alpha/V\beta$ TCR and TCRpMHC complexes. Along this analysis, we compare the results obtained using either an energy or a structure based selection criterion (i.e., C_E or C_R, respectively, as defined in the Methods section). This comparison intends to state if an atomistic force field energy based criterion could be used for future blind homology modeling of TCRpMHC complexes. We finally suggest

further possible routes of improvement for our methodology, based on the analysis of the few cases in which the remodeling process did not lead to a satisfactory structure.

Glutamine orientation correction

The interface between the $V\alpha$ and $V\beta$ domains of TCRs is characterized by the interaction of two highly conserved glutamine (Q) residues [43]. While this Q-Q interaction appears to be of critical importance, the flip state of these residues is often wrongly assigned in experimental crystal structures, due to the ambiguous character of the carboxamide group electron density [50–53]. In the context of this work, we analyzed the flip state of the Q residues among the crystal structures contained in our set of 75 $V\alpha/V\beta$ complexes (i.e., data set DS_T). Only 72.7% of the structures present a correct assignment of the Q-flip state. The details of this analysis are listed in Table S2 of the Supporting Information (Additional file 5: Table S2).

As discussed in the Background section, many modeling tools exist to correct the orientation of glutamine and asparagine residues in a given crystal structure. Among those, we tested Reduce [50, 54, 55] and Protoss [60, 61] on our DS_T set. The application of the Reduce and Protoss programs leads to an improvement of the glutamine flip state in our set, reaching 94.6 and 97.3% of correctly assigned Q-flip states, respectively. Analysis of the failed cases showed that they featured an interaction of the Q residues in an initial bifurcated orientation (i.e., associated in a perpendicular manner). Manual inspection showed that in these cases the perpendicular orientation allowed for optimal interactions with the rest of the domains and should therefore be the most stable in the functional receptor (i.e., not a further artifact of the carboxamide assignments in the experimental structures). As the Reduce and Protoss programs only allow parallel orientations, these cannot successfully predict such particular interactions. Because of this limitation and the below discussed observation that the Q-flip state can change upon the association of the $V\alpha/V\beta$ domains, we decided to implement an independent Q-flip correction approach using our already implemented rigid body operators, such that it can directly be included into our pipeline. This represents a first probing of the modular character of our implementation, which we shall follow towards the future inclusion of local flexibility.

We evaluated the performance of our method for the Q-flip correction in the crystal structure of the DS_T set. We present here the most relevant findings of our analysis, while a more detailed discussion can be found as Supporting Information (Additional file 6: Text S3), together with the entirety of our observations per structure (Additional file 5: Table S2, Additional file 7: Table

S3, and Additional file 8: Table S4). Using the DynaDom correction module, 100.0% of the structures could be assigned in the good Q-flip state. An example of a successful Q-flip correction is depicted in Fig. 3b for the 2f53 crystal structure. This higher performance obtained by our method with respect to the other programs comes from the optimization-based methodology that we implemented. While standard tools only consider two possible parallel conformations per residue (i.e., the original and the flipped state), DynaDom performs an explicit sampling of the carboxamide group using adjustable angular step sizes, followed by an energy minimization step during which the atomistic environment of the residue is taken into account. Such a

protocol allows the system to escape local minima in order to find the most favorable conformations of the Q residues in their environment. For this reason, the Dyna-Dom approach can also lead to a proper paring in the case of the two systems that present a bifurcated Q-Q interaction.

In a second step, we assessed the steps of our pipeline at which this correction should be performed. For this we applied our tool on the two TCR variable domains independently and compared the results with the same calculation performed in the Vα/Vβ complex environment (Additional files 7: Table S3 and Additional file 8: Tables S4). In this analysis, we observed different predicted Q-flip states if the corrections were applied on

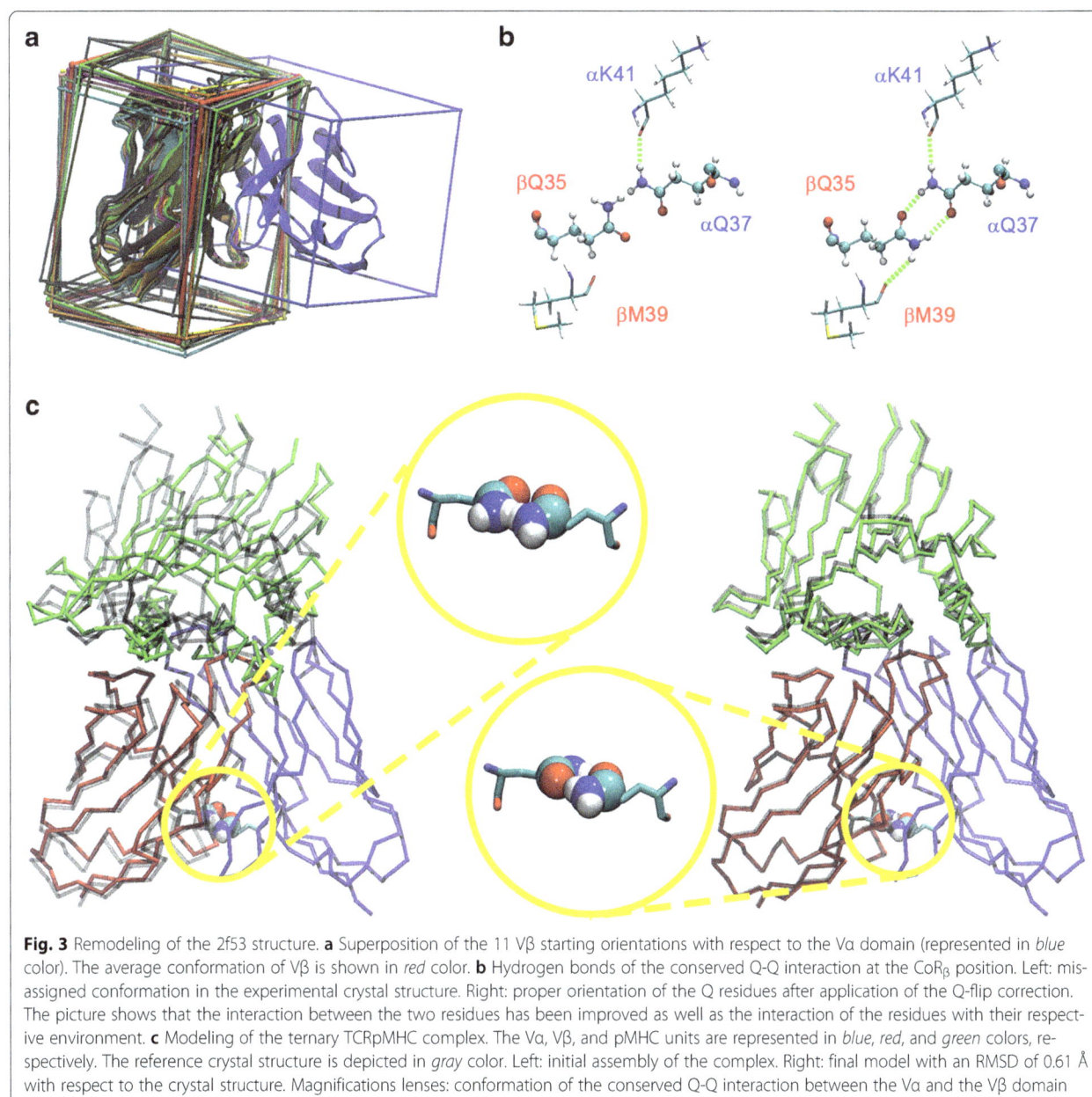

Fig. 3 Remodeling of the 2f53 structure. **a** Superposition of the 11 Vβ starting orientations with respect to the Vα domain (represented in *blue* color). The average conformation of Vβ is shown in *red* color. **b** Hydrogen bonds of the conserved Q-Q interaction at the CoR_β position. Left: misassigned conformation in the experimental crystal structure. Right: proper orientation of the Q residues after application of the Q-flip correction. The picture shows that the interaction between the two residues has been improved as well as the interaction of the residues with their respective environment. **c** Modeling of the ternary TCRpMHC complex. The Vα, Vβ, and pMHC units are represented in *blue*, *red*, and *green* colors, respectively. The reference crystal structure is depicted in *gray* color. Left: initial assembly of the complex. Right: final model with an RMSD of 0.61 Å with respect to the crystal structure. Magnifications lenses: conformation of the conserved Q-Q interaction between the Vα and the Vβ domain

the single domains or on the final complexes. This observation emphasizes the impact of the environment of the glutamine residues on their respective conformation. For this reason, our standard pipeline algorithm performs the Q-flip state correction at two different steps. First, the glutamine residues are optimized accounting only for their respective domain environment (i.e., $V\alpha$ or $V\beta$). This allows for a proper orientation of the glutamine residue prior to the complex assembly, thus eliminating potential errors originating from the experimental structures. It is noteworthy that this feature will be particularly relevant if homology modeled structures are built. Then, further optimizations are performed in the complete environment during the $V\alpha/V\beta$ complex optimization to adapt the Q residues to the final relative orientation of the $V\alpha$ and $V\beta$ TCR domains.

Modeling of the TCR variable domain complexes without pMHC

We tested our DynaDom methodology first for the re-modeling of the $V\alpha/V\beta$ association angles in the complete DS_T set. In particular, we performed a series of evaluations to assess the relevance of two criteria in our algorithm pipeline: the use of the Q-flip correction and the application of a distance restraint between $V\alpha$ and $V\beta$. Furthermore, we analyzed the performance of our method according to two selection criteria. For a given remodeling experiment, DynaDom produces 11 models, which are ranked by their RMSD or energy score (see Additional file 9: Figure S1 Fig. and the description in the Methods section). The remodeling experiment is then counted as successful if the RMSD of the selected model with respect to the original crystal structure is lower than 2 Å.

Our results are summarized in Table 2 and the different evaluation settings that we considered are generally labeled as M_T plus a bit string, which encodes for the use or not of the Q-flip correction and the restraint (e.g., M_T10 labels the remodeling of a TCR complex by applying the Q-flip correction but no distance restraint). The last M_T^*11 test corresponds to the M_T11 settings performed for the DS_T^* data set, which contains only the first BU of each experimental structure.

Considering the C_R criterion, the remodeling procedure already reaches a very high positive prediction rate of 94.7%, even if no Q-flip corrections or restraints are used (M_T00). This rate further increases to 96.0% if the Q-flip correction is switched on (M_T10), while no change is observed if the distance restraint is used alone (94.7% in the M_T01 case). As a consequence, the final prediction rate, with both parameters switched on, also reaches the remarkable rate of 96.0% (M_T11). Only three outliers are observed, originating from the 3dxa and from the 1mwa crystal structures. The relatively low

Table 2 Prediction accuracy for the $V\alpha/V\beta$ association angles modeled without pMHC

ES[a]	Variants[b]		C_R^c		C_E^d	
	Q	R	#	%	#	%
M_T00	off	off	71	94.7	63	84.0
M_T01	off	on	71	94.7	66	88.0
M_T10	on	off	72	96.0	64	85.3
M_T11	on	on	72	96.0	67	89.3
M_T^*11	on	on	47	97.9	43	89.6

[a]Evaluation setting label
[b]Variants: Q = glutamine carboxamide group orientation correction, R = rigid body position restraint
[c]Absolute and relative prediction rate according to the RMSD based criterion (i.e., C_R) in data set DS_T (75 structures). In the particular case of M_T^*11, the prediction was performed on the DS_T^* set (48 structures, without biological units). For each prediction run, the 11 models are ranked by RMSD and a success is counted if the selected structure has an RMSD value lower than 2 Å
[d] Same as c) using the energy criterion to rank the 11 structures and select the best. The prediction is considered as successful if the selected structure has an RMSD value lower than 2 Å

resolution of the 3dxa structure (i.e., 3.5 Å) can partially explain this failure. Furthermore, our modeling process only considers the $V\alpha$ and $V\beta$ domains of the TCR complex. It is possible that the two constant domains of the complex play an important role in these three outlier cases. Regarding the additional experiment performed on the DS_T^* data set (M_T^*11), the results in Table 2 show that the differences in the achieved accuracies with respect to M_T11 are only marginal. This confirms that the inclusion of the BUs does not bias the overall results and demonstrates the robustness of our algorithm with respect to small variations in the structures, thus highlighting the suitability of the approach in a future homology modeling pipeline.

In the perspective of a blind homology modeling experiment of TCR complexes, no reference crystal structure would be available and only an energy-based criterion could be considered for structure selection (i.e., C_E). Based on such a C_E criterion, our remodeling attempt reaches a prediction rate of 84.0% even if no Q-flip corrections or restraints are applied (M_T00). The prediction rate increases with both, the independent use of the Q-flip correction and the distance restraint to 85.3 and 88.0% for M_T10 and M_T01, respectively. If both parameters are used (M_T11), the prediction reaches the remarkable rate of 89.3% and even 89.6% for the M_T^*11 data set. This last result is very promising for the further applications of the DynaDom method in a real structure prediction setting.

It appears that the use of the distance restraint has a stronger impact on the prediction rate obtained according to the C_E criterion than it has for the C_R criterion. This could be attributed to the observation that without distance restraint, the algorithm can yield structures in which the two TCR domains are placed in an unrealistic conformation, which nevertheless has a lower interaction energy (see Additional file 10: Figure S2). Such unphysical associations are far from the original crystal structure and are intrinsically discriminated by an RMSD based selection criterion.

Next to the analysis of the best conformations according to the C_E and C_R criteria we also analyzed the overall performance of the algorithm regarding the quality of all predicted conformations. In Fig. 4, we present the percentage of structures having an RMSD value lower than 1, 2 and 3 Å, depending on the algorithm settings (i.e., M_T00, M_T01, M_T10, and M_T11) among all 75*11 models produced by our DynaDom procedure. In this context we also further analyzed the impact of the Q-flip correction by classifying the resulting models into two groups, with respect to their original Q-flip state in the experimental structures as paired (51*11) and mispaired (20*11). Notice that 4*11 structures lack the presence of Q residues at CoR_β and were therefore not included in the respective analysis. The histograms (Fig. 4) show that an overall percentage of about 80% of the models feature an RMSD lower than 2 Å and thus fulfill our success criterion. This demonstrates the robustness of the presented algorithm and thus its relevance as one step in a comprehensive structure prediction pipeline. By further analyzing the influence of the Q-flip correction on the prediction rates, it can be observed that the overall prediction success is higher for structures in which the Q-Q orientation is already correct in the X-ray structure (paired structures). For these structures 85% of the models have an RMSD value lower than 2 Å, whereas the rate drops to 75% for the mispaired structures. This might be due to the relatively smaller size of the latter data set, as an investigation of a possible correlation between the crystal structure resolution and the quality of the final models did not yield any significant outcome.

Regarding the percentage of structures having an RMSD value lower than 2 and 3 Å for both sets, paired and mispaired, the results are practically independent on the defined settings and only a slight trend towards an improved performance can be observed if the Q-flip correction is applied. This low impact on the overall structures is most likely due to the large surface area of the total TCR domain interface and thus the high number of other interactions, which drive the overall optimization of the domains orientations. The use of the Q-flip correction has, however, a remarkable effect on the quality of the resulting structures once the cutoff is lowered to 1 Å. The percentage of models featuring such a low RMSD indeed increases from 47 to 60% for the mispaired structure set, if the correction is switched on. These observations reveal the importance of a correct orientation of the conserved Q residues at the Center of Rotation CoR_β for an accurate modeling of the TCR variable domain association and the need for their correction if they are wrongly assigned in the template structure.

Overall, this series of remodeling essays highlights the quality of our methodology. It also further emphasizes the applicability of a force field interaction energy-based criterion, which is very promising in the perspective of a homology modeling setting, as it shows that high-quality structures can be identified by this means.

Modeling of the pMHC position with simultaneous TCR variable domain placement

Regarding the successful results obtained for the remodeling of TCR Vα/Vβ assemblies discussed in the

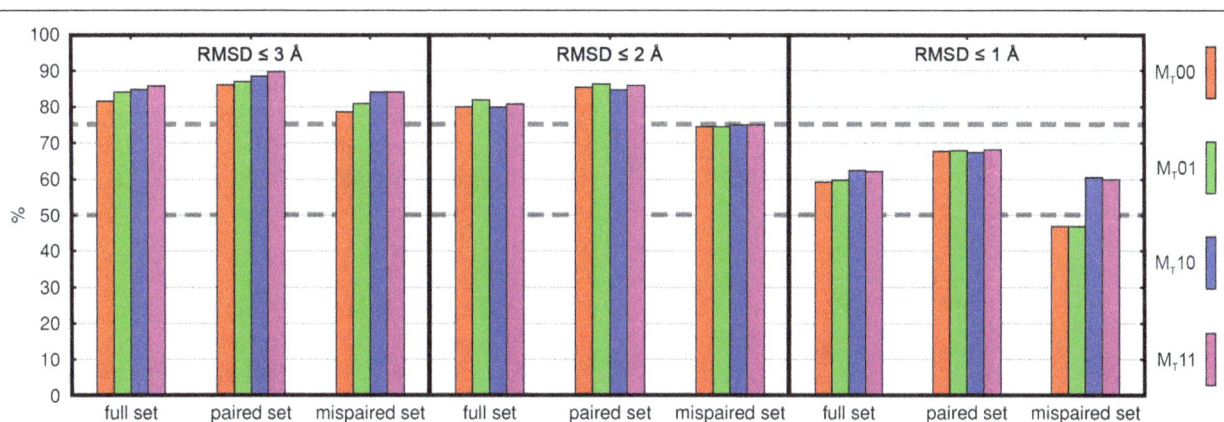

Fig. 4 Percentage of structures among the 75*11 models with an RMSD value lower than 3, 2 and 1 Å. The total set of 75*11 structures is separated into structures for which the Q residues were originally paired or mispaired within their corresponding crystal structure. Each histogram box corresponds to a different setting of the modeling procedure, i.e. with only distance restraint (M_T01), only Q-flip correction (M_T10), both (M_T11), or none of them (M_T00). The percentage of structures with an RMSD value lower than 3, 2 and 1 Å are presented on the left, middle, and right plots, respectively. The right plot shows that for the structures presenting an originally wrong orientation of the Q residues, the Q-flip correction significantly improves the quality of the resulting model (i.e. M_T10 and M_T11)

previous subsection, we shall now assess the performance of the DynaDom method to remodel TCRpMHC complexes. Here, the calculations were performed on the smaller set of structures DS_C, which contains a total of 53 biological units. Our results are listed in Table 3 and the labeling of the test settings follows the nomenclature introduced above (i.e., M_C label and a bit string for the use or not of Q-flip correction and distance restraint).

An example of successfully predicted complex is depicted in Fig. 3c for the structure 2f53. In the figure, the Vα, Vβ, and MHC units are respectively colored in blue, red, and green. The two images represent the complex before and after optimization (on the left and on the right hand side of the figure, respectively). The magnifying glass shows that the Q-flip state is efficiently corrected and one can observe that the final model successfully fits the reference crystal structure (depicted with gray color in the picture).

Regarding the prediction rates according to the C_R and C_E criteria in Table 3, the percentages obtained for the remodeling of TCRpMHC complexes reach a less striking prediction rate, though still relatively high (i.e., 73.6 and 71.7% according to the C_R and C_E criteria, respectively). For the prediction based on the C_R criterion, the success rate appears to be independent on the use of Q-flip correction and distance restraints. A similar trend is observed with the C_E criterion. In this case, the use of one or both of the parameters only marginally increases the prediction rate. As for the modeling of TCR variable domains alone, the use of an energy based criterion yields very satisfactory results compared to a structure based one. This point also confirms the robustness and thus the suitability of our method for a blind homology modeling of TCRpMHC complexes.

Table 3 Prediction accuracy for the combined prediction of the Vα/Vβ and TCR/pMHC association angles

ES^a	Variants[b]		C_R^c		C_E^d	
	Q	R	#	%	#	%
M_C00	off	off	39	73.6	37	69.8
M_C01	off	on	39	73.6	38	71.7
M_C10	on	off	39	73.6	38	71.7
M_C11	on	on	39	73.6	38	71.7

[a]Evaluation setting label
[b]Variants: Q = glutamine carboxamide group orientation correction, R = rigid body position restraint
[c]Absolute and relative prediction rate according to the RMSD based criterion (i.e., C_R) in data set DS_C (53 structures). For each prediction run, the 11 models are ranked by RMSD and a success is counted if the selected structure has an RMSD value lower than 2 Å
[d]Same as c) using the energy criterion to rank the 11 structures and select the best. The prediction is considered as successful if the selected structure has an RMSD value lower than 2 Å

Detailed performance analysis for the TCRpMHC prediction

Regarding the overall, nearly equal, performances of the different settings in Table 3, it clearly appears that the drop of the prediction rate for the remodeling of TCRpMHC complexes with respect to the modeling of only the TCR variable domains is barely dependent on the use of Q-flip correction and distance restraints. The former parameters only affect the relative orientation of the Vα and Vβ domains. This observation indicates that the lower performance observed for the remodeling of TCRpMHC complexes might originate from an incorrect placement of the MHC molecule. We thus performed additional analyses to gain more insights into the shortcomings of the current approach and to identify potential routes for future improvement of our algorithm. To further confirm that the issues encountered in the remodeling of TCRpMHC complexes are solely due to the prediction of the pMHC positions with respect to the TCR, we analyzed the impact of the initial placement of the pMHC ligand on the remodeling of TCRpMHC complexes. In this context, we shall only consider the models obtained according to the RMSD based criterion (C_R).

In the following series of test evaluations (T), we only consider the initial orientation of the TCR domains as found in their crystal structure (i.e., the remodeling procedure does not start from the 11 starting conformations, but only one). This was done to eliminate any potential biasing errors originating from the TCR domain modeling. Next to that, we used the settings of the final modeling pipeline: i.e.,Vβ optimization, Q-flip correction, and position restraints were systematically applied during these tests. As we described in the Methods section and depicted in Fig. 2, our modeling protocol includes a translation of the pMHC unit along a given axis to separate pMHC from the TCR, thus avoiding strong initial forces due to unphysical steric hindrance. This feature is one parameter that we shall analyze in the following tests (i.e., by switching it on or off). For each test setting, a rigid body optimization of the pMHC around its starting position was performed. Finally, for the first two test evaluations (T_1 and T_2), the MHC rigid body was initially placed in its crystal structure orientation, while for the last test (T_3), this unit was oriented according to the general zero conformation discussed in the Method section (i.e., the orientation used in the standard pipeline). The results and details of each test evaluation are presented in Table 4.

In the first test evaluation (T_1) in which the pMHC units are oriented according to their respective crystal structure orientation and no initial translation is performed, the prediction rate reaches 100.0%. Although such a result could be expected as we start from the

Table 4 Prediction rates of the test evaluations

ES[a]	MHC initial orientation[b]	MHC translation[c]	C_R^d	
			#	%
T_1	crystal	no	53	100.0
T_2	crystal	yes	38	71.7
T_3	general	yes	31	58.5

For each test the Q-flip correction as well as the use of distance restraint are systematically applied. The TCR Vβ domain is placed in its original crystal structure orientation and is optimized. The tests are performed for each of the 53 structures present in the DS_C data set and the MHC rigid body position is optimized in each case

[a] Evaluation setting label

[b] The initial orientation of the MHC unit is chosen either according to the original crystal structure or using the general zero orientation as for the standard version of our pipeline (see Methods Section for more details)

[c] Initial translation of the MHC unit to avoid steric hindrance, necessary if the MHC rigid body is not placed according to the crystal structure orientation (see Methods Section for more details)

[d] Absolute and relative prediction rate according to the RMSD based criterion (i.e., C_R) in data set DS_C (53 structures). For each prediction run, a success is counted if the resulting model has an RMSD value lower than 2 Å with respect to the crystal structure

experimental conformations, it proves that our algorithm does not lead to any conformational artifacts. The additional application of the initial translation step for the pMHC ligand (T_2) results in a drastic decrease of the prediction rate to 71.7%, slightly lower than the result obtained using our standard protocol (i.e., 73.6% in the M_C11 case). For the final test (T_3), in which the translated pMHC was placed according to our standard protocol, the prediction rate of our algorithm dramatically drops to 58.5%, which is significantly lower than for the final pipeline setting (73.6%).

These results show that the translation procedure and the preplacement of the pMHC ligand in a single general starting position constitute the accuracy limiting steps of our pipeline. In addition, we confirm here that the use of various starting positions for the Vβ domain clearly outperforms the case in which a single conformation is considered, even if the latter corresponds to the experimental crystal structure (i.e., 73.6% versus 58.6% for the M_C11 and T_3 cases, respectively). At first glance this is a surprising result. However, it clearly appears that the simultaneous optimization of both the TCR domains and the pMHC molecule as performed for the M_C11, but not the T3 setting, is highly beneficial for the performance of the algorithm as it allows for an alternating adaption of the flexible units with respect to each other (Additional file 3: Movie S1). This leads to a smoother optimization path, thus lowering the probability for being trapped in a local minimum. Different starting positions further lower this probability as multiple paths are sampled.

Consequently, one straightforward way to improve our results should be to use multiple starting conformations for the pMHC ligand, in accordance with the 11 Vβ preplacement orientations. To evaluate this procedure, we

chose one structure (PDB-ID 1oga) for which the modeling process failed in the last T_3 test settings. For this structure, the three Euler angle components defining the pMHCs CoR_μ-based rotational center were systematically varied by 5° and all 27 resulting starting poses were constructed. In accordance with the other test settings, the Vβ domain was here again placed in its crystal structure orientation. The results for the 27 resulting models are listed in the Supporting Information (Additional file 11: Table S5). The results improved considerably as this time five structures were obtained with an RMSD lower than 2 Å, thus satisfying our success criterion. Notably, these five models also show the lowest interaction energy among the 27 predicted structures. This last test clearly confirms the necessity of more advanced sampling protocol for the MHC molecule orientation in our modeling strategy to avoid the complex geometry to fall in an unfavorable local minimum. This is in agreement with the observations made by Pierce et al. [33] and demonstrates once again the importance of starting from multiple initial conformations. However, a straightforward combination of the 11 starting conformations of the TCR Vβ domain together with the 27 initial orientations of the pMHC unit would lead to a total of 297 structures to optimize per TCRpMHC complex, thus resulting in a dramatic increase of the computational cost.

Therefore, the presented algorithm provides excellent results and can readily be used for the optimization of the Vα/Vβ association angles. It also yields a fairly good prediction rate for the prediction of TCRpMHC complexes association. However, for the simultaneous optimization of both, the TCR domains and the placement of the pMHC in the latter case, further improvements and evaluations will be necessary prior to its practical use as one step in a real structure prediction pipeline. Considering the general, modular character of our implementation, also different approaches could be combined with the current method to tackle this issue. Among those, basin-hoping techniques [71] have proven to provide good results for the rigid body optimization of tryptophan zippers [72], and Monte Carlo-based rigid body sampling was recently applied by Pierce et al. for the placement of MHC like ligands alone [33, 34]. Despite the numerous tests that would be required for the combination of such techniques, this route represents a promising strategy for the future of our methodology.

Conclusions

In this work we presented a new procedure, DynaDom, for the optimization of protein domain-domain orientations, which was designed for and evaluated on the special case of remodeling T-cell-receptor-peptide-MHC complexes. The approach is based on several rigid body optimization and restraining routines, and uses atomistic force field-based energy calculations. The individual optimization functions are combined in an application-specific pipeline. The method yields remarkable results for the remodeling of TCR Vα/Vβ association angles with prediction rates of 89–96% (RMSD < 2 Å) depending on the evaluation criterion.

The present study shows that it is possible to predict the TCR Vα/Vβ association angles on the basis of structural modeling only, without the need for a specially tailored experimental data dependent scoring function. It also demonstrates that the previously identified Center of Rotation concept [43] can be readily used for the structural prediction of the association angles.

Another striking result arising from this work is the observation that, by simply considering the best-energy conformation for each structure, high prediction rates of 89.3% for the Vα/Vβ association angles could be obtained. This is only marginally lower than the prediction rates obtained for the models with the smallest RMSD. This shows that ranking the modeled structures solely by their force field-based interaction energy allows the identification of high quality structures and demonstrates not only the robustness of the method, but also its suitability as part of a general structure prediction pipeline for TCRpMHC structures.

In a second step, we applied the concept to the simultaneous optimization of the TCR Vα/Vβ association angles and the pMHC positions on the TCR. However, due to efficiency considerations we used a simplified placement method for the pMHC, which resulted in lower prediction rates of 72–74%. This result is still in the predictive range, but not as high as for the TCR domain optimization. Additional preliminary investigations showed that the main reason lies indeed in the initial placement method of the pMHC ligand and that by simply using multiple initial conformations for the pMHC placement, already significant improvements in the placement accuracy are possible. However, a systematic optimization of the method for pMHC placement would require further significant evaluation studies, which would go beyond the scope of this manuscript and which will be the topic of future studies together with the application of DynaDom to the blind homology modeling of TCRpMHC complexes. In general, the presented approach is very well suited to serve as basis for the development of such a method for the prediction of atomistic models of TCRs or TCRpMHC complexes taking inter-domain angles into account. Due to the modular design of our program, a straightforward combination and concurrent optimization of multiple features is possible, as already demonstrated in this work by the concurrent optimization of the Vβ orientation, the pMHC orientation, and the adaption of the glutamine residues connecting the two TCR chains. Thus, the future implementation of partial or full flexibility of side chains or protein backbone regions,

which then could be simultaneously optimized while the rigid body positions are adapted, should be straightforward. In addition, including other domains of the complex, such as the TCR constant domains would also be possible. This could help to study e.g. scissoring effects observed for the constant domains [73] or to investigate TCR signaling, which was elsewhere discussed to be induced by conformational changes in the constant domains [74].

Finally, it is worth noting that the DynaDom strategy is not limited to TCRpMHC assemblies. The combination of the different modules can indeed be easily modified to fit the requirement of other rigid body based predictions of a large variety of biomolecular assemblies.

Additional files

Additional file 1: Text S1. Detailed discussion of important methodological aspects.

Additional file 2: Text S2. Detailed description of the operators.

Additional file 3: Movie S1. Example for the prediction pipeline: Remodeling of the structure with the PDB-ID 1ao7.

Additional file 4: Table S1. Structural Dataset DS$_T$ and the subset DS$_C$.

Additional file 5: Table S2. Performance of the Q-Q interaction optimization.

Additional file 6: Text S3. Glutamine correction and adaption (detailed Results and Discussion).

Additional file 7: Table S3. Per residue flip states using Reduce, Protoss and DynaDom comparing single domains and TCR complexes.

Additional file 8 Table S4: Angular deviations with respect to the crystal structures after DynaDom glutamine refinement.

Additional file 9 Figure S1: Discrimination of the models.

Additional file 10: Figure S2: Influence of the restraint operator.

Additional file 11: Table S5: pMHC optimization for the structure 1oga with different pMHC start conformations.

Abbreviations

BFGS: Broyden-Fletcher-Goldfarb-Shanno; BU: Biological Unit; crystallographically independent molecule in the asymmetric unit; CDR: Complementary Determining Region; C$_E$, C$_R$: ranking Criterion based on the Energy or the RMSD, respectively; CoR$_β$: CoR$_μ$, Center of Rotation in repect to the TCR Vβ and the pMHC, respectively; DS$_C$: Data Set TCRpMHC Complex; DS$_T$*: Reduced DS$_T$ containing only one BU per structure; DS$_T$: Data Set TCR; MHC: Major histocompatibility complex; M$_T$, M$_C$: Modeling run for the TCR test set or the complex test set, respectively; PDB: Protein Data Bank; pMHC: Peptide presented in a Major Histocompatibility Complex molecule; Q: Glutamine; RMSD: Root Mean Square Deviation; SO: Sub-process Operator; T1, T2, and T3: Test evaluations with different conditions; TCR: T Cell Receptor; V$_H$, V$_L$: antibody Variable domains of the Heavy and the Light chain, respectively; Vα/Vβ: Variable domain of the TCR α- and the β-chain, respectively

Acknowledgements

We thank Hanieh Mirzaei and Dima Kozakov for providing their rigid body optimization code, Stefan Bietz for the provision of a Protoss license, Angela Krackhardt, Atanas Patronov, and Ilke Ugur for helpful discussions and critical reading of the manuscript. We thank Antonia Stank for her preparative work on the used MHC complexes.

Funding
TH was supported by the Deutsche Forschungsgemeinschaft through the SFB1035 project A10, AM was supported by the Deutsche Forschungsgemeinschaft through the SFB749, further financial support was provided by the CIPS^M cluster of excellence.

Authors' contributions
TH entirely programmed the code related to this work, performed all the evaluation studies, and contributed to the analysis, to the design of the study, to the development of the algorithms as well as to the writing of the manuscript. AM and IA contributed to the analysis, to the development of the algorithms and to the writing of the manuscript. IA also initiated and contributed to the general design of this work. All authors read and approved the final manuscript.

Competing interests
The authors declare that they have no competing interests.

References
1. Arstila TP, Casrouge A, Baron V, Even J, Kanellopoulos J, Kourilsky P. A direct estimate of the human alphabeta T cell receptor diversity. Science. 1999;286(5441):958–61.
2. Casrouge A, Beaudoing E, Dalle S, Pannetier C, Kanellopoulos J, Kourilsky P. Size estimate of the alpha beta TCR repertoire of naive mouse splenocytes. J Immunol. 2000;164(11):5782–7.
3. Warren RL, Freeman JD, Zeng T, Choe G, Munro S, Moore R, Webb JR, Holt RA. Exhaustive T-cell repertoire sequencing of human peripheral blood samples reveals signatures of antigen selection and a directly measured repertoire size of at least 1 million clonotypes. Genome Res. 2011;21(5):790–7.
4. Tonegawa S. Somatic generation of antibody diversity. Nature. 1983;302(5909):575–81.
5. Garcia KC. Reconciling views on T cell receptor germline bias for MHC. Trends Immunol. 2012;33(9):429–36.
6. Davis MM, Bjorkman PJ. T-cell antigen receptor genes and T-cell recognition. Nature. 1988;334(6181):395–402.
7. Bernstein FC, Koetzle TF, Williams GJ, Meyer Jr EF, Brice MD, Rodgers JR, Kennard O, Shimanouchi T, Tasumi M. The Protein Data Bank: a computer-based archival file for macromolecular structures. J Mol Biol. 1977;112(3):535–42.
8. Michielin O, Luescher I, Karplus M. Modeling of the TCR-MHC-peptide complex. J Mol Biol. 2000;300(5):1205–35.
9. Morgan RA, Dudley ME, Wunderlich JR, Hughes MS, Yang JC, Sherry RM, Royal RE, Topalian SL, Kammula US, Restifo NP, et al. Cancer regression in patients after transfer of genetically engineered lymphocytes. Science. 2006;314(5796):126–9.
10. Zoete V, Irving M, Ferber M, Cuendet MA, Michielin O. Structure-based, rational design of T cell receptors. Front Immunol. 2013;4:268.
11. Rueckert C, Guzman CA. Vaccines: from empirical development to rational design. PLoS Pathog. 2012;8(11):e1003001.
12. Flower DR, Macdonald IK, Ramakrishnan K, Davies MN, Doytchinova IA. Computer aided selection of candidate vaccine antigens. Immunome Res. 2010;6 Suppl 2:S1.
13. Chakraborty AK, Weiss A. Insights into the initiation of TCR signaling. Nat Immunol. 2014;15(9):798–807.
14. Backert L, Kohlbacher O. Immunoinformatics and epitope prediction in the age of genomic medicine. Genome Med. 2015;7:119.
15. Lundegaard C, Lund O, Nielsen M. Predictions versus high-throughput experiments in T-cell epitope discovery: competition or synergy? Expert Rev Vaccines. 2012;11(1):43–54.
16. Atanasova M, Patronov A, Dimitrov I, Flower DR, Doytchinova I. EpiDOCK: a molecular docking-based tool for MHC class II binding prediction. Protein Eng Des Sel. 2013;26(10):631–4.
17. Schueler-Furman O, Altuvia Y, Sette A, Margalit H. Structure-based prediction of binding peptides to MHC class I molecules: application to a broad range of MHC alleles. Protein Sci. 2000;9(9):1838–46.
18. Khan JM, Ranganathan S. pDOCK: a new technique for rapid and accurate docking of peptide ligands to major histocompatibility complexes. Immunome Res. 2010;6 Suppl 1:S2.
19. Antes I, Siu SW, Lengauer T. DynaPred: a structure and sequence based method for the prediction of MHC class I binding peptide sequences and conformations. Bioinformatics. 2006;22(14):e16–24.
20. Rigo MM, Antunes DA, Vaz de Freitas M, Fabiano de Almeida Mendes M, Meira L, Sinigaglia M, Vieira GF. DockTope: a Web-based tool for automated pMHC-I modelling. Sci Rep. 2015;5:18413.
21. Liu T, Pan X, Chao L, Tan W, Qu S, Yang L, Wang B, Mei H. Subangstrom accuracy in pHLA-I modeling by Rosetta FlexPepDock refinement protocol. J Chem Inf Model. 2014;54(8):2233–42.
22. Liu Z, Dominy BN, Shakhnovich EI. Structural mining: self-consistent design on flexible protein-peptide docking and transferable binding affinity potential. J Am Chem Soc. 2004;126(27):8515–28.
23. Almagro JC, Vargas-Madrazo E, Lara-Ochoa F, Horjales E. Molecular modeling of a T-cell receptor bound to a major histocompatibility complex molecule: implications for T-cell recognition. Protein Sci. 1995;4(9):1708–17.
24. Garcia KC, Degano M, Stanfield RL, Brunmark A, Jackson MR, Peterson PA, Teyton L, Wilson IA. An alphabeta T cell receptor structure at 2.5 A and its orientation in the TCR-MHC complex. Science. 1996;274(5285):209–19.
25. Garboczi DN, Ghosh P, Utz U, Fan QR, Biddison WE, Wiley DC. Structure of the complex between human T-cell receptor, viral peptide and HLA-A2. Nature. 1996;384(6605):134–41.
26. Sali A. Comparative protein modeling by satisfaction of spatial restraints. Mol Med Today. 1995;1(6):270–7.
27. Kessler B, Michielin O, Blanchard CL, Apostolou I, Delarbre C, Gachelin G, Gregoire C, Malissen B, Cerottini JC, Wurm F, et al. T cell recognition of hapten. Anatomy of T cell receptor binding of a H-2kd-associated photoreactive peptide derivative. J Biol Chem. 1999;274(6):3622–31.
28. Leimgruber A, Ferber M, Irving M, Hussain-Kahn H, Wieckowski S, Derre L, Rufer N, Zoete V, Michielin O. TCRep 3D: an automated in silico approach to study the structural properties of TCR repertoires. PLoS One. 2011;6(10):e26301.
29. Haidar JN, Pierce B, Yu Y, Tong W, Li M, Weng Z. Structure-based design of a T-cell receptor leads to nearly 100-fold improvement in binding affinity for pepMHC. Proteins. 2009;74(4):948–60.
30. Pierce BG, Hellman LM, Hossain M, Singh NK, Vander Kooi CW, Weng Z, Baker BM. Computational design of the affinity and specificity of a therapeutic T cell receptor. PLoS Comput Biol. 2014;10(2):e1003478.
31. Simons KT, Kooperberg C, Huang E, Baker D. Assembly of protein tertiary structures from fragments with similar local sequences using simulated annealing and Bayesian scoring functions. J Mol Biol. 1997;268(1):209–25.
32. Gray JJ, Moughon S, Wang C, Schueler-Furman O, Kuhlman B, Rohl CA, Baker D. Protein-protein docking with simultaneous optimization of rigid-body displacement and side-chain conformations. J Mol Biol. 2003;331(1):281–99.
33. Pierce BG, Weng Z. A flexible docking approach for prediction of T cell receptor-peptide-MHC complexes. Protein Sci. 2013;22(1):35–46.
34. Pierce BG, Vreven T, Weng Z. Modeling T cell receptor recognition of CD1-lipid and MR1-metabolite complexes. BMC Bioinformatics. 2014;15:319.
35. Ding YH, Baker BM, Garboczi DN, Biddison WE, Wiley DC. Four A6-TCR/peptide/HLA-A2 structures that generate very different T cell signals are nearly identical. Immunity. 1999;11(1):45–56.
36. Ding YH, Smith KJ, Garboczi DN, Utz U, Biddison WE, Wiley DC. Two human T cell receptors bind in a similar diagonal mode to the HLA-A2/Tax peptide complex using different TCR amino acids. Immunity. 1998;8(4):403–11.
37. Chen J-L, Stewart-Jones G, Bossi G, Lissin NM, Wooldridge L, Choi EML, Held G, Dunbar PR, Esnouf RM, Sami M, et al. Structural and kinetic basis for heightened immunogenicity of T cell vaccines. J Exp Med. 2005;201(8):1243–55.
38. Dunn SM, Rizkallah PJ, Baston E, Mahon T, Cameron B, Moysey R, Gao F, Sami M, Boulter J, Li Y, et al. Directed evolution of human T cell receptor CDR2 residues by phage display dramatically enhances affinity for cognate peptide-MHC without increasing apparent cross-reactivity. Protein Sci. 2006;15(4):710–21.
39. Sami M, Rizkallah PJ, Dunn S, Molloy P, Moysey R, Vuidepot A, Baston E, Todorov P, Li Y, Gao F, et al. Crystal structures of high affinity human T-cell receptors bound to peptide major histocompatibility complex reveal native diagonal binding geometry. Protein Eng Des Sel. 2007;20(8):397–403.
40. McBeth C, Seamons A, Pizarro JC, Fleishman SJ, Baker D, Kortemme T, Goverman JM, Strong RK. A new twist in TCR diversity revealed by a forbidden alphabeta TCR. J Mol Biol. 2008;375(5):1306–19.
41. Dunbar J, Fuchs A, Shi J, Deane CM. ABangle: characterising the VH-VL orientation in antibodies. Protein Eng Des Sel. 2013;26(10):611–20.
42. Dunbar J, Knapp B, Fuchs A, Shi J, Deane CM. Examining variable domain orientations in antigen receptors gives insight into TCR-like antibody design. PLoS Comput Biol. 2014;10(9):e1003852.
43. Hoffmann T, Krackhardt AM, Antes I. Quantitative analysis of the association angle between T-cell receptor valpha/vbeta domains reveals important features for epitope recognition. PLoS Comput Biol. 2015;11(7):e1004244.

44. Knapp B, Dunbar J, Deane CM. Large scale characterization of the LC13 TCR and HLA-B8 structural landscape in reaction to 172 altered peptide ligands: a molecular dynamics simulation study. PLoS Comput Biol. 2014;10(8):e1003748.

45. Rudolph MG, Stanfield RL, Wilson IA. How TCRs bind MHCs, peptides, and coreceptors. Annu Rev Immunol. 2006;24:419–66.

46. Sun R, Shepherd SE, Geier SS, Thomson CT, Sheil JM, Nathenson SG. Evidence that the antigen receptors of cytotoxic T lymphocytes interact with a common recognition pattern on the H-2Kb molecule. Immunity. 1995;3(5):573–82.

47. Knapp B, Deane CM. T-cell receptor binding affects the dynamics of the peptide/MHC-I complex. J Chem Inf Model. 2016;56(1):46–53.

48. Bujotzek A, Dunbar J, Lipsmeier F, Schafer W, Antes I, Deane CM, Georges G. Prediction of VH-VL domain orientation for antibody variable domain modeling. Proteins. 2015;83(4):681–95.

49. Bujotzek A, Fuchs A, Qu C, Benz J, Klostermann S, Antes I, Georges G. MoFvAb: modeling the Fv region of antibodies. mAbs. 2015;7(5):838–52.

50. Word JM, Lovell SC, Richardson JS, Richardson DC. Asparagine and glutamine: using hydrogen atom contacts in the choice of side-chain amide orientation. J Mol Biol. 1999;285(4):1735–47.

51. Weichenberger CX, Sippl MJ. Self-consistent assignment of asparagine and glutamine amide rotamers in protein crystal structures. Structure. 2006;14(6):967–72.

52. McDonald IK, Thornton JM. Satisfying hydrogen bonding potential in proteins. J Mol Biol. 1994;238(5):777–93.

53. Hooft RW, Sander C, Vriend G. Positioning hydrogen atoms by optimizing hydrogen-bond networks in protein structures. Proteins. 1996;26(4):363–76.

54. Chen VB, Arendall 3rd WB, Headd JJ, Keedy DA, Immormino RM, Kapral GJ, Murray LW, Richardson JS, Richardson DC. MolProbity: all-atom structure validation for macromolecular crystallography. Acta Crystallogr D Biol Crystallogr. 2010;66(Pt 1):12–21.

55. Davis IW, Leaver-Fay A, Chen VB, Block JN, Kapral GJ, Wang X, Murray LW, Arendall WB, 3rd, Snoeyink J, Richardson JS et al. MolProbity: all-atom contacts and structure validation for proteins and nucleic acids. Nucleic Acids Res. 2007; 35(Web Server issue):W375-383.

56. Weichenberger CX, Byzia P, Sippl MJ. Visualization of unfavorable interactions in protein folds. Bioinformatics. 2008;24(9):1206–7.

57. Weichenberger CX, Sippl MJ. NQ-Flipper: recognition and correction of erroneous asparagine and glutamine side-chain rotamers in protein structures. Nucleic Acids Res. 2007;35(Web Server issue):W403-406.

58. Li X, Jacobson MP, Zhu K, Zhao S, Friesner RA. Assignment of polar states for protein amino acid residues using an interaction cluster decomposition algorithm and its application to high resolution protein structure modeling. Proteins. 2007;66(4):824–37.

59. Labute P. Protonate3D: assignment of ionization states and hydrogen coordinates to macromolecular structures. Proteins. 2009;75(1):187–205.

60. Lippert T, Rarey M. Fast automated placement of polar hydrogen atoms in protein-ligand complexes. J Cheminformatics. 2009;1(1):13.

61. Bietz S, Urbaczek S, Schulz B, Rarey M. Protoss: a holistic approach to predict tautomers and protonation states in protein-ligand complexes. J Cheminformatics. 2014;6:12.

62. Bayden AS, Fornabaio M, Scarsdale JN, Kellogg GE. Web application for studying the free energy of binding and protonation states of protein-ligand complexes based on HINT. J Comput Aided Mol Des. 2009;23(9):621–32.

63. Antes I. DynaDock: a new molecular dynamics-based algorithm for protein-peptide docking including receptor flexibility. Proteins. 2010;78(5):1084–104.

64. Mirzaei H, Beglov D, Paschalidis IC, Vajda S, Vakili P, Kozakov D. Rigid body energy minimization on manifolds for molecular docking. J Chem Theory Comput. 2012;8(11):4374–80.

65. Jones LL, Colf LA, Stone JD, Garcia KC, Kranz DM. Distinct CDR3 conformations in TCRs determine the level of cross-reactivity for diverse antigens, but not the docking orientation. J Immunol. 2008; 181(9):6255–64.

66. Gough B. GNU Scientific Library Reference Manual. 3rd ed. 2009.

67. Jorgensen W, Tirado-Rives J. The OPLS [optimized potentials for liquid simulations] potential functions for proteins, energy minimizations for crystals of cyclic peptides and crambin. J Am Chem Soc. 1988;110(6):1657–66.

68. Jorgensen WL, Maxwell DS, Rives T. Development and Testing of the OPLS All-Atom Force Field on Conformational Energetics and Properties of Organic Liquids. J Am Chem Soc. 1996;118(45):11225–36.

69. Hess B, Kutzner C, van der Spoel D, Lindahl E. GROMACS 4: algorithms for highly efficient, load-balanced, and scalable molecular simulation. J Chem Theory Comput. 2008;4(3):435–47.

70. Almagro JC, Teplyakov A, Luo J, Sweet RW, Kodangattil S, Hernandez-Guzman F, Gilliland GL. Second antibody modeling assessment (AMA-II). Proteins. 2014;82(8):1553–62.

71. Wales D, Doye J. Global optimization by basin-hopping and the lowest energy structures of Lennard-Jones clusters containing up to 110 atoms. J Phys Chem A. 1998;101(28):5111–6.

72. Kusumaatmaja H, Whittleston C, Wales D. A local rigid body framework for global optimization of biomolecules. J Chem Theory Comput. 2012;8(12):5159–65.

73. Gagnon SJ, Borbulevych OY, Davis-Harrison RL, Turner RV, Damirjian M, Wojnarowicz A, Biddison WE, Baker BM. T cell receptor recognition via cooperative conformational plasticity. J Mol Biol. 2006;363(1):228–43.

74. Beddoe T, Chen Z, Clements CS, Ely LK, Bushell SR, Vivian JP, Kjer-Nielsen L, Pang SS, Dunstone MA, Liu YC, et al. Antigen ligation triggers a conformational change within the constant domain of the alphabeta T cell receptor. Immunity. 2009;30(6):777–88.

75. Reiser JB, Darnault C, Guimezanes A, Grégoire C, Mosser T, Schmitt-Verhulst AM, Fontecilla-Camps JC, Malissen B, Housset D, Mazza G. Crystal structure of a T cell receptor bound to an allogeneic MHC molecule. Nat Immunol. 2000;1(4):291–7.

76. Hennecke J, Carfi A, Wiley DC. Structure of a covalently stabilized complex of a human alphabeta T-cell receptor, influenza HA peptide and MHC class II molecule, HLA-DR1. EMBO J. 2000;19(21):5611–24.

77. Hennecke J, Wiley DC. Structure of a complex of the human alpha/beta T cell receptor (TCR) HA1.7, influenza hemagglutinin peptide, and major histocompatibility complex class II molecule, HLA-DR4 (DRA*0101 and DRB1*0401): insight into TCR cross-restriction and alloreactivity. J Exp Med. 2002;195(5):571–81.

78. Reiser JB, Grégoire C, Darnault C, Mosser T, Guimezanes A, Schmitt-Verhulst AM, Fontecilla-Camps JC, Mazza G, Malissen B, Housset D. A T cell receptor CDR3beta loop undergoes conformational changes of unprecedented magnitude upon binding to a peptide/MHC class I complex. Immunity. 2002;16(3):345–54.

79. Kjer-Nielsen L, Clements CS, Purcell AW, Brooks AG, Whisstock JC, Burrows SR, McCluskey J, Rossjohn J. A structural basis for the selection of dominant alphabeta T cell receptors in antiviral immunity. Immunity. 2003;18(1):53–64.

80. Luz JG, Huang M, Garcia KC, Rudolph MG, Apostolopoulos V, Teyton L, Wilson IA. Structural comparison of allogeneic and syngeneic T cell receptor-peptide-major histocompatibility complex complexes: a buried alloreactive mutation subtly alters peptide presentation substantially increasing V(beta) interactions. J Exp Med. 2002;195(9):1175–86.

81. Reiser J-B, Darnault C, Grégoire C, Mosser T, Mazza G, Kearney A, van der Merwe PA, Fontecilla-Camps JC, Housset D, Malissen B. CDR3 loop flexibility contributes to the degeneracy of TCR recognition. Nat Immunol. 2003;4(3):241–7.

82. Stewart-Jones GBE, McMichael AJ, Bell JI, Stuart DI, Jones EY. A structural basis for immunodominant human T cell receptor recognition. Nat Immunol. 2003;4(7):657–63.

83. Maynard J, Petersson K, Wilson DH, Adams EJ, Blondelle SE, Boulanger MJ, Wilson DB, Garcia KC. Structure of an autoimmune T cell receptor complexed with class II peptide-MHC: insights into MHC bias and antigen specificity. Immunity. 2005;22(1):81–92.

84. Colf LA, Bankovich AJ, Hanick NA, Bowerman NA, Jones LL, Kranz DM, Garcia KC. How a single T cell receptor recognizes both self and foreign MHC. Cell. 2007;129(1):135–46.

85. Hoare HL, Sullivan LC, Pietra G, Clements CS, Lee EJ, Ely LK, Beddoe T, Falco M, Kjer-Nielsen L, Reid HH, et al. Structural basis for a major histocompatibility complex class Ib-restricted T cell response. Nat Immunol. 2006;7(3):256–64.

86. Deng L, Langley RJ, Brown PH, Xu G, Teng L, Wang Q, Gonzales MI, Callender GG, Nishimura AI, Topalian SL, et al. Structural basis for the recognition of mutant self by a tumor-specific, MHC class II-restricted T cell receptor. Nat Immunol. 2007;8(4):398–408.

87. Tynan FE, Reid HH, Kjer-Nielsen L, Miles JJ, Wilce MCJ, Kostenko L, Borg NA, Williamson NA, Beddoe T, Purcell AW, et al. A T cell receptor flattens a bulged antigenic peptide presented by a major histocompatibility complex class I molecule. Nat Immunol. 2007;8(3):268–76.

88. Mazza C, Auphan-Anezin N, Gregoire C, Guimezanes A, Kellenberger C, Roussel A, Kearney A, van der Merwe PA, Schmitt-Verhulst A-M, Malissen B. How much can a T-cell antigen receptor adapt to structurally distinct antigenic peptides? EMBO J. 2007;26(7):1972–83.

89. Feng D, Bond CJ, Ely LK, Maynard J, Garcia KC. Structural evidence for a germline-encoded T cell receptor-major histocompatibility complex interaction 'codon'. Nat Immunol. 2007;8(9):975–83.

90. Ishizuka J, Stewart-Jones GBE, van der Merwe A, Bell JI, McMichael AJ, Jones EY. The structural dynamics and energetics of an immunodominant T cell receptor are programmed by its Vbeta domain. Immunity. 2008;28(2):171–82.

91. Dai S, Huseby ES, Rubtsova K, Scott-Browne J, Crawford F, Macdonald WA, Marrack P, Kappler JW. Crossreactive T Cells spotlight the germline rules for alphabeta T cell-receptor interactions with MHC molecules. Immunity. 2008; 28(3):324–34.

92. Piepenbrink KH, Borbulevych OY, Sommese RF, Clemens J, Armstrong KM, Desmond C, Do P, Baker BM. Fluorine substitutions in an antigenic peptide selectively modulate T-cell receptor binding in a minimally perturbing manner. Biochem J. 2009;423(3):353–61.

93. Archbold JK, Macdonald WA, Gras S, Ely LK, Miles JJ, Bell MJ, Brennan RM, Beddoe T, Wilce MCJ, Clements CS, et al. Natural micropolymorphism in human leukocyte antigens provides a basis for genetic control of antigen recognition. J Exp Med. 2009;206(1):209–19.

94. Gras S, Burrows SR, Kjer-Nielsen L, Clements CS, Liu YC, Sullivan LC, Bell MJ, Brooks AG, Purcell AW, McCluskey J, et al. The shaping of T cell receptor recognition by self-tolerance. Immunity. 2009;30(2):193–203.

95. Gras S, Saulquin X, Reiser J-B, Debeaupuis E, Echasserieau K, Kissenpfennig A, Legoux F, Chouquet A, Gorrec ML, Machillot P, et al. Structural bases for the affinity-driven selection of a public TCR against a dominant human cytomegalovirus epitope. J Immunol. 2009;183(1):430–7.

96. Borbulevych OY, Piepenbrink KH, Gloor BE, Scott DR, Sommese RF, Cole DK, Sewell AK, Baker BM. T cell receptor cross-reactivity directed by antigen-dependent tuning of peptide-MHC molecular flexibility. Immunity. 2009;31(6):885–96.

97. Macdonald WA, Chen Z, Gras S, Archbold JK, Tynan FE, Clements CS, Bharadwaj M, Kjer-Nielsen L, Saunders PM, Wilce MCJ, et al. T cell allorecognition via molecular mimicry. Immunity. 2009;31(6):897–908.

98. Burrows SR, Chen Z, Archbold JK, Tynan FE, Beddoe T, Kjer-Nielsen L, Miles JJ, Khanna R, Moss DJ, Liu YC, et al. Hard wiring of T cell receptor specificity for the major histocompatibility complex is underpinned by TCR adaptability. Proc Natl Acad Sci U S A. 2010; 107(23):10608–13.

99. Yoshida K, Corper AL, Herro R, Jabri B, Wilson IA, Teyton L. The diabetogenic mouse MHC class II molecule I-Ag7 is endowed with a switch that modulates TCR affinity. J Clin Invest. 2010;120(5):1578–90.

100. Gras S, Chen Z, Miles JJ, Liu YC, Bell MJ, Sullivan LC, Kjer-Nielsen L, Brennan RM, Burrows JM, Neller MA, et al. Allelic polymorphism in the T cell receptor and its impact on immune responses. J Exp Med. 2010;207(7):1555–67.

101. Borbulevych OY, Piepenbrink KH, Baker BM. Conformational melding permits a conserved binding geometry in TCR recognition of foreign and self molecular mimics. J Immunol. 2011;186(5):2950–8.

102. Newell EW, Ely LK, Kruse AC, Reay PA, Rodriguez SN, Lin AE, Kuhns MS, Garcia KC, Davis MM. Structural basis of specificity and cross-reactivity in T cell receptors specific for cytochrome c-I-E(k). J Immunol. 2011;186(10):5823–32.

16

Binding of undamaged double stranded DNA to vaccinia virus uracil-DNA Glycosylase

Norbert Schormann[1], Surajit Banerjee[2], Robert Ricciardi[3] and Debasish Chattopadhyay[1*]

Abstract

Background: Uracil-DNA glycosylases are evolutionarily conserved DNA repair enzymes. However, vaccinia virus uracil-DNA glycosylase (known as D4), also serves as an intrinsic and essential component of the processive DNA polymerase complex during DNA replication. In this complex D4 binds to a unique poxvirus specific protein A20 which tethers it to the DNA polymerase. At the replication fork the DNA scanning and repair function of D4 is coupled with DNA replication. So far, DNA-binding to D4 has not been structurally characterized.

Results: This manuscript describes the first structure of a DNA-complex of a uracil-DNA glycosylase from the poxvirus family. This also represents the first structure of a uracil DNA glycosylase in complex with an undamaged DNA. In the asymmetric unit two D4 subunits bind simultaneously to complementary strands of the DNA double helix. Each D4 subunit interacts mainly with the central region of one strand. DNA binds to the opposite side of the A20-binding surface on D4. Comparison of the present structure with the structure of uracil-containing DNA-bound human uracil-DNA glycosylase suggests that for DNA binding and uracil removal D4 employs a unique set of residues and motifs that are highly conserved within the poxvirus family but different in other organisms.

Conclusion: The first structure of D4 bound to a truly non-specific undamaged double-stranded DNA suggests that initial binding of DNA may involve multiple non-specific interactions between the protein and the phosphate backbone.

Keywords: Protein-DNA structure, Non-specific DNA, Early DNA recognition complex, Uracil-DNA glycosylase, Poxvirus

Background

Repair of damages in DNA is an essential cellular process regulated by specialized molecular machinery. Base excision repair (BER) pathway [1] for repair of small lesions is initiated by monofunctional DNA glycosylases. These enzymes use a water molecule as a nucleophile to cleave the N-glycosidic bond between the target base and deoxyribose, releasing the damaged base and leaving an apurinic/apyrimidinic (AP) site [2]. The BER pathway for the removal of uracil (Ura), which arises in DNA from deamination of cytosine (Cyt) or incorporation of dUTP during DNA synthesis, is initiated by uracil-DNA glycosylases (UDGs) [3]. UDGs are divided in different families based on their substrate specificity [4]. Members of family I-V share similar

structure and motifs. Family I UDGs, which are also known as UNGs, specifically excise uracil from single-stranded DNA (ssDNA) and double-stranded DNA (dsDNA) with the preference ssU > dsU:G > dsU:A [5]. UNGs are ubiquitous enzymes that use highly conserved motifs for DNA binding, uracil recognition and excision.

The reaction mechanism for UNGs has been elucidated from a series of elegant structural studies in which wild-type and catalytically inactive mutant UNGs were captured in the DNA-bound form at various stages of action [6–12]. For detection of uracil in DNA, UNGs use a 'pinch-push-pull' mechanism, which involves a multi-step base flipping process. In the first step, serine and proline residues in three different loops (the 'Pro-rich loop', the 'Gly-Ser loop' and the 'Leu-intercalation loop') lead to a slight bending of the DNA through compression ('pinch') of the backbone. In the second step, a conserved leucine residue of the 'Leu-intercalation loop' penetrates into the minor groove and the DNA becomes

---

* Correspondence: debasish@uab.edu
[1]Department of Medicine, University of Alabama at Birmingham, Birmingham, AL 35294, USA
Full list of author information is available at the end of the article

fully bent and kinked leading to flipping of the uracil base ('push'). The uracil nucleotide interacts with the 'uracil recognition pocket' where cleavage of the glycosidic bond takes place, and in the final step, the leucine residue is retracted ('pull'). Two catalytic residues, an aspartic acid and a histidine, are found invariant in all UNGs (Table 1).

UNGs of poxviruses are, however, exceptional and most diverse members of this family. The motifs used by these enzymes are fully conserved in all orthopoxviruses but differ significantly from their counterparts in other organisms (Additional file 1: Table S1) [4, 13–15]. More importantly, UNGs of poxviruses assume a novel and essential role in viral replication and serve as an intrinsic component of the replication machinery. In the prototypic poxvirus Vaccinia, UNG (known as D4) binds to another viral protein A20 to form the heterodimeric processivity factor. A20 also binds to the DNA polymerase E9 and thereby tethers D4 to E9 [15–20] and assembles the heterotrimeric (D4:A20:E9) core of the processive polymerase complex. In the absence of D4, Vaccinia E9 synthesizes only short (<10 nucleotides) stretches of DNA [18, 21]. This essential role of D4 in supporting processive DNA synthesis is not dependent on its catalytic (glycosylase) activity [15].

Crystal structures of D4 have been reported in the free form [13,22; 2owr[1], 2owq, 3 nt7, 4dof, 4dog] and as a non-productive[2] uracil complex [14; 4lzb]. Three dimensional structure of D4 retains the overall fold of UNG proteins- a central 4-stranded parallel β-sheet with α-helices on both sides (Fig. 1). There are two additional β-sheets in D4 (residues 2-6/12-16 and 107-109/215-217) not seen in any other UNG (Figs. 1a, b). Although the UNG-specific motifs are very different, the architecture of the uracil-binding pocket in D4 is remarkably similar to other UNGs and five of the six residues forming the pocket are identical. These residues include the catalytic Asp68 and His181, which are appropriately positioned for binding uracil and cleavage of the glycosidic bond [14]. Considering the unique properties of poxvirus UNGs, it should be highly interesting to understand how these proteins adapt to utilizing the altered motifs for binding DNA, and recognizing and excising uracil.

All crystal structures of free-D4 and the structure of the uracil complex feature a characteristic dimeric packing. Although specific protein-protein interactions between D4 subunits vary to some extent the dimer interface is very similar in these structures [13, 14, 22]. This homodimer interface is formed by residues located in the C-terminal β-strand of the central β-sheet in each subunit, a flexible loop/short helix (residues 161–173) and a long helix (187–206) located near the C-terminus. The first site, which is located in a loop connecting the two C-terminal β-strands, represents one of the most flexible areas in the D4 structures (Fig. 1a). At high protein concentration D4 utilizes the same interface for dimerization in solution. However, when A20 is present, very specific interaction at this surface leads to the formation of a D4:A20 heterodimer (Fig. 1c) [19, 20]. Cumulative structural and biochemical evidences indicate that in the processive DNA polymerase complex, D4 remains catalytically active and possibly continuously scans and removes uracil from newly synthesized DNA [18–20]. Details of D4's ability to bind and scan DNA, its enzymatic function and substrate specificity have been reported [22, 23]. However, structural characterization of the DNA-binding site(s) on D4 is missing.

Here, we report the first detailed view of D4-DNA interactions from a crystal structure of D4 in complex with a non-specific undamaged *ds*DNA. The structure of this complex for the first time reveals the DNA-binding interactions by an UNG from the poxvirus family. It also represents the first three-dimensional structure of a UNG bound to a truly undamaged *ds*DNA and provides a representative snapshot of the initial interaction between UNG and DNA. We also present a structural comparison of D4 and human UNG (hUNG) for which crystal structures are available in the DNA-free state, in complex with uracil-containing DNA and in complex with a DNA containing an abasic site [6–8, 10].

Results and discussion
Overall quality of the D4-DNA complex structure
The D4-DNA structure was refined to 2.89 Å resolution with R_{cryst} and R_{free} values of 21.6 and 26.5 %, respectively. The overall quality of the structural model is good

Table 1 Comparison of motifs for DNA binding and catalysis

	1FLZ (EcUNG)	1SSP (hUNG)	4QCB (vUNG)
Catalytic water-activating loop	62-GQDPYH-67	143-GQDPYH-148	66-GIDPYP-71
Pro-rich loop	84-AIPPS-88	165-PPPPS-169	84-FTKKS-88
Uracil specificity β-strand	120-LLLN-123	201-LLLN-204	117-IPWN-120
Gly-Ser loop	165-GS-166	246-GS-247	160-KT-161
Leu intercalation loop	187-HPSPLSVYR-195	268-HPSPLSAHR-276	181-HPAARDR-187
Active site residues	D64, Y66, F77, N123, H187, L191	D145, Y147, F158, N204, H268, L272	D68, Y70, F79, N120, H181, R185

Motifs as previously listed [13] are updated based on structural superimposition using the new D4-DNA complex structure

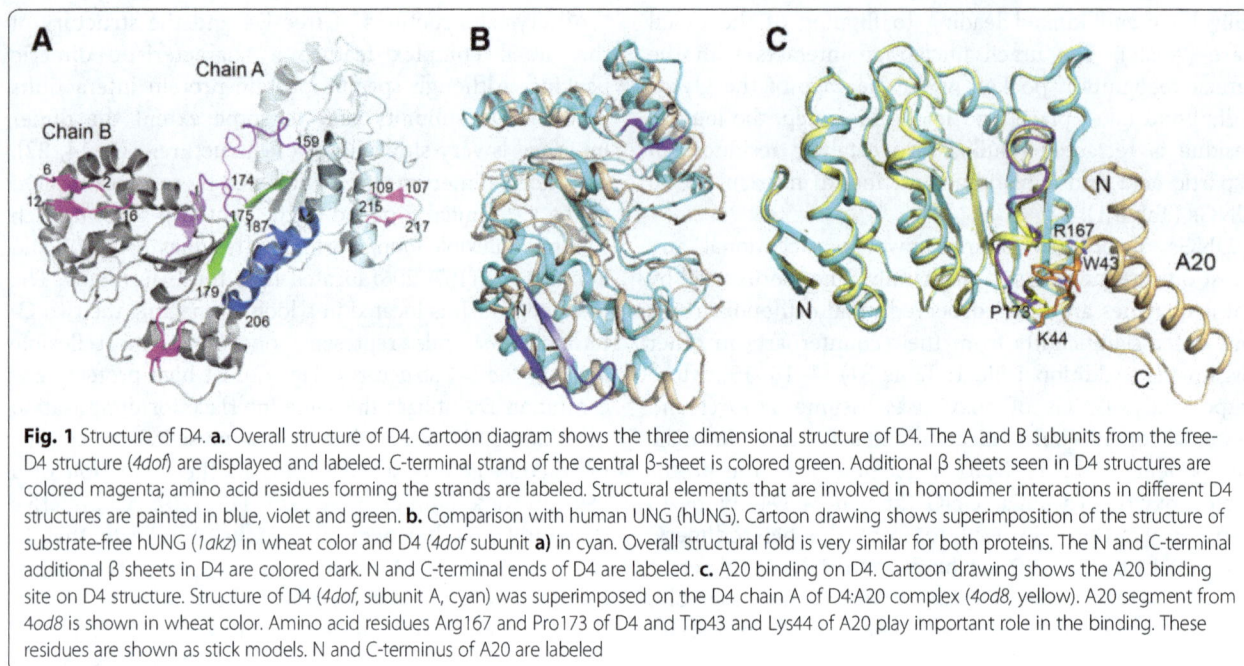

Fig. 1 Structure of D4. **a.** Overall structure of D4. Cartoon diagram shows the three dimensional structure of D4. The A and B subunits from the free-D4 structure (*4dof*) are displayed and labeled. C-terminal strand of the central β-sheet is colored green. Additional β sheets seen in D4 structures are colored magenta; amino acid residues forming the strands are labeled. Structural elements that are involved in homodimer interactions in different D4 structures are painted in blue, violet and green. **b.** Comparison with human UNG (hUNG). Cartoon drawing shows superimposition of the structure of substrate-free hUNG (*1akz*) in wheat color and D4 (*4dof* subunit **a)** in cyan. Overall structural fold is very similar for both proteins. The N and C-terminal additional β sheets in D4 are colored dark. N and C-terminal ends of D4 are labeled. **c.** A20 binding on D4. Cartoon drawing shows the A20 binding site on D4 structure. Structure of D4 (*4dof*, subunit A, cyan) was superimposed on the D4 chain A of D4:A20 complex (*4od8*, yellow). A20 segment from 4od8 is shown in wheat color. Amino acid residues Arg167 and Pro173 of D4 and Trp43 and Lys44 of A20 play important role in the binding. These residues are shown as stick models. N and C-terminus of A20 are labeled

as indicated by MolProbity scores [24] and map correlation coefficients (Table 2). The final model consists of two D4 subunits (chains A and B), a DNA double helix with 12 nucleotides in chain C and 10 in chain D, 4 glycerol (GOL) molecules of which one is in each active site and 118 solvent atoms. Glycerol molecules have been previously observed in the active site of D4 (*2owq, 2owr*) and *E. coli* UNG (*3eug*). Quality of electron density for residues 184–188 of the 'Leu-intercalation' loop was poor. Therefore, these five residues were not modeled. As a result, the discussion of the 'Leu-intercalation' loop for this structure is limited to interactions of loop residues 180–183. Also, side chain atoms of residues 189 and 190 (except for CB) are not included in the final model due to lack of electron density. These residues are located in a flexible area in crystal structures of D4 [13, 14].

In chain C all 12 nucleotides were modeled but only 10 could be modeled in chain D due to insufficient electron density. Of the 10 base pairs forming the DNA double helix, 9 show regular Watson-Crick (WC) base pairing interactions and one non-WC interaction (Fig. 2; Additional file 1: Table S2). In general electron density for the DNA is good for the central regions (C: nucleotides 5–8; D: nucleotides 23–26) that interact with protein residues as indicated by map correlation coefficients (0.92-0.96; average 0.94) and temperature factors (30.5-46.3 Å²; average 37.6 Å²). The overall quality of the electron density for the DNA double helix is displayed in Additional file 1: Figure S1.

A comparison of map correlation coefficients of the DNA nucleotides and the interface residues in D4 is

presented in Additional file 1: Table S3. An electron density map (2mFo-DFc map contoured at 1.5σ level) for both D4-DNA interfaces (chains A and D; chains B and C) is displayed in Additional file 1: Figure S2.

As expected the DNA represents a B-form right-handed double helix. Puckering of the sugar ring for several nucleotides differs from the predominant C2'-endo conformation seen in B-DNA (Additional file 1: Table S4). Similar deviations in sugar puckering conformation have been observed in various protein-DNA complexes.

Assembly of D4-DNA complex

DNA parameters derived from the analysis using the w3DNA web server (http://w3dna.rutgers.edu/) [25] are presented in Additional file 1: Tables S2 and S4. Analyses of DNA-DNA, protein-DNA and protein-protein interfaces using the PDBePISA web server (http://www.ebi.ac.uk/msd-srv/prot_int/pistart.html) [26] are shown in Table 3 and Additional file 1: Table S5.

Conformations of the two D4 chains in the complex are also very similar with an *rmsd* value of 0.15 Å for superposition of all 213 residues in A and B. Each D4 chain binds only to one DNA strand. D4 chain A binds to DNA strand D and D4 chain B binds to DNA strand C (Figs. 2a, b). The electrostatic potential distribution shown in Fig. 2c illustrates the charge complementarity of the DNA binding surface of D4 and the negatively charged phosphate backbone of the DNA strands.

D4 chain B interacts with nucleotides 5–8 (ACGT) in the central part of chain C while D4 chain A interacts

Table 2 Data collection and refinement statistics for *4qcb*

Wavelength [Å]	0.97918
Space group	$P2_12_12_1$
Unit cell parameters [Å]	a = 39.40, b = 92.32, c = 142.88
Data collection statistics	
Resolution limit [Å]	56.50-2.89 (3.05-2.89)[a]
$R_{merge}^{b,c}$	0.158 (0.505)
$R_{meas}^{b,c}$	0.190 (0.609)
$R_{pim}^{b,c}$	0.105 (0.335)
Total number of observations	35850 (5327)
Total number unique	11991 (1750)
Mean I/σ (I)	6.1 (2.3)
$CC_{1/2}$	0.986 (0.855)
CC*	0.996 (0.960)
Completeness [%]	97.9 (99.3)
Multiplicity	3.0 (3.0)
Refinement statistics	
Resolution range (Å)	56.50 - 2.89 (2.97 - 2.89)[a]
Number of unique reflections	11966 (886)
Completeness (%)	97.3 (99.1)
R_{cryst} (%)[d]	21.6 (29.7)
R_{free} (%)[d]	26.5 (32.6)
No. of protein residues	426
No. of DNA nucleotides	22
No. of GOL molecules	4
No. of water molecules	118
Wilson B-factor (Å^2)	29.4
Average B-factors (Å^2)	
Overall	31.4
Protein	28.5
DNA	56.5
GOL	36.6
Water	18.6
Coordinate error (maximum likelihood)	0.39
Correlation coefficient Fo-Fc	0.93
Correlation coefficient Fo-Fc free	0.88
Overall map CC (Fc, 2mFo-DFc)	0.82[e,f]
Ramachandran allowed (%)	99.5

Table 2 Data collection and refinement statistics for *4qcb* (Continued)

Ramachandran disallowed (%)	0.5 [2 outliers, 1 in each subunit]
MolProbity clash score	8.7 [97th percentile]
MolProbity score	1.9 [99th percentile]

[a]Values in parentheses represent highest resolution shell
[b]$R_{meas}$and R_{pim}were calculated with SCALA [38] in the CCP4 program suite [39] using unmerged and not scaled data preprocessed by XDS [36, 37]. $R_{meas}$is a merging R-factor independent of data redundancy while R_{pim} provides the precision of the averaged measurement, which improves with higher multiplicity [48]

$$R_{merge} = \sum_{hkl}\sum_{i}|I_i(hkl) - <I(hkl)>|/\sum_{hkl}\sum_{i}I_i(hkl)$$
$$R_{meas} = \sum_{hkl}\sqrt{N/(N-1)}\sum_{i}|I_i(hkl) - <I(hkl)>|/\sum_{hkl}\sum_{i}I_i(hkl)$$
$$R_{pim} = \sum_{hkl}\sqrt{1/(N-1)}\sum_{i}|I_i(hkl) - <I(hkl)>|/\sum_{hkl}\sum_{i}I_i(hkl)$$

[c]R values for the low resolution shell of 56.50-9.14 Å are: R_{merge} 0.053; R_{meas} 0.065; R_{pim} 0.037
[d]The data included in the R_{free} set (5 %) were excluded from refinement
[e]Final R and R_{free} values based on map calculation in PHENIX [45, 46] are 20.8 % and 24.7 %, respectively
[f]Final R and R_{free} values based on comprehensive validation in PHENIX [45, 46] are 21.5 % and 26.8 %, respectively

with nucleotides 23–26 (AAAC) of chain D. D4 residues at the protein-DNA interface are Ile67, Pro71, Gly128, Glu129, Thr130, Lys131, Gly159, Lys160, Thr161, Asp162, Tyr180, His181 and Ala183 (Table 3 and Additional file 1: Table S5). Polar protein-DNA interactions with the phosphate backbone involve hydrogen-bonds (2.6-3.2 Å) with Lys131, Lys160, Thr161, Tyr180 and His181 from each D4 chain (Table 3; Fig. 2). DNA binding residues in the two D4 chains superimpose very well (*rmsd* 0.13 Å).

Distortion in the DNA chains C and D such as bending and kinking are small compared to those observed in complexes of hUNG with specific uracil-containing DNA [6–8, 10]. The length of the DNA (~37-38 Å) is comparable to the value of 39–41 Å expected for an ideal extended B-form DNA; the widths of the major and minor groove are also similar to those for ideal B-form DNA.

Incidentally, the DNA double helix extends in the unit cell through non-covalent interactions between the ends of the DNA helices generated by the space group symmetry in a head-to-tail fashion (Additional file 1: Figure S3). Similar extended packing through DNA-DNA contacts of *ds*DNA with sticky or blunt ends (head-to-head and head-to-tail fashion) has been observed in other protein-DNA structures [27, 28]. On the other hand this arrangement of DNA chains may be a result of crystal packing.

Both protein-DNA interfaces in the complex have similar interface areas and interface residues (Additional file 1: Table S5). Interactions between the two D4 chains in the complex are minimal and confined to residues Glu32, Val33, Ser35, Trp36, Arg39, Ser132, Ile135, Try136 and Lys139. There is a salt bridge between

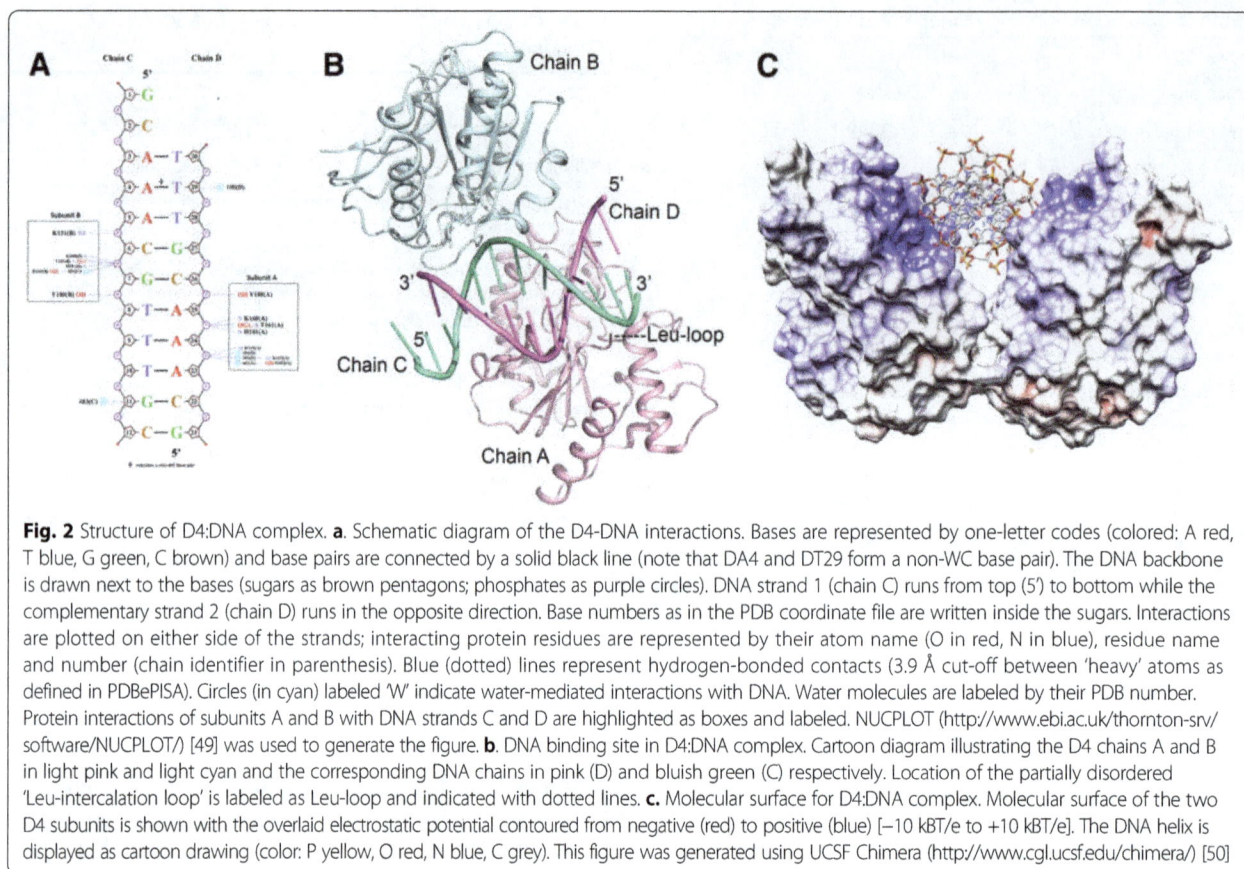

Fig. 2 Structure of D4:DNA complex. **a**. Schematic diagram of the D4-DNA interactions. Bases are represented by one-letter codes (colored: A red, T blue, G green, C brown) and base pairs are connected by a solid black line (note that DA4 and DT29 form a non-WC base pair). The DNA backbone is drawn next to the bases (sugars as brown pentagons; phosphates as purple circles). DNA strand 1 (chain C) runs from top (5') to bottom while the complementary strand 2 (chain D) runs in the opposite direction. Base numbers as in the PDB coordinate file are written inside the sugars. Interactions are plotted on either side of the strands; interacting protein residues are represented by their atom name (O in red, N in blue), residue name and number (chain identifier in parenthesis). Blue (dotted) lines represent hydrogen-bonded contacts (3.9 Å cut-off between 'heavy' atoms as defined in PDBePISA). Circles (in cyan) labeled 'W' indicate water-mediated interactions with DNA. Water molecules are labeled by their PDB number. Protein interactions of subunits A and B with DNA strands C and D are highlighted as boxes and labeled. NUCPLOT (http://www.ebi.ac.uk/thornton-srv/software/NUCPLOT/) [49] was used to generate the figure. **b**. DNA binding site in D4:DNA complex. Cartoon diagram illustrating the D4 chains A and B in light pink and light cyan and the corresponding DNA chains in pink (D) and bluish green (C) respectively. Location of the partially disordered 'Leu-intercalation loop' is labeled as Leu-loop and indicated with dotted lines. **c**. Molecular surface for D4:DNA complex. Molecular surface of the two D4 subunits is shown with the overlaid electrostatic potential contoured from negative (red) to positive (blue) [−10 kBT/e to +10 kBT/e]. The DNA helix is displayed as cartoon drawing (color: P yellow, O red, N blue, C grey). This figure was generated using UCSF Chimera (http://www.cgl.ucsf.edu/chimera/) [50]

residues Glu32 and Arg39 and a stacking interaction between Trp36 of two subunits.

Conformational changes induced by DNA binding
Upon formation of productive DNA complexes, UNGs undergo an 'open to close' conformational transition [6–10]. Such a conformational transition is not expected in the present complex since the DNA in the complex does not contain uracil. D4 residues involved in binding DNA are localized in three structural areas: the extended DNA-binding loop (residues 126–132), the Gly-Ser loop (residues 159–162) and the Leu-intercalation loop (residues 180–187). These three regions are labeled in Fig. 3. Involvement of these sites in DNA-binding and the resulting structural changes in the protein are discussed below. Largest deviations in the DNA complex are noticed in the extended helix/loop segment (residues 164–174,s1 site) joining the two C-terminal strands of the central β-sheet, and in the Leu-intercalation loop (s2 site). These two regions show varying degrees of disorder in D4 structures and portions of these areas are missing from several final models. Therefore, we examined the electron density maps for D4 structures especially focusing on these areas. In the following discussion we used reference free-D4 structures in which

completeness of the model in relevant regions of the protein was good and supported by electron density. We also used the structure of the D4:uracil complex (*4lzb*) for comparison [14]. This structure contains 12 molecules in the asymmetric unit and the s1 and s2 sites could be modeled in 11.

Extended DNA binding loop (residues 126–132)
In the DNA complex the side chain amino group of Lys131 of each D4 subunit forms a strong hydrogen

Table 3 Detailed hydrogen bonding information for D4-DNA interactions in *4qcb* (PISA analysis)

1. D4-DNA (chains A, D)	2. D4-DNA (chains B, C)
#1 D:DA24 [OP2] 2.69 A:LYS131 [NZ]	#1 C:DA5 [OP1] 3.76 B:THR130 [N]
#2 D:DA25 [OP1] 3.10 A:LYS160 [N]	#2 C:DC6 [OP2] 2.59 B:LYS131 [NZ]
#3 D:DA25 [OP2] 3.20 A:LYS160 [N]	#3 C:DG7 [OP1] 3.03 B:LYS160 [N]
#4 D:DA25 [OP2] 2.97 A:THR161 [N]	#4 C:DG7 [OP2] 3.22 B:LYS160 [N]
#5 D:DC26 [OP1] 3.13 A:TYR180 [OH]	#5 C:DG7 [OP2] 2.95 B:THR161 [N]
#6 D:DA25 [OP1] 2.88 A:HIS181 [N]	#6 C:DT8 [OP1] 2.76 B:TYR180 [OH]
	#7 C:DG7 [OP1] 2.91 B:HIS181 [N]

3. D4-DNA (chains A, C)
#1 C:DC12 [OP1] 3.00 A:LYS87 [NZ]

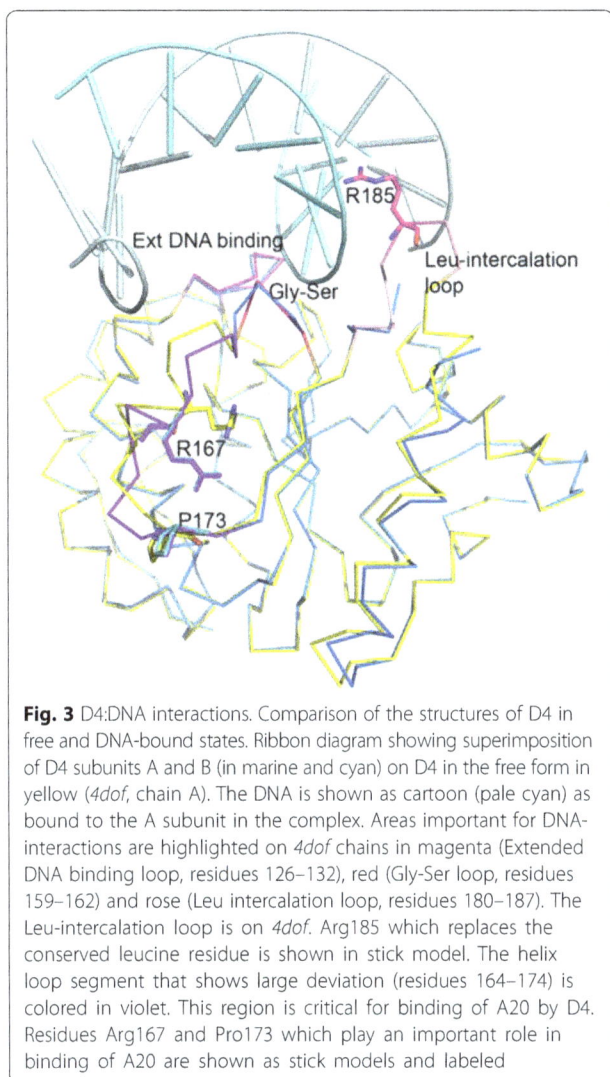

Fig. 3 D4:DNA interactions. Comparison of the structures of D4 in free and DNA-bound states. Ribbon diagram showing superimposition of D4 subunits A and B (in marine and cyan) on D4 in the free form in yellow (4dof, chain A). The DNA is shown as cartoon (pale cyan) as bound to the A subunit in the complex. Areas important for DNA-interactions are highlighted on 4dof chains in magenta (Extended DNA binding loop, residues 126–132), red (Gly-Ser loop, residues 159–162) and rose (Leu intercalation loop, residues 180–187). The Leu-intercalation loop is on 4dof. Arg185 which replaces the conserved leucine residue is shown in stick model. The helix loop segment that shows large deviation (residues 164–174) is colored in violet. This region is critical for binding of A20 by D4. Residues Arg167 and Pro173 which play an important role in binding of A20 are shown as stick models and labeled

bond (2.6 and 2.7 Å) with a phosphate oxygen atom of the partner DNA strand. However, there is no movement in this region of D4 upon DNA binding. Main chain and side chain atoms of Lys131 and neighboring residues superimpose very well with corresponding atoms in free-D4 structures 4dof, 4dog and 4lzb. In all structures the Lys131 side chain extends towards the spatially close Gly-Ser loop and engages in hydrogen bonding with the carboxyl oxygen atoms of Asp162. As a result, the extended DNA-binding loop tilts slightly towards the Gly-Ser loop (Fig. 4). Quality of electron density for Lys131, Asp162 and areas adjacent to these residues is usually very good in these crystal structures (Additional file 1: Figure S4). In a previously published study, the Lys131Val mutant was found to be defective in processivity function [22]. Since this mutant did not show glycosylase activity it was assumed to have a major conformational defect [22]. Both Lys131 and Asp162 are

strictly conserved in UNGs of the orthopoxviruses [15] and thus may play an important role in specialized functions of these viral UNGs, such as a component of the processivity factor. Since neither of these residues is conserved in other UNGs they are less likely to play any direct role in catalysis. On the other hand interactions of Lys131 with Asp162 on the spatially adjacent helix/loop may be important for the stability of the local structure. Therefore the loss of glycosylase activity in the valine mutant may be due to structural impairment. At the same time, D4 surface near Asp162 is critical for its interactions with A20 [22] and therefore, structural changes due to the loss of interactions between Lys131 and Asp162 may affect the processivity function.

In a study combining H/D exchange mass spectroscopy and computational docking Roberts et al. [29] identified two extended non-sequence-specific DNA-binding surfaces in hUNG (residues 210–220 and 251–264) that did not show contacts with DNA in the crystal structures. In D4 the corresponding regions are residues 126–136 and 165–178. The latter segment in D4 is engaged in homodimer formation in the absence of A20 and in heterodimer formation in the presence of A20 [20]. Thus this (residue 165–178) may not represent a potential DNA-binding site in D4.

Gly-Ser loop (residues 159–162)

The s2 site encompasses the Gly-Ser loop. This region shows varying degrees of flexibility in different D4 structures but could be modeled in both subunits in the DNA complex. Quality of electron density in this area of D4 was sufficient for unambiguously placing main chain and side chain atoms of all residues. For comparison of this site we selected the free-D4 structure 4dof (chains A, B) and the D4:uracil complex (4lzb, chain D, F). Comparison of the s2 site in the DNA-complex with free-D4 structures reveals a slight reorganization in this area of the molecule (Fig. 5). Two amino acid residues, Lys160 and Thr161, are involved in hydrophilic interactions with the DNA. Main chain nitrogen atom of Lys160 interacts with two oxygen atoms on one phosphate group (Table 3). Thr161 hydroxyl oxygen atom is hydrogen bonded to one of the oxygen atoms of the same phosphate also through the peptide nitrogen atom. Notably, Lys160 and Thr161 are unique residues in poxvirus UNGs and are replaced by glycine and serine respectively in both E. coli and human enzymes. We showed that D4 mutant Lys160Val was deficient in processivity function but retained catalytic activity and DNA binding ability [22]. In all D4 structures, side chain of Lys160 extends toward the neighboring β-strand and its NZ atom forms a hydrogen bond with the peptide oxygen atom of Val178 on this strand. Loss of this contact in the Lys160Val mutant may impact the structure

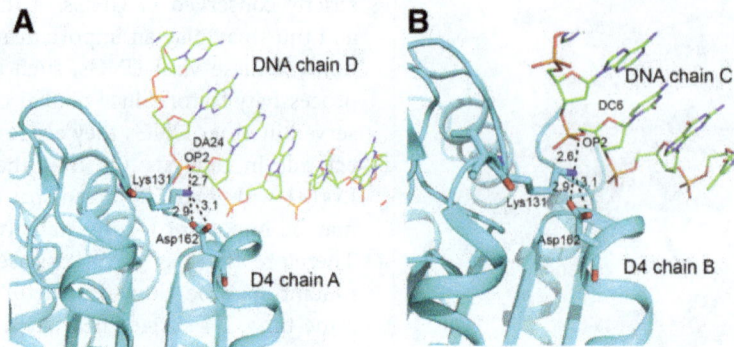

Fig. 4 Protein-DNA interactions at the extended DNA-binding site. **a** and **b** shown in left and right panel are hydrogen bonding interactions (identified by PISA analysis) involving D4 chains A and B, respectively. Relevant D4 residues are shown in stick models and are labeled. DNA chains are shown as lines. Hydrogen bonding distances shown are in a. Interaction of Lys131 side chain NZ atom with Asp162 carboxyl group may be important for local conformational stability. Electron density for this region of DNA complex is shown in Additional file 1: Figure S4

of this area of D4. Notably, a hydrophobic pocket at this site has been shown to be important for D4's interaction with A20 [20].

Leu-intercalation loop (residues 180–187)

The s3 area, which includes the 'Leu-intercalation loop', is important for the catalytic mechanism. Electron density for this segment of D4 structure in *4dog* and *3 nt7* was good [22]. In the DNA-complex quality of electron density for this loop was inadequate for modeling the entire loop. Therefore, residues 184–188 were omitted from the final model. Two amino acid residues immediately preceding the disordered region of D4, namely Tyr180 and His181, directly interact with DNA. Side chain hydroxyl group of Tyr180 and the main chain nitrogen atom of His181 form contact with oxygen atoms

of two phosphate groups from the DNA (Fig. 6). Tyr180 is conserved in UNGs of poxviruses but not in other UNGs. His181 is a critical catalytic residue and is conserved in UNGs across the species. In *4zlb* we noticed some deviations in the position of His181 in different subunits. Generally, in a productive uracil complex, the distance between the NE2 atom of histidine and the O2 atom of uracil is significantly shorter than in nonproductive complexes [6–8]. In *4lzb*, the distance between O2 atom of uracil and His181 NE2 atom varied from 2.8-4.8 Å in different subunits [14]. In the DNA complex His181 side chain moved farther away from the active site area. The missing segment includes Arg185, which is equivalent to the conserved leucine residue (272 in hUNG) of the Leu-intercalation loop. In the mutational analysis mentioned above D4 Arg187Val mutant

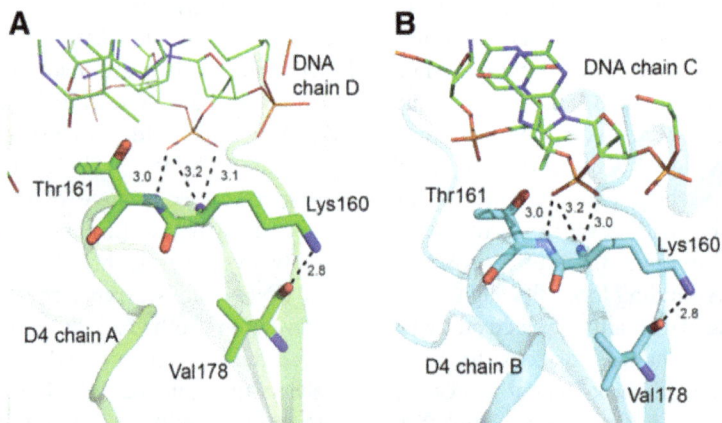

Fig. 5 Protein-DNA interactions at the Gly-Ser loop. Hydrogen bonding interactions between the D4 residues in A and B subunits with partner DNA strands are shown in left and right panel. Relevant D4 residues are shown in stick models and are labeled. DNA chains are shown as lines. Hydrogen bonding distances shown are in Å. Also NZ atom of Lys160 forms hydrogen bonding interactions with the peptide oxygen atom of Val178 located on the C-terminal strand of the central β-sheet

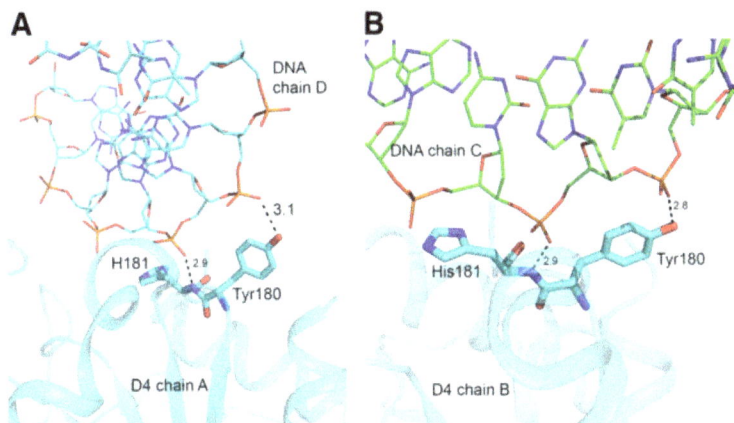

Fig. 6 Protein-DNA interactions at the Leu-intercalation loop. Cartoon and stick drawing displaying hydrogen bonding interactions at the Leu-intercalation loop region between the D4 residues in A and B subunits with partner DNA strands. Relevant D4 residues are shown in stick models and are labeled. DNA chains are shown as lines. Hydrogen bonding distances shown are in Å. Residues 184–188 were missing from both subunits

was also shown to be unable to support processive DNA synthesis but retained binding to A20 and DNA [22]. Unfortunately, due to disorder in this loop detailed comparison of this functionally significant region is not possible from the present structure.

Movement of Arg167 and implications for A20 binding

Most significant changes in the DNA complex were noticed in residues Arg167 and Pro173 although neither of these residues is involved in direct interaction with DNA. In *4dof* and *4lzb*, orientation of the Arg167 side chain is stabilized by a strong hydrogen bond between its NH2 atom and the main oxygen atom of Thr176. In both D4 subunits in the DNA complex, Arg167 side chain is directed away from Thr176 (Fig. 7a). The Arg167 side chain is oriented such that nitrogen atoms of the terminal amino groups form hydrogen bonds with the peptide oxygen atom of Val174. Position of Arg167 side chain is further stabilized by polar interactions of its amino groups with the hydroxyl oxygen atom of Thr176 side chain on one side and the main chain oxygen atom of Ser172 on the other (Fig. 7b).

In the crystal structure of the D4:A20 complex, the aromatic side chain of Trp43 of A20 occupies a hydrophobic pocket referred to as groove by Contesto-Richefeu et al. [22] and inserts into the space between Pro173 and Arg167 side chain. Movement of Arg167 side chain and Pro173 in the A20 complex (as compared to free D4 structures) favors this interaction. On the other hand in the DNA-complex, orientation of Arg167 side chain would interfere with binding of A20 at this site. Interestingly, movement of Pro173 in the DNA-complex also seems to create an unfavorable steric environment for binding of A20. While in the D4:A20 complex, Pro173 of D4 moves in a direction that allowed

packing of Lys44 side chain of A20, in the DNA complex the carbonyl oxygen Pro173 would be in close contact (~1.5 Å) with the CB atom of Lys44 of A20 (Fig. 7c). Thus the structure of the present DNA complex suggests that DNA binding induces conformational changes in D4 that may interfere with its binding to A20. While D4 can independently bind both DNA and A20, perhaps in the formation of the heterotrimeric complex, D4-A20-interaction precedes the DNA-binding event.

Comparison of hUNG and D4

Crystal structures of hUNG have been reported in the free-enzyme form [*1akz*; 6] and in complex with damaged DNA containing uracil [*1ssp*; 6]. Structure of mutant hUNG in which the leucine residue (272) of the Leu-intercalation loop was altered to alanine was described in complex with DNA containing an abasic site [*2ssp*; 6]. Superimposition and comparison of the structure of hUNG with D4 in their DNA complexes demonstrate the divergences and general overlap in the areas important for DNA binding and UNG catalytic mechanism. The three DNA binding sites described above are shown in Fig. 8. Structural comparison of hUNG in the free-state and in complex with different DNA constructs demonstrated that major conformational changes in UNGs are promoted by the steps involved in generating and stabilizing the flipped out nucleotide upon recognition of uracil [10]. Upon binding an undamaged DNA, these conformational changes are not expected in D4. Superimposition of the structure of the D4:DNA complex and the DNA complex of hUNG (*1ssp*) reveals that the Gly-Ser loop and the Leu-intercalation loop provide the main interactions with DNA. Part of the Leu-intercalation loop is disordered in the D4:DNA complex. In hUNG, the catalytic histidine (His268 in *1ssp*) clearly

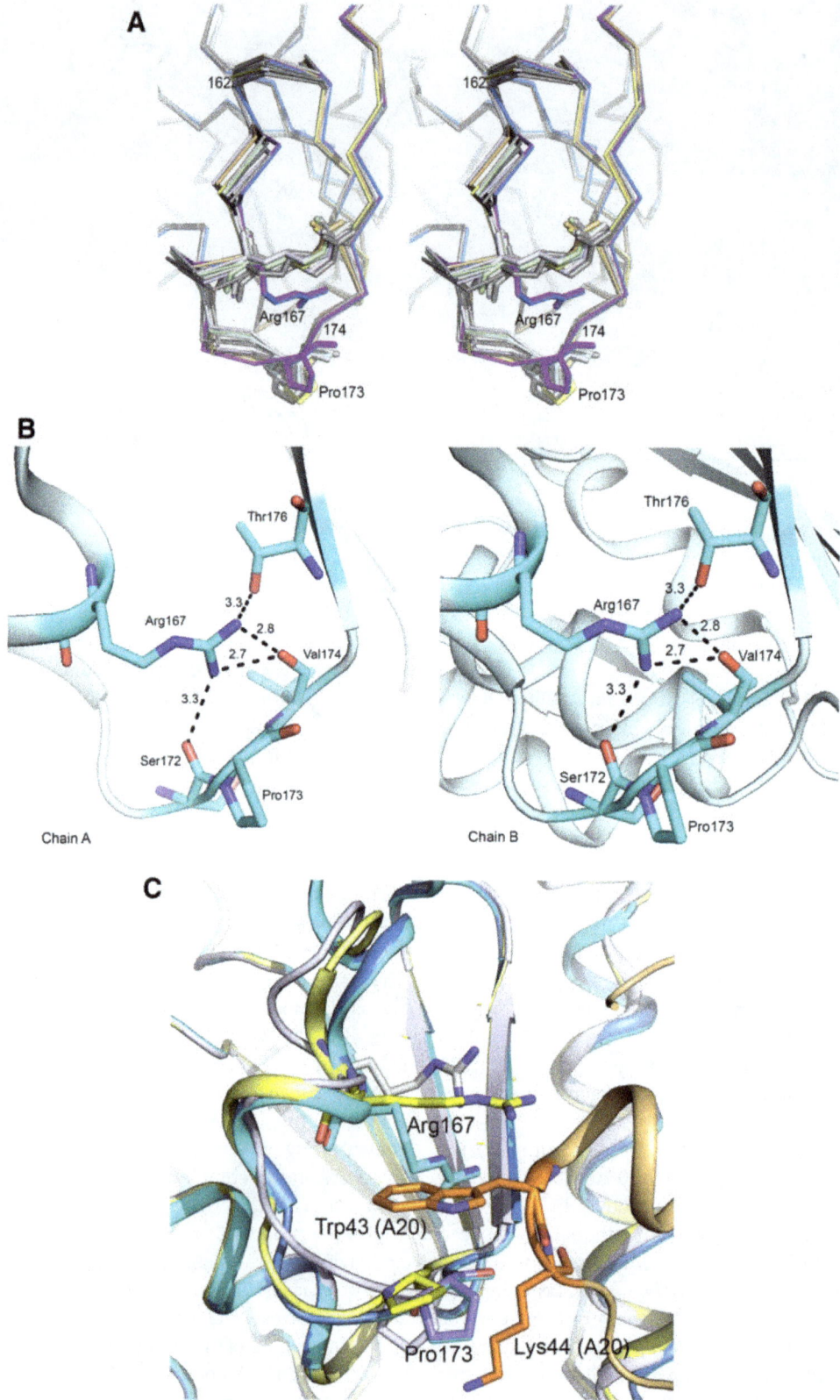

Fig. 7 (See legend on next page.)

(See figure on previous page.)

Fig. 7 Movement of Arg167 in the D4:DNA complex and A20 binding. **a.** Arg167 side chain of D4 in the DNA complex and in DNA-free D4. Stereo diagram showing movement of Arg167 and Pro173 residues of D4 chains A and B in the DNA complex (shown in blue and purple violet colors) superimposed on D4 subunits in different subunits of the uracil complex (*4lzb*) shown in various shades of yellow and green. Arg167 and Pro173 residues are shown in stick model. **b.** Hydrogen bonding interactions of Arg167 in the DNA complex. Close up view of the above region showing hydrogen bonding interactions of Arg167 of the two D4 chains A and B of the DNA complex in the left and right panel, respectively. **c.** Orientations of Arg167 and Pro173 in the DNA complex. Cartoon drawing showing close up view of the A20 binding site on D4. D4 subunits from free-D4 (*4dof* subunit A, white), and D4 subunits A and B from the DNA complex (cyan and marine) are superimposed on the D4 subunit of the D4:A20 complex (*4od8*, yellow). The A20 molecule in *4od8* is colored in wheat. Trp43 and Lys44 residues of A20 are shown in stick model (carbon atoms in orange). Arg167 residues of D4 chains are also shown in stick models and carbon atoms of each stick model are colored according to the chain. Orientations of Arg167 and Pro173 of the D4 subunits in the DNA complex are unfavorable for binding of A20 in the groove

moves slightly closer towards the uracil binding pocket as compared to the D4:DNA complex. It should be noted that in DNA-free state this histidine residue superimposes well in D4 and hUNG (see Fig. 8). The interactions between the DNA and D4 at the extended DNA-binding site may be facilitated because of the larger size of the bound DNA as compared to hUNG:DNA complexes.

Conclusions

The D4-DNA complex described here provides the structural framework for recognizing the DNA binding residues and motifs in poxvirus UNGs. The process of sequence-specific recognition and binding of DNA includes non-specific interaction as an early or intermediate step [30–33]. The structure with undamaged *ds*DNA may represent a snapshot of an early DNA-protein interaction in preparation for the recognition complex for

UNG [34]. It remains to be seen if the Leu-intercalation loop transitions from its disordered state to an ordered state upon recognition of a damage in the DNA. Some of the unique structural features of D4 have evolved for undertaking its novel role in viral replication. Structure presented here provides potential understanding of DNA-binding by D4 in the poxvirus replication machinery.

Methods
Preparation and crystallization of D4-DNA complex

Custom DNA oligos purified by standard desalting method were ordered from Integrated DNA Technologies (IDT). The 12mer DNA oligo was designed to be self-complementary: 5'-GCA AAC GTT TGC-3'. The DNA oligo was dissolved in annealing buffer (10 mM Tris, pH 7.5, 50 mM NaCl, 1 mM EDTA) at a

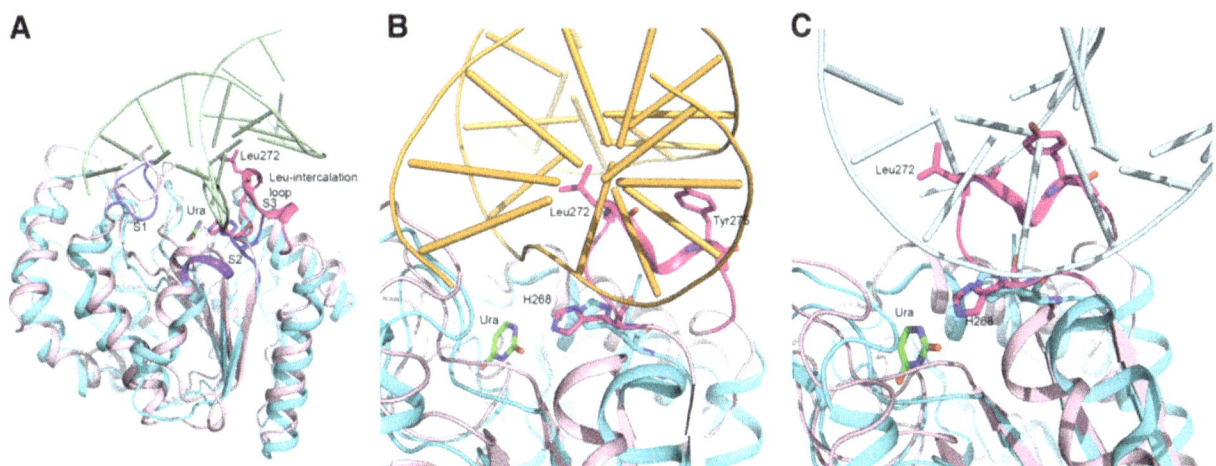

Fig. 8 Comparison of D4:DNA complex with hUNG:DNA complex. **a.** Overall comparison. Structure of the A subunit of the D4:DNA complex (cyan) is superimposed on the structure of hUNG:DNA (*1ssp*, light pink). DNA chain from *1ssp* is shown in cartoon drawing (pale green). Uracil molecule cleaved and bound at the uracil binding pocket in hUNG is shown in stick model (carbon in green). His181 of D4 and the corresponding His268 of hUNG are shown in stick models. Leu272 of the Leu-intercalation loop (magenta) in hUNG is also shown in magenta sticks. DNA-binding areas on D4 are colored darker: S1 site, residues 126–132, S2 site residues 159–162 and S3 site residues 180–187. The corresponding residue of Leu272 in D4 is disordered. **b.** Close up view 1. Close up view showing superimposition of D4 subunit A of D4:DNA complex (cyan) on hUNG:DNA complex structure (light magenta). DNA bound to hUNG is shown in cartoon (light orange). Uracil molecule (Ura) in hUNG uracil binding pocket is shown as stick model (carbon in green). Catalytic His268 of hUNG and the corresponding H181 of D4 are also shown in stick model. **c.** Close up view 2. Cartoon diagram shows superimposition of the hUNG chain (light magenta) from the DNA complex *1ssp* on the D4 chain A (cyan) of the D4:DNA complex. DNA shown in cartoon model is from the D4:DNA complex (chain D, light cyan). Unless specified otherwise figures were generated using PyMOL (http://pymol.org/) [51]

concentration of 5 mM. Annealing using a Perkin Elmer GeneAmp PCR System 2400 resulted in a blunt-end 12mer double-stranded DNA construct with 12 base pairs (final concentration 2.5 mM):

5'-GCA AAC GTT TGC-3' (forward strand)
3'-CGT TTG CAA ACG-5' (reverse complementary strand)

The following annealing protocol was used:

1. Heating to 368 K in 5 min;
2. Cooling to 338 K in 5 min;
3. Cooling to 328 K in 5 min;
4. Cooling to 318 K in 5 min;
5. Cooling to 298 K in 5 min;
6. Cooling to 277 K in 40 min.

DNA oligos and annealed dsDNA were kept at 253 K until use.

Tag-free D4 (tfD4) was purified as previously described [14, 35]. The purified protein contained three vector encoded residues (–GSH) after the hexa-histidine tag was cleaved off.

Protein solution at 3 mg/mL (~120 μM in 25 mM HEPES buffer, pH 7.3, 60 mM KCl, 1 mM TCEP) was incubated with previously annealed self-complementary non-specific dsDNA in a 1:1.2 molar ratio at 277 K for 1 h. High-throughput crystallization screening using the PEGs and Protein Complex Suites from Qiagen was conducted on a Crystal Gryphon (Art Robbins Instruments). Thin plate-like crystals were obtained in hanging drop vapor diffusion at 293 K. The drops contained 1 μL protein plus 0.5 μL reservoir solution containing 16 % PEG6K, 0.08 M TRIS buffer, pH 8.5, 20 % glycerol. For cryo-freezing, crystals were directly plunged into liquid nitrogen.

Data collection and refinement

Intensity data were collected on a Pilatus 6 M detector at the Advanced Photon Source on NE-CAT beamline 24-ID-C. Data were processed with XDS [36, 37] and SCALA [38] of the CCP4 program suite [39] as part of the RAPD data-collection strategy at NE-CAT (https://chem.cornell.edu/rapd/). A total of 120 images (1° width) were collected. While the first 75 frames showed overall good statistics frames 76–120 showed increased mosaicity and R_{merge} values. The processed data from frames 1–75 were indexed in the orthorhombic crystal system. Probabilities for Laue group P222 and space group $P2_12_12_1$ based on systematic absences were 0.856 and 0.967, respectively. Diffraction data extending to 2.89 Å resolution were used for molecular replacement and refinement. $CC_{1/2}$ and CC^* values at 2.89 Å resolution

were 0.86 and 0.96, respectively [40, 41]. Data-collection statistics are listed in Table 1.

The unit-cell parameters (a = 39.4 Å, b = 92.3 Å, c = 142.9 Å; space group $P2_12_12_1$) suggested two protein subunits in the asymmetric unit (52 % solvent; Matthews coefficient V_M = 2.6). The crystal structure was solved by molecular replacement with Phaser (version 2.5.6) [42] using the coordinates of one subunit from the native D4 structure (*4dof*) as search model. The observed pseudo-translation vector was used to position the second subunit in the asymmetric unit. Of the alternative space groups in Laue group P222 only space group $P2_12_12_1$ provided solutions for the translation function. The R value of the refined solution was 45.2 % (RFZ = 5.8, TFZ = 9.3, LLG = 1803). After 1 round of refinement (10 cycles) of the protein coordinates (R_{cryst} and R_{free} dropped to 29.2 % and 32.9 %, respectively) in REFMAC (version 5.8.0073) [43] electron density for the DNA double helix became visible in sigmaA weighted $2mF_o$-DF_c and mF_o-DF_c difference electron density maps (m is the figure of merit and D is the sigmaA weighting factor). Using modeling tools in Coot (version 0.8-pre) [44] B-form DNA nucleotides were fitted into the electron density maps. The forward strand (chain C) of the dsDNA is numbered from 1–12 (5' to 3') while the reverse complementary strand (chain D) is numbered 21–32 (5' to 3'). Base pairing occurs between the base of nucleotide 1 and the base of nucleotide 32 and similarly between the nucleotide 12 and 21 (and so forth). For structure refinement we used automatically generated local non-crystallographic symmetry (NCS) restraints in REFMAC. Structure validation was accomplished using PHENIX (version 1.9-1692) [45, 46], MolProbity [24], QualityCheck (http://smb.slac.stanford.edu/jcsg/QC/) [47], and the new wwPDB X-ray validation server (http://wwpdb-validation.wwpdb.org/validservice/). PHENIX and REFMAC were used for final map calculations. Refinement statistics are listed in Table 1. The structure has been deposited in PDB under *4qcb*. The difference between R and R_{free}, although within reasonable limit for structures in similar resolution, is somewhat higher and may result from incompleteness of the models (for DNA and protein).

Endnotes

[1] PDBID's are shown in italics
[2] Uracil in this complex is not a product of glycosylase activity

Additional files

Additional file 1: **Table S1.** Sequence alignment of six specific regions in UNG enzymes that include the motifs for DNA binding and catalysis. **Table S2.** Detailed hydrogen bonding information for DNA base pairing (w3DNA analysis) in 4QCB. **Table S3.** Map correlation coefficients and average B values for protein interface residues and DNA nucleotides. **Table S4.** Conformational sugar parameters in 4QCB (w3DNA analysis). **Table S5.** Analysis of protein-protein, protein-DNA and DNA-DNA interfaces in *4qcb* (PISA analysis). **Figure S1.** SigmaA weighted 2mFo-DFc and mFo-DFc omit maps for DNA region. DNA double helix with chains C and D. A. 2mFo-DFc map (contoured at 1.5σ level). B. mFo-DFc omit map (contoured at 2.5σ level). C. 2mFo-DFc simulated annealing omit map (contoured at 1.0σ level). **Figure S2.** SigmaA weighted 2mFo-DFc map for D4-DNA interfaces. A. Stick model representation of D4-DNA interface 1 (chains A and D). B. 2mFo-DFc map of D4-DNA interface 1 (chains A and D). C. Stick model representation of D4-DNA interface 2 (chains B and C). D. 2mFo-DFc map of D4-DNA interface 2 (chains B and C). **Figure S3.** Packing of D4-DNA complex. **Figure S4.** Electron density map.

Abbreviations

D4: Vaccinia uracil-DNA glycosylase; BER: Base excision repair; AP: Apurinic/apyrimidinic; Cyt: Cytosine; Ura: Uracil; UDG: Uracil-DNA glycosylase; UNG: Family I uracil-DNA glycosylase; ssDNA: Single-stranded DNA; dsDNA: Double-stranded DNA; hUNG: Human UNG; GOL: glycerol; EDTA: Ethylenediaminetetraacetic acid; PEG: Polyethylene glycol.

Competing interests

The authors declare that they have no competing interests.

Authors' contributions

NS was involved in crystallization, structure determination, structure refinement and manuscript preparation. SB performed data collection and data processing. RR and DC were involved in manuscript preparation. DC designed and participated in the experiments, and provided technical and scientific guidance. All authors read and approved the final manuscript.

Acknowledgements

This work was supported by the National Institutes of Health grant 5U01-A1-082211.
The work is based upon research conducted at the Advanced Photon Source on the Northeastern Collaborative Access Team beamlines, which are supported by a grant from the National Institute of General Medical Sciences (P41 GM103403) from the National Institutes of Health. Use of the Advanced Photon Source, an Office of Science User Facility operated for the U.S. Department of Energy (DOE) Office of Science by Argonne National Laboratory, was supported by the U.S. DOE under Contract No. DE-AC02-06CH11357.

Author details

[1]Department of Medicine, University of Alabama at Birmingham, Birmingham, AL 35294, USA. [2]Northeastern Collaborative Access Team and Department of Chemistry and Chemical Biology, Cornell University, Argonne, Chicago, IL 60439, USA. [3]Department of Microbiology, School of Dental Medicine, Abramson Cancer Center, University of Pennsylvania, Philadelphia, PA 19104, USA.

References

1. Nilsen H, Krokan HE. Base excision repair in a network of defense and tolerance. Carcinogenesis. 2001;22:987–98.
2. Jacobs AL, Schär P. DNA glycosylases in DNA repair and beyond. Chromosoma. 2012;121:1–20.
3. Visnes T, Doseth B, Sahlin Pettersen H, Hagen L, Sousa MML, Akbari M, et al. Uracil in DNA and its processing by different DNA glycosylases. Phil Trans R Soc Lond B. 2009;364:563–8.
4. Schormann N, Ricciardi R, Chattopadhyay D. Uracil-DNA glycosylases – Structural and functional perspectives on an essential family of DNA repair enzymes. Protein Sci. 2014;23:1667–85.
5. Pearl LH. Structure and function in the uracil-DNA glycosylase superfamily. Mutat Res. 2000;460:165–81.
6. Parikh SS, Mol CD, Slupphaug G, Bharati S, Krokan HE, Tainer JA. Base excision repair initiation revealed by crystal structures and binding kinetics of human uracil-DNA glycosylase with DNA. EMBO J. 1998;17:5214–26.
7. Parikh SS, Walcher G, Jones CD, Slupphaug G, Krokan HE, Blackburn GM, et al. Uracil-DNA glycosylase-DNA substrate and product structures: conformational strain promotes catalytic efficiency by coupled stereoelectronic effects. Proc Natl Acad Sci U S A. 2000;97:5083–8.
8. Werner RM, Jiang YL, Gordley RG, Jagadeesh GJ, Ladner JE, Xiao G, et al. Stressing-out DNA? The contribution of serine-phosphodiester interactions in catalysis by uracil DNA glycosylase. Biochemistry. 2000;39:12585–94.
9. Jiang YL, Stivers JT. Mutational analysis of the base-flipping mechanism of uracil DNA glycosylase. Biochemistry. 2002;41:11236–47.
10. Slupphaug G, Mol CD, Kavli B, Arvai AS, Krokan HE, Tainer JA. A nucleotide-flipping mechanism from the structure of human uracil-DNA glycosylase bound to DNA. Nature. 1996;384:87–92.
11. Savva R, McAuley-Hecht K, Brown T, Pearl L. The structural basis of specific base-excision repair by uracil-DNA glycosylase. Nature. 1995;373:487–93.
12. Parker JP, Bianchet MA, Krosky DJ, Friedman JI, Amzel LM, Stivers JT. Enzymatic capture of an extrahelical thymine in the search for uracil in DNA. Nature. 2007;449:433–8.
13. Schormann N, Grigorian A, Samal A, Krishnan R, DeLucas L, Chattopadhyay D. Crystal structure of vaccinia virus uracil-DNA glycosylase reveals dimeric assembly. BMC Struct Biol. 2007;7:45.
14. Schormann N, Banerjee S, Ricciardi R, Chattopadhyay D. Structure of the uracil complex of vaccinia virus uracil DNA glycosylase. Acta Cryst F. 2013;69:1328–34.
15. De Silva FS, Moss B. Vaccinia virus uracil DNA glycosylase has an essential role in DNA synthesis that is independent of its glycosylase activity: catalytic site mutations reduce virulence but not virus replication in cultured cells. J Virol. 2003;77:159–66.
16. Ishii K, Moss B. Mapping interaction sites of the A20R protein component of the vaccinia virus DNA replication complex. Virology. 2002;303:232–9.
17. Stanitsa ES, Arps L, Traktman P. Vaccinia virus uracil DNA glycosylase interacts with the A20 protein to form a heterodimeric processivity factor for the viral DNA polymerase. J Biol Chem. 2006;281:3439–51.
18. Boyle KA, Stanitsa ES, Greseth MD, Lindgren JK, Traktman P. Evaluation of the role of the vaccinia virus uracil DNA glycosylase and A20 proteins as intrinsic components of the DNA polymerase holoenzyme. J Biol Chem. 2011;286:24702–13.
19. Sèle C, Gabel F, Gutsche I, Ivanov I, Burmeister WP, Iseni F, et al. Low-resolution structure of vaccinia virus DNA replication machinery. J Virol. 2013;87:1679–89.
20. Contesto-Richefeu C, Tarbouriech N, Brazzolotto X, Betzi S, Morelli X, Burmeister WP, et al. Crystal structure of the vaccinia virus DNA polymerase holoenzyme subunit D4 in complex with the A20 N-terminal domain. PLoS Pathog. 2014;10:e1003978.
21. McDonald WF, Traktman P. Vaccinia virus DNA polymerase. In vitro analysis of parameters affecting processivity. J Biol Chem. 1994;269:31190–7.
22. Druck Shudofsky AM, Silverman JE, Chattopadhyay D, Ricciardi RP. Vaccinia virus D4 mutants defective in processive DNA synthesis retain binding to A20 and DNA. J Virol. 2010;84:12325–35.
23. Scaramozzino N, Sanz G, Crance JM, Saparbaev M, Drillen R, Laval J, et al. Characterisation of the substrate specificity of homogenous vaccinia virus uracil-DNA glycosylase. Nucleic Acids Res. 2003;31:4950–7.
24. Davis IW, Leaver-Fay A, Chen VB, Block JN, Kapral GJ, Wang X, et al. MolProbity: all-atom contacts and structure validation for proteins and nucleic acids. Nucleic Acids Res. 2007;35:W375–83.
25. Zheng G, Lu X-J, Olson WK. Web 3DNA – A web server for the analysis, reconstruction, and visualization of three-dimensional nucleic-acid structures. Nucleic Acids Res. 2009;37:W240–6.
26. Krissinel E, Henrick K. Inference of macromolecular assemblies from crystalline state. J Mol Biol. 2007;372:774–97.
27. Rice PA, Yang S, Mizuuchi K, Nash HA. Crystal structure of an IHF-DNA complex: A protein-induced DNA U-turn. Cell. 1996;87:1295–306.
28. Dürr H, Körner C, Müller M, Hickmann V, Hopfner KP. X-ray structures of the Sulfolobus solfataricus SWI2/SNF2 ATPase core and its complex with DNA. Cell. 2005;121:363–73.
29. Roberts VA, Pique ME, Hsu S, Li S, Slupphaug G, Rambo RP, et al. Combining H/D exchange mass spectroscopy and computational docking reveals extended DNA-binding surface on uracil-DNA glycosylase. Nucleic Acids Res. 2012;40:6070–81.

30. Cao C, Jiang YL, Stivers JT, Song F. Dynamic opening of DNA during the enzymatic search for a damaged base. Nat Struct Mol Biol. 2004;11:1230–6.

31. Kalodimos CG, Biris N, Bonvin AMJJ, Levandoski MM, Guennuegues M, Boelens R, et al. Structure and flexibility adaptation in nonspecific and specific protein-DNA complexes. Science. 2004;305:386–9.

32. Viadiu H, Aggarwal AK. Structure of BamHI bound to nonspecific DNA: a model for DNA sliding. Mol Cell. 2000;5:889–95.

33. Zharkov DO, Mechetin GV, Nevinsky GA. Uracil-DNA glycosylase: Structural, thermodynamic and kinetic aspects of lesion search and recognition. Mutat Res. 2010;685:11–20.

34. Schonhoft JD, Kosowicz J, Stivers JT. DNA translocation by human uracil DNA glycosylase: Role of DNA phosphate charge. Biochemistry. 2013;52:2526–35.

35. Sartmatova D, Nash T, Schormann N, Nuth M, Ricciardi R, Banerjee S, et al. Crystallization and preliminary X-ray diffraction analysis of three recombinant mutants of Vaccinia virus uracil DNA glycosylase. Acta Cryst F. 2013;69:295–301.

36. Kabsch W. XDS. Acta Cryst D Biol Cryst. 2010;66:125–32.

37. Kabsch W. Integration, scaling, space-group assignment and post-refinement. Acta Cryst D Biol Cryst. 2010;66:133–44.

38. Evans P. Scaling and assessment of data quality. Acta Cryst D Biol Cryst. 2006;62:72–82.

39. Winn MD, Ballard CC, Cowtan KD, Dodson EJ, Emsley P, Evans PR, et al. Overview of the CCP4 suite and current developments. Acta Cryst D Biol Cryst. 2011;67:235–42.

40. Karplus PA, Diederichs K. Linking crystallographic model and data quality. Science. 2012;336:1030–3.

41. Diederichs K, Karplus PA. Better models by discarding data? Acta Cryst D Biol Cryst. 2013;69:1215–22.

42. McCoy AJ, Grosse-Kunstleve RW, Adams PD, Winn MD, Storoni LC, Read RJ. *Phaser* crystallographic software. J Appl Cryst. 2007;40:658–74.

43. Murshudov GN, Skubák P, Lebedev AA, Pannu NS, Steiner RA, Nicholls RA, et al. REFMAC5 for the refinement of macromolecular crystal structures. Acta Cryst D Biol Cryst. 2011;67:355–67.

44. Emsley P, Cowtan K. Coot: Model-building tools for molecular graphics. Acta Cryst D Biol Cryst. 2004;60:2126–32.

45. Adams PD, Afonine PV, Bunkoczi G, Chen VB, Davis IW, Echols N, et al. PHENIX: a comprehensive Python-based system for macromolecular structure solution. Acta Cryst D Biol Cryst. 2010;66:213–21.

46. Afonine PV, Grosse-Kunstleve RW, Echols N, Headd JJ, Moriarty NW, Mustyakimov M, et al. Towards automated crystallographic structure refinement with phenix.refine. Acta Cryst D Biol Cryst. 2012;68:352–67.

47. Yang H, Guranovic V, Dutta S, Feng Z, Berman HM, Westbrook JD. Automated and accurate deposition of structures solved by X-ray diffraction to the Protein Data Bank. Acta Cryst D Biol Cryst. 2004;60:1833–9.

48. Weiss MS. Global indicators of X-ray data quality. J Appl Cryst. 2001;34:130–5.

49. Luscombe NM, Laskowski RA, Thornton JM. NUCPLOT: a program to generate schematic diagrams of protein-nucleic acid interactions. Nucleic Acids Res. 1997;25:4940–5.

50. Pettersen EF, Goddard TD, Huang CC, Couch GS, Greenblatt DM, Meng EC, et al. UCSF Chimera – A visualization system for exploratory research and analysis. J Comput Chem. 2004;25:1605–12.

51. DeLano WL. PyMOL, vol. version 1.7.x. Schrödinger, LLC: Open-Source PyMOLTM; 2002.

PERMISSIONS

The contributors of this book come from diverse backgrounds, making this book a truly international effort. This book will bring forth new frontiers with its revolutionizing research information and detailed analysis of the nascent developments around the world.

We would like to thank all the contributing authors for lending their expertise to make the book truly unique. They have played a crucial role in the development of this book. Without their invaluable contributions this book wouldn't have been possible. They have made vital efforts to compile up to date information on the varied aspects of this subject to make this book a valuable addition to the collection of many professionals and students.

This book was conceptualized with the vision of imparting up-to-date information and advanced data in this field. To ensure the same, a matchless editorial board was set up. Every individual on the board went through rigorous rounds of assessment to prove their worth. After which they invested a large part of their time researching and compiling the most relevant data for our readers.

The editorial board has been involved in producing this book since its inception. They have spent rigorous hours researching and exploring the diverse topics which have resulted in the successful publishing of this book. They have passed on their knowledge of decades through this book. To expedite this challenging task, the publisher supported the team at every step. A small team of assistant editors was also appointed to further simplify the editing procedure and attain best results for the readers.

Apart from the editorial board, the designing team has also invested a significant amount of their time in understanding the subject and creating the most relevant covers. They scrutinized every image to scout for the most suitable representation of the subject and create an appropriate cover for the book.

The publishing team has been an ardent support to the editorial, designing and production team. Their endless efforts to recruit the best for this project, has resulted in the accomplishment of this book. They are a veteran in the field of academics and their pool of knowledge is as vast as their experience in printing. Their expertise and guidance has proved useful at every step. Their uncompromising quality standards have made this book an exceptional effort. Their encouragement from time to time has been an inspiration for everyone.

The publisher and the editorial board hope that this book will prove to be a valuable piece of knowledge for researchers, students, practitioners and scholars across the globe.

List of Contributors

Minky Son, Chanin Park and Keun Woo Lee
Division of Applied Life Science (BK21 Plus), Systems and Synthetic Agrobiotech Center (SSAC), Plant Molecular Biology and Biotechnology Research Center (PMBBRC), Research Institute of Natural Science (RINS), Gyeongsang National University (GNU), 501 Jinju-daero, Jinju 660-701, Republic of Korea

Seul Gi Kwon, Sam Woong Kim and Chul Wook Kim
Swine Science and Technology Center, Gyeongnam National University of Science & Technology, Jinju 660-758, Korea

Woo Young Bang
National Institute of Biological Resources, Environmental Research Complex, Incheon 404-708, Korea

William R. Taylor
Computational Cell and Molecular Biology Laboratory, Francis Crick Institute, Midland Road, NW1 1AT London, UK

Jonathan P. Stoye
Retrovirus-Host Interactions Laboratory, Francis Crick Institute, Midland Road, NW1 1AT London, UK

Ian A. Taylor
Macromolecular Structure Laboratory, Francis Crick Institute, Midland Road, NW1 1AT London, UK

Sandrine Moreira and Gertraud Burger
Department of Biochemistry and Robert-Cedergren Centre for Bioinformatics and Genomics, Université de Montréal, Montreal, QC, Canada

Emmanuel Noutahi
Department of Biochemistry, currently Département d'informatique et de recherche opérationnelle (DIRO), Université de Montréal, Montreal, QC, Canada

Guillaume Lamoureux
Department of Chemistry and Biochemistry, Centre for Research in Molecular Modeling (CERMM), Groupe d'étude des protéines membranaires (GÉPROM), Regroupement québécois de recherche sur la fonction, l'ingénierie et les applications des protéines (PROTEO), Concordia University, Montreal, QC, Canada

Madhulata Kumari
Department of Information Technology, Kumaun University, SSJ Campus, Almora, Uttarakhand 263601, India
School of Computational and Integrative Sciences, Jawaharlal Nehru University, New Delhi 110067, India

Subhash Chandra
Department of Botany, Kumaun University, SSJ Campus, Almora, Uttarakhand 263601, India

Neeraj Tiwari
Department of Statistics, Kumaun University, SSJ Campus, Almora, Uttarakhand 263601, India

Naidu Subbarao
School of Computational and Integrative Sciences, Jawaharlal Nehru University, New Delhi 110067, India

Martha Brennich
ESRF, The European Synchrotron, 71 Avenue des Martyrs, 38000 Grenoble, France

Alexander Andriatis
ESRF, The European Synchrotron, 71 Avenue des Martyrs, 38000 Grenoble, France
MIT, 77 Massachusetts Ave., 02139 Cambridge, MA, USA

Luca Costa
ESRF, The European Synchrotron, 71 Avenue des Martyrs, 38000 Grenoble, France
CBS, Centre de Biochimie Structurale, CNRS UMR 5048- INSERM UMR 1054, 29, Rue de Navacelles, 34090 Montpellier, France

Jean-Marie Teulon, Shu-wen W. Chen and Jean-Luc Pellequer
Univ. Grenoble Alpes, 71 Avenue des Martyrs, 38044 Grenoble, France
CNRS, IBS, 71 Avenue des Martyrs, 38044 Grenoble, France
CEA, IBS, 71 Avenue des Martyrs, 38044 Grenoble, France

Adam Round
European Molecular Biology Laboratory, 71 Avenue des Martyrs, 38000 Grenoble, France
Unit for Virus Host-Cell Interactions, Univ. Grenoble Alpes-EMBL-CNRS, 71 Avenue des Martyrs, 38000 Grenoble, France
Faculty of Natural Sciences, Keele University, Keele, Staffordshire, UK
European XFEL GmbH, Holzkoppel 4, 22869 Schenefeld, Germany

Steffen Grunert and Dirk Labudde
Hochschule Mittweida, University of Applied Sciences, Technikumplatz 17, 09648 Mittweida, Germany

Alexander V. Popov
SB RAS Institute of Chemical Biology and Fundamental Medicine, 8 Lavrentieva Ave., Novosibirsk 630090, Russia

Anton V. Endutkin, Yuri N. Vorobjev and Dmitry O. Zharkov
SB RAS Institute of Chemical Biology and Fundamental Medicine, 8 Lavrentieva Ave., Novosibirsk 630090, Russia
Novosibrsk State University, 2 Pirogova St., Novosibirsk 630090, Russia

Ishfaq A. Sheikh, Ghazi A. Damanhouri and Mohd A. Beg
King Fahd Medical Research Center, King Abdulaziz University, PO Box 80216, Jeddah 21589, Kingdom of Saudi Arabia

Muhammad Abu-Elmagd and Mohammed Al-Qahtani
Centre of Excellence in Genomic Medicine Research, King Abdulaziz University, Jeddah, Kingdom of Saudi Arabia

Rola F. Turki
KACST Innovation Center in Personalized Medicine, King Abdulaziz University, Jeddah, Kingdom of Saudi Arabia
Department of Obstetrics and Gynecology, King Abdulaziz University Hospital, Jeddah, Kingdom of Saudi Arabia

Subhomoi Borkotoky and Ayaluru Murali
Centre for Bioinformatics, School of Life Sciences, Pondicherry University, Puducherry 605014, India

Anne D. Rocheleau, Thong M. Cao, Tait Takitani and Michael R. King
Meinig School of Biomedical Engineering, Cornell University, Ithaca, NY, USA

Manuela Gorgel, Jakob Jensen Ulstrup, Andreas Bøggild, Poul Nissen and Thomas Boesen
Department of Molecular Biology and Genetics, Aarhus University, Gustav Wieds Vej 10c, Aarhus C 8000, Denmark

Nykola C Jones and Søren V Hoffmann
ISA, Department of Physics and Astronomy, Aarhus University, Ny Munkegade 120, building 1525, Aarhus C 8000, Denmark

Dorota Latek
Faculty of Chemistry, University of Warsaw, Pasteur St. 1, 02-093 Warsaw, Poland

Canan Ozgur
Computational Science and Engineering Program and Polymer Research Center, Bogazici University, Istanbul, Turkey

Pemra Doruker
Department of Chemical Engineering and Polymer Research Center, Bogazici University, Istanbul, Turkey

E. Demet Akten
Department of Bioinformatics and Genetics, Faculty of Natural Sciences and Engineering, Kadir Has University, Cibali, 34083 Istanbul, Turkey

Lasse Karhu, Ainoleena Turku and Henri Xhaard
Division of Pharmaceutical Chemistry and Technology, Faculty of Pharmacy, University of Helsinki, P.O. Box 56, 00014, Helsinki, Finland

Thomas Hoffmann, Antoine Marion and Iris Antes
Department of Biosciences and Center for Integrated Protein Science Munich, Technische Universität München, Emil-Erlenmeyer-Forum 8, 85354 Freising, Germany

Norbert Schormann and Debasish Chattopadhyay
Department of Medicine, University of Alabama at Birmingham, Birmingham, AL 35294, USA

Surajit Banerjee
Northeastern Collaborative Access Team and Department of Chemistry and Chemical Biology, Cornell University, Argonne, Chicago, IL 60439, USA

Robert Ricciardi
Department of Microbiology, School of Dental Medicine, Abramson Cancer Center, University of Pennsylvania, Philadelphia, PA 19104, USA

Index

www.ingramcontent.com/pod-product-compliance
Lightning Source LLC
Chambersburg PA
CBHW061243190326
41458CB00011B/3565